MORPHOLOGY *of the* ANGIOSPERMS

Arthur J. Eames
Professor of Botany, Emeritus
Cornell, University

1st Reprint in India, 2015
(Limited Copies)

Reprinted by:

United Book Prints
4806/24, Ansari Road,
Daryaganj,
New Delhi - 110 002
Tel.: 011-41511055

Print: 2021

ISBN: 978-93-83692-24-8

Printed in India

Preface

Progress in the understanding of the morphology of the angiosperms has been rapid in the twentieth century; much new factual information has been obtained and many conceptions and interpretations have been changed. These changes have bren brought about by the studies of many botanists scattered throughout the world in the several fields of morphological study—descriptive, comparative, anatomical, ontogenetic—and these botanists have often brought to bear on their conclusions evidence from the allied botanical fields—taxonomy, cytology, paleobotany, serology, plant geography, palynology. The trend toward the use of a broader basis for the drawing of conclusions and for the proposal of new theories is apparent. At the end of the nineteenth century, Celakovsky emphasized the necessity for using "all evidence" in the interpretation of structure and in drawing phylogenetic conclusions; in the twentieth century, the need for a broad base for all interpretations has been repeated by I. W. Bailey and many others. The importance of this emphasis is being slowly recognized.

This book has been prepared to bring together, in some measure, the results of these many scattered studies for the use especially of advanced students and teachers. It reviews much of the new factual material and many of the theories, old and new, related to the morphology and phylogeny of the angiosperms. Limitations in size of the book have restricted detailed descriptions and discussions of hypotheses, but the author believes that the important aspects of description and hypothesis are covered. In the manuscript, the names of plants that serve as examples are placed at the end of the sentence, set apart by a dash. The taxa cited are not necessarily the only examples.

The viewpoint of the treatment is that of comparative rather than descriptive morphology, with emphasis on evolutionary modifications and phyletic implications. Velenovsky placed on the front page of his excellent textbook, "Vergleichende Morphologie," (1905–1913), these sentences as a maxim: "Zur morphologische Lösung werden wir—wie immer—die vergleichende Methode in Anwendung bringen. Auf diese Wege werden wir zu dem richtigen und einheitlichen Ausschauung gelangen." The author agrees heartily with this point of view.

The book assumes, on the part of the reader, an acquaintance with botany equivalent to that obtained from a general elementary course, although some elementary descriptions and discussions are, of necessity, included.

During the preparation of this book, the author has been impressed with the importance of the earlier morphological literature, that of the middle of the nineteenth century especially, to the research of the twentieth century. Facts and theories discussed in the earlier period, overlooked or forgotten during the succeeding decades of increasing specialization, have been presented again in the twentieth century. The understanding of the earlier students of morphology is remarkable in the light of the small amount of factual information available to them; interpretations made a century ago often appear to be sounder than some later ones; today many botanists, in their highly specialized fields, lose sight of the broader aspects in the maze of details in their own particular area.

In Chap. 11, the morphology of a few families is discussed in some detail, as examples of the use of morphology in determining relative advance in evolutionary modification, and to provide examples of structure important in the interpretation of specialized form throughout the angiosperms. The families selected for discussion are those that possess the most primitive characters—the lowest dicotyledons and monocotyledons. The inclusion of the Amentiferae and Proteaceae—taxa sometimes considered primitive dicotyledons—was planned for this chapter, but they were omitted because of restriction in book size.

A bibliography for general consultation follows the last chapter, and a bibliography following each chapter covers the subject matter in the chapter. The bibliographies contain only a small part of the references consulted in the preparation of this book; the citation of all of these would fill another book. Selections were made on the basis of general importance from the viewpoint of the treatment in this book and of the size and excellence of their bibliographies. Also cited are works from which illustrations have been borrowed.

The author is indebted to all the many botanists who, over a century and more and throughout the world, have laid the foundation for the treatment in this book. He is grateful to the many botanists from whom he has borrowed published illustrations and to Miss Elfriede Abbe who drew Figs. 28, 37, 38, and 97; to Mr. A. List who drew Fig. 3A to C; to Dr. A. T. Hotchkiss who provided the photographs of *Eupomatia bennettii* for the frontispiece and for Fig. 144; to Dr. L. J. Edgerton for Fig.

133; to Prof. K. D. Brase who contributed the proliferated apple shown in Fig. 97; and to Dr. Roger Gauthier for Fig. 49*D*. He is also deeply indebted to his wife, Rita Ballard Eames, for extensive and continuing aid in many ways, especially in the preparation of the manuscript and the illustrations.

Arthur J. Eames

Contents

Chapter 6

Chapter 7

Chapter 8

Chapter 9

Chapter 10

Chapter 11

Chapter 12

Chapter 1

The angiosperms are the dominant seed-bearing plants of the present day, a vast and varied assemblage estimated to consist of 300,000 species. They are commonly considered the "modern" seed plants, geologically young. But, structurally, they do not appear to be of recent origin; even those families now considered most primitive have some well-advanced characters. Evidence is accumulating from the fossil record and from critical morphological studies that the angiosperms are an old group in which there was early differentiation along several lines and that the group had already become diverse and complex by the early Cretaceous period, long considered the time of their origin. The angiosperms, in their dominance, have not, however, "crowded the gymnosperms from the face of the earth," as is often stated or implied in elementary textbooks; large areas of conifer forest exist in the tropics and in both northern and southern temperate lands; the cycads are dominant in some small areas in Australia and South Africa.

In all characters, gross and minute, external and internal, sporophytic and gametophytic, the angiosperms show great diversity of form, a diversity that is clearly the result of adaptive specialization over a very long period of time and under great climatic changes. This specialization leads to both increasing complexity and simplification of structure. Simplicity of form has long been considered evidence of primitiveness; the part played in specialization by reduction, "retrogression" or "suppression," has been commonly overlooked. The realization that simplicity is often secondary rather than primitive has played a prominent part in the interpretations of comparative morphology and phylogeny in the twentieth century. The results of reduction are prominent in the morphology of every part of the plant—in the body of the sporophyte, embryonic and mature, and in gametophyte origin and form.

The angiosperms are commonly set apart from other seed plants by enclosure of the seeds, in contrast with the naked seeds of the gymnosperms; by the presence of vessels in the wood; and by the possession of a complex reproductive structure, the flower. But these characters do not sharply limit the angiosperms. The carpel is open at the time of pollination in some taxa, and some conifers (*Araucaria*) have the seed enclosed; the vessel is present in *Selaginella, Equisetum, Pteridium, Ephedra, Welwitschia,* and *Gnetum* and is absent in many angiosperms.

The carpel is still in process of closing; some stages are to be seen in living taxa. The vessel has arisen independently many times within the angiosperms, and all stages of its development are found in living angiosperms. A supposed basic difference in leaf-trace number between angiosperms and gymnosperms—angiosperms with an odd, gymnosperms with an even number—has been shown to be invalid. But the angiosperms, though not set apart by any single character, clearly represent a well-defined stock, distinct from other seed-bearing plants.

The sporophytic body of angiosperms shows greater variety in form and size than that of any other major plant taxon, ranging in habit from tree to herb; in size, from the minute *Wolffia* to the tallest *Eucalyptus;* in form, from the simplicity of a thallus to the complex branching of trees and giant vines; in flower structure, from the simplicity of a single, naked sporophyll to the complexity of organs of four types, with much connation and adnation. The gametophytes, although consisting of rather few cells, likewise vary a great deal, especially in cell number and arrangement.

Structurally, the plant body consists of an axis, branched or unbranched, with lateral appendages.* The axis is commonly divided on structural and functional grounds into *stem,* with appendages and endarch primary xylem, and *root,* without appendages and with exarch primary xylem. Stem and root may form a continuous axis, as in most seedlings, with a structurally transitional section, the hypocotyl; or roots may develop as appendages of the stem. Continuity of root and stem in the embryo and seedling of the higher plants has been, in part, the basis for the theory that these two organs represent the specialized parts of an original primitive axis which constituted the entire sporophyte, as in simpler members of the Psilophytales.

The stem alone has been considered to represent the entire primitive axis, with the root a secondarily developed organ. This concept is based on the endogenous origin of the radicle and on the common origin, in some embryos and seedlings, of roots as major appendages of the base of the stem. Evidence in support of this view has been seen in the absence or nonfunctioning of the taproot in some supposedly primitive Liliales, where the entire root system is formed by adventitious roots. (Under this theory, the absence of a taproot has been considered more primitive than its presence.) But absence or nonfunctioning of an embryonic taproot seems to be a derived condition. The vascular structure of the hypocotyl, in its symmetry and relations to the primary vascular skeleton of stem and root, supports the view that the primary

* Anatomical details, already discussed in Eames and MacDaniels, *Introduction to Plant Anatomy,* are herein largely omitted.

root is a basic part of the axis; adventitious roots, with simple, crude attachment to the mother axis, are secondary organs.

Variety in growth habit ranges from tree and shrub to herb and vine, woody and herbaceous. The diverse habital forms represent many types and stages of specialization along parallel and convergent lines. Distinction between trees and shrubs is based chiefly on differences in height, woodiness, and permanence, but all habital types merge with others. Some herbs are taller than some trees. Many herbs, even annuals, have strong, woody stems, and some tropical trees have weak, "fleshy" trunks. Vines are both woody and herbaceous, with major and minor structural adaptations to the climbing habit. The tree habit is ancient among vascular plants. Trees of good size were present in the Devonian period and have been present in major taxa since that time. Because of their large size, trees are prominent among angiosperms: it has been estimated that there are 20,000 to 25,000 living species. Great height—over 100 feet—is attained in many families; 200 feet and over in a few. The Australian gums are, without doubt, the tallest angiosperms; *Eucalyptus regnans* reaches a maximum height of 326 feet. (Taller individuals reported are probably mythical.*) Herbs greatly outnumber woody plants in genera and species. Herbs are prominent in some areas because of the vast numbers of individuals. Their early evolutionary history is unknown; their softer structure is not so likely to be preserved in the fossil record as that of woody plants.

THE PLANT BODY

The gross structure of the plant body has been variously interpreted, morphologically. In early days, it was considered to be made up of several "fundamental parts" or organs—root, stem, leaf, floral organs, ovules, trichomes; in recent years, the number of these basic organs has been reduced to three—*root, stem,* and *leaf;* in present usage, the stem and its appendages are commonly considered as a unit, the *shoot.*

Recognition of the shoot as a fundamental part has been a gradual process. The concept that stem and leaf together form an entity was probably first proposed at the beginning of the twentieth century. In succeeding years, three interpretations of the make-up of the shoot were proposed: (1) that the stem is the fundamental part, and the leaves appendages; (2) that the leaf is the basic part, and the stem consists wholly or in large part of proximal parts of the leaves; (3) that the shoot consists of units, "segments," called *phytons.* The second of these theories was later expanded—without the addition of better

* Forest Products Department, Australian Council for Scientific and Industrial Research Organization.

morphological evidence—into the leaf-skin theory (p. 21). The third theory took several forms, including the phyton theory and "phytonism" (p. 21). (The term phytonism has also been applied to the theory that the stem consists of leaf bases.)

If the origin of the complex sporophyte of the higher vascular plants is to be seen in the simple body of the Psilophytales, the description of the body of seed plants as made up of two basic parts, root and shoot, seems reasonable. In general method of development—development by apical meristems—they are alike, but they differ in basic vascular structure, in method of later increase in length, and in structure of the outer tissues. The intercalary growth of stems and leaves (origin of the petiole) is absent in roots. Histological elaboration of the outer tissues in stems is chiefly in the formation of a complex cortex; in roots, of a specialized epidermis with root hairs. In the angiosperms, the presence of an endodermis has been considered a root character, distinguishing root from stem, but an endodermis is present in stems of many taxa—in rhizomes, in seedlings, in aquatic plants, and in some parts of the stems of the majority of herbs. And there is excellent histological evidence in other taxa that the endodermis—characteristic of the entire plant body in many pteridophytes and probably present in the stems and leaves of ancestral angiosperms—has largely been lost. The endodermis ties together root and stem as parts of a continuous plant body, and its presence is not evidence that an organ is a root.

Embryo structure has been seen to demonstrate the unity of the axis in embryos that have, in the hypocotyl, the radial vascular structure of the root and continuity of the endodermis in the hypocotyl and the epicotyl. This continuity of the endodermis is considered a recapitulation of the evolutionary history of the plant body; the axis of the embryo represents the simple, primitive plant body. In the embryo, in the seedling, and persisting (except in some monocotyledons) in the mature plant, the hypocotyl is the section of the axis transitional anatomically from root to shoot; its origin goes back to the early elaboration of the sporophyte as shoot and root.

The description of the axis of the angiosperm embryo as, in structure, a continuous organ, root at one end, stem at the other, has been considered morphologically inaccurate; the embryos of most monocotyledons and those of some dicotyledons cannot be so interpreted. The plumule of these embryos is commonly described as "lateral," but the lateral position is only apparent; the plumule is morphologically terminal, as in the dicotyledons (see Chap. 9).

The Root

In the angiosperm embryo, root and shoot form a continuous structural axis; in the pteridophytes, the embryonic root is the first of a series of lateral structures, which are temporary. (In the lepidodendrids, *Pleuromeia,* and *Isoëtes,* the functional counterpart of the root system of seed plants is the rhizomorph, which, without true root structure, bears adventitious, temporary roots.) Continuity of shoot and root, with a transition region, the hypocotyl, is characteristics of the embryos of dicotyledons and most of the monocotyledons. In some of the arborescent monocotyledons, the embryo has no root or has only an abortive structure at the end of the stem. Absence of a radicle has been considered by some morphologists the primitive condition in monocotyledons, but the presence of briefly functioning and nonfunctioning primary roots in some monocotyledonous taxa is evidence that the taproot has been lost in these highly specialized taxa and adventitious roots on the hypocotyl and stembase have formed a root system. The arborescent shoot system in monocotyledons is a specialized type and is accompanied by a specialized, "secondary" root system.

The Shoot

Recent studies of apical meristems and the anatomy of the node have shown that stem and leaf are closely interrelated and have emphasized the concept of the shoot as, morphologically, a fundamental part of the plant body. Anatomically, no structural line, external or internal, separates stem and leaf. Ontogenetically, also, no separation can be made. There is no constancy in details of meristematic origin of these organs; histological limits of tunica and corpus vary even in the same plant. The region of union of leaf and stem, the nodal region, is one of merging tissues, cauline and foliar. Yet the node is commonly called a part of the stem and the region of attachment of the leaf is termed the leafbase. Tissues of leaf and stem merge in the nodal region, and limits cannot be set; this region is a part of the shoot, neither leaf nor stem. Though stem and leaf cannot be sharply delimited—and the leafbase should not be called one of the basic parts of the leaf—stem and leaf are units of body structure that must be treated as parts of the shoot.

The term shoot is an old one, going back as far as the late nineteenth century. Stem and leaf were, in early studies, called "correlative parts" of the shoot, but "shoot" has long been suppressed, because of the supposedly important distinction between leaf and stem. Now this old, largely abandoned term has been revived and is slowly coming into use, because it has been found to be useful and morphologically valid. Although stem and leaf are convenient descriptive terms, necessary for use in most fields of botanical study, they should not be considered primary categories of body structure. The flower, a reproductive shoot, is treated in later chapters.

The Stem. The stem, as an axis, is described as divided into parts, the *internodes,* by more or less well-defined areas where appendages of leaf rank, the *nodes,* are attached. Nodes are often defined as levels of attachment of leaves, sometimes as levels of departure of leaf traces. Nodes and internodes are continuous parts of the stem; there is no structural delimitation. A description of "nodal anatomy," for example, necessarily covers much of the structure of internodes. Significant morphological differences in the shoot are chiefly those of internal structure; they are discussed under The Anatomy of the Plant Body later in this chapter.

Ecological modifications of the stem and shoot lie outside the scope of this book, but the *short shoot* should have brief mention, because, in its extreme form, it involves the morphology of "terminal" leaves. In the gymnosperms, the slow-growing, lateral short shoots of *Larix* and *Ginkgo* and the determinate, deciduous short shoots of *Pinus* are well known. In *Pinus monophylla,* the short shoot with a solitary pseudoterminal leaf resembles a simple leaf. In the angiosperms, similar, but less highly specialized, slow-growing short shoots, with crowded nodes, are characteristic of species of *Betula, Pyrus,* and other genera. Stems terminating in reduced leaves are occasional in monocotyledons—*Polygonatum, Streptopus, Disporum,* some bamboos. The species of *Uvularia* show stages in the loss of the stem apex and origin of a terminal leaf.

The Leaf. The leaf, commonly called an appendage of the stem, consists of two fundamental and more or less clearly distinct parts, the *blade,* or *lamina,* and the *leafstalk,* or *petiole.* A third part, the *leafbase* —"leaf cushion," "leaf buttress," "leaf foundation"—has long been recognized by many morphologists. (The term leafbase has also been applied to the basal, sheathing part of many monocotyledon leaves, both descriptively, to the winged base, and morphologically, to the part developed from the base of the leaf primordium.)

The leafbase, as a basic part of the leaf connecting petiole and stem, is not clearly defined, structurally. Some interpretations of leaf morphology call the leafbase a definite part of the leaf, the part that connects petiole and stem, "the base on which the leaf stands." In other interpretations, this base is merely the connecting or transitional region between leaf and stem and not a fundamental part of the leaf itself. The limits of the leafbase are not recognizable, externally or internally. In Europe and Asia, the leafbase has been generally recognized as an important part of the leaf; in America, it has received little attention. Definitions of the leafbase are loose and various, and often seem remarkable as examples of definitions of a morphological entity: the leaf primordium, the basal part of the leaf primordium, the ontogenetic buttress or foundation of the leaf, the hypopodium, the part of the apical meristem of a vegetative shoot where a leaf arises that has no share in the development of the leaf itself, the connecting tissues of leaf and stem, the region in the stem and leaf where lateral leaf traces change course toward the median trace, the segment of the axis which subtends a leaf initial and surrounds the leaf trace as it differentiates. The failure of American morphologists to recognize the leafbase as an important part of the leaf cannot be understood by some Indian morphologists, "because a leaf without a base cannot be conceived."

Ontogeny provides the best basis for an understanding of the nature of the leafbase. The leaf primordium arises as a lateral, more or less crescent-shaped mound or ridge on the apical meristem of the shoot. From this early primordium, as a base, arise a median conical or cylindrical projection—from which the blade and petiole develop—and, if stipules are present, lateral lobes. If a leafbase is to be recognized, it is recognizable only in shoot ontogeny as the base of the leaf primordium; it cannot be distinguished in the mature leaf or node.

The leafbase has been well described as a primordial region in the young node, consisting of two components—a proximal cauline or axial part, and a distal foliar part. Together, these components form the part of the shoot transitional from stem to leaf, a part not limited internally but sometimes limited externally as decurrent areas of epidermal and cortical tissues set apart by unusual color or vesture. This description sets no limit internally in the stem for the leafbase; others set as the limit "to the pith" or "sometimes to the pith"; still others "to the center of the pith." Interpretations of the radial extent of the leafbase conflict with morphological distinctions within the stem. If the leafbase tissues extend only to the pith, the leafbase, as a part of the leaf, clothes the pith, or "core," as a "mantle," and the shoot axis consists of the pith only; if the leafbase extends to the center of the pith, no axis, as such, exists in the nodal region. Structure of the apical meristem is reported to show that the view that the leafbase tissues extend to the center of the axis is incorrect, because "two different zones of the meristem form mantle and core." The leafbase, if such a structure is worthy of distinction, is a segment of the shoot, not of the leaf. (The extent of the leafbase as a unit of structure of the shoot is further discussed under the leaf-skin theory, p. 21.)

If the shoot is accepted as one of the two basic parts of the plant body, the leaf can well be called a "partial shoot arising from a parent whole shoot." There is then no necessity to distinguish a part of the shoot, partly stem and partly leaf, as a fundamental part of the leaf. The term *partial shoot* is good in its implication that the leaf is a lobe of the axis, but not because it has "lost its adaxial part in the development of the dorsiventral form," as suggested under one theory. The leafbase is not recognized in this book as a part of the leaf; it is considered a part of the shoot, a nodal region transitional between stem and leaf—a transition region. Because the term leafbase is used not only in this way but also, loosely, for the proximal part of leaves—which is morphologically various--the term *leaf buttress* would distinguish the part that forms the nodal transitional region from that which forms petiole and blade.

The angiosperm *leaf blade,* or *lamina,* is remarkable for its extraordinary variety of form. Descriptively, two chief types are distinguished as *simple* and *compound.* From the phylogenetic viewpoint, each of these has been considered primitive, but strongest support has come for the view that the simple leaf is primitive.

The leaves of the woody ranalian families—now recognized as showing many primitive characters and probably the most primitive living dicotyledons—are simple. The simple form is apparently more common in the fossil leaves of the Cretaceous and Tertiary periods. Comparative study shows many series leading from simple, through increasing dissection, to compound—in *Acer,* in *Rubus,* in the Vitaceae. These series can hardly be read in the other direction. Chief support for the compound leaf as primitive has come from families where reduction of leaflets to one is shown by comparison of many taxa and by the presence of vestigial structure in leaflets, petiole, and rachis—Leguminosae (Fig. 1), Rosaceae, Rutaceae. But the woody ranalian families, with their highly primitive flowers and wood, have simple leaves. The decompound leaves of some genera in families with primitive flowers—especially the fernlike leaves of some genera of the Ranunculaceae and Fumariaceae—have been regarded as possible evidence that these taxa retain in some measure the leaf form of an ancestral fernlike stock.

Under the Durian theory of the habit of the primitive angiosperms, the large, pinnately compound leaf is considered the form characteristic of early angiosperms. The basis for this conclusion is that leaves of this type accompany the growth habit believed primitive and that they inhabit tropical forests, where today the most primitive living angiosperms are found. But the woody ranalian families with simple flowers and anatomy also inhabit these forests, and their leaves are simple.

Though the simple leaf seems undoubtedly at least one primitive type, the simple leaves of many taxa clearly represent modified compound leaves; the lamina of the compound leaf has been reduced by loss or fusion, or by both loss and fusion, of leaflets. Unquestioned reduction of these types is present in the Leguminosae, Rosaceae, Oleaceae, Rutaceae, Proteaceae. Evidence of loss and fusion of leaflets comes from comparison of closely related taxa, with structural evidence in venation and in the presence of an articulation between rachis and petiole below the solitary leaflet (Fig. 1*A*). The genus *Citrus*

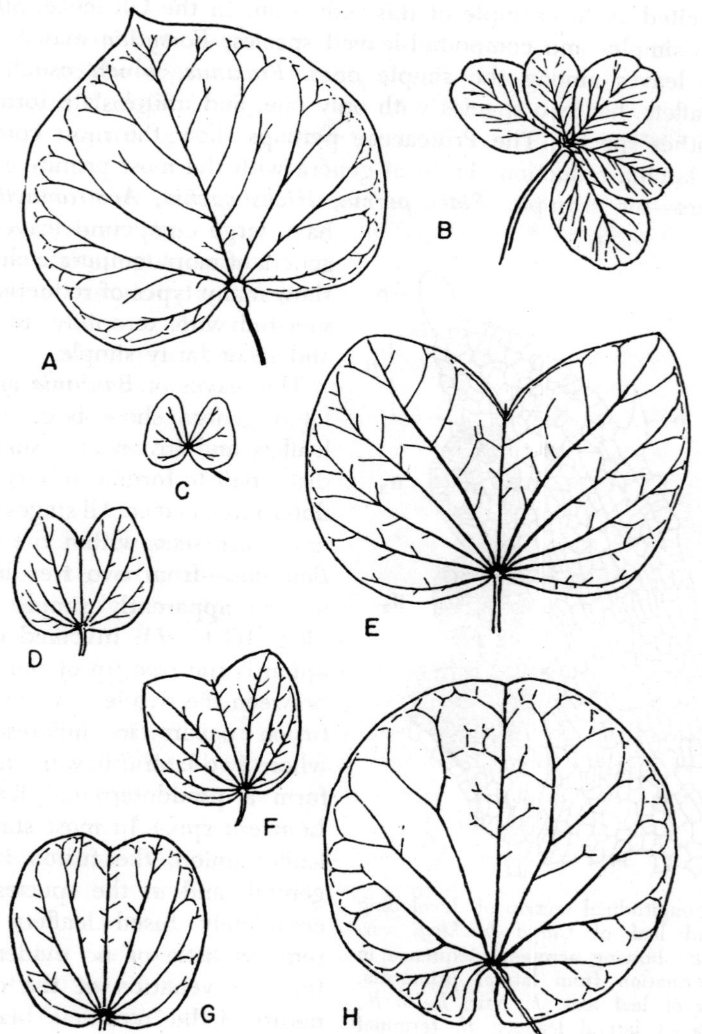

Fig. 1. Sketches of leguminous leaves showing evidence of reduction in compound leaves. *A,* in *Cercis canadensis,* survival of a terminal leaflet, articulation prominent; *B,* in a 5-foliolate leaf of *Trifolium arvense,* reversion of normal trifoliolate to ancestral multifoliolate; *C* to *H,* in leaves of *Bauhinia* (*C, B. saigonensis; D, B. acuminata; E, B. reticulata; F, B. mollicella; G, B. malabarica; H, B. retusa*), a series in connation of two lateral leaflets to form one pseudoterminal leaf, the tip of the rachis still evident in *D* to *G.*

is often cited as an example of this reduction. In the Oleaceae, *Syringa* has both simple- and compound-leaved species; *Forsythia* may have 3-foliolate leaves among the simple ones. *Fraxinus,* which usually has many leaflets, has two species with only one, and unifoliolate forms occur in other species. The Proteaceae perhaps show the most complete story of lamina reduction. Tropical genera with the most primitive floral characters—for example, *Placosperma, Hicksbeachia, Austromuellera*—have large compound leaves; the genera of more temperate climates show many types of reduction, associated with xerophily, to small and secondarily simple.

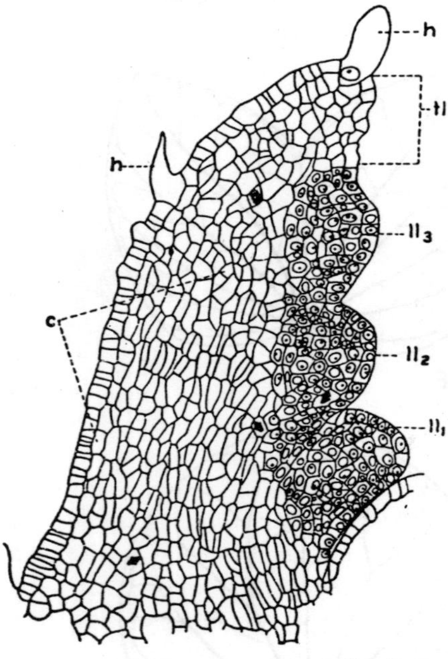

Fig. 2. Longitudinal section of developing compound leaf of *Carya buckleyi* var. *arkansana,* showing acropetal sequence in leaflet formation from lateral primordia. *c,* cortex of leaf axis; *h,* hair; ll_1 to ll_3, primordia of lateral leaflets; *tl,* terminal leaflet. (*After Foster.*)

The leaves of *Bauhinia* and related genera show both loss of leaflets and fusion of a surviving distal pair to form a solitary pseudoterminal leaflet. All stages of the union are seen within the genus *Bauhinia*—from two free leaflets to one apparently simple blade (Fig. 1*C* to *H*). Involved in this union is the free tip of the rachis between the two leaflets. (Similar fusion occurs in inflorescences where two lateral flowers unite to form a pseudoterminal flower—*Lonicera* spp.) In most stages of leaflet union, the fusion is congenital, and in the species with completely fused leaflets, there remains little or no evidence in form or venation of the double nature of the "terminal" blade.

The compound leaf, as an advanced type, is believed to have arisen by the evolutionary dissection of the simple leaf. In ontogeny, the leaflets develop as do the lobes of a simple leaf—by the development of lateral primordia on the median axis (Fig. 2). The development of the compound leaf of the palms is wholly different—by an ontogenetic splitting of the primordium (Fig. 18).

In gross *venation,* pinnate venation appears to be primitive, the palmate derived. Many transitional forms occur. Evidence that pinnate venation is primitive is found in the anatomy, especially in the leaves of the woody Ranales. The pattern of minor vein arrangement is also of two major types—*reticulate* or *netted*, and *parallel* or *striate*—but there are intermediate types. There are many exceptions to the characterization of the leaf venation of dicotyledons as netted and that of the monocotyledons as parallel; the leaves of many of the lower monocotyledons have netted veins, and some of the dicotyledons—the Epacridaceae, for example—have parallel veins. The venation of the Epacridaceae contrasts strongly with that of the closely related Ericaceae and has clearly been derived from the reticulate type. (Modification of nodal structure has accompanied the change in venation; a multilacunar node in the Epacridaceae has replaced the unilacunar node in the Ericaceae.)

Perhaps the most important feature in the arrangement of the lesser veins is the presence of free vein endings within the vein eyelets in some taxa and the absence of free endings in other taxa. The view that significant changes in size of the vein eyelets occur with increasing age of the plant was shown to be without foundation. Vein eyelets are usually of irregular form, but, in some tropical genera, they are rectangular and remarkably uniform in size.

The *petiole* arises late in the ontogeny of the leaf, by intercalary growth of a region at the base of the lamina. Increase in length is probably chiefly by cell enlargement. The delayed development of the petiole has been considered evidence that the petiole is a recent development in the specialization of the leaf. But this seems unlikely, because sessile leaves can often be shown, by comparison with the leaves of related taxa, to have lost their petioles. The leaves of the primitive woody dicotyledons are petiolate, and the leaflike carpels of many of them are stipitate. (The homology of leaf and carpel has been strongly supported in recent years by evidence from ontogeny and anatomy.)

Stipules are lateral parts or appendages of the leaf, usually borne one on each side at, or near, the base of the leaf. When borne on the stem free from the leaf itself, they appear to be independent organs and have been called "cauline stipules," an unfortunate term which implies that they are part of the stem. But anatomical structure shows that these stipules are a part of the leaf; their vascular supply is derived from the lateral leaf traces. Although this anatomical relationship was shown in 1880, its morphological significance was not recognized until nodal anatomy was critically studied in the early decades of the twentieth century.

Stipules have been called "appendages of the leafbase, not of the leaf" because they "arise from the leafbase." But only independently attached stipules arise in this manner; other stipules arise congenitally fused with the leaf primordium as this meristem develops on the leaf buttress. In some compound leaves, appendages—*stipels*—occur at the bases of the leaflets. Stipels are surely a part of the leaf, and are evidence that stipules also are a part of the leaf. Stipules rarely occur on cotyledons. They are occasionally present on perianth parts in some of the more primitive monocotyledons—*Melanthium, Zygadenus*—and on stamens—*Ornithogalum* and other liliaceous genera and many of the Bromeliaceae. *Allium* shows in its many species great variety in the stipules of stamens, from well-developed to vestigial, and some species have none.

Stipules have been rather loosely classified on the basis of form and relation to the leaf and stem. They are called *lateral* when they seem to be lateral parts of the leaf, appearing like lobes of the lamina or wings of the petiole. Lateral stipules that are borne on the stem independently of the leaf are termed *free* (Fig. 3*A, B, C*)—*Begonia, Vitis, Hydrocotyle, Liriodendron;* they are termed *adnate* where fused along one margin to the petiole for all or part of their length—many Rosaceae (Fig. 3*D*), Leguminosae. The term adnate is sometimes restricted to stipules that are fused to the body of the leaf throughout their length, as in the sheathing leaves of many monocotyledons; this type is also called "vaginal." *Ventral* stipules are those whose margins meet on the ventral side of the petiole—*Artocarpus, Magnolia,* some species of *Begonia.* In *axillary* stipules, the margins are fused above the petiole, forming a single leaflike pseudoaxillary structure. Axillary stipules have also been called *intrapetiolar*. Intrapetiolar stipules are formed by the union of stipules of two different leaves at the same node—*Fuchsia, Elatine.* In the whorls of leaves of *Galium* (Fig. 3*C*) and related genera, there are only two true leaves; the other "leaves" are intrapetiolar stipules or, in part, individual simple stipules. Free stipules arise early on the leaf buttress, close to the primordium of the blade and petiole, but are isolated from the rest of the leaf during the enlargement and maturation of the buttress.

Stipules, common in most of the less advanced angiosperms, are absent in most of the Sympetalae; they are present in some of the Rubiaceae. Some families in the Archichlamydeae—Rosaceae, Leguminosae—have stipulate and estipulate genera. In the woody Ranales, now considered very primitive, there are estipulate families—Winteraceae, Degeneriaceae, Eupomatiaceae, Annonaceae, Himantandraceae, Monimiaceae, Calycanthaceae, Lauraceae; and stipulate families—Magnoliaceae, Chloranthaceae. Stipules commonly accompany the woody habit; 40 per cent of woody dicotyledons are stipulate, as contrasted with 20 per cent of herbs.

In the monocotyledons, stipules, as paired, free appendages of the leaf, are rare; they occur in the Hydrocharitaceae, Butomaceae, Najadaceae, perhaps some of the Dioscoreaceae. The monocotyledons have sometimes been described as lacking stipules, but there is little agreement in the interpretation of the sheathing base of the monocotyledon leaf. The thin margins are commonly interpreted as representing

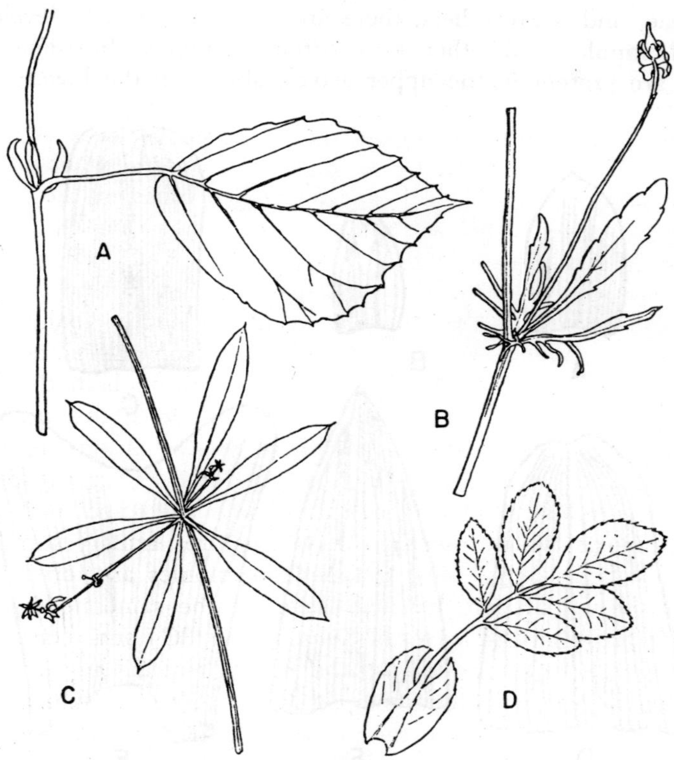

Fig. 3. Sketches of stipules showing variations in type. *A, Hamamelis,* simple, free; *B, Viola,* deeply dissected, free; *C, Galium,* two leaflike stipules with each leaf, free; *D, Rosa cinnamomea,* leaflike, adnate to petiole. (*A–C, drawn by A. List.*)

adnate stipules that have lost identity as such in adaptation to the general monocotyledon habit of "telescoped shoot." The leaf sheath is also sometimes considered an elaboration of the primordial leafbase, without distinction of blade and stipules. The loose use of the term *ligule* for the wings of the sheathing leaf adds to the difficulty of interpreting the monocotyledon leaf as a whole.

Evidence in support of the concept that the stipules form the marginal parts of the sheathing base has been found in those monocotyledons that show paired adnate stipules, especially *Potamogeton.* In this genus, the leaves of seedlings show stages in the development of the sheathing base by the merging of adnate stipules with the base of the petiole (Fig. 4).

In many families, there is evidence of reduction of the stipules, associated with the loss of their function as protective structures for meristems and young leaves; in the Leguminosae, Droseraceae, Onagraceae, Cornaceae, and many others, there are taxa with poorly developed or vestigial stipules and other taxa without stipules. In some species, stipules are present in the upper leaves, absent in the lower; in other

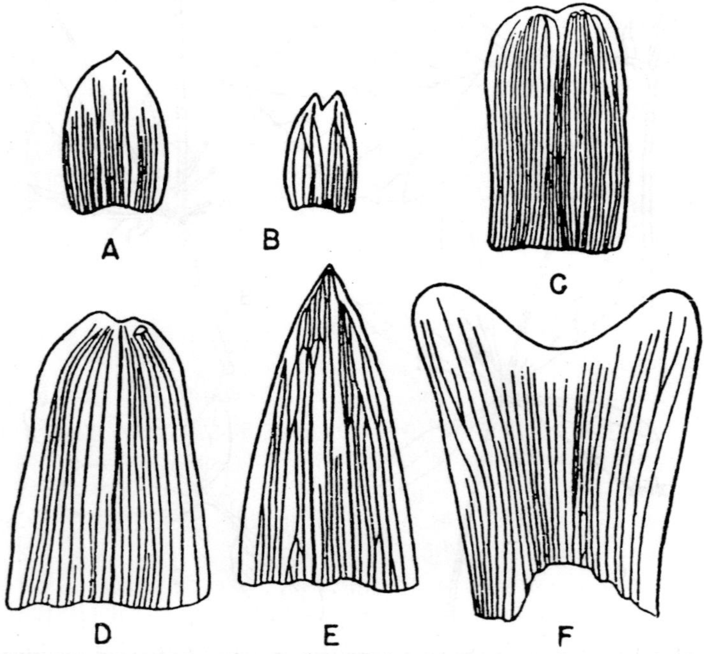

Fig. 4. Sketches of stipules of *Potamogeton* showing stages in connation of a pair of stipules to form a stipular sheath. *A, P. perfoliatus; B, C, P. lucens; D, E, P. natans; F, P. crispus.* (*After Monoyer.*)

species, present in the lower, absent in the upper leaves. In *Tropaeolum,* they are present only in seedlings. Stipules appear to be a disappearing feature of the angiosperm leaf. In dicotyledons, they have apparently been lost in most of the higher families; in the monocotyledons, they may survive as a part of the sheathing leaf, an adaptation to the dominant monocotyledon habit.

Anatomical evidence for the primitiveness of the stipulate condition seems contradictory. Stipulate taxa commonly have trilacunar nodes; they rarely have unilacunar nodes with an odd number of traces. If the unilacunar node with two traces is primitive (Fig. 6*A*), as now seems probable, the primitive angiosperm leaf should be stipulate. The primitive Helobiales fit into this picture, as do the Magnoliaceae and a few other woody-ranalian taxa, but other woody-ranalian lines are without stipules.

Stipules have been interpreted as basal leaflets of a compound leaf and as remnants of an ancestral whorl of leaves, but neither ontogeny nor anatomy supports these views, and, in angiosperms, whorled leaves represent advanced phyllotaxy. The stipules of the gymnosperms and ferns (except perhaps the Eusporangiatae) are not nodal appendages.

Prolongations of the stipulate margins beyond the median part of the sheathing base of monocotyledon leaves form a more or less free structure, the *ligule*. The ligule may be a two-lobed or two-toothed structure, or a simple structure where the two parts are united as a ventral "collar." (This collar may become greatly enlarged and form an important part of the leaf.) Formed in this way, a ligule is obviously stipular in nature, but the interpretation of the ligule has been expanded to cover the basal sheath, with which the ligule is continuous, and the entire sheath called the ligule. This understanding of the ligule has brought about great confusion in the interpretation of the monocotyledonous embryo (Chap. 9). The ligule is not morphologically distinct from the stipules, of which it usually represents the distal part. (A descriptive distinction becomes necessary only in the interpretation of the coleoptile of the grasses and some other families where the coleoptile is called "ligular." The coleoptile is a part of the sheathing leafbase, not of the ligule.) Ligules are present not only in many monocotyledons but also in some dicotyledonous families—Droseraceae, Saxifragaceae, Araliaceae, Piperaceae. The ligule, or hastula, of the palms represents a part of the telescoped rachis (Fig. 148).

Whether or not the stipules are recognized as parts of the primitive angiosperm leaf, they show great variety in form and relation to other parts of the leaf. Evolutionary modification is obvious, and there are two very different concepts of the direction of the series. According to one concept, paired stipules, free or partly adnate, represent the primitive leaf structure. With specialization, adnation became complete, with the stipule tips becoming auriculate at the base of the blade. Reduction and loss of the auricles left a simple sheathing leafbase, consisting of the united "leafbase" and the stipules, an apparently simple but really complex structure.

According to the other concept, the primitive angiosperm leaf had a prominent sheathing base and a blade, sometimes separated by a petiole. In specialization, this sheath became auriculate and, with shortening of the sheath, became auricles. Under this concept, free stipules are the higher type; under the other concept, free stipules are the most primitive. Under one theory, the monocotyledons, under the other, the dicotyledons have the most primitive stipules. Evidence from nodal structure, from seedlings, and from comparison in related taxa indicates that the free stipules are the more primitive form. The many traces of the monocotyledon-sheath margins suggest advanced structure. (The multilacunar node is generally recognized as an advanced type.) Seedlings in some monocotyledons show progressive stages in their early leaves from paired free stipules to sheathing leafbases—*Potamogeton.* Free paired stipules are present in *Potamogeton densus,* in species of *Ruppia, Althenia, Najas, Hydrocharis, Smilax.* Increasing adnation of free stipules is seen in many dicotyledonous families—Rosaceae, Leguminosae. Fused adnate stipules and sheathing leafbases are frequent in dicotyledons, rare in monocotyledons—*Zannichellia palustris.* The theory that free basal stipules are the primitive type seems best supported by evidence of all types. In stipular type, the leaves of the dicotyledons are more primitive than those of the monocotyledons.

On the basis of the theory that paired lateral stipules are primitive, two views have been held as to the relationship of the free and the adnate. It is claimed that the free stipules stood at first at the base of the petiole; that a change to a free (cauline) position came later as the leaf buttress matured and displaced them. The other view is that the position of the free stipules is the primitive one in the shoot, that a closer relation to the leaf is an advanced character. Ontogenetic evidence can be found to support each of these views; to determine the ancestral position of the stipules, evidence from ontogeny alone is not sufficient.

Small, reduced leaves borne close to the base of lateral branches have long been termed *prophylls.* They have also often been called *bracteoles,* though this term is usually restricted to inflorescences and flowers. Commonly, there are two lateral prophylls in the dicotyledons and one in the monocotyledons. Where there are two prophylls, they may stand "opposite," forming a pair, or one may be distal to the other. Definitions of prophylls emphasize their position in relation to leaves on the branch; they do not fit into the phyllotactic spiral of the branch. It is apparent that they do not when the pairs are opposite but when one prophyll is above the other, they commonly seem to fit into the spiral.

The numerical distribution of the prophylls—two in the dicotyledons and one in monocotyledons—has sometimes been considered significant, because it seems to parallel cotyledon number, but there are many exceptions. In inflorescences, prophylls may be prominent organs, especially on the ultimate branchlets, and even on the pedicles, as in the Cyperaceae. (Here they are commonly called bracteoles.)

The solitary prophyll is usually superaxillary, with its ventral side toward the lateral branch. The form and anatomy of some solitary prophylls suggest that the solitary organ represents two fused prophylls. The solitary prophyll may have two apices and is frequently two-keeled. Two to seven vascular bundles are present. Where there are two bundles, one lies in each keel and one is larger than the other; there is no median bundle. The older interpretation of this asymmetry is that compression of the organ in ontogeny has distorted it, so that the middle part has become lateral. Form, position, and structure support the view that the solitary prophyll clearly represents a fused pair; the asymmetry results from closeness to the mother branch and consequent difference in development. Location and number of buds axillant to the prophylls should be further evidence of number of organs present, but it is claimed that in the prophylls of some monocotyledons there is only one bud. This is perhaps the condition to be expected if prophylls are the reduced first leaves of a shoot; the distal node is normally the better developed and the only one to have an axillary bud. Anatomy demonstrates that the prophylls are not stipules of the subtending leaf; their vascular supply is derived from the lateral branch, not from the mother stem.

Prophylls seem to have no significance as unique appendages in the shoot; they are merely leaves of reduced form, sometimes in apparently peculiar positions. They seem to represent the first appendages of a lateral shoot, weakly developed and sometimes displaced by closeness to the mother shoot. The prophyll of monocotyledons is undoubtedly a pair of leaves.

The *terminal leaf* has been described as present in some reduced and highly specialized shoots, as in some grass inflorescences. In these shoots, the apical meristem has been transformed into a leaf primordium, and the leaf is, ontogenetically, strictly terminal. Stages in the assumption of a pseudoterminal leaf position in determinate shoots are seen in both dicotyledons and monocotyledons. In sympodial growth of woody twigs, the stem apex aborts and the dead tip may be abscised, as is a leaf, either in a terminal position or after crowding to one side into an apparently lateral position (*Tilia, Cladrastis*). In the monocotyledons, evidence of the true position of an apparently terminal leaf may be obscure or lacking. The genera *Streptopus, Disporum, Polygonatum,* and *Uvularia* show various stages in the loss of the stem tip. A vestigial apex may be enclosed in the sheath of the distal leaf. In *Uvularia,* one species shows no trace of a stem tip; another, a vestigial cone; another, an obvious stem tip. The "terminal" position of leaves in herbaceous monocotyledons is obviously secondary; the shoot apex is abortive or lost, and a leaf assumes its position. Determinate leaves have somewhat different origins in monocotyledons and dicotyledons. Similarly, a solitary carpel or stamen and a basal ovule (in a syncarpous ovary) may acquire a terminal position in a floral shoot. The "terminal" cotyledon of many monocotyledon embryos is also secondary (Chap. 9). Change of position of these organs, congenitally established, does not make such a leaf, carpel, ovule, or cotyledon morphologically cauline, as has been sometimes claimed. The error of interpretation of basal ovules as cauline because of their apparently terminal position has brought about suggested changes in the classification of major angiosperm taxa that are morphologically unsound: the division of angiosperms into Phyllosporeae and Stachyosporeae, and the removal of the Casuarinaceae from the angiosperms.

The *monocotyledon leaf* is characteristically simple; compound forms are rare—Dioscoreaceae, some of the Liliaceae and Araceae. Parallel venation is dominant; the compound leaves and some simple leaves have parallel-reticulate, palmate-reticulate, or pinnate-reticulate venation—*Trillium, Smilax, Colocasia, Arisaema,* Butomaceae, Pontederiaceae, Alismataceae. The simple leaf is elongate, commonly linear, and consists of a sheathing base—the leaf sheath or leafbase—and a distal "limb"; these parts may merge, or be more or less clearly delimited.

The resemblance of the simple, linear monocotyledon leaf to the phyllodes of some dicotyledons suggested the *phyllode theory,* the conception that this leaf represents morphologically the basal sheath or petiole of an ancestral leaf. The sheathing base is looked upon as an adaptation in the development of the characteristic habit in monocotyledons—leaves close-packed on a greatly shortened stem. In the parallel venation of the "limb" is seen the vascular pattern of a petiole.

Morphological support for this theory is supposedly given by the close resemblance of the monocotyledon leaf to the sheathing bases of some dicotyledon leaves—Umbelliferae—and of the resemblance of the terete "limb" to petioles in such genera as *Triglochin* and species of *Allium, Sagittaria, Sisyrinchium.* Vascular anatomy is considered to support the interpretation of the monocotyledonous leaf as homologous with the proximal part of the dicotyledonous leaf. Parallel venation is characteristic of sheathing leafbases and petioles. Species of *Sagittaria* are believed to show stages in the flattening of terete leaves. The flattened leaves show two series, dorsal and ventral, of vascular bundles oriented with phloem toward the dorsal and ventral surfaces, respectively. Bifacial leaves, so formed, show stages in the merging of these series, with, in the completely flattened leaves of some taxa, all the vascular bundles in one plane and alternating irregularly in orientation. This vascular structure is present also in the leaves of the Pontederiaceae, Hydrocharitaceae, and other families. An apparently dorsiventral leaf is shown by its anatomy to be a modified cylindrical leaf.

It has been argued that the concept of the monocotyledon leaf as a phyllode cannot be supported by the presence of inverted bundles, because similar bundles occur in petioles of dicotyledon leaves, but the leaves cited—*Eryngium, Ranunculus, Plantago*—are also phyllodes—flattened petioles, rachises, or midribs.

As interpreted under the phyllode theory, the modification of the monocotyledon leaf by reduction has gone beyond the loss of the blade. The leaves of many genera of the Liliaceae and Amaryllidaceae—*Hyacinthus, Doryanthes, Dracaena*—have petioles reduced to vestigial, "solid" apices (the leafbase forms the photosynthetic structure). In *Hemerocallis* and other taxa, even this tip is considered lost, and the leafbase forms the entire leaf.

The phyllode theory also holds that, in contrast with this reduction, specialization has brought about the development of a new leaf part, a secondary blade, by the expansion of the tip of the phyllode to form a distal lamina. The palms, Scitamineae, Alismataceae, *Smilax, Eichornia, Pontederia,* are cited as examples. In such blades, parallel venation is considered evidence that the "blade" is petiolar in nature, but the venation is commonly intermediate between parallel and reticulate.

The theory that the simplicity of the typical monocotyledon leaf is the result of reduction was discussed at the beginning of the twentieth century as a part of the idea that the monocotyledons arose by the "self-adaptation" of dicotyledons "to a moist or aquatic habitat."

The blade-bearing monocotyledon leaf has been considered by some to be the primitive, by others, the highly developed type. The simple, ligulate leaf has been called "rudimentary," but it is associated with highly specialized habit, as in the grasses. A reticulate lamina in monocotyledon leaves may represent survival of an ancestral character or a secondary structure developed as an elaboration of a part of a parallel-veined leaf. The evidence that the monocotyledon leaf is, in large measure, a phyllode is convincing, and there seems no doubt that the "blade" of such leaves as those of *Eichornia* and *Pontederia* represents the modified tip of a phyllodelike leaf.

The interpretation of the blade of the palm leaf as an example of a lamina secondarily acquired by modification of the tip of the petiole is not supported by its structure or its ontogeny. Early stages of its ontogeny show that this leaf is a "complete leaf"—sheath, petiole, and blade. The development of petiole and blade follows the pattern of a typical dicotyledon leaf; the primordium of the blade is formed on the leaf buttress. This is followed by development of the petiole from an intercalary meristem at its base. If the blade is to be interpreted as petiolar in nature, the presence of an unusually long petiole and a prominent, well-defined leafbase must be explained.

There is little evidence to support the selection of an ancestral type for the monocotyledon leaf. On the assumption that the monocotyledons arose from dicotyledonous stock, the primitive monocotyledon leaf probably resembled that of the dicotyledon in gross structure. But, if the angiosperms are polyphyletic, the ancestral monocotyledon leaf may have differed greatly from that of the dicotyledon and comparisons with modern types would therefore be valueless.

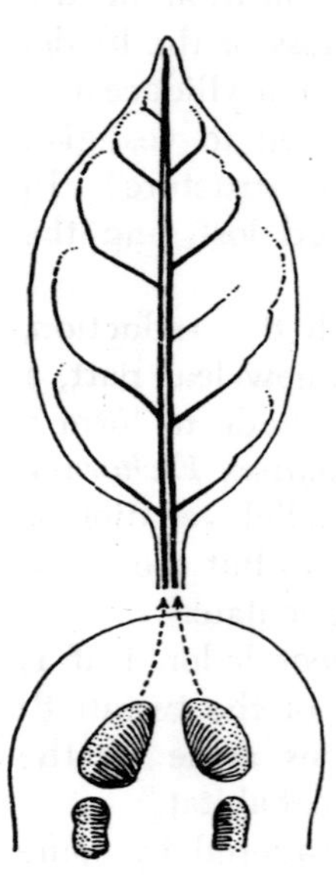

Fig. 5. Diagram showing vascularization of node and leaf of *Austrobaileya.* Unilacunar nodal structure with two traces which continue as two bundles through petiole and leaf blade. (*After Bailey and Swamy.*)

In the morphological study of the angiosperm leaf, recognition of shoot and root as the basic structural categories in the plant body raises the question of the fundamental nature of the leaf and its relation to the stem. The lower vascular plants suggest that there are two morphologically distinct types of leaves: *microphylls,* which are lobes or outgrowths of the stem; and *megaphylls,* which are modified minor branch systems. Leaf gaps are associated with the traces of megaphylls but not with those of microphylls. The angiosperms are megaphyllous.

As a part of the development of the *telome theory* of the nature of the plant body, considerable attention has been given to the interpretation of the make-up of the angiosperm leaf and sporophylls. These appendages have been interpreted as systems of dichotomizing axis tips, united laterally. But nodal structure—number, position, and relation of traces and gaps—and ontogeny of the appendages, especially development by apical and marginal meristems, do not support a telomic make-up.

All appendages in the angiosperms have long been interpreted as having basically an odd number of traces, but it has been shown recently that, in the primitive node, the appendage has two traces associated with one gap (Fig. 5). Branches of the shoot also have two traces with a single gap; leaf and branch are alike in the origin of their vascular supply. The nature of the leaf as a basic part of the shoot system is apparent, and the term "partial shoot" well expresses its fundamental nature.

A two-part trace system throughout the shoot may suggest ancestral dichotomy. Evidence in support of this view has been seen in veinlet branching, cotyledonary lobing, and the forking of some simple stigmas, but phyllotaxy and branching habit do not show dichotomy. (The "dichotomous" branching of a palm is probably not true dichotomy. See Chap. 11.) It has been pointed out in studies of apical meristems that leaf and branch primordia have different origins; leaf initials arise less deeply in the meristem than do branch initials but this difference is apparently not constant. (Little attention has been given to the origin of branch primordia.)

The various interpretations of the make-up of the shoot fall into three morphological categories: that in which the stem is regarded as an axis, with lateral lobes or appendages, the leaves; that in which the shoot is considered a multiple structure, made up of segments or units called phytons; and that in which the stem is considered a secondary structure, built up partly or wholly of leaves, which are the fundamental units. On the basis of the second and third interpretations, the phyton theory and the leaf-skin theory of the nature of shoot have been proposed.

Akin to the interpretation that the leafbase is, in part, cauline, is the theory that the outer tissues—"skin"—of the shoot axis consist of decurrent leafbases. This "skin" has been described for some taxa, both dicotyledons and monocotyledons, and regarded as present in most, probably in all, angiosperms. Formation of the "leaf skin" is considered to be by downward continuation and development of the leafbase, as the internode elongates. Limitation of the "skin," both externally and internally, has been variously interpreted; according to one description, this is "a question solely of definition." Obviously, the theory meets major morphological difficulties where structural limits are not set, and it seems of little importance.

Under the phyton theory, the shoot consists of "units of growth" that are renewed by a type of terminal "budding." The make-up of these units has been variously defined: as an internode with its attached leaf; a leaf with a root (the internode being the base of the leaf); a segment of the stem, limited by nodes, with or without a leaf; a leaf primordium with its base incorporated in the axis. (A root, as a part of the phyton, is lacking in angiosperms.) Concepts of the shoot as consisting of a series of structural units are old and have been obscured by the dominance of the stem-and-leaf theory. Anatomically, units like those described under the leaf-skin and phyton theories do not exist; the shoot is the basic unit.

The Anatomy of the Plant Body*

Structurally, the basic plant body is that built up by the embryo and its apical meristems. This is the *primary body*. All fundamental body parts are represented in the primary body. To this primary body is

* Anatomy of the flower in detail is discussed in Chap. 3.

commonly added a secondary body, by the activity of secondary meristems of the cambium type. The *secondary tissues* so formed may obscure, distort, or destroy parts of the primary body. The various organs may consist entirely of primary tissues (formed by primary meristems) or partly of primary and partly of secondary tissues. The primary body is complete in itself in so far as the presence of all basic organs is concerned—root and shoot (stem and leaves, sterile and fertile). In all these parts, vascular tissues—forming the *vascular skeleton*—are prominent structural features. In the axis (root and shoot), a central core, the *stele* or *central cylinder*, is more or less clearly set apart from the surrounding cortical and epidermal tissues. Limiting the stele is the *endodermis,* a uniseriate layer of specialized cells. Morphologically, the endodermis is probably best considered the outermost layer of the stele. In the angiosperms, as compared with lower vascular plants, the endodermis is greatly reduced in distribution and in histological structure.

The anatomical structure of the primary body of the angiosperms, especially that of the shoot, is complex. The modifications of high specialization, with reduction, accompanying the herbaceous, aquatic, epiphytic, and parasitic habits, have brought about vascular structure that may be difficult to interpret.

A basis for the interpretation of the vascular structure of the stele was provided by the *stelar theory,* late in the nineteenth century. This theory has been somewhat modified and new terms added, but satisfactory definitions are difficult to make and there has been much looseness in use of the terms applied to the stele and its vascular prolongations. The terms *protostele* and *siphonostele* were applied to two basic types: protostele, to a central cylinder with a solid (pithless) vascular core; and siphonostele, to a cylinder with a tubular vascular core and central pith. The siphonostele, when broken by openings (gaps) where traces pass out to lateral appendages, is *phyllosiphonic;* where gaps are formed by branch traces alone, it is *cladosiphonic.* The term *solenostele* has been applied to siphonosteles in which the leaf gaps in the internode above are closed before the gaps associated with traces next above appear, and *dictyostele* where the vascular tube is dissected into a meshwork by overlapping leaf gaps. But, in general use, the term dictyostele has been incorrectly applied to angiosperm steles; by original definition, the vascular bundles of a dictyostele are concentric, and this is a type unknown in angiosperms. The newer term *eustele* is now in general use for steles with a vascular skeleton of collateral or bicollateral bundles which anastomose more or less freely. This is the common type in angiosperm shoots. Steles in which phloem is present only external to the xylem are *ectophloic;* those in which there is also phloem internal to the xylem are *amphiphloic.*

The ectophloic stele is the common type, but amphiphloic steles are frequent—Apocynaceae, Asclepiadaceae, Campanulaceae Compositae, Convolvulaceae, Cucurbitaceae, Gentianaceae, Myrtaceae. Internal phloem is especially well developed in the Solanaceae and Cucurbitaceae, where, in herbaceous genera, it may play a prominent part in conduction. The phylogenetic relations of these stelar types in the angiosperms are uncertain. The sometimes apparently vestigial or degenerate condition of the internal phloem and its late ontogenetic development suggest that the amphiphloic stele may be more primitive than the ectophloic, but the absence of internal phloem in primitive families and its presence in some advanced families does not support this concept. If the angiosperms are polyphyletic, both stelar types may have been present in the stock from which they arose.

Anomalous vascular structure is frequent in angiosperm roots and stems. The peculiarities are various and are chiefly in the secondary tissues and in their method of origin and development. The original cambium of the young stem or root may cease to function and may be replaced by other secondary cambia which arise outside of the phloem. A cambium of normal type and persistence may form xylem and phloem of unusual distribution and arrangement. Excessive multiplication of parenchyma in restricted regions may break the original cylinder into strips. A cambium that develops tissue only centripetally may form complete vascular bundles embedded in "interfascicular tissue." Anomalous steles are doubtless in large part adaptations to unusual function, habit, or habitat; they occur in many woody vines and parasites, and in food-storage organs. They may characterize families—Chenopodiaceae, Amaranthaceae.

Anatomy of the Root. The root is typically protostelic; in specialized types, a pith may be present. Angiosperm root steles range from *monarch* to *polyarch*, but are most commonly *diarch* and *tetrarch*. The basic types are clearly the diarch and tetrarch, and each of these has been claimed to be the most primitive. The diarch type is characteristic of large groups of herbaceous genera—members of the Ranales, Rhoeadales, and Urticales, for example. Woody taxa commonly have tetrarch roots. The less common stelar types seem more likely to have been derived, phylogenetically, from the tetrarch than from the diarch. Diarch and monarch are probably reduction types; monarch steles are found in greatly reduced roots. The roots of the monocotyledons are typically polyarch and usually have a pith. It seems probable that the tetrarch stele is the primitive type; it best fits, structurally, into hypocotyls and into the pattern of the two-trace node; it is characteristic of most of the Archichlamydeae and is rare in the Metachlamydeae. The polyarch stele with a pith seems to be highly specialized.

Anatomy of the Shoot. The vascular structure of the shoot axis in the angiosperms is complex and has a pattern that is repeated from node to node. The nodal sections are more complicated than the internodes, because of the presence of divergent vascular strands—*leaf traces*—which extend from the internodal system to the leaf. *Branch traces,* always two, arise just above the leaf traces and similarly connect the vascular stele of the branch with the stele of the mother axis. The number of leaf traces ranges from one to many and is characteristic of taxa; it is phylogenetically significant.

The term leaf trace was first applied to the traces of a leaf, collectively, but is now applied to the individual strand, because number of strands per leaf is important. A leaf trace cannot be delimited rigidly, because it is continuous from stele to leaf. Distally, it is considered to end where it passes into the cortex. In the cortex, it may fork or unite with other traces. (The base of the petiole may show a number different from that of the stele.) Proximally, the trace unites with the stelar cylinder or with bundles of the internodal steles at various levels, according to patterns related to trace number and phyllotaxy. The stele may be siphonostelic, with an essentially unbroken vascular cylinder, or eustelic, with "free" bundles of various sizes. The eustelic type has been incorrectly considered characteristic of herbs, but many herbaceous genera—*Nicotiana, Salvia, Aster, Hypericum*—have siphonostelic stems. Woody angiosperms have both types of steles. The eustele, with reduced vascular tissue, is perhaps the advanced stelar form.

Areas where cortex and pith are continuous are termed *leaf* and *branch gaps,* respectively. Where the primary vascular stele has the form of a more or less unbroken cylinder, breaks are present above and often lateral to the leaf and branch traces where they depart from the cylinder. Each trace may have its own gap, or two or more traces may have a common gap. Nodes are termed unilacunar, trilacunar, and multilacunar, where the number of gaps is one, three, and many, without regard to number of traces per gap. In a unilacunar node, more than one trace may be associated with the gap.

The primary vascular cylinder is often weakly developed, and the gaps may not be obvious, becoming prominent only after secondary thickening has begun. In typical eusteles, where the cylinder consists of more or less isolated strands, gaps are usually not apparent; they have been considered as merged with the spaces between the bundles. In the floral shoot, nodal anatomy is like that in the vegetative shoot. This similarity is strong evidence that floral appendages are of leaf rank, that they are not organs *sui generis,* not mere lobes of the axis, not mere spore-bearing areas (Chap. 3).

The trilacunar node, with three traces (Fig. 6C), has long been looked upon as the primitive type in angiosperms, a type associated with the leaf type considered primitive—simple and pinnately veined. Support for the primitiveness of the trilacunar node was found in the dominance of this nodal structure in the Archichlamydeae, with elaboration to multilacunar types (Fig. 6*D*) in the Epacridaceae, Platanaceae, Araliaceae, Umbelliferae, Polygonaceae, and Meliaceae, and to reduction types in the Centrospermae, Myrtiflorae, and most of the Metachlamydeae. Some Archichlamydeae show reduction in trace number

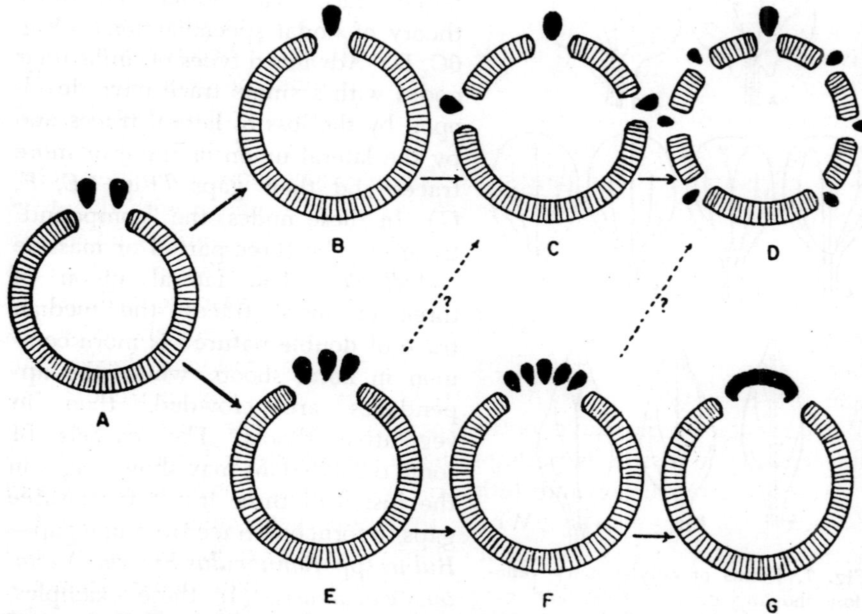

Fig. 6. Diagrams showing evolutionary development in nodal angiosperm anatomy from a primitive type. *A*, unilacunar with two traces; *B*, unilacunar with two traces fused; *C*, *trilacunar*, with three traces, the median a double trace; *D*, multilacunar with seven traces; *E*, ***unilacunar*** with three traces from one gap; *F*, five traces from one gap; *G*, unilacunar with one massive trace of five to seven fused traces. (*After Canright.*)

within the family—Leguminosae, Anacardiaceae; and some Metachlamydeae, multiplication of traces—Epacridaceae. Further evidence supporting the primitiveness of the three-trace node was seen in the presence of these nodes in seedlings of dicotyledons that have many traces in the mature plant. The monocotyledons are commonly multilacunar, but seedlings frequently have three-trace leaves, and their carpels, usually multitrace, may have, as reduction forms, three traces or only one.

The two-trace, unilacunar node (Fig. 6*A*)—a "fourth type" of node —was long overlooked. Certain two-trace organs—cotyledons, stamens in a few taxa, carpels with double midribs, set aside as abnormalities or examples of fused organs—are now seen to be persisting examples of early nodal structure in angiosperms. The odd-numbered types, so characteristic of angiosperms in general, have been derived by the union of the two traces of the primitive node, as a part of the specialization of stem and leaf. Fusion of the two traces of a unilacunar node gives a one-trace unilacunar node; the addition of lateral traces in pairs, with their individual gaps gives tri- and multilacunar nodes with odd-numbered traces (as under the older theory of nodal specialization) (Fig. 6*C*, *D*). Advanced types of unilacunar nodes with a single trace have developed by the loss of lateral traces and by the lateral union of three or more traces and their gaps (Fig. 6*E*, *F*, *G*). In these nodes, the "compound" trace may be three-parted or massive —*Asclepias*. The lateral union of three or more traces—the median trace of double nature—is more common in floral shoots, where the appendages are crowded, than in vegetative shoots. The carpels of some dicotyledons may show stages in the fusion of three traces from three gaps to form one trace from one gap—*Rubus* spp. *Ranunculus Ficaria*, *Anemone canadensis*. (In these examples, the reduction and simplification of nodal structure accompanies the evolutionary modification of a follicle in achene development. The carpels appear to be one-trace organs until the origin of their vascular supply is determined.)

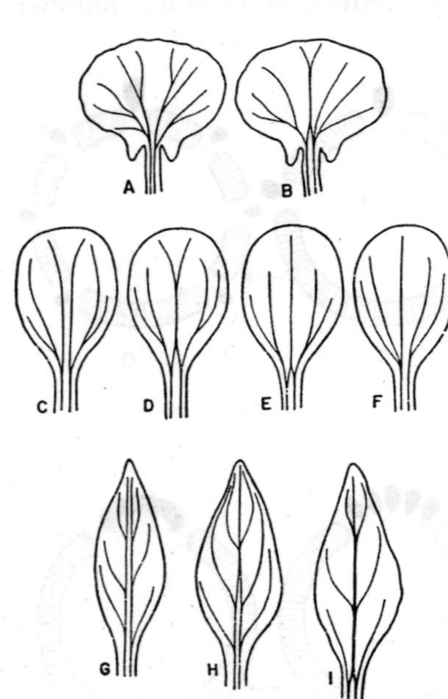

Fig. 7. Forms of cotyledonary venation showing double traces and extent of independence of the two vascular systems within the cotyledon. (*After Bailey.*)

Two-trace unilacunar nodes are known in the stems of the Austrobaileyaceae (Fig. 5), Chloranthaceae, Lactoridaceae, Trimeniaceae, and some Verbenaceae, Labiatae, and Solanaceae. In most of these taxa, the traces are independent well down in the eustele. In cotyledons and sporophylls also, there are frequent examples of paired or fused median traces. Paired traces are probably more numerous in cotyledonary than in vegetative nodes (Figs. 7 and 8). In reproductive nodes, they are doubtless more frequent than is known; their presence cannot be so readily determined as in leafy stems, but two-trace floral organs are occasional. Double median bundles are frequently found in the carpels of taxa of widely scattered families—Ericaceae (*Clethra, Pyrola, Epigaea*), Caryophyllaceae (*Anagallis*). In some free carpels, the stigmas are two-lobed and the vascular supply forked. The double-bundle was described in many taxa in demonstration of the carpel polymorphism theory.

Two-trace stamens are uncommon but are present in both advanced and primitive taxa: *Austrobaileya, Sarcandra, Victoria, Nuphar, Casuarina, Cyrtandra, Eranthemum, Doryanthes* (two pairs), several genera in the Betulaceae, Fagaceae, and Proteaceae. Origin of the single trace in stamens is seen in *Hakea,* where, in a single species, even a single flower, some stamens have two independent traces and others have obviously double traces, which may fork distally.

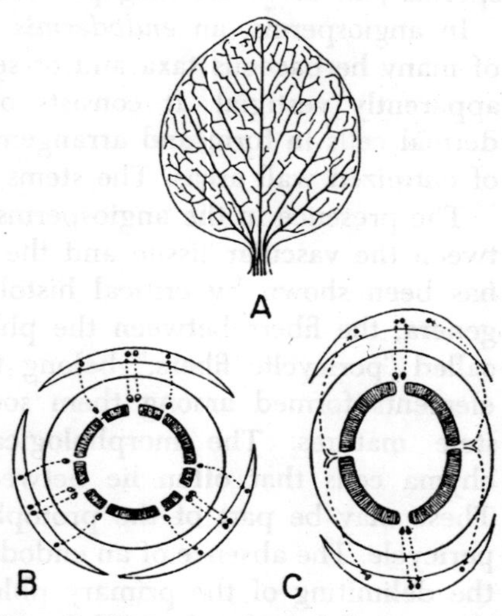

Fig. 8. Diagrams of cotyledonary nodes and cotyledon. *A, B,* of *Degeneria vitiensis* showing four traces in petiole and venation in *A* and a trilacunar node with three cotyledons with four traces each from three gaps each; *C,* of *Magnolia grandiflora* showing a trilacunar node with two cotyledons with four traces each. The four lateral traces "derived by the bifurcation of two traces which arise from two gaps in the stele." (*After Swamy.*)

Sporophylls with double vascular supplies have doubtless not been recorded because they were considered abnormal. The frequency of occurrence of these sporophylls with two traces cannot be compared with that of similar leaves, because trace origins in the flower are often concealed by fusion.

The presence of two traces in cotyledons has long been known and at one time aroused interest as possible evidence of ancestral dichotomy persisting in the embryo. This formed part of the basis for a "theory of the double leaf trace"—a theory that two-trace appendages characterize most major vascular taxa. This theory received little attention, because of interest in the problem of the phylogenetic relation of the cotyledons of the dicotyledons and the monocotyledons—does the single cotyledon of the monocotyledons, with two traces, represent two fused cotyledons, or one of a pair, the other lost (Chap. 9)? The two-trace cotyledon was considered primitive for the Liliaceae, but from the viewpoint that it represents retention of the vascular supply of the two cotyledons of dicotyledonous ancestors, rather than as basic angiosperm structure.

Reduction in the cotyledon of the two traces to one—as in stems—has been described in *Scilla* (Chap. 9).

Recognition of the two-trace node as primitive for angiosperms constitutes a major forward step in the use of anatomy in interpreting the phylogeny of the angiosperms (although odd-numbered traces will doubtless continue to be described as characterizing angiosperms). In the possession of a nodal trace system based on two units, the angiosperms join the other megaphyllous taxa.

In angiosperms, an *endodermis* is present in roots and in the stems of many herbaceous taxa and of seedlings. In these locations it is often apparently vestigial. It consists of a layer of cells resembling endodermal cells in form and arrangement but lacking the critical character of cutinized wall areas. The stems of woody plants lack an endodermis.

The presence in the angiosperms of a *pericycle*, a sheath of tissue between the vascular tissue and the endodermis, has been questioned. It has been shown by critical histological studies that, in at least some genera, the fibers between the phloem and the endodermis, commonly called "pericyclic fibers," belong to the primary phloem; protophloem elements formed among them soon degenerate and disappear as the stele matures. The morphological problem here relates to parenchyma cells that often lie between these fibers and the endodermis. These may be part of the protophloem or of the more or less distinct pericycle. The absence of an endodermis in most angiosperm stems makes the delimiting of the primary phloem, with its parenchyma and fibers, uncertain. A pericycle is characteristic of vascular cryptogams and of the roots of seed plants. In the angiosperm stem, the pericycle, together with the endodermis, is in process of reduction and loss.

The primitive angiosperm leaf, under the interpretation of the primitive nodal structure as trilacunar, was seen as palmately veined, perhaps three-lobed, with three major veins united at the base of the blade. The present recognition of the unilacunar, two-trace node as apparently the basic nodal structure also supports the concept of the palmately veined leaf as primitive. Leaves with two traces continued independently through petiole and blade (as a double midvein) are rare—*Austrobaileya* (Fig. 5), *Chloranthus, Sarcandra, Ascarina.* Pairs of traces that continue through the organ are more frequent in cotyledons and sporophylls than in leaves (Fig. 7).

The more advanced three-trace system is formed by the addition of two lateral traces with separate gaps (Fig. 6*C*). The three traces—the median double in nature fundamentally—unite at various points: in the cortex, at the base of the petiole, or in the petiole. From this three-trace system, a common type, have been derived other still more specialized types.

There is strong correlation between the simple, pinnate leaf and the unilacunar node, but there are also extreme exceptions, as in *Eupomatia*, where there are 7 to 11 traces from as many gaps. Leaves with sheathing bases usually have several traces and gaps. Though the fundamental

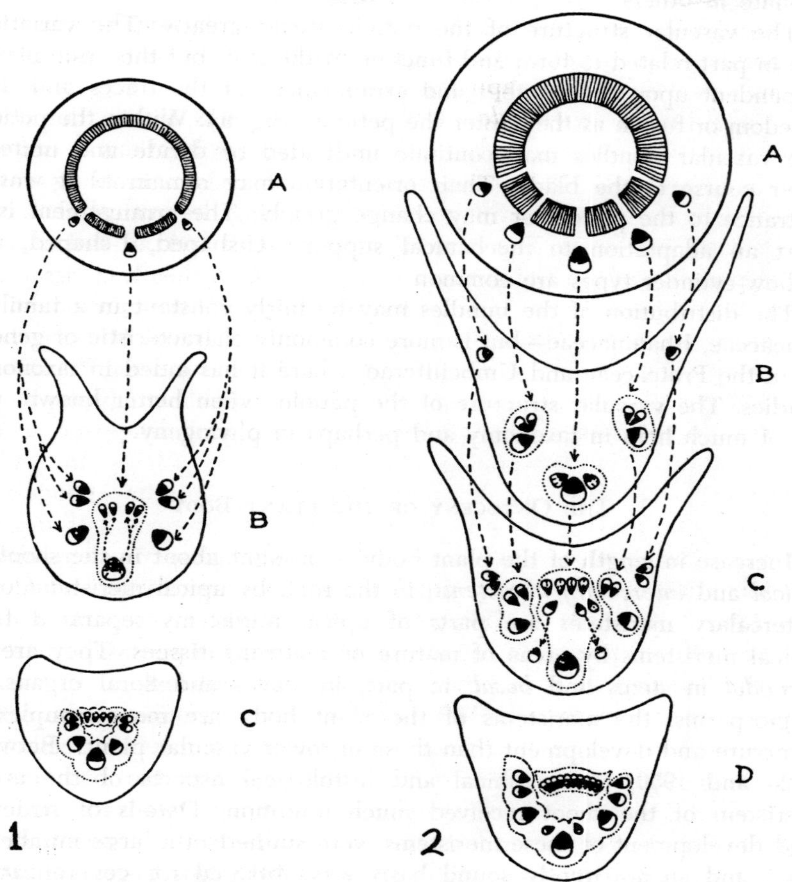

Fig. 9. Two series of diagrams showing vascularization of node and successively higher levels of the petiole of *Degeneria vitiensis:* series 1 from a seedling leaf and series 2 from a mature leaf. (*After Swamy.*)

vascular relationship of node and leaf now seems clear, variations related in part to leaf form are many. For example, highly complex nodal structure accompanies interpetiolar stipules and other stipules of unusual form and function.

Pinnate venation may have been derived in two ways: directly from a simple, two-trace leaf with a double midvein, by a strengthening of lateral veinlets; from a three-trace leaf with three veins by a weakening or loss of the lateral traces and veins. Among leaves of the woody Ranales, which, by association in this order with so many other highly primitive characters, have been looked upon as simple, there are both palmate and pinnate types—palmate in *Austrobaileya* and *Tetracentron,* pinnate in others.

The vascular structure of the petiole varies greatly. The variations are in part related to form and function of the leaf, but the basic plan is dependent upon the number and arrangement of the traces and their freedom or fusion as they enter the petiole (Fig. 9). Within the petiole, the vascular bundles may continue undivided or divide and unite in their course to the blade. Their orientation may remain as it was at entrance to the petiole or may change greatly. The arrangement is in part an adaptation to mechanical support; U-shaped, I-shaped, and hollow-cylinder types are common.

The distribution of the bundles may be fairly constant in a family—Ericaceae, Rhamnaceae—but is more commonly characteristic of genera, as in the Proteaceae and Umbelliferae, where it has aided in taxonomic studies. The vascular structure of the petiole, when better known, will be of much help in taxonomy and perhaps in phylogeny.

The Ontogeny of the Plant Body

Increase in length of the plant body is brought about in the shoot by *apical* and *intercalary meristems*; in the root, by apical meristems only. Intercalary meristems are parts of apical meristems separated from apical meristems by areas of mature or maturing tissues. They are *internodal* in stems and *basal,* in part, in leaves and floral organs. In angiosperms, the meristems of the plant body are more complex in structure and development than those of lower vascular plants. Between 1920 and 1950, topographical and histological aspects of the apical meristem of the shoot received much attention. Details of structure and development of these meristems were studied in a large number of taxa, and an apparently sound basis was obtained for generalizations on the morphological value and significance of their structure.

Ontogeny of the Shoot Apex. The *apical-cell theory,* which arose in the earlier days of interest in the ontogeny of the plant body, assumed development from a solitary apical cell or group of apical initials. This theory was replaced in the 1860s by the *histogen theory,* which holds that development is by meristems that build up individually the various tissues or parts of the axis. The histogen theory dominated interpretation of apical meristems for more than fifty years, but in the early decades of the twentieth century, with increasing interest in the anatomy of the shoot, it was found to have little morphological value and was replaced by the *tunica-corpus theory* (Fig. 10). This new theory stimulated research in shoot apices of many taxa during the following decades.

According to the tunica-corpus theory, the shoot apex consists of meristematic tissues arranged more or less distinctly in two major parts

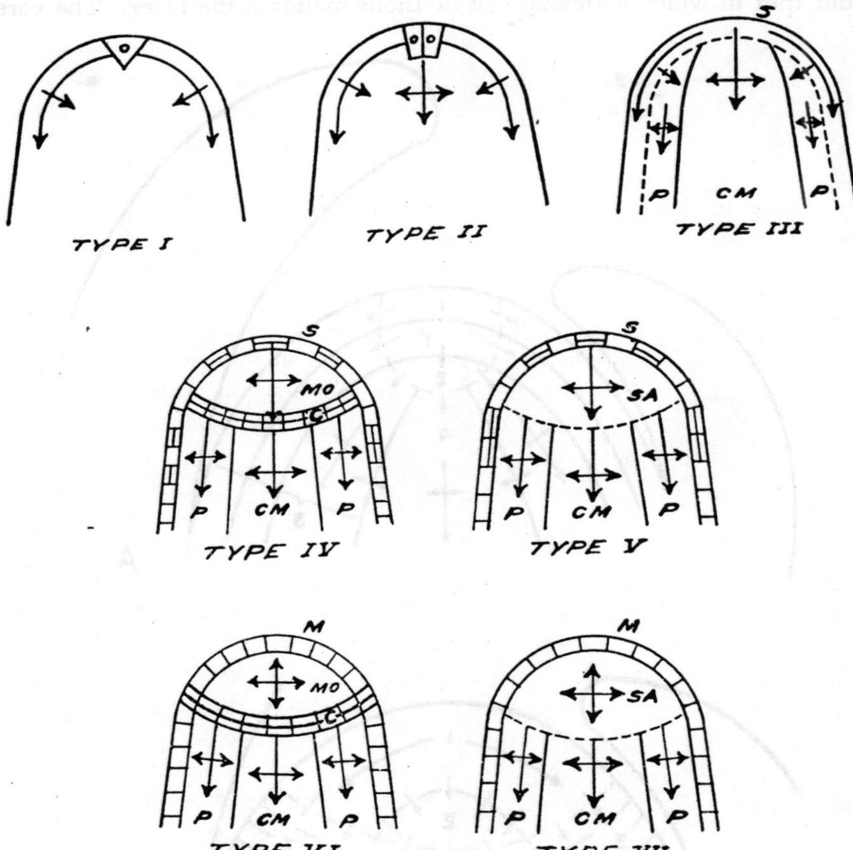

Fig. 10. Diagrams of longitudinal sections of shoot apices showing types of organization in vascular plants. S, surface meristem; *M*, mantle; *MO*, central mother cells; *C*, cambium-like zone; *SA*, subapical initials; *CM*, central meristem; *P*, peripheral meristem. (*After Popham.*)

—a central core, or *corpus*, sheathed by one or more external layers, the *tunica* (Fig. 11). In descriptions of the earlier years under this theory, interpretations were made rather rigidly. But it soon became evident that limitation of zones is not always clear and was sometimes "found" in different places by different investigators; that the number of layers in the tunica, considered characteristic of a species, varies with position in the plant, vigor of the shoot, and season. To improve interpretations, it was suggested that the terms tunica and corpus be replaced by *mantle* and *core*—so defined as to avoid the rigidity of the earlier use and to allow considerable variation in some characters. Mantle is defined as consisting of all the distinct peripheral layers of the apex in which anticlinal cell divisions maintain the layers. The core

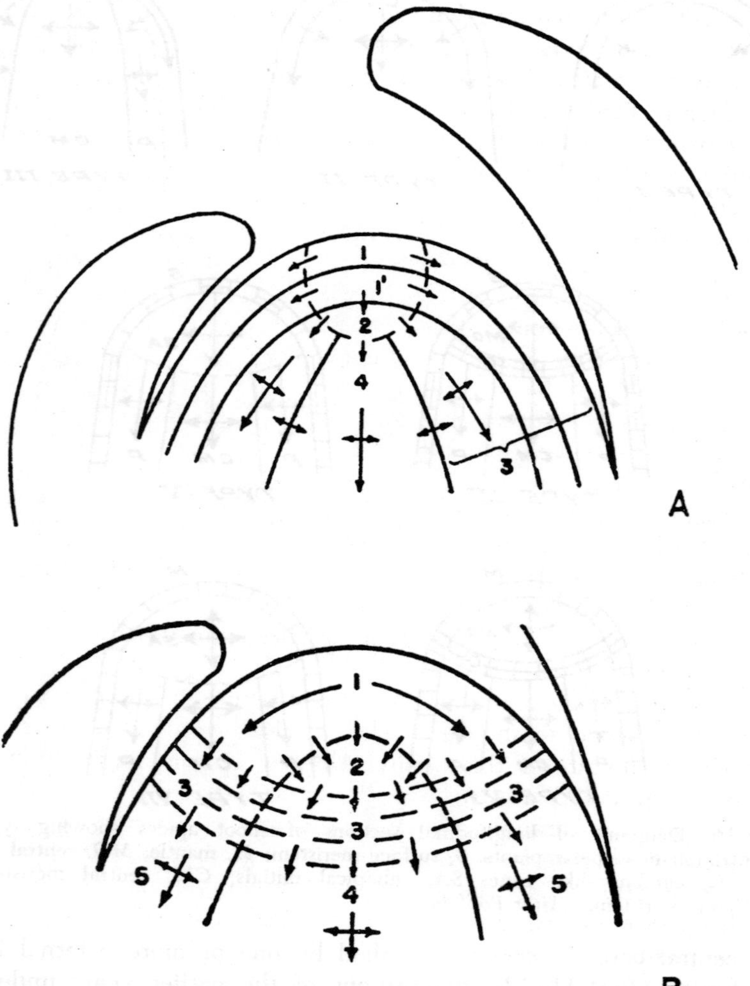

Fig. 11. Diagrams to illustrate two concepts of growth in dicotyledonous shoot apex. *A*, the tunica corpus theory: zone 1 and 1′, initials of tunica layers; zone 2, corpus initials; zone 3, peripheral zone; zone 4, rib meristem. *B*, the mantle-core concept: zone 1, mantle layers; zone 2, central-mother-cell zone; zone 3, cambium-like zone; zone 4, rib meristem; zone 5, peripheral zone. (*After Gifford.*)

consists of the remaining central tissues in which cell divisions are in various planes and little or no peripheral layering is present. The mantle may add to the core.

In somewhat modified definitions, the corpus is the central core of the apex, consisting of larger cells, varying in form, without definite peripheral layering, and with cell divisions in many planes; the tunica consists of one to several outer uniseriate layers of smaller cells, uniform in size, with divisions wholly or largely anticlinal (Fig. 12). In the corpus, cell divisions are infrequent, and the protoplasts stain lightly; in the tunica, divisions are more frequent and the cytoplasm stains deeply. Tunica and corpus, though not always sharply delimited, are independent, self-perpetuating meristems. The chief variations are in number of layers in the tunica—at first considered constant for a species—and in the clearness of separation of tunica and corpus. The major difficulty in delimiting tunica and corpus has been the interpretation of the outer layer of the corpus, a layer, with frequent anticlinal divisions, which may be more or less distinct as a uniseriate sheath and appears transitional between corpus and tunica.

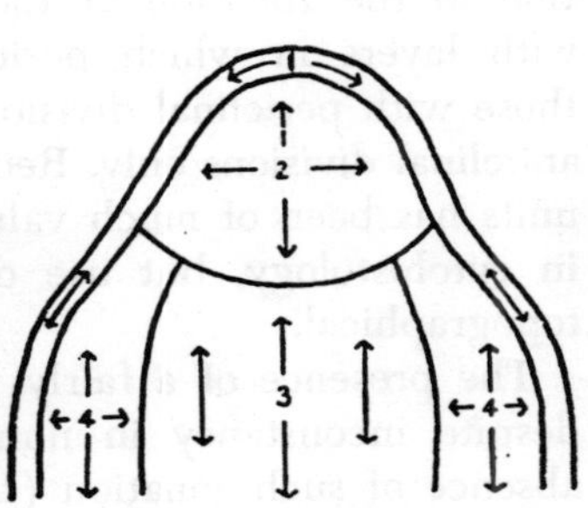

Fig. 12. Zonation diagram of a hypothetical shoot apex showing direction of cell divisions. 1, tunica; 2, corpus; 3, rib meristem; 4, flank meristem. (*After Stant.*)

Below the tunica-corpus zone in the apex, there may be distinguishable a region of transition between the highly meristematic distal cells and the maturing cells, in which enlarging cells of like type stand in long rows or files. This zone has been called a third meristem, a *rib meristem* (Fig. 12).

A rather broad survey of shoot-apex structure in various tribes of the Rosaceae seems to show that clear tunica-corpus zonation is not a characteristic feature of this family. Absence of definite zonation here has been considered a surprising departure from the apex structure described for taxa throughout angiosperms and as perhaps the result of differences in interpretation. But delimitation of tunica and corpus may be weak or obscure. In the monocotyledons, rather few shoot apices have been studied, perhaps because of greater complexity of structure in these plants.

The tunica is probably most commonly two-layered. A one-layered tunica is common in monocotyledons and occasional in dicotyledons. A few major taxa have mostly several-layered tunicas—Compositae, Caprifoliaceae, Rosaceae, Guttiferae. Variability in number of layers is more frequent in dicotyledons than in monocotyledons. According to the information now available, number of tunica layers seems to have no phylogenetic significance. A comparison of doubtful importance has been made with the reduction in number of layers in the nucellus, where the larger number of cell layers seems to represent primitive structure (Chap. 7). Morphological significance has been seen in manner and place of origin of leaf initials, but a leaf primordium may arise wholly from cells of the tunica or from initials in both tunica and corpus.

Angiosperm shoot apices seem to have two types of organization: one, in which the cells of the tunica divide anticlinally except in the center of the apex, where some periclinal divisions occur; the other, in which no periclinal divisions occur. Evolutionary progress in specialization in the zonation of the apex has apparently been from a tunica with layers in which periclinal divisions are general throughout, to those with periclinal divisions restricted to the apex, and to those with anticlinal divisions only. Recognition of tunica and corpus as descriptive units has been of much value in studies of growth activities, especially in cytohistology, but use of the terms must be considered primarily topographical.

The presence of a fairly well-defined zonation—tunica and corpus—despite inconstancy in number of layers in the tunica and perhaps absence of such zonation (as reported for some rosaceous taxa), seems to characterize the angiosperms. The zonation of apical meristems should not be interpreted too rigidly, as it was in the earlier years of its study; the apices must be recognized as dynamic structures, responding within limits to various growth conditions.

The shoot apex from which a flower will develop (the *reproductive* or *floral apex*) has been claimed to differ fundamentally from the vegetative apex, to be a completely different structure and, like the sporophylls borne upon it, a structure *sui generis*. But this concept has been shown by several critical studies to be without foundation. In the transformation of a vegetative into a floral apex, the vegetative axis undergoes, as it elongates, a gradual change in zonal pattern. The change is in proportion of zones and is, in large part, the result of the bringing close together of the nodes and the many appendages of the flower.

The interpretation of the floral apex as morphologically distinct from the vegetative apex was based largely on the following claims. First, the vegetative apex has a tunica and a corpus; the flower axis has a "parenchymatous core"—not highly meristematic—and a thick, "heavy," meristematic "mantle." But differences in gross structure between the two types of apices are in degree only and are associated with basic structure. In the floral apex, the nodes are crowded together, and the core is "parenchymatous" (not meristematic), because the shoot is determinate and cell divisions have largely ceased in the central part. The mantle is wide and complex in structure, because in it are developing the primordia of many appendages. The mantle corresponds to the tunica and the outer part of the corpus. Secondly, the floral appendages are claimed to have arisen from more superficial layers than do leaves. This difference does not exist; the wider mantle and the contrast between mantle and core—greater than in vegetative apices, because of more abundant primordia in the outer part and few cell divisions in the center—makes the origin appear to be more superficial. The third claim was that there are no foliar buttresses in the floral apex. But the buttresses develop later and are smaller, because the nodes are crowded. A fourth claim was that development of procambial strands is strictly acropetal in the floral apex, whereas in the vegetative axis development is in both directions from the base of an appendage. But development of the procambium is alike in both kinds of apices. The small size of the appendages at this stage and their crowded arrangement make direction of development difficult to determine. And, finally, it was claimed that carpellary primordia are not crescent-shaped and do not embrace the shoot axis as leaf primordia do. But the carpel primordia of many taxa, those in which carpel closure is postgenital, are crescent-shaped; those in which closure is congenital are not crescent-shaped. The floral apex is morphologically like the vegetative apex; differences are of degree only, associated with differences in function. It is a vegetative axis gradually transformed, not a new structure. Under some growth conditions, the floral axis may be transformed into a vegetative axis, as in terminally proliferated flowers.

Ontogeny of the Root Apex. The anatomy of the root apex was well known long before critical studies of the shoot apex were made, doubtless because of the greater simplicity of the root apex and better distinction of histogenetic zones. In the root apex was found, in large part, the basis for the histogen theory, but greater interest in the more complex shoot apex has placed acquaintance with this growing point ahead of that of the root.

Early descriptions of the apical meristem of the root set apart a central core, the *plerome;* an outer sheath, the *periblem;* and a uniseriate, outer layer, the *dermatogen.* Distinction of these layers, with implication of restriction in function and morphology in the parts developed by each, formed the *histogen theory.* Though better acquaintance with anatomy showed that the terms do not have the morphological value assigned to them, they have continued in use as of topographical value. Tunica and corpus are terms not applicable to the root apex, because of its markedly different morphology. The apical meristem of the root is shorter than that of the shoot and differs from it in its clearer and more constant zonation, in the definition of layers close to the apex, in the presence of a root cap, and in the absence of primordia of appendages.

The meristematic layers of the root apex of angiosperms are usually formed by three, rarely four, groups of initiating cells (Fig. 13). In the dicotyledons, the distal group forms the cap and the dermatogen; the median group, the periblem; the innermost, the plerome. In the monocotyledons, the distal group forms the cap; the median, the dermatogen

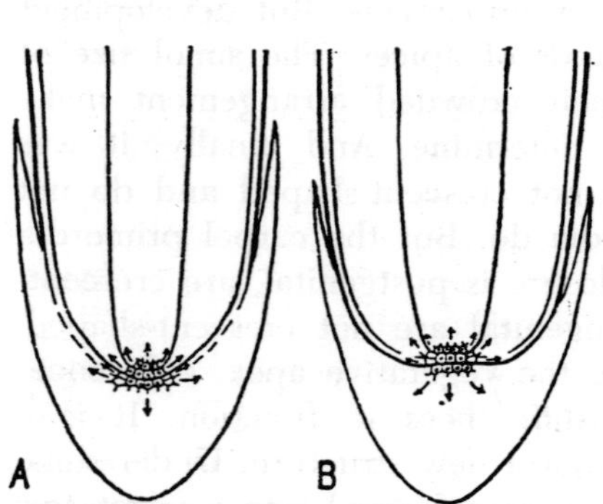

Fig. 13. Diagrams of root-apex types in angiosperms. *A*, initials in three groups, cap not distinct, formed by same initials as dermatogen; *B*, initials in three groups, cap distinct in structure and independent in origin. (*After Eames and MacDaniels.*)

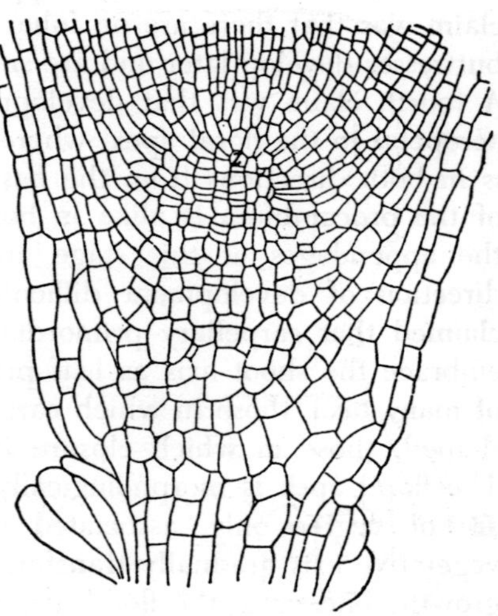

Fig. 14. Median longitudinal section of the basal end of a mature embryo of *Tropaeolum majus*, showing initial cell (Z) and suspensors attached at the base. (*After von Guttenberg.*)

and periblem; the innermost, the plerome. The outstanding characteristic of the apex of the dicotyledon root is the common origin of cap and dermatogen, a resemblance to the ancient type of root-apex origin, where both cap and epidermis are formed by a solitary apical cell (Fig. 14).

The root apex, lacking the complexities of appendage development, is simple, though the cap adds a different "appendage." Its simplicity of structure suggests that it is more primitive than the shoot apex. The simplicity of the root apex adds to the evidence (lack of appendages, protostelic structure, and exarch xylem) that the root is a more primitive organ than the shoot.

The *root cap* is an important morphological feature of the root. It is absent in only a few monocotyledons, where it seems to have been lost as a part of adaptation to an aquatic habitat.

Ontogeny of the Leaf. The development of the leaf follows patterns related as a whole to the type of leaf that is formed. Variety in details of meristem activity and tissue specialization is great, and only general features of form and anatomy are discussed here.

In the dicotyledons, the leaf primordium is initiated close to the shoot apex in the tunica or in an area involving cells of both tunica

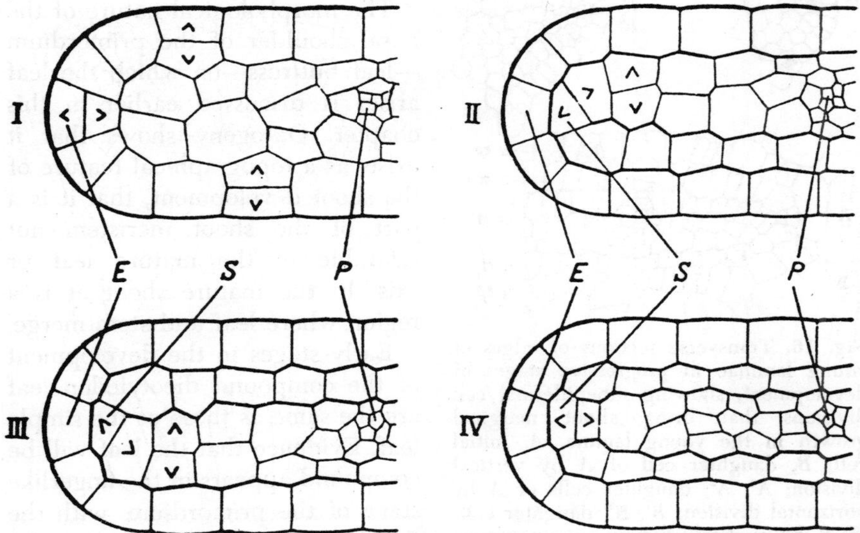

Fig. 15. Diagrams showing types of marginal growth in the leaf blade. *E*, **epidermal initial; S, subepidermal initial;** *P*, **procambium.** (*After Hara.*)

and outer corpus. A lateral shoulder, or "buttress," increases in size by apical growth and by lateral expansion around the apical meristem. The lateral extent of the more or less crescent-shaped mound so formed varies with form of leaf and with stipule form and position. Median growth builds up a somewhat flattened, fingerlike projection, on which marginal meristems soon appear. Apical growth continues, but increase in size is soon largely a result of the activity of the marginal meristems (Figs. 15 and 16), which early lay down the foundation of the blade, often outlining its general form very early. Below the marginal meristems a basal region, which shows less activity, later becomes an "intercalary" or "basal" meristem, which builds the petiole. If the leaf is stipulate, stipule meristems arise below this meristem on the shoulders of the buttress, or, where adnate to the petiole, on the petiole meristem when it becomes active. The petiolar meristem may remain inactive for a long period, or may continue cell division, with little cell enlargement. The petiole may develop either from a transverse meristematic plate, or, more commonly, by the rapid enlargement of small cells.

The midrib region of the leaf is formed by the central part of the early elongate primordium, the remainder of the blade by specialization of the tissues formed by the marginal meristems. The history of the various tissue layers of the blade is complex but follows a general pattern.

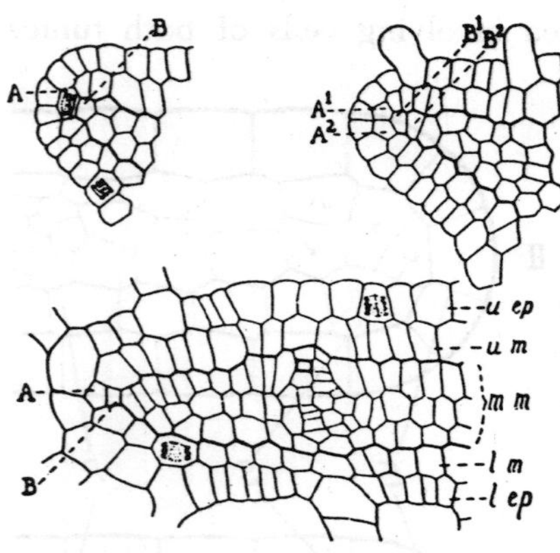

Fig. 16. Transverse sections of edges of young laminae in progressive stages of development, showing subepidermal cell divisions that bring about marginal growth in the young lamina. *A*, initial cell; *B*, daughter cell of *A* by vertical division; A^1, A^2, daughter cells of *A* by horizontal division; B^1, B^2, daughter cells of *B* by vertical division; *u ep*, upper epidermis; *u m*, upper mesophyll; *m m*, middle mesophyll; *l m*, lower mesophyll; *l ep*, lower epidermis. (*After Avery.*)

The morphological nature of the basal shoulder of the primordium —leaf buttress—on which the leaf arises is discussed earlier in this chapter. Ontogeny shows that it exists as a topographical feature of the shoot development, that it is a part of the shoot meristem not referable to the mature leaf or axis. In the mature shoot, it is a region where leaf and stem merge.

Early stages in the development of the compound dicotyledon leaf are the same as those of the simple leaf. Evidence that the leaf will be compound appears in the fingerlike stage of the primordium, with the development of leaflet primordia along the marginal ridges. Sequence in development in the leaflet primordia varies with the taxon—acropetal, basipetal, or "divergent." Apparently little is known of the origin of these new apical and marginal meristems from the margins of the mother primordium—whether they appear before or after the mother marginal meristems are established. Leaflet development follows the same course as that in the simple leaf. Ontogenetic fusion between leaflet primordia apparently may occur, as between floral-organ primordia, but is infrequent or rare and little is known about it. (Leaflet fusion in *Bauhinia* and related genera is apparently largely congenital.)

The history of development of the leaf of the monocotyledons differs only in detail from that of the dicotyledons. The leaf buttress is less prominent than in dicotyledons and may be absent. (In many monocotyledons, the leaves are borne more closely than in dicotyledons, and buttresses are inconspicuous or late in development, as in floral axes where the organs are crowded.) Buttress and primordium may be indistinguishable. The primordium has a broader base than in the dicotyledons; the crescent margins extend far, or completely (in some sheathing leaves), around the axis—an obvious relation to the sheathing bases of so many monocotyledons. Marginal meristems are absent. The activities of an apical meristem are brief, and the leaf is built up chiefly

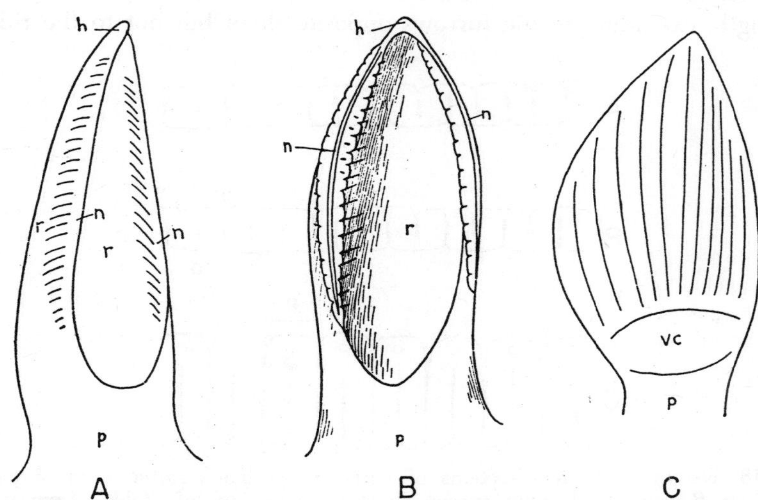

Fig. 17. Diagrammatic sketches, showing position and relationship of leaf parts in primordia of the two leaf types in palms. *A, B,* pinnate leaf, *B,* showing detail of furrows and ridges of pinnae primordia; *C,* palmate leaf; *h,* hook; *n,* rein; *p,* petiole; *r,* rachis; *vc,* ventral crest (hastula). (*From Eames,* 1953.)

by an intercalary meristem. Correlations can perhaps be found here with the supposed petiolar or leafbase nature of this leaf (Chap. 12).

Compound leaves are rare in the monocotyledons. Details of development of compound monocotyledon leaves are little known. The ontogeny of the compound palm leaf is described here because it is probably a type unique among angiosperms.

In the palms, the simple leaf, present in seedlings of most taxa and mature plants of a few, is primitive; the compound leaf has been derived from the simple. The derivation has been described as ontogenetic—by a folding of the blade brought about by alternate dorsal and ventral invaginations and a later splitting along the lines of the folds. The dissection is indeed ontogenetic, but the process is much more complex than this. The leaf is a "complete leaf," with sheathing base, petiole, and blade. The primordium is similar to that of other monocotyledonous leaves. But very early, while the primordium is only 1 or 2 mm long, a pattern of dissection is laid down which later determines the lines of folding and splitting of the blade into "leaflets" (Fig. 17).

The pattern of folding is laid down on both dorsal and ventral surfaces of the blade primordium by a series of shallow ridges and furrows (Fig. 18). Simultaneously, minute, needle-shaped openings, formed by separation of cells, appear in close rows in the tissue below each ridge, dorsal and ventral (Fig. 18*A*). These openings, at first only 10 to 20μ in diameter, do not extend to the surfaces of the blade; later, they increase in length, extending to the furrows opposite them but not to the ridges

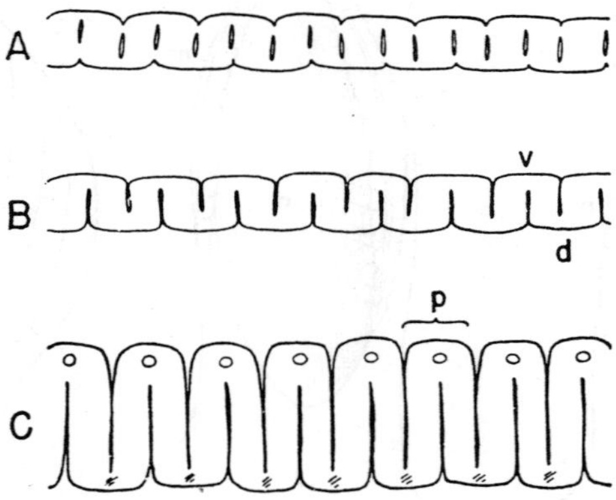

Fig. 18. Diagrams of cross sections of parts of leaf-blade primordium of palms, based on *Roystonea*, showing successive stages in origin of "folds" from which pinnae arise. *A*, early stage of differentiation, acicular slits below each ridge; *B*, somewhat advanced stage, the slits extended to the furrows opposite them; *C*, later stage, blade increased in thickness and in area, procambium of midveins of pinnae developed, beginnings of abscission-tissue development in dorsal ridges which separate pinnae. *d*, dorsal; *p*, pinna; *v*, ventral. (*From Eames*, 1953.)

(Fig. 18*B*). Rows of these needlelike perforations then unite, splitting the ridges longitudinally from below, and forming a series of low, compressed "folds" (Fig. 18*C*). Increase in tissues of the blade in area and in thickness builds up the ridges (folds), which alternate on the upper and lower sides and form the typical plicate structure (Fig. 19) of the immature and, in many palmate types, the mature leaf. Lateral separation into leaflets comes about as the leaf unfolds from the bud, through division by abscission layers or by "disorganization." The folding is initiated by dissection, and the folds are built up by differential growth—"invagination," in one sense. The folds are not formed by compression within the sheaths of outer leaves, though, in some genera, there may be a crinkling in other planes.

This is only the first part of the story of the formation of the compound leaf in palms. The major veins of the blade extend pinnately or palmately from a midrib or basal area, distally running parallel with the margin, and the lowermost veins form a narrow, strongly vascular band along the margin of the blade (Fig. 20). The early ridges and later folds do not extend into this band. When the folds are separated,

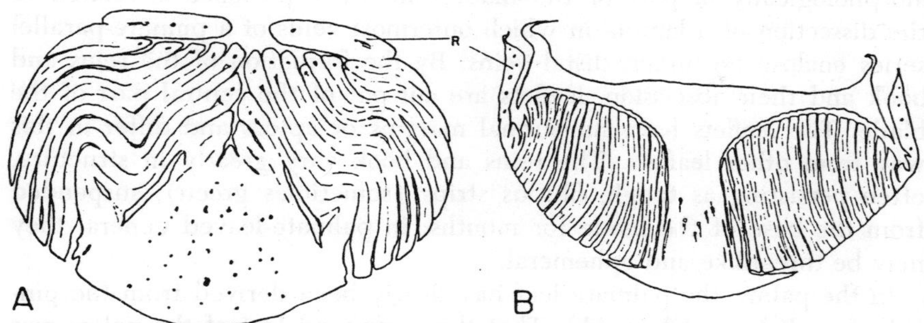

Fig. 19. Sketches of transverse sections of young leaves showing pinnate and palmate development, reins *R* at margins of blades. *A, Roystonia regia,* a pinnate type, rachis below developing pinnae; *B, Livistona chinensis,* a palmate type, hastula above united with dorsal crest (lobe of rachis) below. (*From Eames,* 1953.)

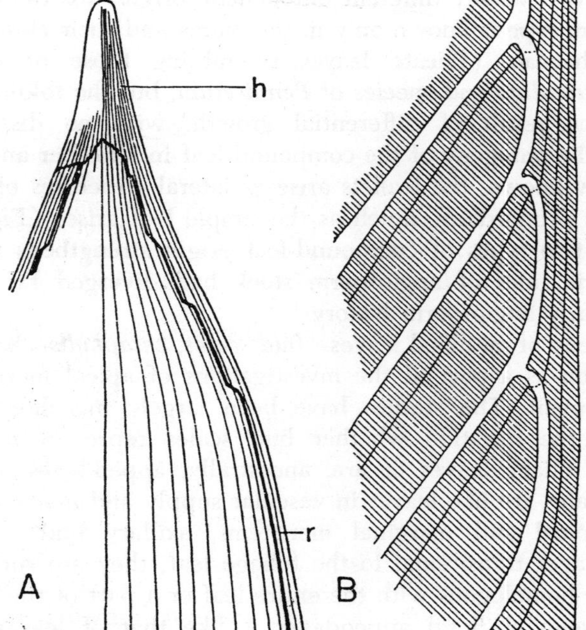

Fig. 20. Diagrams showing details of reins and hook in palm leaves. *A,* showing lines of abscission and relation of reins and hook to blade; *B,* showing course of vascular bundles and lines of abscission between pinnae and between pinnae and rein. *h,* hook; *r,* rein. (*From Eames,* 1953.)

as the leaf opens, the leaflets or pinnae so formed are at first held together at their tips by the marginal band but are soon freed by the abscission of the band, together with the apex of the blade where the veins meet. From each leaf hang two straplike structures, the *reins*, with the apex, the *hook*, attached to one of them. These structures are, morphologically, a part of the blade, and their presence is related to the dissection of a lamina in which outermost veins of a pinnate-parallel series enclose the inner, distal veins. By the formation of the reins and hook and their abscission, leaflets are cut out of the central part of the blade. The leaflets have no normal margins or apices and differ in this way from other leaflets. The reins and hook vary greatly in structure, often persisting as tough, fibrous strips (sometimes green), suspended from the base of the blade for months; in palmate-leaved genera, they may be threadlike and ephemeral.

In the palms, the palmate leaf has clearly been derived from the pinnate (see Palmae, Chap. 11). That the compound leaf of the palms was derived from the simple was accepted by morphologists in the middle of the nineteenth century, but its method of origin was not understood for one hundred years.

There are two wholly different ontogenetic origins for the compound leaf. The palm type is known only in the palms and their close relatives, the Cyclanthaceae. (Plicate leaves resembling those of palms are present in *Curculigo* and species of *Pennisetum*, but the folding in these genera is the result of differential growth, with no dissection involved.) In the ontogeny of the compound leaf in all other angiosperms, so far as now known, the leaflets arise as lateral structures of the elongating rachis primordium, much as the simple leaf arises (Fig. 2). The existence of two types of compound-leaf origin strengthens the theory that the basic lines of angiosperm stock had diverged far from one another in early angiosperm history.

The ontogeny of modified leaves—*bud scales, cataphylls*—has received some attention as a part of the investigations of apical meristems and leaf development. The studies have been largely morphogenetic and have raised the question whether bud scales represent modified or transformed foliage leaves or are, ancestrally, appendages of different rank. Bud scales are like leaves in vascular supply and major features of ontogeny—apical and marginal meristems, axillary buds—and many taxa show stages transitional to the foliage leaf; they are surely of leaf rank and are homologous with the entire leaf or a part of the leaf.

The ontogeny of floral appendages is like that of leaves at many and at all important stages. Differences believed to exist, which have been used to demonstrate that the floral shoot apex is morphologically unlike that of the vegetative apex, are those of degree only, or do not exist. (These differences have already been discussed earlier in this chapter.) It has been reported that floral organs are initiated in the inner tunica layers only, but carpel initials in some taxa arise in part in the outer corpus, as do some leaf initials. In stamens, marginal meristems are present in those that have laminar form and in some of those with broad filaments. Ontogeny strongly supports, as does vascular anatomy, the homology of all floral appendages with leaves.

Secondary Vascular Tissues and the Cambium

Xylem. The xylem of angiosperms ranges widely in structure from simple to highly complex and to a secondary simplicity derived by reduction. The evolutionary story of the two cell systems in xylem—vertical and transverse—and their cell types has been worked out in detail. In the most primitive woods, the vertical system consists of *tracheids* and a small amount of *wood parenchyma;* the transverse system consists of parenchyma cells in the form of *wood rays.* In specialization, the tracheids gave rise to *vessel elements* and various types of *fibers,* and the wood parenchyma was increased in amount and changed in distribution. In the transverse system, specialization changes were in form, size, and structure of the rays.

The primitive angiosperm tracheid is well shown in form and structure by vesselless genera and those with very primitive vessels. This tracheid is very long, with long, tapering, overlapping ends and numerous pits. The pits of these tracheids are all scalariform—*Trochodendron, Eupomatia.* Such tracheids are remarkably fernlike. In other primitive genera, some of the pits of the tracheids are round. Where both forms of bordered pit are present, as in the Winteraceae, the scalariform type occurs in the early-formed part of the annual ring. Scalariform pitting, the undoubtedly primitive type of tracheary pitting in angiosperms, is widely distributed among the less specialized families but is absent in the higher families.

The most primitive type of vessel in the angiosperms differs from the primitive angiosperm tracheid only in the absence of closing membranes in some of the scalariform pits of the end walls. In these vessel elements, the number of perforated pits is high—in *Eupomatia,* from twenty to one hundred. They are arranged in ladderlike series along the flattened, oblique end walls (Fig. 21*A*). In evolutionary modification, this primitive vessel element became progressively shortened; accompanying the shortening, the advanced vessel element became rounded in cross section, with the wall usually thinner than that of the primitive element and often irregularly thick. The characteristic long-tapering ends became less acute, until the end walls stood at right angles to the

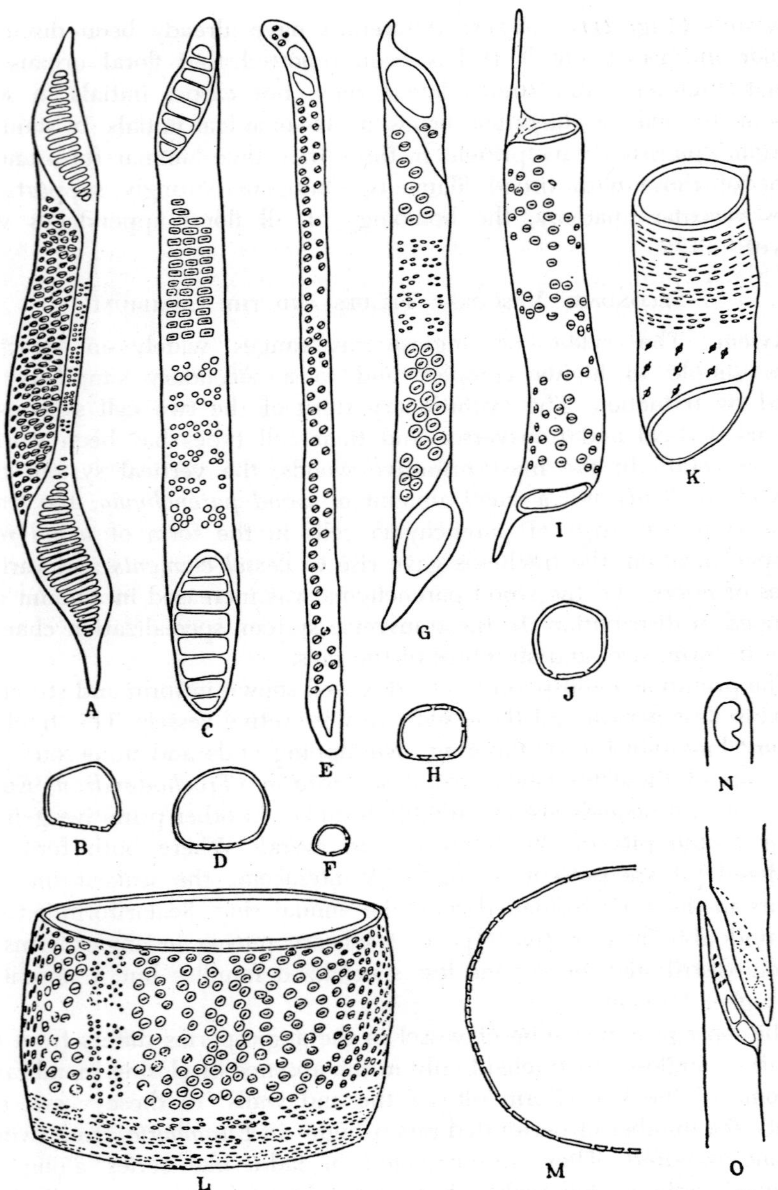

Fig. 21. Vessel elements in side view and cross section. *A, B, Betula alba; C, D, Liriodendron; E, F, Lobelia cardinalis; G, H, Quercus alba; I, J, Pyrus Malus; K, Acer Negundo; L, M, Quercus alba; N,* end of vessel element from *Lobelia* showing perforation indicating derivation of porous vessel from scalariform; *O,* ends of vessel elements from *Lobelia* showing method of union of elements in a series. (*From Eames and MacDaniels.*)

side walls (Fig. 21*L*). The perforations were reduced in number, enlarged, and fused, by loss of the ladderlike bars between them, to form a single opening. Vessel elements with one large pore in each end wall are termed *simply perforate* (Fig. 21*G* to *L*); elements of less specialized type, with two or more perforations on the end walls, are termed *scalariform* or *multiperforate* (Fig. 21*A* to *C*). Modification in form of vessel element is phylogenetic, but perforation of the wall is ontogenetic.

In primitive, vessel-bearing woods, the vessels are solitary or chiefly so (Fig. 22); in advanced types, they tend to be aggregated. Where the vessels, solitary or clustered, are scattered through the annual ring—as seen in cross section—the wood is *diffuse-porous;* where the vessels, especially the largest ones, are chiefly in the early wood of the annual rings, the wood is *ring-porous.* Ring porosity is an advanced character associated usually with simply perforate vessels.

Scalariform vessels occur in about 110 families—exclusively in some, dominantly in others, infrequently or sporadically in others. Except for the Paeoniaceae, which includes both herbaceous and woody species, families in which scalariform vessels occur exclusively are all woody—Eupomatiaceae, Himantandraceae, Annonaceae, for example. Families in which part of the vessels are scalariform are scattered through the dicotyledons—Cornaceae, Betulaceae, Ericaceae, Bixaceae, Magnoliaceae, Theaceae. Scalariform vessels are most common in families recognized as primitive in other characters, especially those that seem to be primitive members of more or less isolated lines—Magnoliaceae, Betulaceae, Ericaceae. They sometimes accompany advanced flower structure—Betulaceae, Caprifoliaceae. Simple vessels are also broadly distributed among angiosperms. The vessels of herbs, even in primitive families like the Ranunculaceae, are simply porous. (*Paeonia,* with scalariform vessels, has been removed from the Ranunculaceae.)

Vessels are absent in some aquatics—Ceratophyllaceae, Nymphaeaceae, Podostemonaceae—parasites, saprophytes—*Monotropa;* in these taxa, absence clearly represents phylogenetic loss. Throughout angiosperms, in highly specialized secondary wood, vessel number tends to be greatly reduced. Some annual dicotyledonous herbs have few or no vessels, as do some larger plants of unusual habit, such as the Cactaceae and larger Crassulaceae, which possess strong vascular cylinders.

The vessel has arisen independently in most of the major lines of vascular plants: Selaginellales—*Selaginella;* Equisetales—*Equisetum;* Gnetales (*sensu stricto*)—*Gnetum;* Welwitschiales—*Welwitschia;* Ephedrales—*Ephedra;* Dicotyledoneae—apparently at least several times; Monocotyledoneae—many times. The vessel cannot be considered a distinguishing character of the angiosperms. That some angiosperms are vesselless was known as early as 1842; in 1900, the "Homoxyleae," a

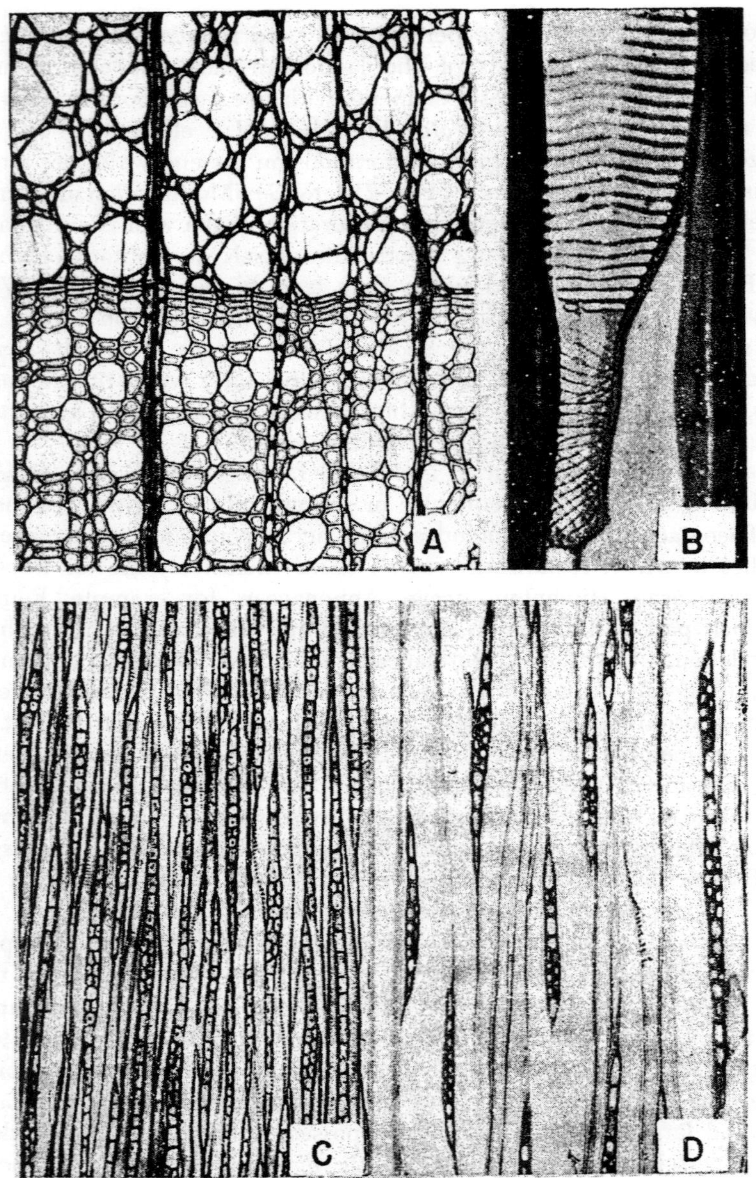

Fig. 22. Secondary xylem of *Cercidiphyllum japonicum*. *A*, transverse section of mature wood; *B*, part of a vessel member showing the spiral thickenings in the tapering end; *C*, tangential section through late-summer wood; *D*, tangential section through early-summer wood. (*From Swamy and Bailey.*)

taxon consisting of the family Winteraceae and the genera *Tetracentron* and *Trochodendron,* were described. Other vesselless genera—*Sarcandra* in the Chloranthaceae (Fig. 23) and *Amborella* (Fig. 24), perhaps constituting a family close to the Monimiaceae—have been added to the

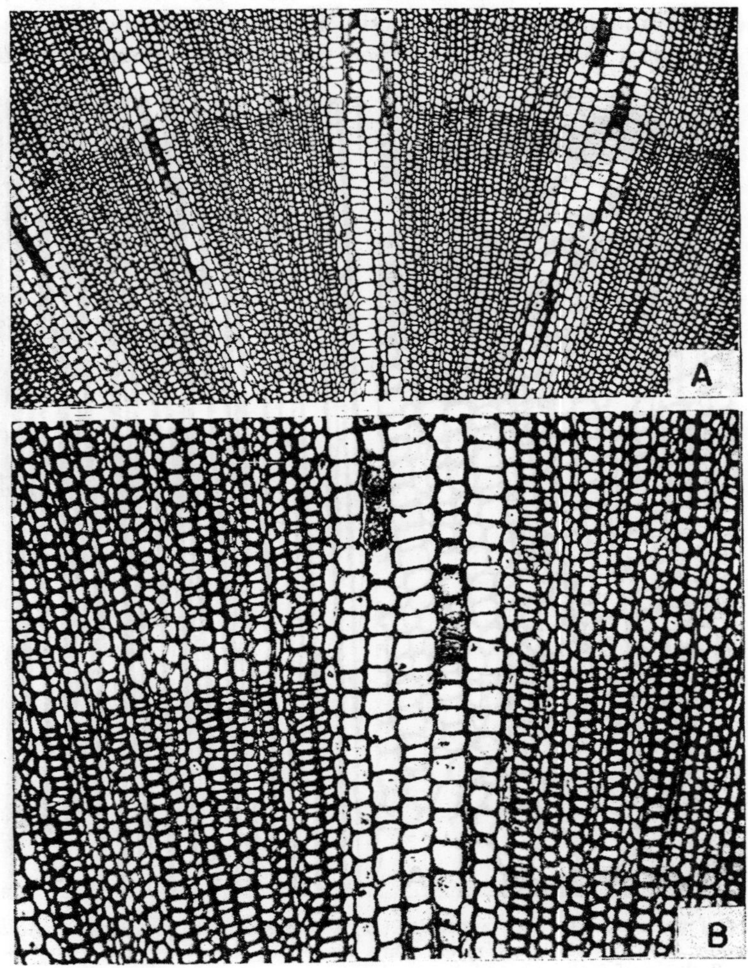

Fig. 23. Transverse section of part of a two-year-old stem of *Sarcandra* showing vesselless xylem with multiseriate and uniseriate rays in the first annual ring. *B*, a highly magnified part of *A* to show detail. (*From Swamy and Bailey.*)

vesselless group. But the group is not phylogenetically related; they constitute surviving remnants of several ancestral stocks.

Critical and extensive study of the distribution of vessels in the plant body shows that in the dicotyledons vessels apparently appeared first in the secondary xylem, then, progressively, in metaxylem and protoxylem. Advance in specialization in wood has "worked backward" from later to earlier and earlier formed tissue. In the monocotyledons, vessels clearly arose first in the later-formed metaxylem, then in earlier-formed metaxylem and protoxylem. Vessels did not appear simultaneously throughout the plant body. In the monocotyledons at least, they arose first in the roots and then developed successively in higher

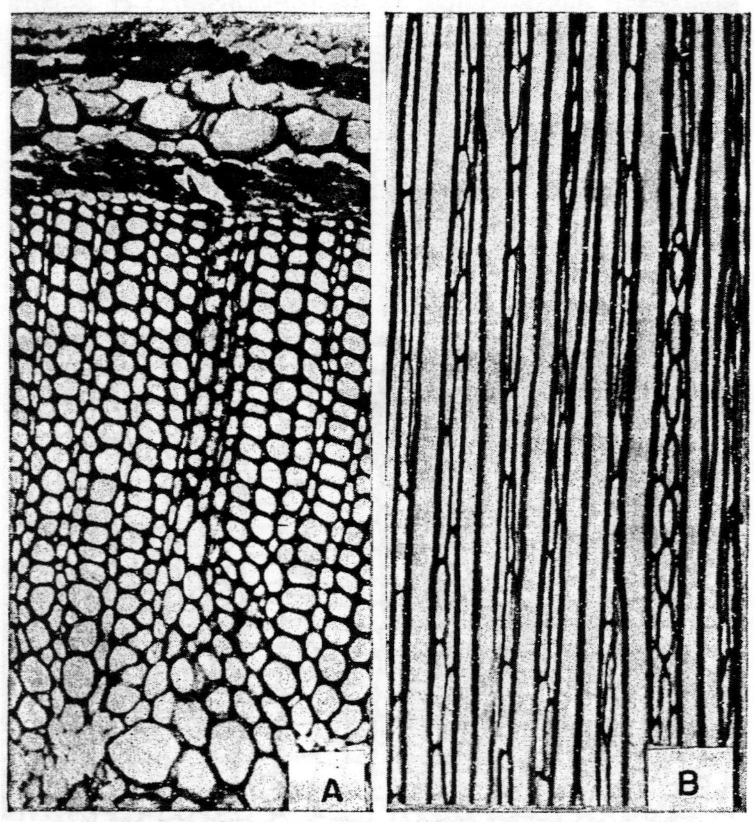

Fig. 24. Transverse (*A*) and tangential (*B*) sections of part of a stem of *Amborella trichopoda* showing vesselless xylem with uniseriate, biseriate, and triseriate rays. (*From Bailey and Swamy.*)

and higher parts of the plant. All steps in this phylogenetic "spread" of the vessel through the plant body are found. In the primitive Alismataceae, they are present only in the roots; in the highly advanced Gramineae, they are present throughout the plant. The origin of the vessel in the monocotyledons is perhaps comparatively recent.

Vessels in angiosperms arose without question from scalariformly pitted tracheids. Such an origin makes impossible the derivation of angiosperms from the Gnetales (*sensu lato*) or from other higher gymnosperms that have tracheids with only round-bordered pits. The evolution of the vessel provides strong evidence that the angiosperms arose from fernlike stock. (The resemblance of the vessels of *Eupomatia* to those of *Pteridium* is remarkable.)

Evidence from vessels supports the theory that dicotyledonous herbs arose from woody ancestors; the vessels of herbs are more highly specialized than those of trees. The obviously recent origin of vessels in monocotyledonous herbs makes impossible the origin of monocotyledons from dicotyledonous herbs, as expressed in some theories of the relationship between monocotyledons and dicotyledons.

The primitive pitting in intertracheary walls is scalariform-bordered. From this primitive pitting has been derived, by reduction in size, by division, and by change in form, the circular-bordered pit. Division of the elongate pit formed a row of shorter rectangular pits; these later became rounded. In a few of the woody dicotyledons, as in *Eupomatia,* all the intertracheary pits are scalariform; in others, as in the Winteraceae, scalariform pits are present only in the early wood, with transitional types and round pits in the late wood. The round pits, at first arranged in transverse rows—*opposite*—become, in higher types, spirally or irregularly arranged—*alternate.* In the various types of fibers—which are derivatives of tracheids—the pits are modifications of the round type, arranged irregularly.

In *wood fibers* of various types—structural modifications of the primitive tracheids—the tracheid has been reduced in diameter and in number and size of its pits, and its wall has been much thickened. No evidence of the primitive scalariform pits persists; all the pits are reduced forms of the circular pit. In length, there is little change. (The fibers of advanced types of wood appear unusually long in contrast with the shortened vessel elements and the great increase in length in development from the short cambium cell of this wood.)

Several types of wood fibers are recognized, distinguished on the basis of degree of modification from the ancestral tracheid. Cells in which the wall is greatly thickened and the lumen nearly occluded, with pits small and reduced in number, are typical *fibers.* Cells intermediate between tracheids and fibers in thickness of wall and size and number of pits—the pits with narrower borders and slitlike apertures—are *fiber tracheids. Libriform fibers* are fibers in which the walls are very thick and the pits very small, often essentially simple. No line separates these types, which represent stages in the elaboration of tracheids as supporting cells, with the conducting function lost. *Substitute fibers* are fiberlike cells with protoplasts. That they are tracheids morphologically is doubtful. The term *fusiform wood-parenchyma cells* perhaps best suggests their nature.

The more highly specialized types of fibers occur in wood that is in other ways highly specialized—often much simplified—especially that of woody herbs, subshrubs, and shrubs. In herbaceous stems with strong cylinders of secondary wood, as in many Compositae, the wood consists largely of fibers.

The cells of *wood parenchyma* stand in vertical rows that extend for long distances in the secondary xylem. The rows are basically uniseriate, consisting of elongate, mostly rectangular cells placed end to end, but two or more rows may stand together, forming clusters or bands of various sizes. Wood-parenchyma cells are formed by the transverse division of daughter cells of fusiform cambium initials, each daughter cell forming a vertical series of a few cells. Each series unites, end to end, with a similar series above and below to form long, continuous columns. Elongate fiberlike parenchyma cells, formed directly by fusiform initials, are called fusiform wood-parenchyma cells. These living cells, with fiberlike shape and wall, were formerly called "substitute fibers."

Wood-parenchyma distribution is termed *diffuse* where single columns are isolated and scattered among the other cells of vertical series; *terminal* where they are present as isolated strands or tangential clusters at the end of a season's growth; *metatracheal* where the strands are aggregated in clusters or bands but not including scattered, isolated strands and are wholly or largely free of contact with vessels; *paratracheal* where aggregates of columns are associated, wholly or largely, with vessels and typical tracheids; *vasicentric* where the aggregates are restricted to sheaths about the vessels. No sharp distinction can be made between the diffuse and metatracheal types, and the term *apotracheal* has been proposed to include paratracheal and vasicentric types.

In wood-parenchyma distribution, the diffuse type seems to be the most primitive, and the vasicentric the most advanced. Terminal distribution appears to be a reduction type derived from diffuse and, therefore, an advanced type.

Wood parenchyma is usually absent in stems of herbs that have discrete vascular bundles and in some vines and subshrubs with anomalous vascular steles where the bundles are separated by wide primary rays. In some shrubs, wood parenchyma may increase greatly in amount where the rays are reduced in the later-formed wood. In wood where rays are reduced and disappearing, the ray cells are upright, and the disappearing ray is apparently transformed into wood parenchyma by the elongation of the ray initials in the cambium.

In highly specialized woods, shorter and wider wood-parenchyma cells accompany shorter fusiform cambium cells. This shortening is closely correlated with a similar shortening in the vessel element.

The types of wood-parenchyma distribution merge to some extent and may differ in seedlings and mature plants. The presence of both apotracheal and paratracheal parenchyma in the same wood is rare; it is known in only six families—among them Tiliaceae, Myrtaceae, Bombacaceae.

Within a family, the amount of wood parenchyma may be fairly constant or may range from little to much, but type of distribution is fairly constant except in families with considerable range in floral structure.

There is little information about the occurrence of fusiform wood-parenchyma cells ("substitute fibers"). They are found in the first annual rings of some genera, especially in tropical families, in woody vines, and in suffrutescent herbs. They are, perhaps, most abundant in woody plants that appear to have been derived secondarily from herbs, as in the Berberidaceae; in these plants, strong-walled, elongate parenchyma cells may furnish the support given by fibers in typical woody plants.

In contrast with the vertical system, the transverse system of cells in the secondary wood of angiosperms—*xylem rays* or *wood rays*—is made up of a single basic cell type, the parenchyma cell. But the rays, radial sheets of these cells, though simple in cellular make-up, show much diversity in form and cell arrangement. As they elongate, they may increase or decrease in size; they may divide or fuse with other rays; they may have different form close to the primary body than farther out in the secondary wood. In tangential section, rays have various shapes—linear, oblong, fusiform, elliptical, oval; the larger types—as seen in section—may have parallel sides with extended, uniseriate wings at top and bottom.

Two major types of wood rays are distinguished: *uniseriate* and *multiseriate,* made up, respectively, of one and of more than one series of cells (Figs. 22 and 25). Uniseriate rays consist of one type of cell only; multiseriate rays may be *homogeneous,* consisting of one type of cells only, or *heterogeneous,* consisting of two types of cells: upright, vertically elongate cells, which form an outer, limiting, enclosing layer, and prostrate inner cells, with their longest diameter usually radial. These two kinds of cells seem to differ cytologically also. The term *aggregate* is applied to clusters of rays that lie more closely together than other rays in the same wood; "aggregate rays" are not a kind of ray. Clustering of rays is characteristic of genera in widely scattered families. The term *primary ray* is applied to rays that are connected with the pith.

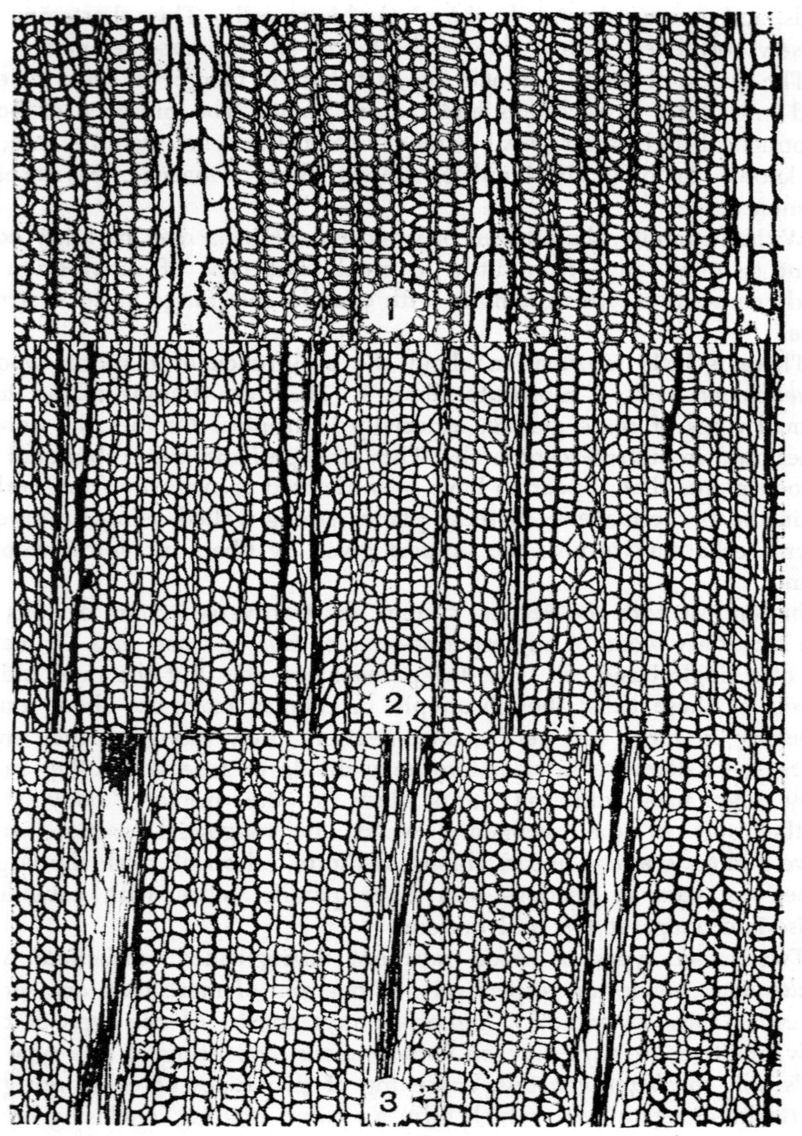

Fig. 25. Transverse and tangential sections of wood showing uniseriate and multi-
3, 6, *Belliolum haplopus.* (*From Bailey.*)

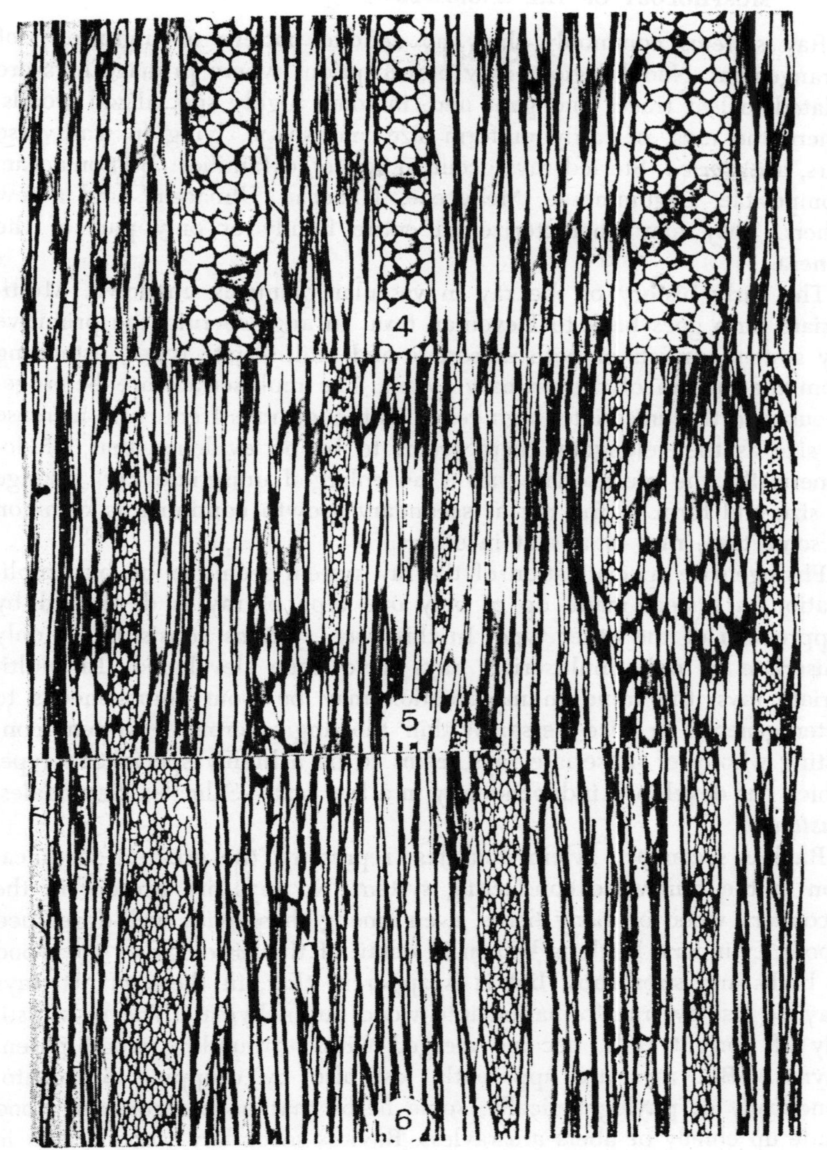

seriate rays in the Winteraceae. 1, 4, *Zygogynum Vieillardi;* 2, 5, *Drimys Winteri;*

Ray systems commonly show no definite pattern or uniformity of arrangement. Nodal regions may be exceptions where primary rays are related to leaf traces and gaps and, in some highly specialized woods, where the rays, of fairly uniform size and shape, lying in transverse tiers, are *storied.* Storied rays occur in scattered families: Bignoniaceae, Compositae, Leguminosae, Meliaceae, Tiliaceae, Ulmaceae, and a few others. They may characterize an entire family or only part of the genera.

The early history of the ray in vascular plants is unknown. Multiseriate form goes back to Devonian taxa. In angiosperms, the primitive ray system consists of both uniseriate and multiseriate types, extending from the margin of the primary xylem. The multiseriate are heterogeneous, and the uniseriate, high celled. Although rays frequently increase in size as they elongate, there seems to be no evidence that, phylogenetically, the multiseriate ray is an enlarged uniseriate ray. Change in size and form of the ray as secondary growth continues is common in some taxa, rare or absent in others.

Phylogenetic specialization of the ray system seems to be by simplification; a system made up of only one type of ray is developed by suppression of the other type. In this way, systems consisting of only uniseriate or only multiseriate rays have been developed. In multiseriate rays, the trend in modification may be from homogeneous to heterogeneous or vice versa. Within the angiosperms, a system consisting wholly of uniseriate rays seems to be a highly specialized type, which has developed independently in a few taxa—Salicales, Sapindales, *Castanea.*

Rayless secondary xylem doubtless represents the greatest modification in the transverse conducting system. No rays are present in the secondary wood of many herbs, even among those with well-developed woody cylinders. Little is known in detail of the histology of the wood of herbs and subshrubs, but it is apparent that, in these plants, rays may be lost by transformation into wood parenchyma. (The rays, usually of upright cells, become merged with surrounding wood parenchyma.) But rays are apparently lost also by dropping out, ontogenetically or phylogenetically; some herbs that have secondary wood made up chiefly of fibers are rayless. Rayless secondary wood occurs in woody herbs and subshrubs in widely scattered families—Geraniaceae, Tremandraceae, Crassulaceae, Empetraceae, Caprifoliaceae.

The loss of wood rays in herbs is associated with reduction in cambial activity and the consequent drop in demand for radial conduction. Histologically, distal termination of rays comes about by transformation in the cambium of ray initials into fusiform initials. As the axis increases in diameter, new cells are added by change in function of cambium initials.

The ray cells of vascular plants were probably derived originally from tracheids—the conifers show the origin of prostrate ray tracheids from erect tracheids—but the origin of ray cells must go far back in the history of vascular plants. The high, upright cells of uniseriate rays appear to be specialized types, which approach wood-parenchyma cells in form and perhaps in function.

The xylem of angiosperms, both as a whole and by its constituent cell types, provides excellent series in evolutionary modification. Advance to greater complexity and reduction to simpler structure are clearly shown. Specialization in wood structure is correlated with advance in flower structure; the wood of sympetalous families is more highly specialized than that of polypetalous families.

Broadly considered, the various types of wood cells advance in specialization together: the higher types of vessels usually accompany the highest types of fibers; scalariform vessels commonly accompany fiber tracheids and are not present with libriform fibers. But there are conspicuous exceptions; for example, fiber tracheids accompany simply perforate vessels in *Quercus*.

Specialization in the ray appears to be chiefly in simplification and reduction. The vesselless woods have both uniseriate and multiseriate rays, and many advanced genera have only one type. Heterogeneous rays, present in some of the more primitive genera, are present in lower seed plants—Pteridospermae, Bennettitales, Cycadales.

In wood structure, generic differences are usually evident; specific differences, rarely; well-marked differences may exist between subgenera or sections, as in *Quercus*. (Some anatomists believe that, in structure, the wood may be as conservative, or even more conservative, than the flower.) Within families, wood structure ranges from remarkably uniform to only fairly uniform and even diverse; in orders, there may be uniformity or an almost complete lack of resemblance (probably evidence that the order is unnatural under the present interpretation). Comparisons of advance in wood structure with specialization in flowers frequently show high correlation in groups of genera commonly assumed to be closely related, but there are examples of major discrepancies in large families and orders. Broadly considered, the structure of wood is one of the most important characters in the determination of natural relationships.

Phloem. The phloem of angiosperms has not been studied as thoroughly as the xylem, but the story of its evolutionary modification is known. It closely parallels that of the xylem. The fundamental conducting cell, the *sieve cell,* is elongate, and has walls with minute perforations, through which extend, in restricted areas, cytoplasmic strands connecting the protoplast with the protoplasts of adjacent similar cells. With specialization, the sieve cell structurally parallels the vessel element; it becomes progressively shorter, with end walls more and more oblique, until they stand finally at right angles to the side walls. Where sharper delimitation sets apart a group of sieve areas, *sieve plates* are formed. These plates indicate lines or courses of conduction in the sieve cell, a first step in the formation of a linear group of these cells, a *sieve tube.* (Sieve cells that have sieve plates and unite to form a sieve tube are termed *sieve-tube elements.*) With increased specialization, the sieve plates become more and more prominent, their pores larger and fewer, and the plates become restricted to the end walls of the cells. Where the end wall is transverse, the plate occupies most of the wall and the pores are few and large. A cylindrical cell of this type is the highest form of sieve-tube element, resembling in form the most highly specialized vessel element. With the elaboration of sieve plates, the other sieve areas remain weakly developed or vestigial —"ghost plates" or *lattices.* Advanced woody families, herbs, vines, and plants of unusual habit usually have sieve-tube elements of the highest type.

The *companion cell,* a new cell type in the phloem of vascular plants, is characteristic of angiosperms. No companion cells have been found in *Austrobaileya* (Fig. 26). Companion cells are parenchyma cells of special structure and function, closely related in position and function to a sieve element. They lie alongside the sieve elements and are pitted only with those cells—evidence that their function is closely related to that of the sieve elements. Companion cells are characteristic of angiosperms but absent in other seed plants.

Usually one or more companion cells accompany each sieve element; rarely, a sieve-tube element has no companion cells. The number of companion cells appears to increase with increase in specialization of the phloem, and the phloem of herbs and many woody taxa may have numerous companion cells. In general, highly specialized phloem accompanies highly specialized xylem. The correlation in advancement between xylem and phloem is often closer than that between flower and vascular tissues.

Phloem is usually a complex tissue, including *phloem parenchyma* and *sclerenchyma* of one or more types, in addition to sieve tubes and companion cells. In the more primitive woody angiosperms, the phloem tends to be simpler than in the more advanced families and has fewer companion cells and much less sclerenchyma. The phloem of some of the woody Ranales is "soft," not "stratified" with alternate bands of sclerenchyma and sieve tubes. Sclerenchyma, where present in phloem of simpler structure, consists chiefly of sclereids; fibers are few or absent. The phloem of *Drimys* is entirely "soft"; that of *Zygogynum,* in the same family, has a few isolated fibers. *Amborella* has sclereids but

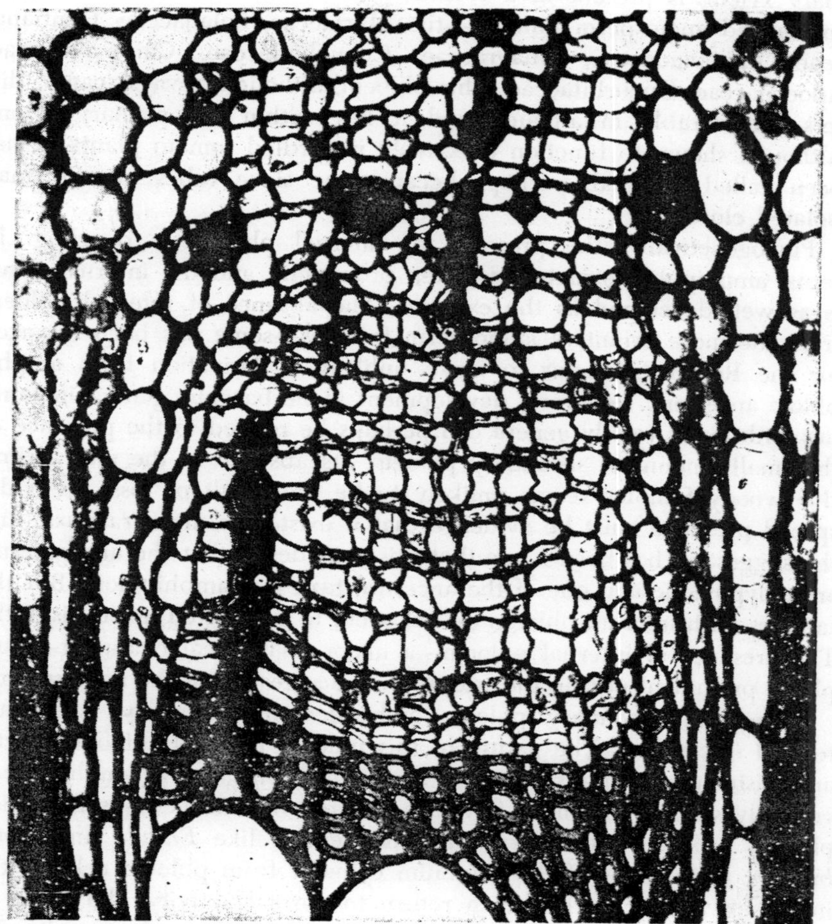

Fig. 26. Transverse section of part of stem of *Austrobaileya scandens* showing secondary phloem with no companion cells. (*From Bailey and Swamy.*)

no fibers. *Calycanthus,* on the other hand, has stratified phloem and sieve tubes of high type. In herbs and subshrubs where there is little phloem, the phloem is simple, usually without sclerenchyma—except as a sheath or cap—and consists sometimes of sieve tubes and companion cells only. The story of evolutionary modification seems to be one of increase in complexity from a simple type, consisting wholly or largely of "soft cells," to one that has also sclerenchyma cells of one or more types variously arranged, and to a secondarily simple type, consisting largely or wholly of sieve tubes and companion cells.

Internal (or *intraxylary*) *phloem,* as well as that external to the primary xylem, is present in a considerable number of the more highly specialized angiosperm families—the Myrtaceae, Solanaceae, Gentianaceae, Asclepiadaceae, Onagraceae, Convolvulaceae, Campanulaceae, Apocynaceae, Cucurbitaceae. This phloem, although always scanty, differs considerably in amount, and it lies within the primary xylem. Although shown to function effectively in girdled tomato plants, it has been called vestigial, perhaps because the sieve tubes are in small isolated clusters.

Phylogenetically, the presence of internal phloem is puzzling. In ferns, amphiphloic structure is often prominent, and the internal tissue is as well developed as the external. The absence of internal phloem from the more primitive angiosperm families—some has been reported for the Rosaceae—seems to make unlikely a derivation from amphiphloic ancestors. Its strong development in herbaceous genera in families with many woody genera can perhaps be related to the presence of the small amount of secondary phloem. Its absence in the most primitive woody families seems unlikely to be the result of loss. Its widespread presence must be evidence that it existed in ancestral taxa, and the suggestion has been made that this is evidence that the angiosperms are polyphyletic and one of the ancestral taxa was amphiphloic. But the families with internal phloem do not seem to be related, even distantly. The presence of internal phloem—a major anatomical character—must play a prominent part in the search for the ancestors of the angiosperms.

Strands of secondary phloem embedded in secondary xylem are termed *interxylary*. The nature of this phloem is chiefly of ontogenetic and histological interest, but its presence within the xylem has been critically studied in only a few taxa. There are two methods by which phloem becomes interxylary. In some genera, like *Entada* and *Combretum,* some strips of the cambium cylinder form phloem cells to the inside for a brief period, then return to normal activity; the strips of phloem are thus embedded in xylem, which was formed normally. By the other method, strands of phloem are normally formed by the cambium, but the initials that formed them cease to function, and new arcs of cambium develop external to the phloem strands and form xylem, which encloses the phloem strands, as in *Strychnos.*

Phloem of peculiar origin is present in the complex tissues formed inwardly by secondary or accessory cambia, as in the Chenopodiaceae. In this family, the cambium forms both xylem and phloem to the inside, mostly in cell aggregates that suggest vascular bundles.

The Cambium. In the axis of typical woody plants, the cambium forms a continuous cylinder, broken only by branch and leaf gaps; in some herbs and subshrubs, the secondary vascular tissue of the stem is small in amount; it consists of a cylindrical network of more or less discrete strands—the stele is a eustele (Fig. 27*B*). Complete cylinders of secondary vascular tissues are present in many herbs (Fig. 27*A*), both annual and perennial; the description of herbs in general as characterized by dissected steles is erroneous. The network of bundles represents a reduced and dissected vascular cylinder. Proof of this interpretation lies in part in the cambium. In the areas between the

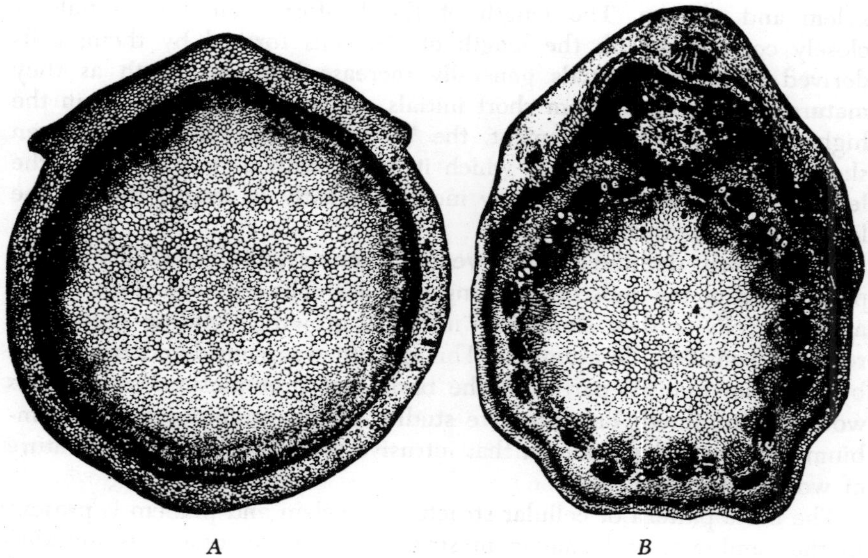

Fig. 27. Transverse sections of herbaceous stems showing two dicotyledonous types. *A*, ***Digitalis,*** stele continuous; *B*, ***Artemisia,*** stele dissected. (*From Eames and MacDaniels after Sinnott and Bailey.*)

vascular bundles, lines of tangential divisions in parenchyma cells connect the strips of cambium in the bundles. All stages in the loss of cambium between the bundles are present in related species and even in the same shoot, which may have the cambium cylinder complete at the base and vascular bundles free at the top, with, in between, strips of vestigial cambium, which are progressively weaker toward the top.

The evolutionary history of the cambium in the angiosperms is one of reduction in activity and in area. All stages of loss are present in dicotyledons; in monocotyledons, the vascular bundles typically have no cambium, but, in many genera, some of the bundles have a weak cambium, which functions only briefly. In some genera, "secondary" cambia—cambia arising late outside the primary vascular tissue—may build up the vascular cylinder, as in many arborescent monocots and some families of herbaceous and subshrubby dicotyledons, such as the Chenopodiaceae. These accessory cambia may continue to function permanently or may soon be replaced by similar, later-formed meristems that arise successively in outer tissues. Many woody vines—lianas especially—increase in diameter in this way.

Histologically, the angiosperm cambium is a sheet of initiating cells of two shapes, fusiform and rectangular or rounded. The fusiform cells build the vertical system; the rectangular, the radial systems of the xylem and phloem. The length of the fusiform cambium initials is closely correlated with the length of the cells formed by them; cells derived from long initials generally increase little in length as they mature. Cells derived from short initials also increase little and, in the higher types of vessel element, the "body" of the cell may be even shorter than the initial from which it arose, with a "tail" showing the length of the initial. Fibers may increase greatly in length beyond the length of the mother initials.

In the early decades of the twentieth century, the possibility of the penetration of the ends of maturing vascular cells between the walls of adjacent cells as they elongate—"intrusive growth," "gliding growth"—received considerable attention. This growth was considered impossible by some anatomists, partly on the basis that protoplasmic connections would be ruptured; but intensive studies of the cambium and of cambium-cell length have shown that intrusive growth is a common feature of wood ontogeny.

The basic pattern of cellular structure in xylem and phloem is present in the cambium, and change in structure of these tissues is initiated there. Changes in type, number, and arrangement of cells formed as secondary growth continues are based on changes in the cambium initials. New cells are added with increase in area of the cambium. Fusiform initials form additional similar cells by radial division or by a transverse or oblique division followed by elongation. New ray initials are formed by transverse divisions of fusiform initials. The fusiform cells of the more primitive dicotyledons are long and long-tapering, as are the tracheids, scalariform vessels, and sieve elements formed by them. Specialization in the cambium has been a progressive shortening of the fusiform initials and the development of uniformity in ray-initial clusters. The cambium initials of the vesselless taxa are very long and long-tapering; a general shortening of the initials accompanies the presence of vessels. Within an individual plant, as secondary growth continues, the fusiform initials tend to increase slightly in length for some years, although vessel elements and wood parenchyma cells may become shorter.

BIBLIOGRAPHY

Root and Stem

Andrews, E. C.: The development of the natural order Myrtaceae, *Proc. Linn. Soc. N.S.W.*, **38**: 529–568, 1913.

———: The development and distribution of the natural order of Leguminosae, *Jour. Roy. Soc. N.S.W.*, **48**: 333–407, 1914.

Arber, A.: The tree habit in angiosperms: its origin and meaning, *New Phyt.*, **27**: 69–84, 1928.

———: Root and shoot in the angiosperms: a study of morphological categories, *New Phyt.*, **29**: 297–315, 1930.

———: The interpretation of root and shoot in the angiosperms, *Biol. Rev.*, **16**: 81–105, 1941.

Bailey, I. W.: Nodal anatomy in retrospect, *Jour. Arnold Arb.*, **37**: 269–287, 1956.

Ball, E.: The development of the shoot apex and of the primary thickening meristem in *Phoenix canariensis* Chaub., with comparisons with *Washingtonia filifera* Wats. and *Trachycarpus excelsa* Wendl., *Am. Jour. Bot.*, **28**: 820–832, 1941.

Bancroft, H. The arborescent habit in angiosperms: A review, *New Phyt.*, **29**: 153–169, 227–275, 1930.

Bews, J. W.: "Plant Forms and Their Evolution in South Africa," London, 1925.

Boke, N. H.: Development of the adult shoot apex and floral initiation in *Vinca rosea* L., *Am. Jour. Bot.*, **34**: 433–439, 1947.

Bower, F. O.: Size, a neglected factor in stelar morphology, *Proc. Soc. Edinburgh*, **41**: 1–25, 1921.

Canright, J. E.: The comparative morphology and relationships of the Magnoliaceae. IV. Wood and nodal anatomy, *Jour. Arnold Arb.*, **36**: 119–140, 1955.

Catesson, A.-M.: Structure, évolution et fonctionnement du point végétatif d'une monocotylédone: *Luzula pedemontana* Boiss. et Reut. (Joncacées), *Ann. Sci. Nat. Bot.*, 11 sér., **14**: 253–291, 1953.

Celakovsky, L. J.: Die Gliederung der Kaulome, *Bot. Zeit.*, **59**: 77–114, 1901.

Corner, E. J. H.: The durian theory or the origin of the modern tree, *Ann. Bot.*, **13**: 367–414, 1949.

———: The durian theory extended. I. *Phytomorph.*, **3**: 465–476, 1953; II. *Phytomorph.*, **4**: 263–274, 1954.

Cotton, A. D.: The megaphytic habit in the tree Senecios and other genera, *Proc. Linn. Soc. London,* **156**: 158–168, 1943.

Davy, J. B.: The suffrutescent habit as an adaptation to environment, *Jour. Ecol.*, **10**: 211–219, 1922.

Eames, A. J., and L. H. MacDaniels: "Introduction to Plant Anatomy," 2d ed., New York, 1947.

Engard, J. A.: Organogenesis in *Rubus, Univ. Hawaii Res. Publ.*, **21**: 1–234, 1944.

Esau, K.: Vascular differentiation in the pear root, *Hilgardia,* **15**: 299–324, 1934.

———: Origin and development of the primary vascular tissues, *Bot. Rev.*, **9**: 125–206, 1943.

Fahn, A., and I. W. Bailey: The nodal anatomy and the primary vascular cylinder of the Calycanthaceae, *Jour. Arnold Arb.*, **38**: 107–117, 1957.

Foster, A. S.: The shoot apex in angiosperms, *Bot. Rev.*, **20**: 477–529, 1954.

Gifford, E. M., Jr.: The structure and development of the shoot apex in certain woody Ranales, *Am. Jour. Bot.*, **37**: 595–611, 1950.

———: The shoot apex in angiosperms, *Bot. Rev.*, **20**: 477–529, 1954.

Griffith, A. M., and M. E. Mallius: The unit of shoot growth in dicotyledons, *Proc. Leeds Phil. Soc.*, **2**: 125–139, 1930.

Guttenberg, H. von: Studien über die Entwicklung des Wurzelvegetationspunktes der Dicotyledonen, *Planta,* **35**: 366–396, 1947.

Jeffrey, E. C., and R. E. Torrey: Transitional herbaceous dicotyledons, *Ann. Bot.*, **35**: 227–248, 1921.

Kumazawa, M.: The medullary system in the Ranunculaceae, Lardizabalaceae, and Berberidaceae, *Jap. Jour. Bot.*, **8**: 19–46, 1936.

Majumdar, G. P.: The shoot of higher plants: Its morphology and phylogeny, *Jour. Asiatic Soc. Letters and Sci.*, **23**: 39–62, 1957.

Markfeldt, O.: Ueber das Verhalten der Blattspurstrange immergrüner Pflanzen beim Dickenwachstum des Stammes oder Zweiges, *Flora,* **68**: 33–39, 81–90, 99–113, 1885.

Marsden, M. P. F., and I. W. Bailey: A fourth type of nodal anatomy in dicotyledons, illustrated by *Clerodendron trichotomum* Thunb., *Jour. Arnold Arb.*, **36**: 1–51, 1955.

Metcalfe, C. R.: The systematic anatomy of the vegetative organs of the angiosperms, *Biol. Rev.*, **21**: 159–172, 1946.

Nast, C. G.: The comparative morphology of the Winteraceae. VI. Vascular anatomy of the flowering shoot, *Jour. Arnold Arb.*, **25**: 454–466, 1944.

Parkin, J.: The durian theory: a criticism, *Phytomorph.*, **3**: 80–88, 1953.

Perry, B. A.: Chromosome number and phylogeny of relationship in the Euphorbiaceae, *Am. Jour. Bot.*, **30**: 527–543, 1943.

Petersen, H. E.: Some preliminary remarks on the origin of isolated vascular bundles in herbaceous dicotyledonous plants, *Dansk Bot. Tidskr.*, **37**: 136–147, 1920.

Philipson, W. R.: The ontogeny of the shoot apex in dicotyledons, *Biol. Rev.*, **24**: 21–50, 1949.

———: Organization of the shoot apex in dicotyledons, *Phytomorph.*, **4**: 70–76, 1954.

Popham, R. A.: Principal types of vegetative shoot apex organization in vascular plants, *Ohio Jour. Sci.*, **51**: 249–270, 1951.

———: "Developmental Plant Anatomy," Columbus, Ohio, 1952.

Posthumus, O.: On some principles of stelar morphology, *Rec. Trav. Bot. Néerl.*, **21**: 111–295, 1924.

Reeve, R. M.: The "tunica-corpus" concept and development of shoot apices in certain dicotyledons, *Am. Jour. Bot.*, **35**: 65–75, 1948.

Rüdiger, W.: Die Sprossvegetationspunkte einiger Monokotylen, *Beitr. Biol. Pfl.*, **26**: 401–443, 1939.

Sargant, E.: A theory of the origin of monocotyledons, founded on the structure of their seedlings, *Ann. Bot.,* **17**: 1–92, 1903.

Saunders, E. R.: The leaf-skin theory of the stem, *Ann. Bot.*, **36**: 135–165, 1922.

Schade, C., and H. von Guttenberg: Über die Entwicklung des Wurzelvegetationspunktes der Monocotyledonen, *Planta,* **40**: 170–198, 1951.

Schoute, J. C.: Die Stammesbildung der Monokotylen, *Flora,* **92**: 32–48, 1903.

———: On phytonism, *Rec. Trav. Bot. Néerl.*, **28**: 82–96, 1931.

Sifton, H. B.: Developmental morphology of vascular plants, *New Phyt.*, **43**: 87–129, 1944.

Sinnott, E. W.: Investigations on the phylogeny of the angiosperms. I. The anatomy of the node as an aid in the classification of angiosperms, *Am. Jour. Bot.*, **1**: 303–322, 1914.

———: The evolution of herbs, *Science,* **44**: 291–298, 1916.

——— and I. W. Bailey: Investigations on the phylogeny of the angiosperms. IV. The origin and dispersal of herbaceous angiosperms, *Ann. Bot.*, **28**: 547–600, 1914.

Stant, M. Y.: The shoot apex of some monocotyledons. I. Structure and development, *Ann. Bot.*, n.s. **16**: 115–128, 1952.

Stebbins, G. L., Jr.: Cytological characteristics associated with growth habits in the dicotyledons, *Am. Jour. Bot.*, **25**: 189–198, 1938.

Thomas, E. N.: A theory of the double leaf-traces founded on seedling structure, *New Phyt.*, **6**: 77–91, 1907.

Tippo, O.: The role of wood anatomy in phylogeny, *Am. Midl. Nat.*, **36**: 362–372, 1946.

Van Tieghem, P.: Sur les diverses sortes de méristèles corticales de le tige, *Ann. Sci. Nat. Bot.*, 9 sér., **1**: 33–44, 1905.

Wagner, N.: Über die Entwicklungsmechanik der Wurzelhaube und des Wurzelrippenmeristems, *Planta*, **30**: 21–66, 1939.

Wardlaw, C. W.: Comparative observations on the shoot apex of vascular plants, *New Phyt.*, **52**: 195–209, 1953.

———: On the organization and reactivity of the shoot apex in vascular plants, *Am. Jour. Bot.*, **44**: 176–185, 1957.

Williams, W. C.: The structure of the meristematic root tip and origin of the primary tissues in the roots of vascular plants, *Am. Jour. Bot.*, **34**: 455–462, 1947.

Worsdell, W. C.: The origin and meaning of medullary (intraxylary) phloem in the stem of dicotyledons. I. Cucurbitaceae, *Ann. Bot.*, **29**: 567–590, 1915.

Zimmerman, W.: Der Baum in seinem phylogenetischen Werden, *Ber. Deutsch. Bot. Ges.*, **48**: 34–49, 1930.

LEAF

Artschwager, E.: Anatomy of the vegetative organs of sugar cane, *Jour. Agr. Res.*, **30**: 197–221, 1925.

Avery, G. S.: Structure and development of the tobacco leaf, *Am. Jour. Bot.*, **20**: 565–592, 1933.

Bailey, I. W., and E. W. Sinnott: Nodal anatomy and the morphology of stipules, *Am. Jour. Bot.*, **1**: 441–453, 1914.

——— and B. G. L. Swamy: The morphology and relationships of *Austrobaileya*, *Jour. Arnold Arb.*, **30**: 211–226, 1949.

Blaser, H. W.: Studies in the morphology of the Cyperaceae. II. The prophyll, *Am. Jour. Bot.*, **31**: 53–64, 1944.

Bugnon, P.: Sur les homologies de la feuille chez les Graminées, *Bull. Soc. Bot. France*, 4 sér., No. 24, **71**: 246–251, 1924.

Corner, E. J. H.: The durian theory extended. II. The arillate fruit and the compound leaf, *Phytomorph.*, **4**: 152–165, 1954.

Cross, G. L.: The origin and development of the foliage leaves and stipules of *Morus alba*, *Bull. Torrey Bot. Club*, **64**: 145–163, 1937.

Domin, K.: Ein Beitrag zur Morphologie des Dikotylenblattes, *Ceská Akad. Umĕné, Prague* (*Bull. Internat. Compt. Rend. Sci.*), **16**: 145–170, 1911.

———: Morphologische und phylogenetische Studien über die Stipularbildungen, *Ann. Jard. Bot. Buitenzorg*, **24**: 117–320, 1911.

———: Phylogenetic evolution of the phyllome, *Am. Jour. Bot.*, **18**: 237–242, 1931.

Eames, A. J.: Neglected morphology of the palm leaf, *Phytomorph.*, **3**: 172–189, 1953.

Foster, A. S.: Leaf differentiation in angiosperms, *Bot. Rev.*, **2**: 349–372, 1936.

———: A histogenetic study of foliar determination in *Carya buckleyi* var. *arkansana*, *Am. Jour. Bot.*, **22**: 88–147, 1935.

Gaisberg, E. von: Zur Deutung der Monokotylenblätter als Phyllodien, unter besonderer Berücksichtigung der Arbeit von A. Arber, *Flora*, N.F., **15**: 177–190, 1922.

Glück, H.: "Die Stipulargebilde der Monokotyledonen," Heidelberg, 1901.

———: "Blatt-und Blütenmorphologische Studien," Jena, 1919.

Hara, N.: On the types of the marginal growth in dicotyledonous foliage leaves, *Bot. Mag. Tokyo*, **70**: 110–114, 1957.

Hare, C. L.: On the taxonomic value of the anatomical structure of the vegetative organs of the dicotyledons. II. The anatomy of the petiole and its taxonomic value, *Proc. Linn. Soc. London*, **155**: 223–229, 1942.

Krauss, B.: Anatomy of the vegetative organs of the pineapple, *Bot. Gaz.*, **110**: 159–217, 333–404, 550–587, 1948.

Lubbock, J.: On the stipules, their form and function. II. *Jour. Linn. Soc. Bot.*, **30**: 465–532, 1894.

MacDaniels, L. H., and F. F. Cowart: The development and structure of the apple leaf, *Cornell Univ. Agr. Exp. Sta. Mem.*, 258, 1944.

Majumdar, G. P.: The complete foliage leaf, *Proc. Indian Acad. Sci.*, **B42**: 65–72, 1955.

Marsden, M. P. F., and I. W. Bailey: See under Root and Stem Bibliography.

Mitra, G. C., and G. P. Majumdar: The leaf-base and the internode: Their true morphology, *Paleobotanist*, **1**: 351–367, 1952.

Money, L. L., I. W. Bailey, and B. G. L. Swamy: The morphology and relationships of the Monimiaceae, *Jour. Arnold Arb.*, **31**: 372–404, 1950.

Monoyer, A.: Sur les stipules des *Potamogeton, Rpt. Assoc. Fran. Adv. Sci.*, **1926**: 1–3.

Petit, L.: Nouvelles recherches sur le pétiole des phanérogames, *Actes Soc. Linn. Bordeaux*, **43**: 11–60, 1889.

Philipson, W. R.: Development and morphology of the ligule in grasses, *New Phyt.*, **34**: 310–325, 1935.

Pijl, L. van der: The leaf of *Bauhinia, Acta Bot. Néerl.*, **1**: 287–309, 1952.

Priestley, J. H.: Cell growth and cell division in the shoot of flowering plants, *New Phyt.*, **28**: 54–81, 1929.

——— and L. I. Scott: Leaf venation and leaf trace in the monocotyledon, *Proc. Leeds Phil. Soc.*, **3**: 305–324, 1937.

Quéva, C.: Contributions à l'anatomie des Monocotylédonées. II. Les Uvulariées rhizomatenses, *Beih. Bot. Centralbl.*, **22**: 30–77, 1907.

Rüter, E.: Über Vorblattbildung bei Monokotylen, *Flora*, N.F., **10**: 193–261, 1918.

Schrödinger, R.: Phylogenetische Ansichten über Scheiden- und Stipularbildungen, *Verhandl. Zool.-Bot. Ges. Wien*, **69**: 162–193, 1931.

Sinnott, E. W., and I. W. Bailey: Investigations on the phylogeny of angiosperms. V. Foliar evidence as to the ancestry and early climatic environment of the angiosperms, *Am. Jour. Bot.*, **2**: 1–22, 1915.

Swamy, B. G. L.: Further contributions to the morphology of the Degeneriaceae, *Jour. Arnold Arb.*, **30**: 10–38, 1949.

——— and I. W. Bailey: *Sarcandra*, a vesselless genus of the Chloranthaceae, *Jour. Arnold Arb.*, **31**: 117–129, 1950.

Troll, W.: Morphologie der schildformigen Blätter, *Planta*, **17**: 231–314, 1932.

———: Vergleichende Morphologie der Fiederblätter, *Nova Acta Leopold.*, N.F., **2**: 315–455, 1935.

Tyler, A. A.: The nature and origin of stipules, *Ann. N. Y. Acad. Sci.*, **10**: 1–49, 1897.

Yarborough, J. A.: *Arachis hypogaea:* The seedling, its cotyledon, hypocotyl, and roots, *Am. Jour. Bot.*, **36**: 722–758, 1949.

Xylem, Phloem, and Cambium

Arber, A.: Studies on intrafascicular cambium in monocotyledons, *Ann. Bot.*, **33**: 459–465, 1919; **36**: 251–256, 1922.

Bailey, I. W.: The cambium and its derivative tissues. II. Size variations of cambial initials in gymnosperms and angiosperms, *Am. Jour. Bot.*, **7**: 355–367, 1920.

———: The problem of differentiating and classifying tracheids, fiber-tracheids, and libriform fibers, *Trop. Woods*, **45**: 18–23, 1936.

———: The comparative morphology of the Winteraceae. III. Wood, *Jour. Arnold Arb.*, **25**: 97–103, 1944.

———: The potentialities and limitations of wood anatomy in the study of the phylogeny and classification of angiosperms, *Jour. Arnold Arb.*, **38**: 211–254, 1957.

——— and C. G. Nast: The comparative morphology of the Winteraceae. VII. Summary and conclusions, *Jour. Arnold Arb.*, **26**: 37–47, 1945.

——— and B. G. L. Swamy: *Amborella trichopoda* Baill.: A new morphological type of vesselless dicotyledon, *Jour. Arnold Arb.*, **29**: 245–254, 1948.

——— and W. W. Tupper: Size variation in tracheary cells. I. A comparison between secondary xylem of vascular cryptogams, gymnosperms, and angiosperms, *Proc. Am. Assoc. Arts and Sci.*, **54**: 149–204, 1918.

Bancroft, H.: The arborescent habit in angiosperms: A review, *New Phyt.*, **29**: 153–169, 227–275, 1930.

Barghoorn, E. S., Jr.: The ontogenetic development and phylogenetic significance of rays in the xylem of dicotyledons. I. The primitive ray structure, *Am. Jour. Bot.*, **27**: 918–928, 1940; II. Modifications of the multiseriate and uniseriate rays, *Am. Jour. Bot.*, **28**: 273–282, 1941; III. The elimination of rays, *Bull. Torrey Bot. Club*, **68**: 317–325, 1941.

Barkley, G.: Secondary stelar structures of Yucca, *Bot. Gaz.* **78**: 433–442, 1924.

Boureau, E.: Sur certaines espèces homoxylées, vivantes et fossiles, á ponctuations aréolées scalariformes de la Nouvelle-Calédonie, 8° *Congr. Bot. Rapp.*, **2**: 231–232, 1954.

Chalk, L.: The phylogenetic value of certain anatomical features of the dicotyledonous woods, *Ann. Bot.*, n.s., **1**: 409–428, 1937.

Cheadle, V. I.: Secondary growth by means of a thickening ring in certain monocotyledons, *Bot. Gaz.*, **98**: 535–555, 1937.

———: The occurrence of types of vessels in the various organs of the plant in the Monocotyledoneae, *Am. Jour. Bot.*, **29**: 441–450, 1942.

———: The origin and certain trends of specialization of the vessel in the Monocotyledoneae, *Am. Jour. Bot.*, **30**: 11–17, 1943.

———: Vessel specialization in the late metaxylem of the various organs in the Monocotyledoneae, *Am. Jour. Bot.*, **30**: 484–490, 1943.

———: Specialization of vessels within the xylem of each organ in the Monocotyledoneae, *Am. Jour. Bot.*, **31**: 81–92, 1944.

———: Independent origin of vessels in the monocotyledons and dicotyledons, *Phytomorph.*, **3**: 23–44, 1953.

———: Research on xylem and phloem: Progress in fifty years, *Am. Jour. Bot.*, **43**: 719–731, 1956.

——— and K. Esau: Secondary phloem of the Calycanthaceae, *Univ. Calif. Publ. Bot.*, **20**: 397–510, 1958.

——— and N. B. Whitford: Observations on the phloem in the Monocotyledoneae. I. The occurrence and phylogenetic specialization in structure of the sieve tubes in the metaphloem, *Am. Jour. Bot.*, **28**: 623–627, 1941.

Church, A. H.: Thalassiophyta and the subaerial transmigration, *Bot. Mem. Oxford Univ. Press,* 1919.

Eames, A. J., and L. H. MacDaniels: See under Root and Stem Bibliography.

Eckardt, T.: Kritische Untersuchungen über das primäre Dickenwachstum bei Monocotylen, mit Ausblick auf dessen Verhältnis zur sekundären Verdickung, *Bot. Archiv,* **42**: 289–334, 1941.

Esau, K., V. I. Cheadle, and E. M. Gifford: Comparative structure and possible trends of specialization of the phloem, *Am. Jour. Bot.*, **40**: 9–19, 1953.

Frost, F. H.: Specialization in secondary xylem of dicotyledons. I. Origin of vessels, *Bot. Gaz.*, **89**: 67–94, 1930.

Gilbert, S. G.: Evolutionary significance of ring porosity in woody angiosperms, *Bot. Gaz.*, **102**: 105–120, 1940.

Gupta, K. M.: On the wood anatomy and theoretical significance of homoxylous angiosperms, *Jour. Indian Bot. Soc.*, **13**: 71–101, 1934.

Helm, J.: Das Erstarkungswachstum der Palmen und einiger anderer Monokotylen, zugleich ein Beitrag zur Frage des Erstarkungswachstums der Monokotylen überhaupt, *Planta,* **26**: 319–364, 1936.

Huber, B.: Die Siebröhrensystem unserer Bäume und seine jahreszeitlichen Veränderungen, *Jahr. Wiss. Bot.*, **88**: 176–242, 1939.

——— and K. Mägdefrau: Zur Phylogenie des heterogenen Markstrahlbaues, *Ber. Deutsch. Bot. Ges.*, **66**: 117–124, 1953.

Kraus, G.: Über Dickenwachstum der Palmenstämme in den Tropen, *Ann. Jard. Bot. Buitenzorg,* **24**: 34–44, 1911.

Kribs, D. A.: Salient lines of structural specialization in the wood rays of dicotyledons, *Bot. Gaz.*, **96**: 547–557, 1935.

———: Salient lines of structural specialization in the wood parenchyma of dicotyledons, *Bull. Torrey Bot. Club,* **64**: 145–163, 1937.

Lemesle, R.: Les divers types de trachéids et leurs principaux modes d'association chez les dicotyledones hétéroxylées, *Rév. Gén. Bot.*, **54**: (643): 326–340, 1947.

Metcalfe, C. R.: The systematic anatomy of the vegetative organs of the angiosperms, *Biol. Rev.*, **21**: 159–172, 1946.

Mohl, H. von: Über den Bau des Palmenstammes in "Vermischte Schriften Botanisches Inhalts," Tübingen, 1845.

Schoute, J. C.: Die Stammesbildung der Monokotylen, *Flora,* **92**: 32–48, 1903.

———: Über die Verdickungsweise des Stammes von *Pandanus, Ann. Jard. Bot. Buitenzorg,* 2 ser., **6**: 115–137, 1907.

Sifton, H. B.: Developmental morphology of vascular plants, *New Phyt.*, **43**: 111–129, 1944.

Stern, W. L.: Comparative anatomy of xylem and phylogeny of Lauraceae, *Trop. Woods,* **100**: 1–72, 1954.

Swamy, B. G. L., and I. W. Bailey: The morphology and relationships of *Cercidiphyllum, Jour. Arnold Arb.*, **30**: 187–210, 1949.

Tippo, O.: See under Root and Stem Bibliography.

See also references for families, Chap. 11.

Chapter 2

THE INFLORESCENCE

Flowers are borne singly and in clusters. Those in clusters, together with the stems and bracts associated with them, form *inflorescences;* those borne singly are termed *solitary flowers.*

Solitary flowers may terminate shoots of major or of minor rank, or may be axillary. In the terminal position, some solitary flowers may represent a simple and primitive condition; for example, in primitive woody genera such as those of *Magnolia, Annona, Eupomatia, Calycanthus.* Others, the great majority, are doubtless surviving members of reduced inflorescences, some derived from terminal, some from axillary clusters. Solitary axillary flowers are commonly surviving units of axillary clusters.

It is difficult to draw a line between a group of solitary flowers and an inflorescence. The limits of inflorescences are often poorly defined; specialized inflorescences of various types pass downward into flowers —solitary or in small clusters—in the axils of foliage leaves—for example, in *Lysimachia, Campanula.* Obviously, a definition for inflorescence that will satisfactorily describe and limit all flower clusters called inflorescences cannot be made. The inflorescence is not a morphological unit; it is a part of the branching system of the stem with more or less definitely segregated flowering tips. In the long course of specialization of the angiosperms, the inflorescence has been morphologically modified in many ways, with the production of similar form in unrelated lines. Under extreme specialization, some inflorescences no longer resemble branch systems but flowers and fruits—Compositae (Figs. 28 and 37), Proteaceae, Urticaceae. Inflorescences may be grouped in larger inflorescences, and these may be reduced, as in the smaller herbaceous species of *Euphorbia.*

Inflorescences range in size from minute to huge, and in flower number up to millions. The multiflowered panicles of the grasses and the heads (often compound) of some Compositae are well known examples. The terminal inflorescence of *Corypha umbraculifera,* a monocarpic palm, is a pyramidal panicle* about 10 m high and about 1 m in diameter at the base, with a flower number estimated at 6,000,000. The

* The term panicle is here used in the broader sense, for any large, compound inflorescence, determinate, indeterminate, or mixed in type.

grass trees, *Xanthorrhoea* (Fig. 29), similarly have millions of flowers in a contracted panicle of cylindrical form, which may be 2 m long. The inflorescence of *Typha* has been estimated to bear 300,000 flowers.

Most inflorescences die as soon as the fruit is mature, and many are soon abscised; rarely, they are biennial or perennial, with annual or continuous flowering. Those of some Bromeliaceae produce flowers for two or three seasons, and those of *Couroupida,* the cannonball tree,

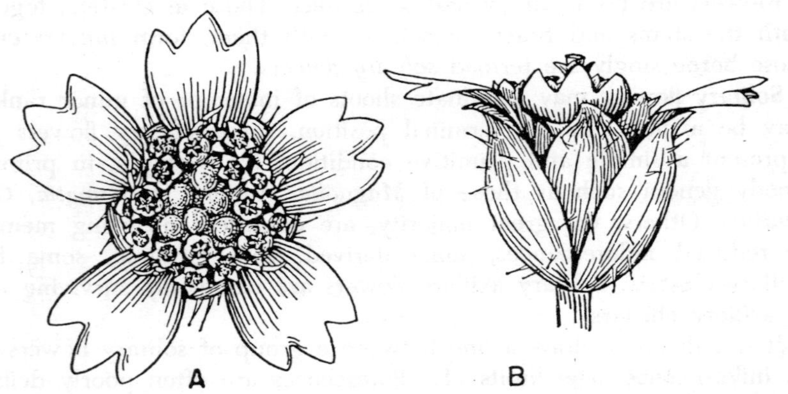

Fig. 28. Sketch of inflorescence of *Galinsoga* simulating a flower. *A,* face view showing ray and disc flowers; *B,* lateral view showing subtending bracts. (*Drawing by Elfriede Abbe.*)

persist on the tree trunk for many years, leafless but woody, and increase in length and diameter like vegetative branches.

Classification

Many descriptive terms have been applied to the varieties of inflorescences, terms the use and meaning of which have often been loose; transitional types are common and rigid definition cannot be made. Inflorescences have been classified on two bases: that of position on the stem system, and that of sequence in flowering within the cluster.

On the basis of position on the stem, they are grouped as terminal, axillary, and intercalary. Strictly terminal clusters terminate branches; axillary clusters are terminal on short axillary branches or represent foliaceous axillary branches which have been reduced to inflorescences; intercalary inflorescences are terminal clusters that have been "left behind" by continuing apical growth of the main axis, which forms, alternately, fertile and sterile sections, or by sympodial growth. Myrtaceous genera, such as *Callistemon,* the bottle brush, and *Melaleuca,* the paperbark tree, are excellent examples of the intercalary type. Inflorescences borne on internodes have sometimes been called intercalary, sometimes adventitious, because believed developed from adventitious buds. Though some adventitious inflorescences may be truly so, the morphological position of the clusters is commonly misinterpreted. The

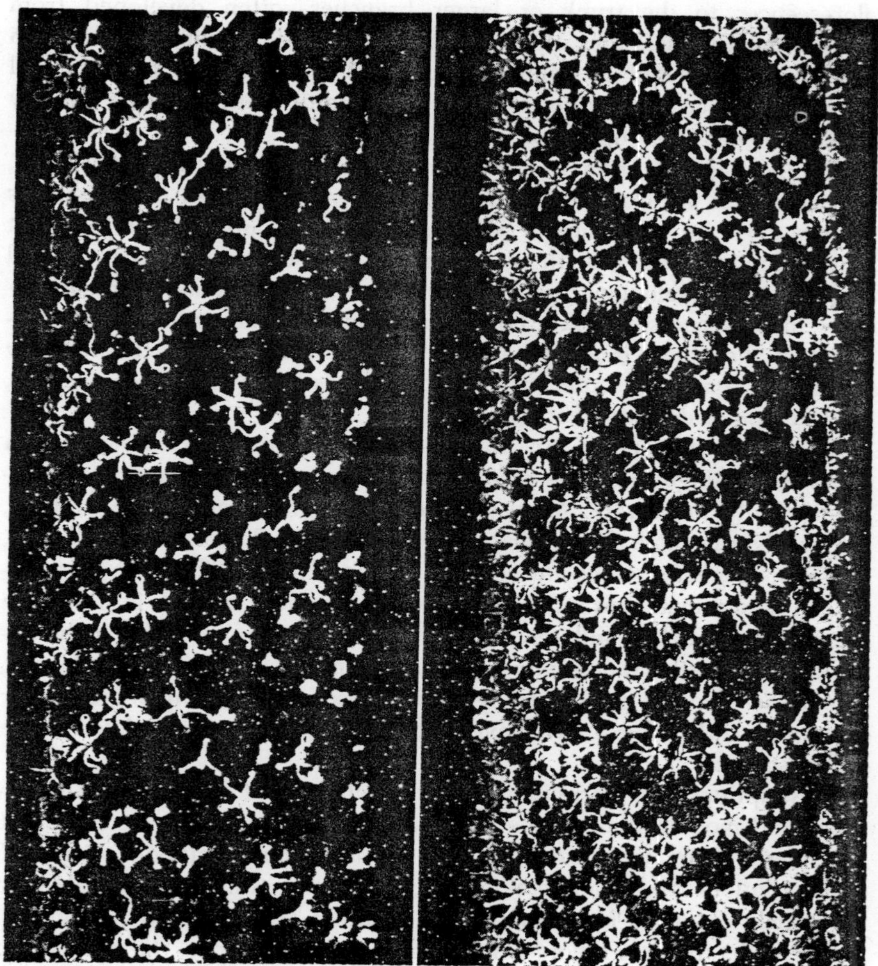

Fig. 29. Photograph of part of an inflorescence of *Xanthorrhoea hastilis,* a condensed cymose panicle, with branches spirally arranged. Left, young stage, with only the first median flower of each cyme open; right, later stage, with secondary flowers also mature.

pseudolateral position may be the result of sympodial growth, as in *Asclepias,* or of adnation of the peduncle to the internode. In *cauliflory,* where inflorescences develop on older branches or tree trunks, various relations of the flower cluster to the mother stem may be present. Long-dormant axillary buds of dwarfed lateral branches may develop flowers, as in some legumes—*Cercis*. Cauliflory is frequent in tropical forests but rare in temperate regions. It has been considered, under the durian theory, a primitive character, but the morphological relation of the inflorescences to the trunk or larger branches—often developed from adventitious buds—does not support this view. Cauliflory has been described as related to pollination or seed dissemination by bats and birds. Evidence that pollination in the primitive angiosperms was by insects seems well supported.

On the basis of order of development of flowers within the cluster, inflorescences are classified as *determinate* (*cymose*) and *indeterminate* (*racemose*). In determinate inflorescences, development of the first flower limits apical growth of the main axis of the inflorescence, and sequence in development of the other flowers is basipetal; in indeterminate inflorescences, the first flower is the lowest, and order of development of the others is acropetal—the number of flowers is theoretically unlimited. Basipetal order of development is often called centrifugal, and acropetal development, centripetal.

It is unsatisfactory to use the order of development of flowers as a basis for distinguishing the major types of inflorescences, and it is morphologically inaccurate. Many inflorescences are intermediate in type, especially racemose forms that have a terminal flower, as in *Clethra, Juglans, Digitalis, Convallaria, Campanula, Pyrola*. That the line between determinate and indeterminate inflorescences is weak is shown by the large, branched inflorescences, which are often *mixed*, partly determinate, partly indeterminate. Some families—the Campanulaceae, Violaceae—have a great variety of inflorescences. A single genus, for example, *Hybanthus*, may have both determinate and indeterminate inflorescences.

Descriptions and morphological interpretations of all the many variations and modifications of the two major types of inflorescences are not considered here; they are described in elementary textbooks. Some of the variations and modifications that are not readily interpretable or are partly obscured by connation or adnation are discussed below.

In the determinate inflorescences, the simple basic type, the *dichasium*, consists of a terminal flower, with other flowers terminal on two lateral axes borne in the axils of bracts below the terminal flower. A similar inflorescence, with more than two lateral branches, is a *pleiochasium;* one with only one lateral branch, a *monochasium*. All these are termed *cymes*, although "cyme" is often used as synonynous with dichasium. Different types of monochasia are formed by differences in the position and sequence of the successive branchings; among these are the *helicoid* and *scorpioid cyme* and the *cincinnus*. The morphological nature of some examples of these has been misinterpreted, because of complexities of nodal structure. Variations in pedicel length and in orientation of flowers produce forms of inflorescences difficult to place in the usual categories. In all the determinate types, branching may continue successively, as in the first cluster, and large, compound, cymose inflorescences be built up. In these clusters, the ultimate three-flower unit is often called a *cymule*. This term is applied also to the few units surviving in a small cluster reduced from a large one, as in the Betulaceae.

The term *panicle* is applied to large, branched, racemose inflorescences, but it is more commonly used for large, loose clusters of any type, determinate, indeterminate, or mixed. The inflorescence of the lilac, *Syringa*, is racemose, but all its branchlets have a terminal flower, and those that have only three flowers form a typical cyme. The large inflorescences of many grasses—for example, the rice, *Oryza*—are cymose in major, racemose in minor branching. In large inflorescences, such as those of many rosaceous genera—*Crataegus* and *Rubus*, for example—the order of development may be irregular and inconstant. The umbel may be determinate or indeterminate. The importance of sequence in development has been overemphasized.

Some, perhaps many, of the apparently simple inflorescences—certain types of monochasia, even some racemes—are not morphologically simple but represent reduced branched types, as is shown by anatomy and critical comparative studies. Some dichasia become simple by the abortion of the terminal flower; the branching then appears dichotomous; compound dichasial inflorescences reduced in this way have been called dichotomous.

Branched inflorescences of any type are called *compound*, as contrasted with *simple*, but this is, morphologically, a valueless distinction. Many apparently simple clusters have been derived from branched ones by reduction, but the origin is more or less obscure. For example, the apparently simple "raceme" of *Claytonia* is not simple but highly complex in nature. The raceme and the dichasium are commonly called the basic simple types, but the dichasium is a branched, not simple, inflorescence.

Reduced Inflorescences

In the evolutionary modification of inflorescences, reduction and condensation have played prominent parts (Fig. 30). Reduction is primarily in flower number, but the stem system is commonly also involved; internodes are shortened or suppressed, and, in reduction to one flower, nearly all the stem system may be lost. The bracts of the reduced stem system may similarly be greatly reduced, but they may be retained where the internodes are lost and form an involucre about the remaining flowers or flower. Involucres of this type may be petaloid and simulate a corolla or perianth—*Actinotis* (Umbelliferae), *Pimelea* (Thymelaeaceae), *Leucadendron* and *Protea* (Proteaceae), *Cornus*.

Reduction in flower number has taken place in many genera throughout angiosperms. Examples of reduction to one flower from racemose clusters are seen in *Drosera, Tulipa, Plantago, Cypripedium, Moneses, Vicia, Vaccinium, Scutellaria, Gentiana, Monotropa, Ornithogalum;* from cymose clusters in *Narcissus, Viola, Potentilla, Silene, Dalibarda, Rosa, Anemone, Philadelphus, Rubus, Vinca.* A solitary flower may represent one surviving member of a raceme, as in *Vicia* (Fig. 31); of an umbel, as in *Xanthosia* (Umbelliferae); and of a head, as in *Corymbium* (Com-

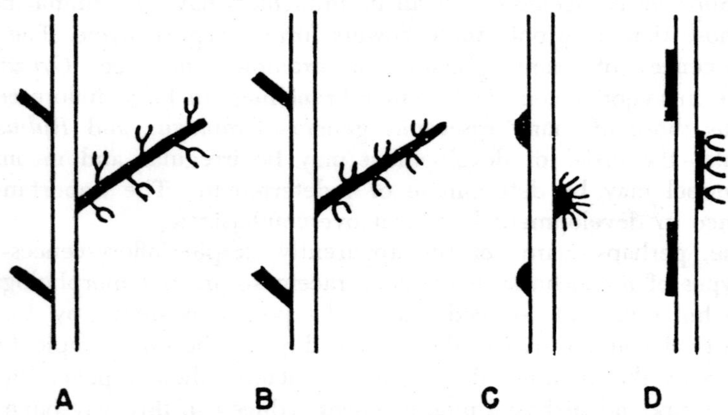

Fig. 30. Diagrams showing progressive reduction of lateral branchlets of the inflorescences of coryphoid palms. *A*, all axes more or less shortened; *B*, ultimate branchlets (pedicels) shortened, flowers sessile; *C*, flower-bearing branchlets reduced to hemispherical bases; *D*, lateral branchlets wholly reduced, the base elongated, flowers borne in a row along main axis. Reduction in flower number continuous from many to three (triads) or two, sometimes sunken in the fleshy main axis. (*After Bosch.*)

positae). Each flower of a head may be the surviving flower from one of a compacted cluster of heads, as in *Echinops, Oedera* (Compositae). Where reduction is to a few-flowered cluster, no satisfactory descriptive term may be applicable, and such phrases as "umbellate cluster" (in Rosaceae) and "contracted cymose cluster" (in Labiatae) are used. The morphological nature of the cluster may be obscure, unless comparisons are made with inflorescences of related taxa.

Solitary flowers like that of the quince bush (*Cydonia*) have been cited as examples of primitive solitary flowers. But the evidence of reduction through a racemose-umbellate series exists in the related genera, *Pyrus, Malus, Chaenomeles. Chaenomeles* shows all the stages in this reduction.

Reduction is usually superficially evident from comparison with related taxa and, in extreme reduction, when externally obscure, can often be detected anatomically, as in *Acer*, where vestiges of the inflorescence axis are present in the clusters of flowers of the soft maples, and in *Vicia*, where the axis of an axillary raceme is buried in the cortex of the mother stem beyond the attachment of the trace of a

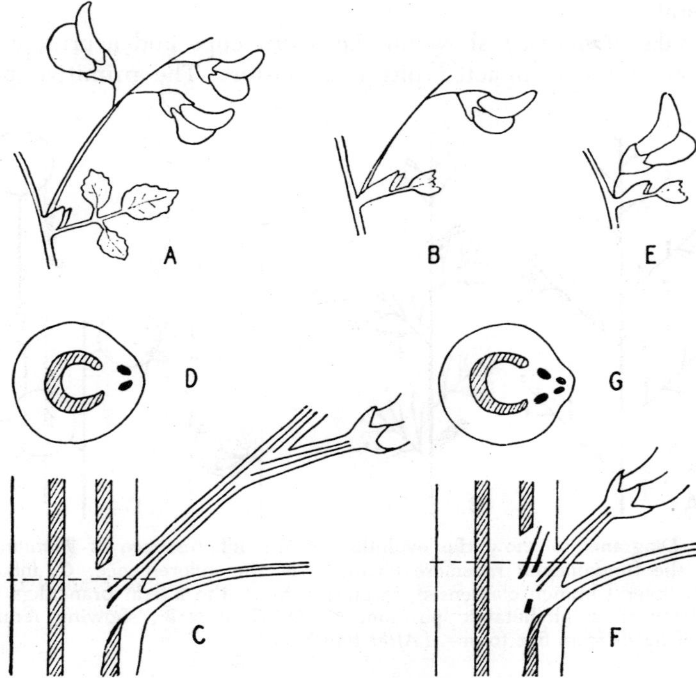

Fig. 31. Sketches of flowers and inflorescences of *Vicia* spp. and diagrams of vascularization of the nodes. *A, B, E,* showing stages in flower reduction in raceme: *E,* the surviving flower apparently borne on the main stem; *C, D,* and *F, G,* diagrams in longitudinal and transverse sections showing the anatomy of the nodes of *B* and *E,* respectively. In *F,* the peduncle of the raceme has disappeared externally, vestiges of its traces are seen buried in the main stem, and traces of the flower are apparently derived independently of the inflorescence stele.

proximal flower (Fig. 31), so that the flower of a secondary axis seems to be borne on a primary axis. Similarly, in the well-known proteaceous genera *Grevillea* and *Banksia,* the flowers that appear to be borne on the main (primary) axis of the inflorescence are borne on greatly reduced, hidden, secondary axes.

A different type of reduction is that where the number of flowers remains high but the axis system is greatly reduced. In *Platanus,* the numerous flowers are crowded into heads; those of each major branch of the inflorescence form a spherical cluster, with its many flowers compacted on the fused branchlets (Fig. 32*C*). Within the genus, the pistillate heads are reduced from a racemose series to one (Fig. 32*E*). The surviving head appears terminal but is the proximal member of the series. The phyllotactic spiral of the branchlets making up a head can be seen indistinctly on the proximal surface of the head and is clear anatomically. (The staminate heads are not reduced in number but are ephemeral.)

The oaks (*Quercus*) show, in the acorn cups and acorns, a similar reduction of the compacted pistillate clusters. The primitive pistillate

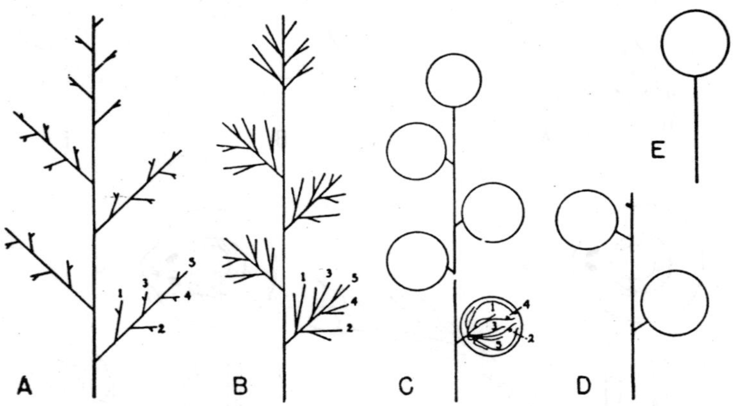

Fig. 32. Diagrams to show the evolution of the inflorescence of *Platanus*. *A*, *B*, showing the fundamental racemose character of the inflorescence; *C*, inflorescence with each lateral branch condensed, forming a head, the lateral branchlets retaining in the head their phyllotactic position; *C*, *D*, *E*, a series showing reduction in number of heads from five to one. (*After Boothroyd.*)

inflorescence of this genus is a large complex panicle, one still present in some tropical species. The ultimate cymose branchlet systems have been reduced to fused woody "cups," surrounding a terminal flower (Fig. 33A). The bracts of the fused branchlets form the scales of the cup. In primitive living species, many acorns are borne laterally on elongate branching axes; in temperate-climate oaks, the cups are reduced to a few, which are borne on shortened, rarely branching axes (white oaks), and to two or one on a very short axis (black oaks). The family Fagaceae shows also a series in reduction in number of flowers within the surviving fertile cluster—*Castanea* has five or three flowers; *Fagus,* two (Fig. 33*B*); *Quercus,* one (Fig. 33A). The lepidocaryoid palms show a similar reduction of the pistillate inflorescence, with the union of the bracts of the sterile branchlets forming an "armored" sheath, the *lorica* (Fig. 34). The lorica surrounds a surviving terminal flower of a lesser branch system (see Palmae). (The scales that make up the lorica have been called emergences of the ovary wall, not bracts, because believed to be borne directly on the wall, but sections of the flowers show the ovary free from the encasing lorica.)

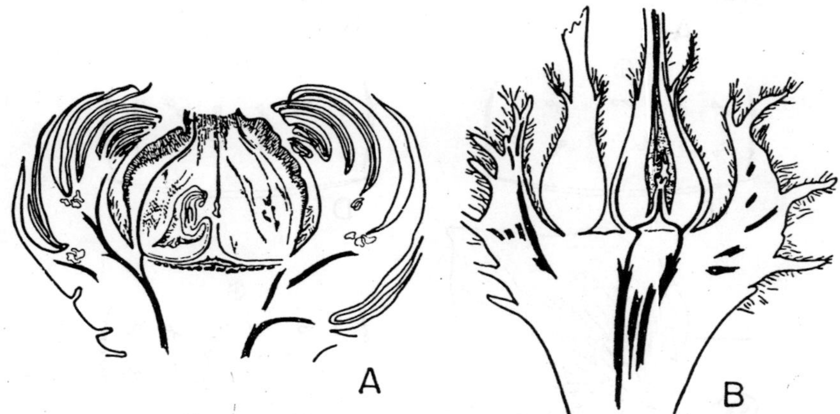

Fig. 33. Sketches of longitudinal sections of young female flowers and inflorescence with involucre prominent. *A, Quercus rubra; B, Fagus americana.* (*After Langdon.*)

The multiflowered catkins of the Amentiferae are often described as racemes and spikes, but probably none of them is of so simple a type. The catkins of the Betulaceae are fine examples of reduction from a complex cymose inflorescence. The "florets" represent, in various genera, a cymule of three flowers, with their subtending bracts in various stages of connation and reduction (Fig. 35).

In the elms, *Ulmus americana* has a compacted panicle; *U. thomasi,* a raceme; *U. pumila,* a head—an obvious reduction series. In *Acer, A. pseudo-Platanus* has a panicle; *A. spicatum* and *A. pennsylvanicum,* a simple raceme or a raceme with a few weak basal branches; *A. saccharum,* an umbellate cluster; *A. rubrum* and *A. saccharinum,* few flowered heads with vestiges of the rachis.

The solitary flower of *Viola* is the surviving median flower of a dichasium. *V. betonicifolia* of Australia and Asia has flowers in the axils of the pair of

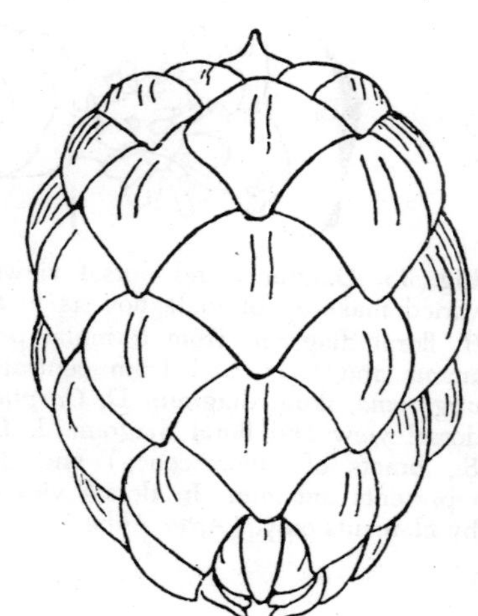

Fig. 34. Loricate fruit of *Metroxylon.* (*After L. H. Bailey.*)

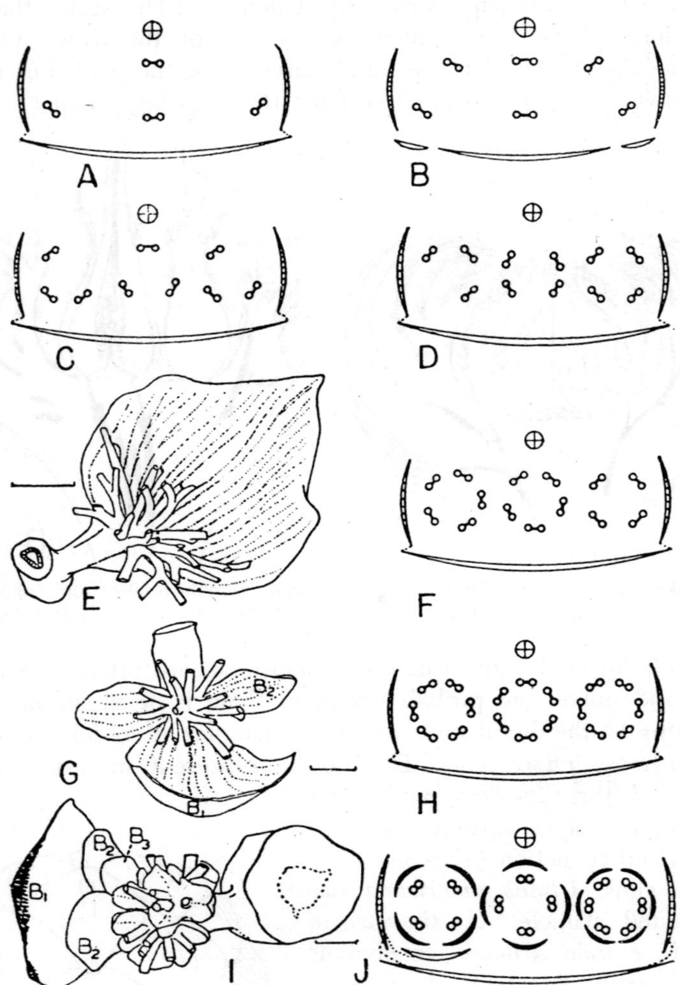

Fig. 35. Diagrams and dorsal views of "flowers," cymules, of Betulaceae showing varied make-up of male flowers in a cymule. *A, B, G, H, Carpinus japonica: A, B, H,* floral diagrams from terminal portion of ament, from transition region between ament and twig, and from central portion of ament; *G,* dorsal view. *C, Ostrya virginiana,* floral diagram. *D, Carpinus laxiflora,* floral diagram. *E, F, C. caroliniana,* dorsal view and floral diagram. *I, J, Alnus firma* var. *hirtella,* dorsal view, (B_1, B_2, B_3, bracts of inflorescence) and floral diagram. Horizontal bar in *E, G,* and *I* represents one mm. In dorsal views, all anthers are removed, stamens represented by filaments only. (*After Abbe.*)

bracts ("prophylls") on the peduncle. Species of the section Nosphinium of *Viola*—woody shrubs of Hawaii—have a cyme with two to four flowers. In other sections cleistogamous or vestigial flowers are described as present in the axils of the bracts. The related genus, *Hybanthus,* with various forms of inflorescence, shows similar reduction to solitary flowers.

The inflorescence of prominent genera of the Proteaceae—*Banksia, Grevillea*—shows flowers arranged in pairs in longitudinal rows, each pair subtended by a single bract. Comparative and anatomical study shows that each pair represents two surviving flowers of a lateral racemose cluster which has been shortened until the axis is buried within the cortex of the main axis. Stages in this reduction are present in the more primitive genera, *Macadamia* and *Hicksbeachia,* which have compound racemose inflorescences with two flowers on each lateral axis.

Extreme inflorescence reduction accompanying vegetative reduction is seen in annual and perennial herbs with solitary terminal flowers and radical leaves—*Cypripedium, Tulipa.*

Inflorescences Resembling Flowers. Many greatly reduced inflorescences resemble flowers; some have passed as flowers in taxonomic treatments. The flowerlike inflorescences of such families as the Com-

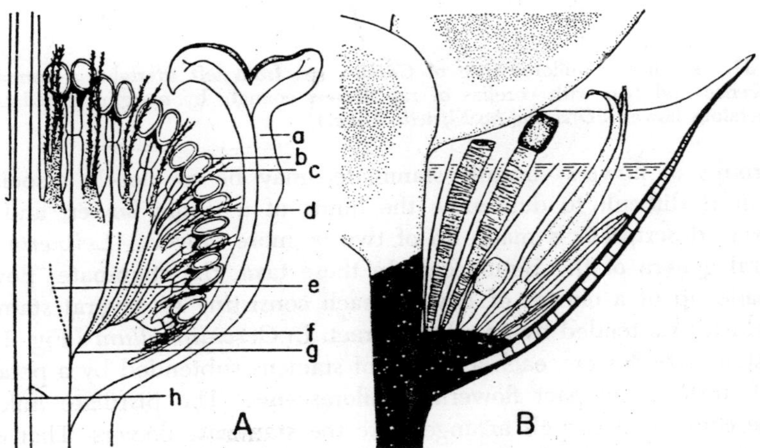

Fig. 36. Reduced inflorescences. *A, Euphorbia pulcherrima,* median, longitudinal section of one half an inflorescence, pistillate flower not shown. *a,* gland; *b,* stamen; *c,* bracteole; *d,* articulation between pedicel and flower consisting of one stamen; *e,* left secondary branch of dichasium; *f,* pedicel of flower; *g,* right secondary branch of dichasium; *h,* two of three traces in stele of pistillate flower. *B, Heliconia bihai* L., portion of inflorescence with side of single bract cut away to show, from right to left, two unopened buds, flower in anthesis, and immature and mature fruits on elongated pedicel. Each structure subtended by bract which subtends a raceme of many flowers not shown. Water level indicated by broken lines. (*A, after Haber; B, after Skutch.*)

positae, Cornaceae, and Euphorbiaceae with petaloid involucres or modified outer flowers are well known. Similar "false flowers" occur in many other families, as in the genera *Leucadendron, Actinotis, Pimelea.* In *Euphorbia,* extreme reduction in flower number and flower structure forms a highly specialized flowerlike inflorescence (Fig. 36*A*); a naked pistillate flower is surrounded by staminate flowers, each reduced to a single stamen; sterile, fused branches of the inflorescence form glands. The outer, petaloid, ray flowers of the Compositae form a pseudocorolla. Where these flowers are connate, as in the "cup *Cosmos,*" the resemblance to a true corolla is close (Fig. 37).

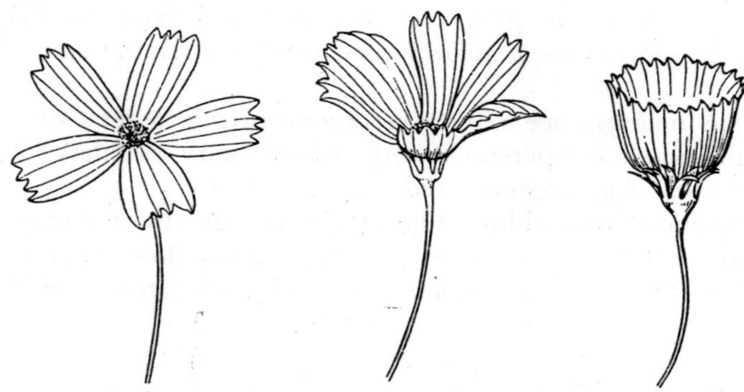

Fig. 37. Sketches of inflorescence of *Cosmos* sp., from left to right, a normal inflorescence and two with corollas of ray flowers connate by margins, simulating a sympetalous flower. (*Drawing by Elfriede Abbe.*)

Groups of flowers—usually staminate—may be so closely associated that it is difficult to determine the limits of a single flower, and the "flower" description is made up of two or more flowers—*Leitneria* and several genera of the Betulaceae. In these taxa, the staminate "flower" is made up of a cluster of flowers, each consisting of several stamens, the cluster subtended by an adnate bract. In *Cercidiphyllum* (Fig. 146), the staminate flowers, each a cluster of stamens subtended by a petaloid bract, form a compact flowerlike inflorescence. The pistillate inflorescence consists of carpels arranged like the staminate flowers. That each carpel, subtended by its bract, constitutes a flower is evident from the orientation of the carpels: the ventral margins appear abaxial. Inflorescences commonly described as flowers are those of *Triglochin* and *Potamogeton.* In *Triglochin,* anatomy demonstrates that the "flower" consists of a whorl of staminate flowers, separated by a whorl of bracts from a whorl of pistillate flowers. The presence of bracts (not staminodes) between the stamens and the carpels is in itself sufficient evidence that this is not a true flower. The "flower" of *Potamogeton,* long considered by a few students an inflorescence, is shown by anatomy to consist of four staminate flowers adnate to four calyxlike bracts, which subtend them, and four distal pistillate flowers.

Extensive Fusion in Inflorescences. The great variety of form in inflorescences is increased and complicated by connation and adnation in the branch system and among parts of the flowers. Simple adnation of inflorescences to adjacent organs is frequent and often apparent, as in

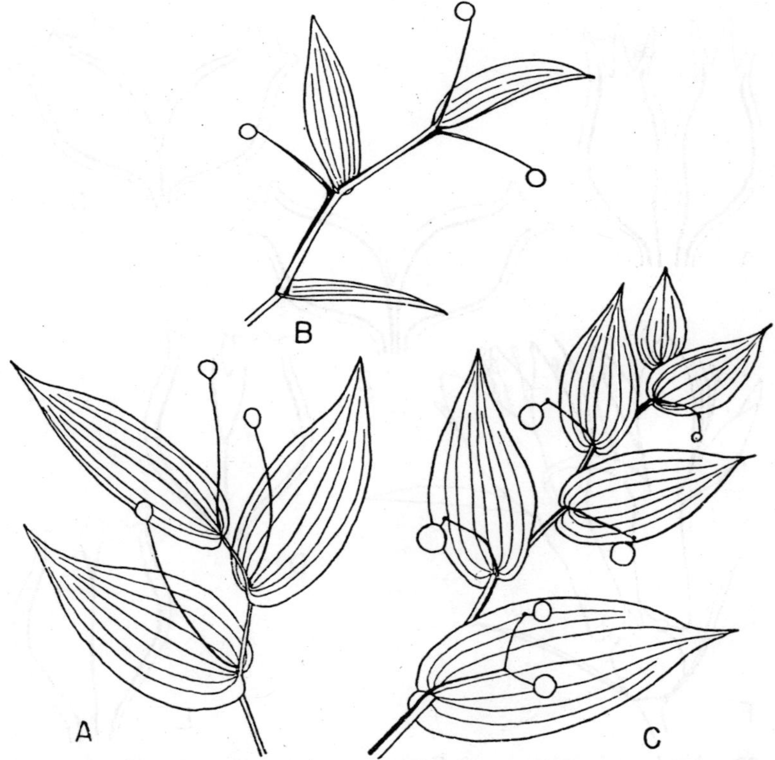

Fig. 38. Sketches of *Streptopus* showing reduction of inflorescence and adnation of peduncle to stem. *A, S. simplex*, inflorescence reduced to solitary flower, peduncle adnate to stem for only a short distance above axil of leaf; *B, S. roseus*, inflorescence as in *A*, peduncle adnate almost up to leaf above the one in whose axil it is borne; *C, S. amplexifolius*, lowest inflorescence showing two flowers with bract indicating a third, and those above showing one flower and a bract indicating another, peduncle adnate to stem up to leaf above, making it appear to be borne below a leaf. (*Drawing by Elfriede Abbe.*)

the adnation of the inflorescence of *Tilia* to its subtending bract and that of the spathe of some aroid inflorescences to the flower cluster—*Phyllocarpus*. More obscure relationships involve fusion of the peduncle to the mother axis, as in *Streptopus, Sparganium,* and other genera, where the flowers appear to be borne below a leaf rather than in its axil (Fig. 38).

The flowers of an inflorescence may be connate laterally either by their basal parts or throughout their length, forming "compound" flowers whose morphological structure is recognizable only anatomically —some of the Betulaceae. Among the species of *Lonicera* with two-flowered inflorescences, there are found stages in the lateral connation of the pairs of flowers (Fig. 39). Some species have the flowers wholly

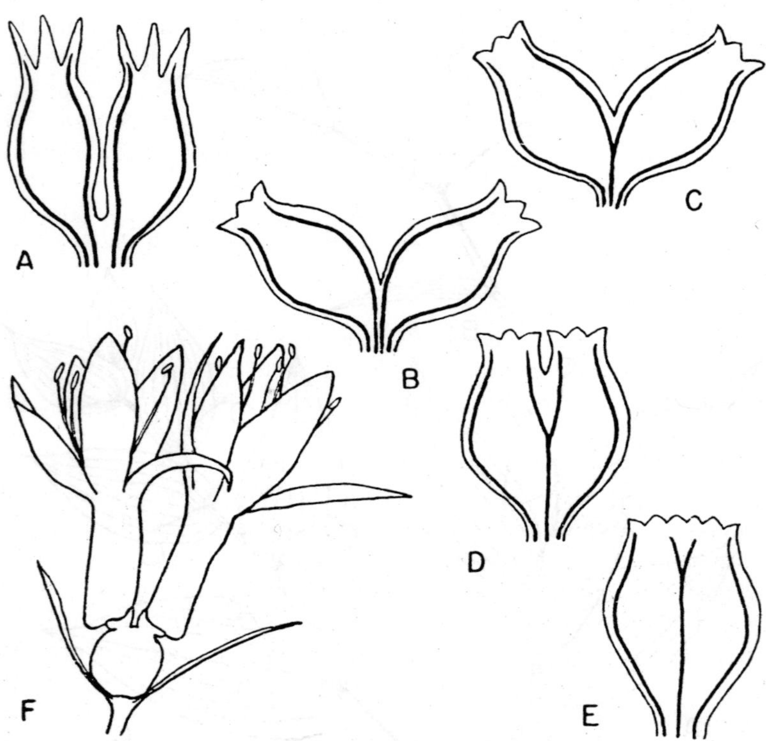

Fig. 39. Sketch and diagrams of *Lonicera* spp. showing an inflorescence of two flowers fused by their ovaries, and, in the diagrams, the fusion of vascular bundles under connation. *A, L. syringantha; B, L. canadensis; C, L. tatarica; D, L. Standishii; E, L. oblongifolia; F, L. caerulea.* (*A to E after Wilkinson.*)

free; other species show various degrees of lateral connation of the flowers by their pedicels and ovaries up to complete union of the flowers to the base of the perianth. The various species show well not merely external fusion but fusion between major elements of the vascular skeleton. The free-flowered species show derivation from a cymose inflorescence like that of the related *Diervilla*. Similar lateral fusion of the several flowers of an inflorescence by their ovaries is seen in *Syncarpea* (Myrtaceae); the united, inferior ovaries form a globose base, on which are attached the perianths and androecia of all the flowers. In *Mitchella* (Rubiaceae), fusion similar to that in *Lonicera* occurs, but the flowers are solitary and belong far apart, morphologically, in the axils of leaves on opposite sides of the stem. The erect flowers are fused laterally to the axis above the node at which they are borne and to one another. The fusion extends to the base of the perianth (sometimes to the top of the corolla) and involves the tip of the mother axis; a falsely terminal, double flower is thus formed. The abortive tip of the stem can sometimes be seen on the double ovary between the perianth tubes. Two solitary axillary flowers become a falsely terminal inflorescence of two flowers. (Similar conditions are seen in *Ephedra*, a gymnosperm, in which ovules on opposite sides of a cone axis become fused, forming a falsely terminal ovule.)

Critical study of other inflorescences reduced to a single flower shows their apparent position to be false. The solitary axillary flower of species of *Vicia* is a lateral flower of a reduced axillary raceme (Fig. 31). The pistillate flower of *Carex*, apparently terminal on a spikelet, is a lateral flower, which, in the reduction of the spikelet, has become adnate to the vestigial tip of the spikelet. The axis tip is seen in related genera and, within the genus, in all stages of reduction and fusion with the ovary.

Fleshiness, associated with fusion in inflorescences, may obscure, in various degrees, form and structure. The entire inflorescence or only parts may be involved in the fleshy transformation; leaflike or platelike structures are formed in this way. Absence of bracts or distortion in their position may increase difficulties of interpretation. In the Urticaceae and Moraceae, fleshy inflorescences are common; the "fruit" of the fig tree (*Ficus*) is a highly specialized fleshy inflorescence. In this genus, the many inflorescence branches are erect and closely approximated, fleshy, and fused to one another to form a hollow structure which is open distally. The major bracts may enter into the fusion or remain free on the outer surface of the "fruit." The greatly reduced flowers cover the inner surface of the hollow structure. Other genera of the Moraceae and Urticaceae (*Dorstenia, Elatostemon*) show phylogenetic steps in the development of the fig type of inflorescence.

Fusion, involving all the parts of an inflorescence and associated with fleshiness, is seen in the fruit of the pineapple plant (*Ananas*), where the inflorescence axis, bracts, and ovaries of all the flowers become intimately fused with one another.

Fusion of sterile branches of the inflorescence, with the development of much woody tissue, and the transformation of bracts into scales or spines form the burrs of such genera as *Castanea* and *Fagus* and the acorn cup of *Quercus* (Fig. 33).

Phylogeny of the Inflorescence

Different views have been held as to the basic phylogeny of the inflorescence: that the solitary flower is primitive and the inflorescences are built up by the addition of flowers from stems below the original flower; that the large compound inflorescence is primitive and that the simpler types and the solitary flower have been derived from this by reduction; and that both these conditions were present in the ancestral stock.

The theory that the solitary flower represents the primitive condition is largely based on the presence of the solitary flower in certain woody genera now known to possess many other primitive features—*Magnolia, Eupomatia, Degeneria, Calycanthus, Dillenia, Himantandra.* In support of the theory that the panicle is the primitive flower-bearing form are the dominance of paniculate inflorescences in woody plants in general; the presence of panicles in herbaceous families considered fairly primitive—Ranunculaceae, Rosaceae, Liliaceae; the prevalence of reduction series that lead to solitary flowers throughout angiosperms; and evidence that at least some of the flowers considered examples of primitive flower position are really surviving members of an inflorescence.

Within inflorescences, the racemose type has frequently been considered the more primitive, doubtless because development in this type is chiefly acropetal, like that of vegetative stems. But flower development limits apical growth, and the determinate condition in itself would seem to be primitive. The solitary flower—such as that of *Magnolia*—which seems to be primitive is determinate, and the majority of large much-branched inflorescences are at least in part determinate. No clear line can be drawn between determinate and indeterminate types; in many genera scattered throughout angiosperms, the racemes have a terminal flower and the acropetal sequence in development may be upset. The racemose arrangement may well be secondary. Changes in the normal ontogenetic sequence that are obviously secondary are the centrifugal development within the androecium and the basipetal development of inflorescences in some palms. The determinate inflorescence seems to be the primitive type.

Evidence is strong that the solitary flowers, which some consider represent the primitive type of flower arrangement, are, in reality, examples of reduction from inflorescences. Part of the basis for the citation of the *Magnolia* flower as primitively solitary is the general acceptance of this genus as one of the most primitive living angiosperms. But the closely related genus, *Michelia,* has clusters of large flowers, and *Raimondea,* in the related Annonaceae, has some flowers in cymes (Fig.

40), others solitary. The solitary flowers in *Annona* have bracts along the peduncle, where abortive flowers may be present. *Degeneria* and *Eupomatia laurina* also have bracts along the peduncle, and, in *Eupomatia*, one of these may subtend a second flower.

Whatever the flower position in the first angiosperms, it is apparent that reduction in flower number and in inflorescence complexity has taken place along many, often parallel and convergent, lines. This reduction is made evident by comparative study in many families and is especially clear when found within generic limits, a common condition. If the solitary flower is primitive, there must have been extensive building up to the paniculate inflorescences, followed frequently by

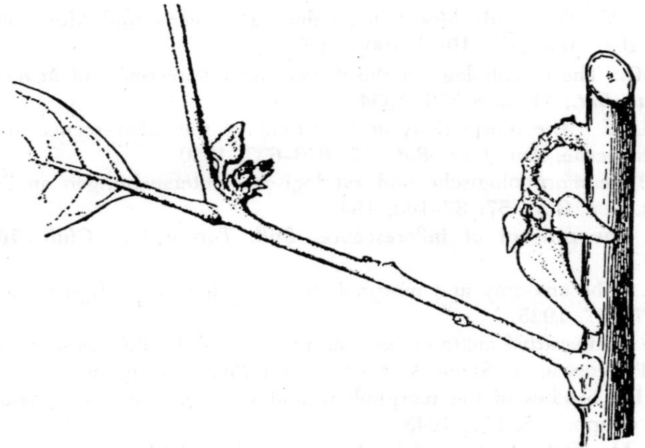

Fig. 40. Sketch of *Raimondea,* Annonaceae, showing inflorescences of large flowers in bud, extra-axillary. (*After Safford.*)

reduction to simple clusters and solitary flowers, which represent a highly specialized condition. Proof of a multiplication of flowers is difficult to obtain. It has been suggested that this multiplication occurred as an accompaniment to the development of microflory, such as that seen in several families in the Australian flora. But the abundance of small flowers in these families is doubtless correlated with xerophily, and the many flowers are not borne typically in inflorescences.

Simplicity in flower position may represent either a primitive or an advanced condition; it probably represents an advanced condition, the result of reduction. The phylogenetic position of the simple inflorescence is not clear; many of them obviously represent modified compound types. If the large compound type is primitive, as it seems to be, no inflorescence is fundamentally simple; all simple inflorescences are reduction types. The apparently simple "catkins" of the Amentiferae, the umbellate clusters of the Rosaceae, the spikes of the Polygonaceae, and the racemes of the Proteaceae are basically complex.

BIBLIOGRAPHY

Abbe, E. C.: Studies in the phylogeny of the Betulaceae. I. Floral and inflorescence anatomy and morphology, *Bot. Gaz.*, **97**: 1–67, 1935.

Bailey, I. W., and C. G. Nast: The comparative morphology of the Winteraceae. VII. Summary and conclusions, *Jour. Arnold Arb.*, **26**: 37–47, 1945.

Bailey, L. H.: Palms and their characteristics, *Gentes Herb.*, **3**: 1–29, 1933.

Becker, W.: Violae asiaticae et australienses, *Beih. Bot. Centralbl.*, **34** (II): 209–216, 1916.

Bergdolt, E.: Morphologische und physiologische Untersuchungen über *Viola, Bot. Abhandl.*, **20**: 1–120, 1932.

Bernbeck, F.: Vergleichende Morphologie der Urticaceen—und Moraceen—Infloreszenzen, *Bot. Abhandl.*, **19**: 1–100, 1932.

Blaser, J. LeC.: The morphology of the flower and inflorescence of *Mitchella repens, Am. Jour. Bot.*, **41**: 533–539, 1954.

Boothroyd, L. E.: The morphology and anatomy of the inflorescence and flower of the Platanaceae, *Am. Jour. Bot.*, **17**: 678–693, 1930.

Bosch, E.: Blütenmorphologische und zytologische Untersuchungen an Palmen, *Ber. Schweiz. Bot. Ges.*, **57**: 37–100, 1947.

Croizat, L.: The concept of inflorescence, *Bull. Torrey Bot. Club*, **70**: 496–509, 1943.

Haber, J. M.: The anatomy and morphology of the flower of *Euphorbia, Ann. Bot.*, **156**: 657–707, 1925.

———: The comparative anatomy and morphology of the flowers and inflorescences of the Proteaceae. I. Some Australian taxa, *Phytomorph.*, **9**: 325–358, 1959.

Hjelmquist, H.: Studies in the morphology and phylogeny of the Amentiferae, *Bot. Not.*, **2** (Supp.): 5–171, 1948.

Langdon, L. M.: Ontogenetic and anatomical studies of the flower and fruit of the Fagaceae and Juglandaceae, *Bot. Gaz.*, **101**: 301–327, 1939.

Manning, W. E.: The morphology of the flowers of the Juglandaceae. II. The pistillate flowers and fruits, *Am. Jour. Bot.*, **27**: 839–852, 1940.

Melchior, H.: Blütenstandbildung bei der Gattung *Viola;* ein Beitrag zur Phylogenie der Violaceen, *Ber. Deutsch. Bot. Ges.*, **50**: 198–204, 1932.

Müller, F.: "The Fertilization of Flowers by Insects," London, 1883.

Nast, C. G., and I. W. Bailey: The comparative morphology of the Winteraceae. VI. Vascular anatomy of the flowering shoot, *Jour. Arnold Arb.*, **25**: 454–466, 1944.

Parkin, J.: The evolution of the inflorescence, *Jour. Linn. Soc.*, **42**: 511–562, 1914.

Pauchet, L.: Recherches sur les Cupulifères, *Thèse. Fac. Sci., Paris*, 1949.

Rickett, H. W.: The classification of inflorescences, *Bot. Rev.*, **10**: 187–231, 1944.

Safford, W. E.: *Raimondia*, a new genus of Annonaceae from Colombia, *Contr. U.S. Nat. Herb.*, **16**: 217–219, 1913.

———: *Annona sericea* and its allies, *Contr. U.S. Nat. Herb.*, **16**: 263–275, 1913.

Skutch, A. F.: The aquatic flowers of a terrestrial plant, *Heliconia bihai* L., *Am. Jour. Bot.*, **20**: 535–543, 1933.

Smith, A. C.: Taxonomic notes on the old world species of Winteraceae, *Jour. Arnold Arb.*, **24**: 119–164, 1943.

Snell, R.: Anatomy of the spikelets and flowers of *Carex, Kobresia*, and *Uncinia, Bull. Torrey Bot. Club*, **66**: 277–295, 1936.

Swamy, B. G. L., and I. W. Bailey: The morphology and relationships of the Cercidiphyllaceae, *Jour. Arnold Arb.*, **30**: 187–210, 1949.

Turpin, P. J. F.: Mémoire sur l'inflorescence des Graminées et des Cyperées comparée avec celle des autres végétaux sexifères, *Mém. Muséum Hist. Nat. Paris*, **5**: 426–492, 1819.

Wilkinson, A. M.: Floral anatomy and morphology of some species of the tribe Lonicereae of the Caprifoliaceae, *Am. Jour. Bot.*, **35**: 261–271, 1948.

Zimmermann, W.: Die Phylogenie der Angiospermen-Blütenstände, *Beih. Bot. Centralbl.*, **53A** (I): 95–121, 1935.

Chapter 3

THE FLOWER

The flower is often considered characteristic of angiosperms; yet no one character can be used to set these plants apart from other seed-bearing plants. Moreover, the flower cannot be so defined as to separate it from similar reproductive structures in the gymnosperms. The morphological basis of the flower is rarely emphasized in definitions; often the flower is "the reproductive structure of the angiosperms." Mor-

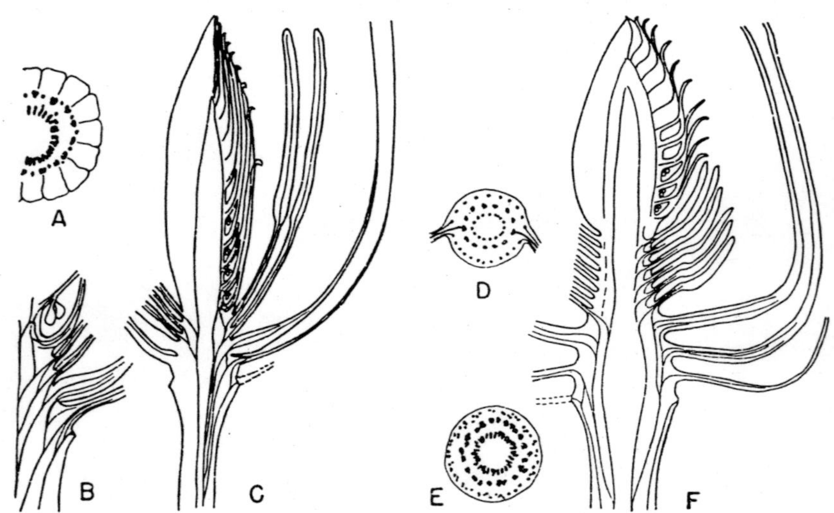

Fig. 41. Diagrams showing primitive structure of flowers of *Magnolia acuminata*, *A* to *C*, and *Liriodendron tulipifera*, *D* to *F*. *C* and *F*, longitudinal sections; *A* and *E*, cross sections through peduncle; *B*, longitudinal section showing vascular supply to appendages; *D*, cross section through androecial region of receptacle showing origin of stamen traces.

phologically, it is a determinate stem tip bearing sporophylls and, commonly, other appendages that are sterile (Fig. 41). But this definition applies equally well to many cones, those of the gymnosperms and even of some of the lower vascular plants. The cone is a characteristic reproductive structure in most vascular plants. If the term flower is to be restricted to the fertile stem tip of the angiosperms and its appendages, it is necessary to compare these structures and their sporangia with those of other groups.

The looseness of use of the term flower by botanists is, in part, responsible for the difficulty in defining the structure. By "the flowering plants" is commonly meant the angiosperms, but such phrases as "the flowering of the conifers" (referring to the period of pollination), "the flowers of the gymnosperms," and "the flowers of the seed plants" are frequently seen.

Basic Structure

The Receptacle. The flower is, first of all, a stem tip, the *receptacle,* resembling in ontogeny and fundamental structure a vegetative tip. It consists of nodes and internodes and bears appendages. The nodes are usually closely crowded by shortening and often brought together by suppression of internodes. Apical growth is limited early in development, but other growth may continue until the fruit is mature. That the flower is only in part a mature structure is commonly overlooked; the receptacle and carpels are still in early stages when the rest of the flower is mature. (Failure to recognize that the vascular supply of the carpels within the receptacle is incomplete at flowering time has led to incorrect interpretations of the nature of the flower and its parts.) The receptacle is often greatly modified and unstemlike in form, size, and structure; and, as it matures with the fruit, it may become still less stemlike.

On the receptacle are borne, typically, both fertile and sterile appendages. The shortening and suppression of internodes bring the appendages close together, either in spirals or whorls. Whorls represent the more specialized arrangement, as in leaf arrangement; a whorl represents one or more "turns" of a flattened spiral. Parts of highly compressed spirals often pass as whorls, as they appear to be from naked-eye and hand-lens inspection (and are, for practical purposes, in taxonomic description), but they can be seen to be spiral by anatomical study. Each organ stands at a level microscopically different from the others, as in the corolla of *Ranunculus.* The appendages may all stand in spirals or all in whorls, or part may be spirally placed and part in whorls. Where all types are spiral, the spiral may run continuously throughout the flower, as in *Paeonia;* more often, there are discontinuities between the different kinds of organs. Where the appendages are all in whorls, the members of successive whorls usually alternate in position with those of the whorls directly above and below; where there are several or many whorls of organs of one kind, as in the androecium of *Aquilegia,* the organs may stand in longitudinal rows. Typically, the different kinds of organs—though separated by breaks in phyllotactic continuity—stand close to one another longitudinally on the receptacle, but there may be prominent, naked internodal areas on the lower part of receptacle, as in *Liriodendron* (Fig. 41*F*). Where the arrangement of organs is spiral, the phyllotactic spiral is often the same as that of the leaves but may be different, especially where the floral organs are numerous.

The Sterile Appendages. The sterile appendages are typically of two kinds: *sepals,* which together form the *calyx;* and *petals,* which make up the *corolla.* These appendages are below the fertile appendages, the calyx below the corolla. (In *Eupomatia,* petaloid staminodia form a pseudocorolla between the stamens and the carpels.) Sepals and petals commonly differ in form, size, and other characters. In some families, they may be closely alike, as in most of the Liliaceae; in others, transitional forms occur, as in the Magnoliaceae. Interpretation of the perianth as calyx or corolla may be difficult and unimportant, as in *Drimys* and *Wintera,* where the organs, ranging from two to several, are spirally placed; and where the perianth is represented by only one or few appendages, which serve as a bud-scale-like cap.

Commonly, sepals are more or less leaflike or bractlike in form and structure, especially in their vascular relations to the stem. Morphologically, they are modified leaves. Typically, they stand in whorls, but, in some primitive families, are spirally arranged—Dilleniaceae, Paeoniaceae. They may be petaloid, but this condition usually accompanies reduction or loss of the petals, as in the Proteaceae. Where greatly reduced, they have the form of minute teeth, scales, bristles, or mere ridges. *Cornus* shows all stages of calyx reduction in both external and internal structure, and some species have no vestige of a former calyx. (Connation and adnation involving sepals are discussed in later pages.)

Petals are typically laminar and larger than the sepals. They, like the sepals, are, morphologically, appendages of leaf rank. In many families, they represent sterile stamens, but, in some primitive families, they are probably modified leaves (like the sepals), as evidenced by transitional forms and anatomical structure in Magnoliaceae. Petals have great range in size and form, from the large, elaborate organs of some orchids to minute structures; under reduction, they may become scales, bristles, or glands of many forms. Accompanying compacting of the flowers in inflorescence specialization, they are reduced to small size and lose their petaloid appearance.

Theories of the nature and development of the perianth are closely bound up with theories of the origin of the flower. One theory is that the perianth, at least in part, was preexistent to the flower. In ancestral forms, sterile appendages were associated with groups of sporophylls, as seen in the bracts below the cones of some conifers and the "flower" of the Bennettitales. These appendages became, with the establishment of angiospermy (enclosure of the ovules), a part of the flower. Elaboration of the perianth, with the development of two series of organs, took place within the angiosperms. Under a second theory, all perianth parts are considered modified sporophylls. This is an old theory that has received little attention since 1900. The petals in most families have been shown to be sterile, petaloid stamens. But all sepals and the petals of some families, for example, Magnoliaceae and Calycanthaceae, have been shown by transitional forms and anatomical structure to be modified leaves. A third theory is based on the view that the primitive flower was naked, and the perianth developed within the angiosperms as a new structure, accompanying the specialization of the flower. Under this theory, the perianth appeared in early forms as bractlike or scalelike organs, protective in function; these organs were increased in number and size, with elaboration in color and complexity of form. The development of a distinct upper series, the corolla, was one of the later steps. The flowers of ranalian families may give evidence of the origin of the perianth from bractlike, protective organs. They show variety in number and form of bud-scale-like outer appendages. *Eupomatia* and *Himantandra* have one calyptralike organ; other genera, a few small, spirally arranged organs. *Trochodendron* has a few appendages below a naked flower. The distinction between bracts and sepals is hardly possible in some of these genera. There are probably no strictly perianthless flowers—with the exception of those that are greatly reduced in structure, as in the inflorescences of *Cercidiphyllum* and many of the Amentiferae. All stages in the evolutionary development of the calyx from bracts are present in the Ranales and Dilleniales.

The corolla has probably arisen in two different ways. A perianth of many spirally arranged organs, with gradual transition from bractlike to larger petaloid members—as in the Magnoliaceae, Calycanthaceae, Himantandraceae, Nymphaeaceae—apparently became separated into proximal and distal parts, the distal more petaloid and otherwise distinct from the proximal. Stages in this specialization are present in the Magnoliaceae. The genus *Magnolia* itself shows reduction in number in the members of the perianth, with gradual change from spiral to whorled arrangement (Fig. 41*A* to *C*). Some species show definitely distinct corolla, as does the allied genus, *Liriodendron* (Fig. 41*D* to *F*). In the corolla, stages in transition from spiral to whorled arrangement are frequent, and there may be two whorls of petals. A second origin lies in the sterilization and petaloid elaboration of stamens. Anatomy demonstrates that the petals of many families are, in vascular structure, unlike the sepals but like the stamens, regardless of extent of superficial resemblance in form to sepals; the number of traces departing to a petal is like that going to the stamen and unlike that to the sepal. The corolla has undoubtedly arisen in two ways. It represents modified stamens in most families; in some families, it represents the distal part of a primitive unspecialized perianth. The effect of reduction of the perianth is primarily on the corolla; the petals are reduced in prominence, to small nonpetaloid organs, often easily overlooked, or to complete external loss. The "apetalous" condition has commonly been considered an early stage in flower development, one in which the perianth is still in primitive form. But, in many taxa, the apetalous condition has been shown to represent a high stage in perianth specialization rather than a low one —Salicaceae, Betulaceae, Juglandaceae, Urticaceae, Aceraceae, Platanaceae, Proteaceae, *Fraxinus*.

The Fertile Appendages. The fertile appendages, also of leaf rank, are of two types: *microsporophylls* (*stamens*), which bear microsporangia; and *megasporophylls* (*carpels*), which bear megasporangia. The stamens constitute the *androecium;* the carpels, the *gynoecium*. (The term *pistil* is applied to a unit of the gynoecium: to a single carpel when the carpels are free from one another; to a group of carpels when they are fused to one another.) Where the flower has only one kind of sporophyll, it is *unisexual;* where both are present, it is *bisexual*. If flowers of both unisexual kinds are borne on the same plant, the taxon is *monoecious;* if staminate and pistillate flowers are borne on separate plants, the taxon is *dioecious*.

The flower varies greatly in number of kinds of parts, from four to one; in number of organs of a kind, from many to one; in extent of fusion of organs to one another, from complete absence of fusion to connation and adnation in extreme form; in elaboration of organs in size and form. The range in flower form and structure is very great— from the (superficially) simplest type, in which the flower consists of a single sporophyll on a receptacle, as in the staminate flowers of *Euphorbia* (Fig. 36A), to those with all kinds of appendages present, extensively fused together and elaborate in form. Variety in form is recognized as largely the result of reduction and of fusion. (Closely similar conditions are seen in the cones of the conifers.)

Broad comparative studies of flower structure show lines of specialization and, in general, the sequence in which changes have occurred. From these, the more primitive structure is determined and the course of phylogenetic modification can be outlined. The structure of the flower has been the chief basis for the classifications, artificial and natural, of angiosperms. And, although it is now recognized that evidence of natural relationships must come from all parts of the plant, vegetative and reproductive, and from all fields of plant study, flower structure provides a foundation upon which theories of evolutionary relationships can be based.

The Primitive Flower

The type of flower now generally recognized as morphologically simple is one that shows the least change, under adaptive evolutionary modification, from the original primitive flower. Evidence now seems to support strongly the theory that the ancestral flower was bisexual, with numerous stamens and carpels, without a perianth, or with a uniseriate perianth of simple, bractlike organs. (This theory differs greatly from that still held by some botanists that the primitive flower was unisexual.) All the appendages were spirally arranged, and the flower was symmetrical and without fusion among its parts. From this theoretically basic flower—essentially the "pattern flower" of preevolutionary taxonomy—have developed many lines of modification, with reduction producing types more simple in kinds and number of organs, and with elaboration producing complexity in form.

The major principles of evolutionary modification, upon which is based the acceptance of this type of flower as morphologically simple and primitive among living forms, are the following advances:

1. From many parts, indefinite in number, to few, definite in number
2. From three or four sets of appendages—perianth, androecium, and gynoecium—to one
3. From spiral to whorled arrangement of appendages
4. From freedom of floral parts to fusion—connation and adnation
5. From radial symmetry (actinomorphy) to bilateral symmetry (zygomorphy)

In these principles, there is important departure from those of the Engler system of classification, which has long formed the foundation of the natural system of plant classification and is still largely in use in herbaria. The major difference lies in the interpretation of simplicity. Under the Engler system, the primitive flower is unisexual; advance is seen to bisexual flowers and to increase in number of sporophylls. In more recent classifications, fewer kinds and smaller number of appendages represent specialization—reduction by loss of parts. This viewpoint has brought major changes in phylogenetic relationships of both large and small taxa; taxa formerly believed primitive are now considered advanced, and new lines of apparent relationship have been drawn. Prominent among these changes is the interpretation of the unisexual flower with few sporophylls as advanced, rather than primitive.

Reduction in the Flower

Reduction in the flower may occur in many or all parts, simultaneously in several parts, or progressively from part to part. Loss may be of individual organs or of entire whorls of organs; loss may be minor, as loss of one petal or one stamen, or may occur in all parts of the flower, so that there remain only the receptacle and one or few sporophylls of one kind. All stages in loss of function and reduction in size can be seen in closely related taxa. Organs in process of evolutionary reduction, "vestigial organs," can be recognized by abortive form and structure and by position in the flower; vascular anatomy may aid in the interpretation of vestigial structures where identity is uncertain. The vascular traces of lost organs are usually present when external form is reduced, and may persist in the receptacle after the organ itself has disappeared.

Under reduction, organs may be reduced in form and structure and changed in function. Transformations of petals and stamens into glands and of stamens into staminodia are probably the commonest changes. Evidence of the change of petals and stamens into glands is usually apparent in position of the glands and type and origin of vascular supply. (Some glands and glandular surfaces represent merely secretory areas, not modified organs.) Plants with nonfunctioning petals are sometimes called apetalous when the petals are still present in vestigial or greatly reduced form, as in the Proteaceae. In this family, some genera have the petals present as laminar scales—*Placospermum, Austromuellera*—or filamentous projections; others have glands in the positions and with the anatomy of petals. In the Salicaceae, glands that undoubtedly represent a lost perianth stand around, behind, or in front of the sporophyll. The glands have the position and vascular supply of perianth parts, and some of them are lobed and somewhat petaloid. The presence of a single gland seems to represent the greatest reduction, because the willows, primitive in other floral features, have several glands, and remnants of the vascular supply to other lost perianth parts are found in species with a single gland.

Reduction in the Stamen. Reduction in the stamen occurs in all stages, from abortion of sporangia only, to complete disappearance of the organ. The abortion of two sporangia—one of each pair—is frequent; abortion of three of the four is rare. Loss of the entire anther is frequent; the stamen survives as a sterile, laminar, or filamentous organ, which may be petaloid. The petals of the majority of families seem to represent completely petaloid stamens. Stages in the loss of the stamen by gradual reduction in size of a staminodium are well shown in the genera of the Scrophulariaceae. In genera where no external remnant of the lost stamen survives, the vascular trace of the stamen is still present in the receptacle.

Reduction in the Carpel. Reduction in the carpel is primarily in size and in number of ovules. Primitively, the carpel contained many ovules.

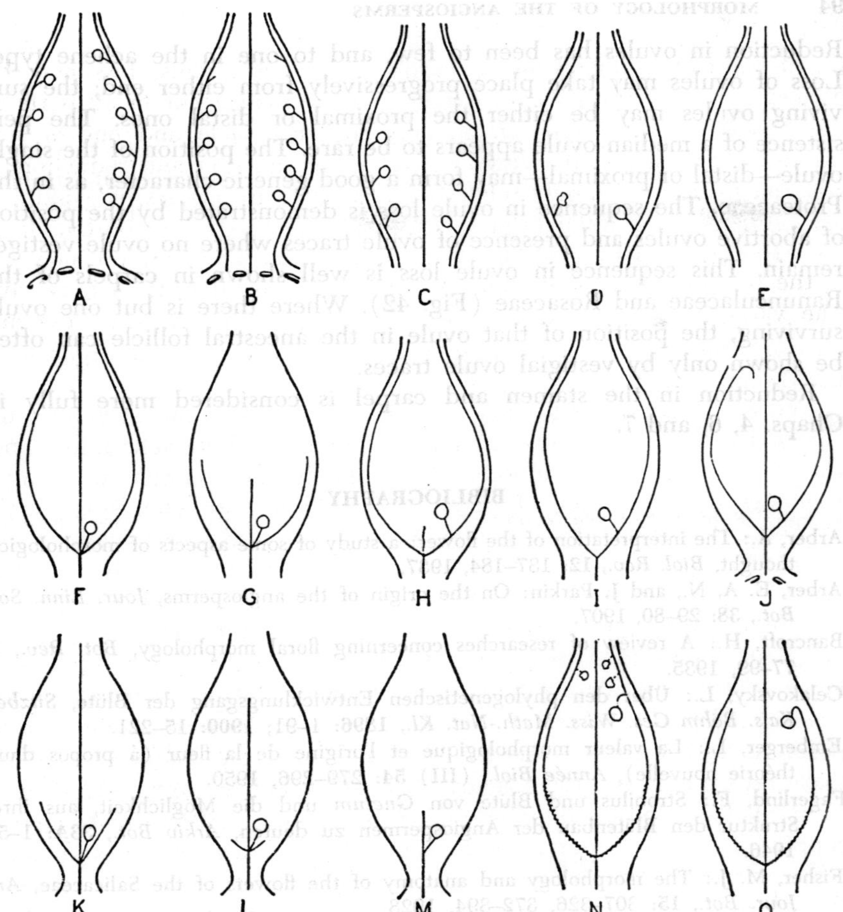

Fig. 42. Diagrams of carpel structure showing reduction of ovules, and fusion and reduction of vascular supply. *A* to *E*, follicles: *A*, *Helleborus viridis*, typical follicle, with many ovules and three traces; *B*, *Trollius laxus*, ovules reduced in number, traces to lost ovules persisting; *C*, *Aquilegia canadensis*, upper ovules and their traces lost; *D*, *Hydrastis canadensis*, all ovules but lower two lost, one of these abortive; *E*, *Waldsteinia fragarioides*, all ovules but one lost, trace to one other persisting. *F* to *O*, achenes: *F* to *M*, with basal ovule surviving; *N*, *O*, with an upper ovule surviving. All with dorsal and ventral traces united at the base, in some as far as the ovule, which then appears to be attached to the dorsal (*F*, *G*, *K*, *L*, *M*). *F*, *Geum rivale*, with distal parts of all bundles present; *G*, *Duchesnea indica*, dorsal and ventral bundles greatly shortened; *H*, *Fragaria vesca*, with dorsal bundle greatly shortened; *I*, *Agrimonia striata*, with dorsal bundle lost beyond its union with the ventral bundles; *J*, *Ranunculus Ficaria*, the ventral bundles recurving, not entering the style; *K*, *R. Flammula*, the ventral bundles greatly shortened; *L*, *R. Cymbalaria*, only vestiges of the ventral bundles persisting, the dorsal bundle greatly shortened; *M*, *R. aquatilis*, the ventral bundles lost beyond the ovule, the dorsal bundle continuing hardly beyond it; *N*, *Potentilla recta*, one ovule surviving, others vestigial, the ventral bundles unreduced; *O*, *P. canadensis*, one ovule surviving, the ventrals lost beyond the ovule. (*From Eames and MacDaniels, adapted from Chute.*)

Reduction in ovules has been to few, and to one in the achene type. Loss of ovules may take place progressively from either end; the surviving ovules may be either the proximal or distal ones. The persistence of a median ovule appears to be rare. The position of the single ovule—distal or proximal—may form a good generic character, as in the Proteaceae. The sequence in ovule loss is demonstrated by the position of abortive ovules and presence of ovule traces where no ovule vestiges remain. This sequence in ovule loss is well shown in carpels of the Ranunculaceae and Rosaceae (Fig. 42). Where there is but one ovule surviving, the position of that ovule in the ancestral follicle can often be shown only by vestigial ovule traces.

Reduction in the stamen and carpel is considered more fully in Chaps. 4, 6, and 7.

BIBLIOGRAPHY

Arber, A.: The interpretation of the flower: a study of some aspects of morphological thought, *Biol. Rev.*, **12:** 137–184, 1937.

Arber, E. A. N., and J. Parkin: On the origin of the angiosperms, *Jour. Linn. Soc. Bot.*, **38:** 29–80, 1907.

Bancroft, H.: A review of researches concerning floral morphology, *Bot. Rev.*, **1:** 77–99, 1935.

Celakovsky, L.: Über den phylogenetischen Entwicklungsgang der Blüte, *Sitzber. Kais. Böhm Ges. Wiss. Math.-Nat. Kl.*, **1896:** 1–91; **1900:** 15–221.

Emberger, L.: La valeur morphologique et l'origine de la fleur (á propos d'une théorie nouvelle), *Année Biol.*, (III) **54:** 279–296, 1950.

Fagerlind, F.: Strobilus und Blüte von *Gnetum* und die Möglichkeit, aus ihrer Struktur den Blütenbau der Angiospermen zu deuten, *Arkiv Bot.*, **33A:** 1–57, 1946.

Fisher, M. J.: The morphology and anatomy of the flowers of the Salicaceae, *Am. Jour. Bot.*, **15:** 307–326, 372–394, 1928.

Glück, H.: "Blatt- und Blütenmorphologische Studien," Jena, 1919.

Grant, V.: The protection of the ovules in flowering plants, *Evol.*, **4:** 179–201, 1950.

Grégoire, V.: Sporophylles et organes floraux, tige et axe florale, *Rec. Trav. Bot. Néerl.*, **32:** 453–466, 1935.

Hallier, H.: Zur morphologischen Deutung der Diskusbildung in der Dikotylenblüte, *Medel. Rijks Herb. Leiden*, **41:** 1–14, 1921.

Henslow, G.: On the vascular systems of floral organs and their importance in the interpretation of the morphology of flowers, *Proc. Jour. Linn. Soc. London Bot.*, **28:** 152–196, 1889.

Heslop-Harrison, J.: The experimental modification of sex expression in flowering plants, *Biol. Rev.*, **32:** 38–90, 1957.

Horne, A. S.: A contribution to the study of the evolution of the flower, with special reference to the Hamamelidaceae and Cornaceae, *Trans. Linn. Soc.*, (II) **8:** 239–309, 1914.

Janchen, E.: Die Herkunft der Angiospermen-Blüte und die systematischen Stellung der Apetalen, *Oesterr. Bot. Zeitschr.*, **97:** 129–167, 1950.

Just, T.: Origine et évolution de la fleur, *Année Biol.*, **28:** 135–148, 1952.

Kasapligil, B.: Morphological and ontogenetic studies of *Umbellularia californica* Nutt. and *Laurus nobilis* L., *Univ. Calif. Publ. Bot.*, **25**: 115–240, 1951.

Kaussmann, B.: Vergleichende Untersuchungen über die Blattnatur der Kelch-, Blumen-, und Staubblätter, *Bot. Archiv*, **42**: 503–572, 1941.

Kozo-Poljanski, B.: On some "third" conceptions in floral morphology, *New Phyt.*, **35**: 479–492, 1936.

Leppik, E. E.: The form and function of numerical patterns in flowers, *Am. Jour. Bot.*, **43**: 445–455, 1956.

Martens, P.: La grain et le tube pollinique: Réflexions sur les caractères propres des phanérogames, *Acad. Roy. Belg. Bull. Cl. Sci.*, 5 sér., **33**: 919–943, 1948.

Mason, H. L.: The concept of the flower and the theory of homology, *Madrono*, **14**: 81–95, 1957.

Mathews, J. R.: Floral morphology and its bearing on the classification of angiosperms, *Trans. Bot. Soc. Edinburgh*, **23**: 69–82, 1941.

Nast, C. G.: The comparative morphology of the Winteraceae. VI. Vascular anatomy of the flowering shoot, *Jour. Arnold Arb.*, **25**: 454–466, 1944.

Neumayer, H.: Die Geschichte der Blüte, *Abhand. Zool.-Bot. Ges. Wien*, **14**: 1–112, 1921.

Ozenda, P.: Recherches sur les dicotylédones apocarpiques, *Publ. Lab. École Norm. Sup.*, sér. biol. II, Paris, pp. 1–183, 1949.

Parkin, J.: The unisexual flower: a criticism, *Phytomorph.*, **2**: 75–79, 1952.

———: The unisexual flower again—a criticism, *Phytomorph.*, **7**: 7–9, 1957.

———: The strobilus theory, *Proc. Linn. Soc. London*, **135**: 51–64, 1922.

Payer, J. B.: "Traité d'Organogénie Comparée de la Fleur," Paris, 1857.

Plantefol, L.: Fondements d'une théorie florale nouvelle: L'ontogénie de la fleur, *Ann. Sci. Nat. Bot.*, 11 sér., **9**: 35–186, 1948.

Rauh, W., and H. Reznik: Histogenetische Untersuchungen an Blüten- und Infloreszenzenachsen. I. Die histogenese becherförmiger Blüten und Infloreszenzenachsen sowie der Blütenachsen der Rosoideen, *Sitzber. Heidelb. Akad. Wiss. Math.-Nat. Kl.*, **1951**: 3–71.

Sargant, E.: The reconstruction of a race of primitive angiosperms, *Ann. Bot.*, **22**: 121–186, 1908.

Sporne, K. R.: Correlation and classification in dicotyledons, *Proc. Linn. Soc. London*, **160**: 40–47, 1948.

———: A new approach to the problem of the primitive flower, *New Phyt.*, **48**: 259–276, 1949.

———: Statistics and evolution of dicotyledons, *Evol.*, **8**: 55–64, 1954.

Stebbins, G. L., Jr.: Natural selection and the differentiation of angiosperm families, *Evol.*, **5**: 299–324, 1951.

Thompson, J. M.: Studies in advancing sterility. VII. The state of flowering known as angiospermy, *Hartley Bot. Lab. Liverpool Publ.* 12, 1934.

———: Towards a modern physiological interpretation of flowering, *Proc. Linn. Soc. London*, **156**: 46–69, 1943–44.

Troll, W.: "Organization und Gestalt im Bereich der Blüte," Berlin, 1928.

Van Tieghem, P.: Recherches sur la structure du pistil et sur l'anatomie comparée de la fleur, *Mém. Sav. Acad. Sci. Imp. France*, **21**: 1–261, 1875.

Vautier, S.: La vascularisation florale chez les Polygonacées, *Candollea*, **12**: 219–341, 1949.

Wilkinson, A. M.: The floral anatomy and morphology of some species of *Cornus* and of the Caprifoliaceae, Thesis, Cornell University, 1945.

Wilson, C. L., and T. Just: The morphology of the flower, *Bot. Rev.*, **5**: 97–131, 1939.

Chapter 4

THE ANDROECIUM

Classification by Arrangement on the Receptacle

The aggregate of the stamens of a flower constitutes the *androecium.* In number, the stamens of a flower range from many to one, from indefinite to definite. In arrangement on the receptacle, they are *spiral, whorled,* or *fasciculate* (clustered)—the fascicles usually in whorls.

The Spiral Androecium. Many stamens, spirally arranged, constitute the primitive androecium, from which have been derived the whorled and the fasciculate types. This androecium is present in many of the lower dicotyledons—Degeneriaceae, Dilleniaceae, Magnoliaceae, Eupomatiaceae, Annonaceae, Calycanthaceae, Nymphaeaceae, and some of the Ranunculaceae, Monimiaceae, and Winteraceae.

The Fasciculate Androecium. The nature of stamen fascicles was much discussed—with little agreement—in the earlier days of taxonomic morphology. It has recently received attention again, chiefly in connection with the theory of the telomic structure of the stamen. The fascicle has been considered both a branching and a compound organ: its branching, the result of radial and tangential splitting (chorisis) of a simple stamen; its compound nature, the result of aggregation of simple stamens, connate in various degrees (Fig. 43). Fascicles consist of various numbers of stamens—from very many, to few, and to one. Within the fascicle, development is often, perhaps always centrifugal. Connation is of various degrees—by the stamen traces within the receptacle and, externally, by the filaments. Stages in the connation are present in several families—Dilleniaceae, Paeoniaceae, Cactaceae. In these families, a few stamens in the cluster may be wholly free, both externally and in their traces. Ontogeny, in showing fascicles arising as a mound, with individual stamens developing on the surface, has been considered to show evidence of branching. But evidence from ontogeny is not in itself conclusive; organs phylogenetically fused arise congenitally united, wholly or in part. A comparable condition exists in the sympetalous corolla where the lobes—morphologically, the petal tips—and the tubular base arise from a simple primordium.

The fascicle has been interpreted as a primitive form of the stamen, a branching system of fertile telomes, reduced and compacted, with the simple stamen a surviving solitary branch of this compound system. Similarity of the fascicle to the stamen of *Ricinus* (Fig. 44) has been cited in support of this theory, and resemblance to the bennettitalian stamen has been noted as a part of the evidence for the origin of angiosperms from the Bennettitales. The fasciculate arrangement seems to

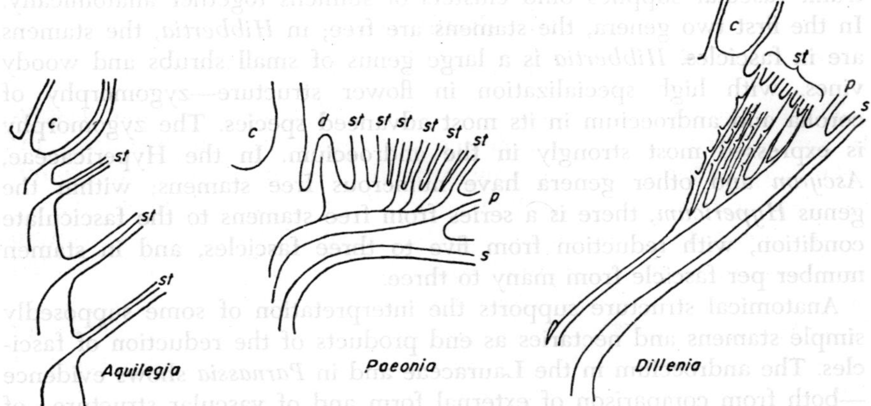

Fig. 43. Diagrams showing vascular supply to the androecium. *Aquilegia*, stamens with independent traces; *Paeonia* and *Dillenia*, "trunk" vascular supply to the stamen fascicles. *c*, carpel; *d*, disc; *p*, petal; *s*, sepal; *st*, stamen. (*From Eames*, 1953. *Paeonia after Brouland; Dillenia drawing by Wilson.*)

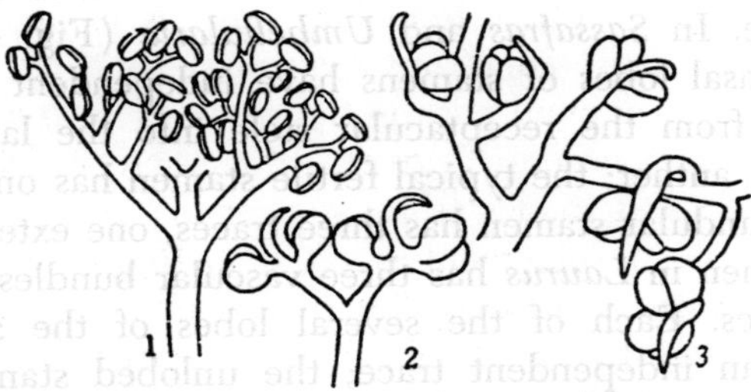

Fig. 44. Sketches of parts of branching stamens of *Ricinus communis*. 2, 3, varieties from Java showing morphology of the anther; anther sacs free, attached to the base of the elongate connective. (*After van der Pijl*; 1, *from Sach's Lehrbuch*.)

have arisen at least twice—among the Dilleniales and among the woody Ranales. A series in elaboration is seen in the Myristicaceae, Monimiaceae, and Lauraceae. In these three families, part or all of the members of a fascicle are transformed into nectaries, which often retain stamen form.

The theory that the fascicle is merely an aggregation of simple stamens is supported by comparison of related taxa and by anatomy. In the Dilleniaceae, Paeoniaceae, Myrtaceae, and Hypericaceae, series in taxa show the clustering of numerous free stamens to form multi-stamen fascicles, and reduction in number of fascicles and number of stamens per fascicle. In the Dilleniaceae, *Wormia, Dillenia,* and *Hibbertia* form an obvious series in evolutionary specialization in flower structure and in habit, from trees to small shrubs and vines. In *Dillenia,* where there is no external grouping among the many spiral stamens, trunk vascular supplies bind clusters of stamens together anatomically. In the first two genera, the stamens are free; in *Hibbertia,* the stamens are in fascicles. *Hibbertia* is a large genus of small shrubs and woody vines, with high specialization in flower structure—zygomorphy of corolla and androecium in its most advanced species. The zygomorphy is expressed most strongly in the androecium. In the Hypericaceae, *Ascyron* and other genera have numerous free stamens; within the genus *Hypericum,* there is a series from free stamens to the fasciculate condition, with reduction from five to three fascicles, and in stamen number per fascicle from many to three.

Anatomical structure supports the interpretation of some supposedly simple stamens and nectaries as end products of the reduction of fascicles. The androecium in the Lauraceae and in *Parnassia* shows evidence —both from comparison of external form and of vascular structure—of strong reduction. Some of the stamens and nectaries (in the Lauraceae, modified stamens) have lateral or basal appendages (Fig. 45). Anatomically, these appendages represent individual stamens connate in a fascicle with a fertile stamen or with other sterile members of an original fascicle. In *Sassafras* and *Umbellularia* (Fig. 45*A* to *C*), the nectariferous basal lobes of stamens have independent vascular traces, two extending from the receptacular stele into the lateral structures, the other to the anther; the typical fertile stamen has only one trace. In *Benzoin,* the glandular stamen has three traces, one extending into each lobe. Each stamen in *Laurus* has three vascular bundles, two supplying the lateral lobes. Each of the several lobes of the staminodium of *Parnassia* has an independent trace; the unlobed stamen has several traces arranged in the filament in a loose cluster. The stamen and the staminodium (nectary) of this genus are homologous structures, representing greatly reduced stamen fascicles with connate members. (The presence of the vascular bundles of several connate stamens within the "filament" of the typical stamen is an excellent example of the persistence of vascular structure after external form has been lost.) The existence of fasciculate stamens in these taxa is of phylogenetic importance. It supports the view—maintained by cytological evidence—that *Parnassia* does not belong in the Saxifragaceae; it aids in determining the relatives of the Lauraceae (Chap. 11).

Though the existence of "trunk" vascular supplies for fascicles would, in itself, seem to support the view that the fascicle is a unit organ, other anatomical evidence shows the evolutionary development of the fascicle. In the Paeoniaceae (Fig. 46), Dilleniaceae, Monimiaceae, and other families, some stamens—especially those on the outside of the fascicles —have simple, independent traces, and other stamens have traces

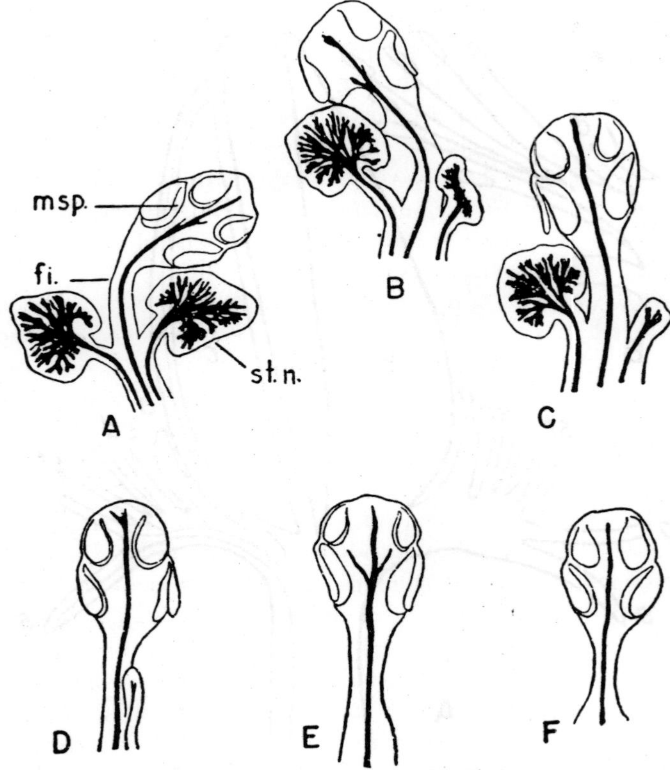

Fig. 45. Stamen fascicles of *Umbellularia* showing lateral members transformed into nectaries and stages in the loss of the nectaries, vascular supply to nectaries independent and like that of the fertile stamen in type and origin. *msp.*, microsporangium; *fi.*, filament; *st.n.*, staminal nectary. (*After Kasapligil.*)

loosely coherent with the "trunk," or connate with it for only a short distance. (The anatomy of the vascular supply of fasciculate stamens, especially the histological structure of the trunk vascular supply, is little known.)

The Whorled Androecium. Whorled arrangement in the androecium represents a modification of spiral arrangement, as does the similar condition in leaves. Where the stamens are in one whorl, the androecium is *haplostemonous;* in two whorls, the members of the outer whorl alternating with the petals, *diplostemonous*, and the members of the outer whorl opposite the petals, *obdiplostemonous;* where in more than two whorls, *polystemonous* (Fig. 47, *A*, *B*). Polystemonous androecia are uncommon [four whorls in Lauraceae (Fig. 47*B*), several in *Aquilegia, Delphinium, Nandina, Trochodendron;* three in some species of *Illicium*]; haplo- and diplostemonous are common. Obdiplostemonous

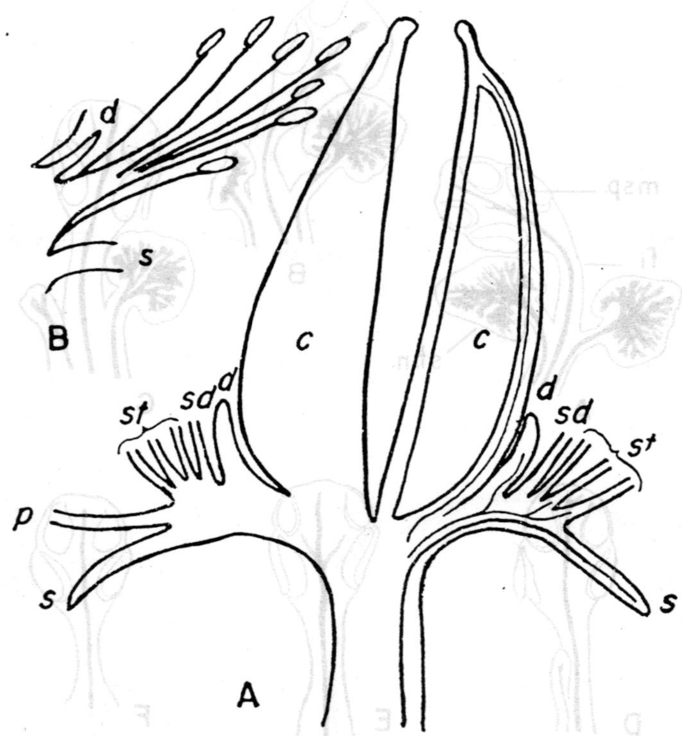

Fig. 46. Diagrams of flower structure of *Paeonia brownii*, showing fascicled stamens. A, showing trunk vascular supply to stamens; *B*, showing various degrees of connation of stamens in fascicle. *c*, carpel; *d*, disc appendage; *p*, petal; *sd*, staminodia; *st*, stamen; *s*, sepal. (*A*, *from Eames*, 1953.)

androecia are characteristic of the Caryophyllaceae, Geraniaceae, Oxalidaceae, Rutaceae, some Saxifragaceae, and some other taxa.

Obdiplostemony represents an interruption in the usual sequence of alternation in floral whorls; its morphological nature was much discussed in the 1870s and 1880s. Three interpretations of obdiplostemony were proposed: that it arose by the addition of a new or extra whorl of stamens (the intercalation theory); by the loss of a whorl between the corolla and the androecium (the reduction theory); and by ontogenetic displacement (the displacement theory). Under the intercalation theory, a new whorl of stamens was considered interposed between the corolla and the lower stamen whorl. Under the *dédoublement* theory, the alternate new members were formed by "dédoublement," a

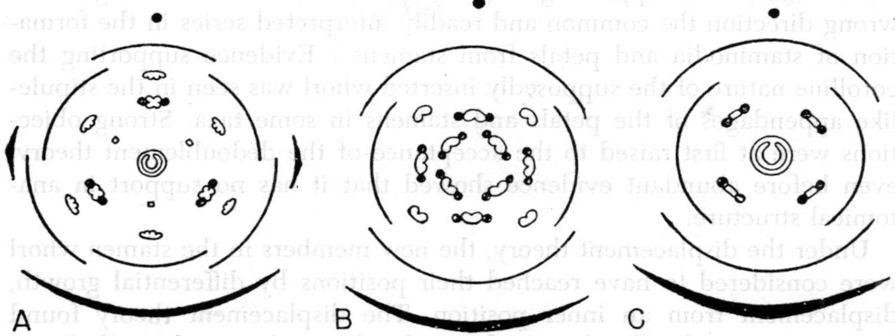

Fig. 47. Floral diagrams of polystemonous androecia showing reduction in the androecium; transformation of lateral members of the stamen fascicles into nectaries and reduction in number of the whorls. *A, Umbellularia californica:* the inner whorl vestigial, the next outer whorl of nectary-bearing fascicles, the two outer whorls of simple stamens. *B, C, Laurus nobilis:* staminate flower *B* with two whorls of nectary-bearing fascicles and one whorl of stamens; pistillate flower *C* with one whorl of staminodes. (*After Kasapligil.*)

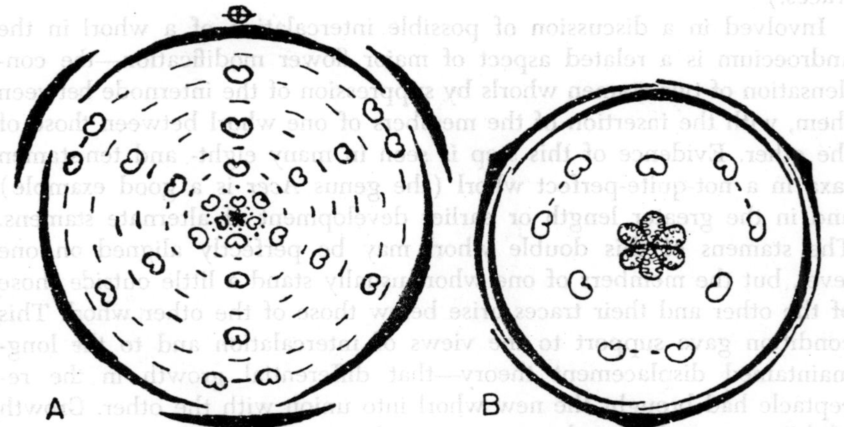

Fig. 48. Floral diagrams showing paired stamens (*dédoublement*) in the outer whorl of androecium. *A, Sagittaria sagittifolia; B, Butomus umbellatus.* (*After Salisbury.*)

doubling by radial splitting of the original outer whorl (Fig. 48). Under the Carpel Polymorphism theory, it has been claimed that obdiplostemony transgresses the law of alternation of successive whorls, but, in androecia with more than two whorls of stamens, there is commonly no alternation (Fig. 48*A*). Under still another theory, that of origin by division, which is supported by the frequent presence of epipetally, the new stamens were considered morphologically petals, formed by tangential splitting of the petals, with the new inner organs transformed into fertile appendages. (This theory arose from reading in the wrong direction the common and readily interpreted series in the formation of staminodia and petals from stamens.) Evidence supporting the corolline nature of the supposedly inserted whorl was seen in the stipule-like appendages of the petals and stamens in some taxa. Strong objections were at first raised to the acceptance of the dédoublement theory, even before abundant evidence showed that it has no support in anatomical structure.

Under the displacement theory, the new members in the stamen whorl were considered to have reached their positions by differential growth, displacement from an inner position. The displacement theory found later support under the theory of carpel polymorphism, where "ballooning," caused by the development of carpellary locules, was considered to have pushed outward the antepetalous stamens, so that, in obdiplostemonous flowers, they stand farther out than the antesepalous stamens. (Displacement of this sort would not be morphologically significant, because it does not affect the level of origin of the stamen traces.)

Involved in a discussion of possible intercalation of a whorl in the androecium is a related aspect of major flower modification—the condensation of two stamen whorls by suppression of the internode between them, with the insertion of the members of one whorl between those of the other. Evidence of this step is seen in many eight- and ten-stamen taxa in a not-quite-perfect whorl (the genus *Acer* is a good example) and in the greater length or earlier development of alternate stamens. The stamens of this double whorl may be perfectly aligned on one level, but the members of one whorl usually stand a little outside those of the other and their traces arise below those of the other whorl. This condition gave support to the views of intercalation and to the long-maintained displacement theory—that differential growth in the receptacle had brought the new whorl into union with the other. Growth of this sort was believed even to move the inner whorl to a position outside the outer whorl and thus to bring about obdiplostemony. Evidence of this was supposedly found in the course of the traces through the cortex, but the position of trace origin was not noted. Union of two whorls is well demonstrated in several families. In the papilionate Leguminosae, the ten stamens may arise as two whorls but, in maturity, form one; differential growth brings the whorls together, with resulting connation. In the Caryophyllaceae, obdiplostemony is incomplete; transitional stages in union of whorls are numerous. In the Saxifragaceae, both diplo- and obdiplostemonous taxa are present, with whorls separate or fused.

Evidence supporting the reduction theory—loss of the outer whorl of stamens—is strong. Polystemony is generally accepted as primitive and diplo- and haplostemony as advanced. (The story of simplification in the flower is one of progressive shortening of the receptacle and dropping out of whorls.) Stages in reduction of whorls in the androecium are readily seen in some families—Lauraceae, Caryophyllaceae, Saxifragaceae, Primulaceae, Sapotaceae, Myrsinaceae, Theophrastaceae. Most genera of the Primulaceae are diplostemonous, but, in *Samolus* and *Naumburgia,* the outer whorl of stamens is represented by rudiments; in some species of *Primula,* the outer whorl, usually absent, may be present in normal form.

Objection has been raised to the interpretation of obdiplostemony as brought about by loss of a whorl of stamens, on the ground that it is merely a matter of spatial and mechanical possibilities and, therefore, has no morphological significance. But the presence of remnants of the lost organs and their traces, as in *Primula,* is evidence of the loss.

Reduction in the Androecium

In reduction in the androecium, individual organs or entire whorls may be lost. In related taxa, the inner whorl may be lost in some families, as in the Iridaceae; the outer whorl in others, as in the Burmanniaceae and Haemodoraceae. The Orchidaceae are an example of members lost from both ancestral whorls.

The origin and course of the vascular traces to the stamens in obdiplostemonous taxa support the theories of the loss of the outer androecial whorl and of the union of the two whorls in those taxa that have only one whorl but have double the number of members present in other floral whorls. Obdiplostemony and the union of two stamen whorls are steps in the shortening of the receptacle in specialization of the flower by reduction.

Ontogeny provides no evidence of value in the interpretation of obdiplostemony. Sequence in development among the various whorls of organs in the flower is variable; it is probably less commonly acropetal than otherwise. In *Oxalis* and *Geranium,* sequence in development of the two whorls may be in either direction; in *Commelina,* the inner whorl of stamens develops before the outer.

Transformation of Fertile Stamens into Staminodia. Reduction of the androecium, whether spiral or whorled, by transformation of fertile stamens into staminodia may occur in either outer or inner members (Fig. 47A) or both; typically the outer members. Transformation of the inner stamens seems to be the more primitive condition, as seen where the petaloid staminodia form a prominent pseudocorolla above and inside the androecium in *Eupomatia* (frontispiece). In other primitive flowers—*Degeneria, Himantandra, Calycanthus,* and some of the Monimiaceae, Nymphaeaceae, and Helobiales—staminodia are present both above and below the stamens. If the upper position is the earlier one in the evolution of the flower, as suggested by these primitive taxa, the first "corolla" was above the stamens, a position which accompanies pollination by beetles.

Reduction in Stamen Number within Whorls. Aside from the reduction of whorls in the androecium, loss of stamens within the whorl is a prominent feature of specialization in many families. This loss accompanies zygomorphy and other adaptations to pollination methods. There is often variety in reduction in stamen number within a family, as from five to four and to two—Scrophulariaceae; even within a genus, as in *Polygonum,* from nine or six to three and two. Reduction in stamen number is in one whorl only or in both whorls and is of all degrees; it ranges from suppression of a single stamen in the androecium to suppression of all but one. Extreme reduction—to a single stamen—is occasional, as in *Euphorbia, Callitriche, Sarcandra, Najas, Casuarina, Hippuris, Lilaea, Zostera, Triglochin, Wolffia, Mangifera,* most orchids. Solitary stamens occur in flowers otherwise greatly reduced, chiefly those of aquatic genera, of elaborate zygomorphic form, or of greatly compacted inflorescences. In unisexual flowers, the solitary stamen has often been claimed to be terminal on the receptacle and therefore cauline in nature, but ontogeny and vascular structure of the entire flower show such a stamen to be appendicular, a true microsporophyll in a pseudoterminal position (Fig. 36). (In this respect, it is similar to the pseudoterminal carpel.) Some so-called terminal stamens are fused stamens; two or more connate sporophylls occupy a central position in the flower—*Salix* (section Diandrae), *Zannichellia, Typha.* Evidence that the solitary, apparently terminal, stamen is not morphologically terminal and cauline is seen in its ontogeny, for it arises off center; in its form, it is not radially symmetrical, because two sporangia are larger than the other two; and the anther is definitely dorsiventral. The interpretation of solitary stamens and carpels as cauline involves the acceptance of sporangium position in the angiosperms as both cauline and appendicular. Indiscriminate sporangial position—cauline or foliar—is found only in the lower plants, and the presence of both types in the highest plants—even within a genus, as is seen under the Phyllosporae-Stachyosporae theory—is morphologically inconsistent.

There is great variety in the form and function of sterile modified stamens: they may be transformed into petaloid organs—*Eupomatia, Trollius;* or into secretory organs—*Coptis.* Frequently they are represented by abortive structures, mere stubs, or vascular traces ending in the receptacle—Scrophulariaceae, Anacardiaceae. Reduction to one fertile stamen and the sterilization of half the anther of that stamen, with the sterile part petaloid, are characteristic of some highly zygomorphic families—Cannaceae, Zingiberaceae, Marantaceae, some Aizoaceae.

Reduction of Stamen Fascicles. The formation of fascicles is probably an early step in androecium reduction. Further steps consist of reduction in number of fascicles and in number of stamens per fascicle. Both these steps are well shown in *Hibbertia,* where reduction of fascicles is from several to one, and of stamens, similarly, from many to very few; and in *Hypericum,* where five fascicles are reduced to three, and stamens per fascicle from many to about three. Extreme reduction in the fascicle is seen in the Lauraceae and in *Parnassia;* some apparently simple stamens are surviving members of reduced fascicles—those of *Sassafras, Benzoin.*

Fusion in the Androecium

Connation of members of a whorl—ontogenetic or phylogenetic—is frequent in both whorled and fasciculate androecia. An androecium with connation by filaments is *monadelphous* where all stamens form a single cluster; *diadelphous* where two clusters are formed; *polyadelphous* where there are more than two clusters. It is *adelphous* where there is no connation. An androecium with anthers united is termed *syngenesious.* Union of anthers is commonly ontogenetic but is congenital in some taxa. Ontogenetic fusion commonly covers *cohesion* as well as connation. (Cohesion implies a loose, not intimate, union of similar structures, brought about by glandular secretions, by close appression with interlocking of epidermal cells, or by cuticular projections; connation implies histological union, with lines of union either evident or obscure. The anthers of the Compositae, Lobeliaceae, Solanaceae, some Gentianaceae are coherent.) Connate anthers are infrequent or rare—the Typhaceae, some Cucurbitaceae. The anthers of the Lobelioideae seem to show stages from cohesion to connation. Fusion by filaments—monadelphous or diadelphous—may involve one, two, or perhaps more whorls. Union of two whorls to form an apparent one is frequent, as in some legumes and the Thymelaeaceae.

Adnation of the Androecium

Fusion of stamens to other organs of the flower, especially to the corolla (this is termed epipetaly), is common. Fusion to the calyx is less common than to the corolla—Proteaceae. Fusion to the carpels, where all the outer organs are together fused to the gynoecium in perigyny and epigyny, is common. Fusion to carpels alone is rare—*Sarcandra,* Monimiaceae. Fusion to styles and stigma, with the formation of a *gynostemium,* is characteristic of the Orchidaceae and Stylidiaceae.

Adnation may be by filaments for part or for all their length, where the anthers are *sessile* on other organs. (The term sessile is unfortunately applied both to anthers where free stamens consist of anthers only, and to those where filaments merge with the uniting organ.) Adnation of the anther to other organs varies in extent, from attachment near the base only to union by the entire dorsal or ventral surface—the dorsal surface to a sepal in many Proteaceae, the entire ventral surface to the gynoecium in the orchids. In *Viscum,* fusion of the entire stamen to the sepal is intimate, and all external evidence of the fusion is lost; the sporangia are apparently borne on the ventral surface of the sepal, which is described as "polliniferous."

Stages in fusion—ontogenetic or phylogenetic—of stamen to adjacent organs can be seen in many taxa. External evidence of the fusion may remain, the filament forming a ridge on the surface of the petal or sepal, or the only evidence may be internal, in the presence of the vascular bundles of the two organs, which lie, radially, side by side. Fusion, when phylogenetically established, may include the vascular tissues, and the bundles of the two organs may be merged and histologically indistinguishable.

The Androecium under Zygomorphy

Number, form, and structure of stamens are greatly modified in the development of zygomorphy. Some of the stamens become sterile, are reduced in size, and lost, leaving little or no evidence in external form, though sometimes internal vascular traces may persist. Often, where there is more than one whorl of stamens, one entire whorl, or some members of either, or of both, whorls are suppressed. Of a whorl of six in *Canna,* one remains fertile, four become petaloid staminodia, and one is lost; in the Bignoniaceae, two stamens are fertile and three sterile. Correlated with sterilization and loss in the androecium under development of zygomorphy, are modifications in size and form, in time of maturation, and in adnation of the stamens to other organs. A number of terms are used in descriptive taxonomy for these variations but they have little morphological significance—for example, *didynamous,* where the androecium consists of two pairs of stamens of unequal length; *tetradynamous,* of four long and two short stamens.

Ontogeny of the Androecium

The stamens commonly develop in acropetal, that is, *centripetal*, sequence. The existence of basipetal, or *centrifugal*, sequence in the androecium in several taxa was noted as early as 1871 but received little attention. Recently, centrifugal sequence in the androecium has been emphasized as probably an important character in the determination of phylogenetic relationships between families and orders. In many families, order of origin and maturation is not on record; it is conspicuous in taxa with massive, spiral, multiorgan androecia; in families with few stamens, it is difficult to determine. A sequence is present within fascicles in the Hypericaceae and Dilleniaceae. Undoubtedly, centrifugal development was derived from centripetal, the normal acropetal sequence. Centrifugal development is known in the following multistaminate families and will doubtless be found in others—Actinidiaceae, Aizoaceae, Bixaceae, Cactaceae, Capparidaceae, Dilleniaceae, Hypericaceae, Loasaceae, Lecythidaceae, Malvaceae, Paeoniaceae, Theaceae; it is unknown in the monocotyledons. The major many-stamen families with centripetal stamens are Annonaceae, Lauraceae, Leguminosae, Lythraceae, Magnoliaceae, Myrtaceae, Nymphaeaceae, Papaveraceae, Punicaceae, Ranunculaceae, Rosaceae. The Geraniales and Centrospermae have been considered probably "referable to the centrifugal series," and, certainly, if the Capparidaceae belong in this group, so must their close relative, the Cruciferae. Examples of the value of sequence of development in determination of relationships can be seen in its use as an important character in the removal of *Paeonia* from the Ranunculaceae and the establishment of the Paeoniaceae as a member of the Dilleniales; and in the transfer of the Crossosomataceae from the Rosales to the Dilleniales.

Centrifugal development seems to characterize certain groups of families commonly recognized as related and is apparently an important character in confirming supposed relationships and in suggesting others. But this does not mean that all taxa with centrifugal stamens belong in the same phylogenetic line; reversal of sequence of development has doubtless appeared more than once, as have all other advanced floral characters. Similar departure from normal sequence in development is seen in the inflorescences of some palms (*Caryota*), where the inflorescences appear successively down the trunk, and in some cauliflorous tropical taxa. Centrifugal sequence in flowering is occasionally seen in inflorescences—*Vallisneria* (staminate), *Dipsacus* (proximal part). This sequence in inflorescences may be apparent only—the result of the condensation of large complex inflorescences.

THE STAMEN

The stamen, or microsporophyll, in contrast with the other floral appendages, is typically a slender organ, unleaflike in form. But it is, like the other organs, basically of leaf rank and is closely similar to all these organs in morphological and anatomical structure, in ontogeny, and in its relation to the stem (floral receptacle). Less important resemblances are in the epidermis: stomata are frequent on the connective —less common on the filament and absent on terete forms; and epidermal appendages like those of other floral organs are occasionally present, even over the anther sacs, as in *Calycanthus fertilis.* With the exception of the primitive laminar types, the stamen shows little superficial resemblance to the megasporophyll, the carpel, but the homology of the two organs is usually unquestioned.

In most families, the stamen is a highly specialized organ, with great variety of elaborate form related to methods of pollination. Typically, it consists of two more or less distinct parts: a proximal, sterile part, the *filament;* and a distal, fertile part, the *anther* (Fig. 49*P*). The anther consists of the microsporangia and the sterile tissues between and within which the sporangia are borne. The term *connective* is usually applied to the strip of tissues that lies between the pairs of sporangia, but this median strip is not separable, histologically or morphologically, from the tissues of the anther-sac wall, which enclose the sporangia. The stamen is usually considered a rather simple organ, but its simplicity is false; it is, in general, more complex and more reduced than the carpel. Filament and anther are usually distinct, but in the more primitive families there may be no clear limitation of fertile and sterile parts. The filament merges into the anther (Fig. 49*E* to *H*). The connective may extend beyond the sporangia, sometimes forming a more or less distinct appendage (Fig. 49*A, B, F, I, K*). The filament, the connective, and the appendage of the connective constitute continuous parts of the microsporophyll and are not morphologically distinct units. The connective is the sterile median part of the anther, connecting and, in various degrees, embedding the sporangia. Morphologically, it represents a median part of the laminar sporophyll.

The Primitive Stamen

Like the primitive carpel, which has no distinction of ovary, style, and stigma, the primitive stamen has no distinction between filament and anther. The most primitive stamens in living angiosperms are probably those of some of the woody Ranales (Fig. 50). These are broad, more or less leaflike organs, without, or with weak, distinction between fertile and sterile parts. The sporangia are borne near the

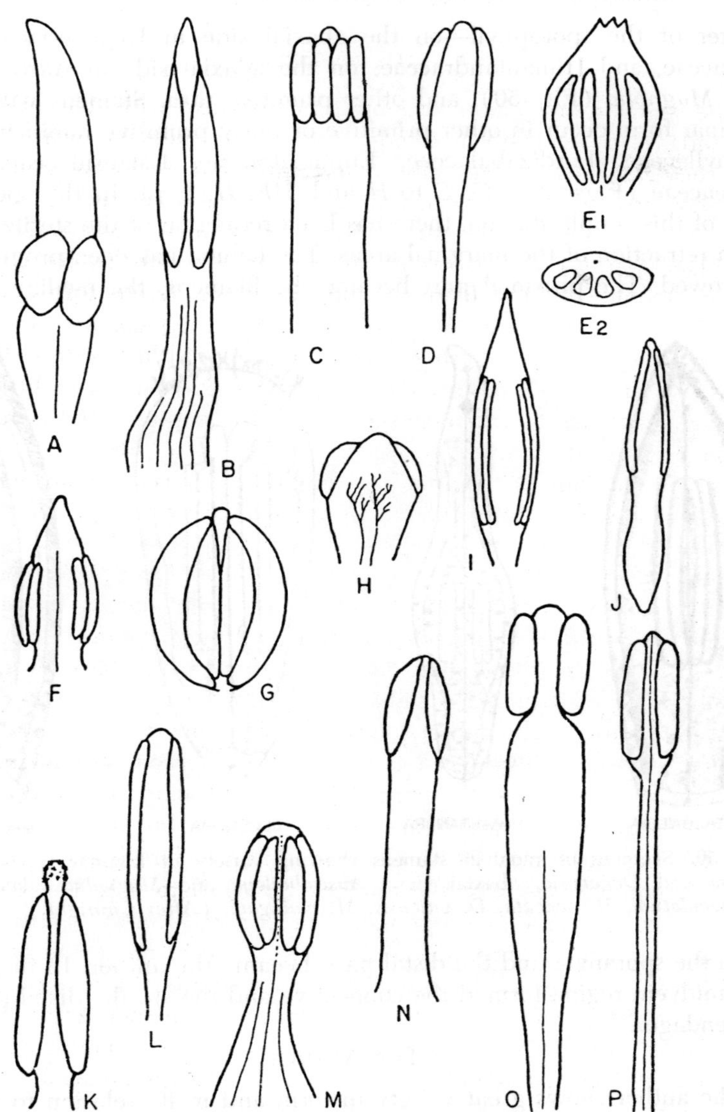

Fig. 49. Semidiagrams showing variety in stamen form from laminar to anther-filament with intermediate types. *A, Belliolum haplopus; B, Eupomatia laurina; C, Illicium; D, Hillebrandia; E, Ceratophyllum; F, Descainea; G, Lactoris; H, Chloranthus glaber; I, Magnolia nitida; J, M. hypoleuca; K, Calycanthus floridus; L, Michelia; M, Euryale; N, Sagittaria; O, Hydrastis; P, Caltha.* (*Adapted* from—*A, Bailey and Nast; C, I, J, L, Canright; D, photograph by R. Gauthier; G, Challenger Exped.; H, Swamy & Bailey; M, Chifflot.*)

center of the sporophyll—on the abaxial side in Degeneriaceae, Annonaceae, and Himantandraceae; on the adaxial side in *Austrobaileya* and *Magnolia* (Fig. 50), and other primitive taxa. Stamens with semilaminar form occur in other primitive or fairly primitive families—Ceratophyllaceae, Lardizabalaceae, Eupomatiaceae, Lactoridaceae, Nymphaeaceae (Figs. 49*B, C, E* to *H* and 54*F, H, I, J*). In the specialization of this simple stamen, there has been reduction of the sterile tissues, with retraction of the marginal areas. The lamina has been progressively narrowed. The proximal part became the filament; the median section,

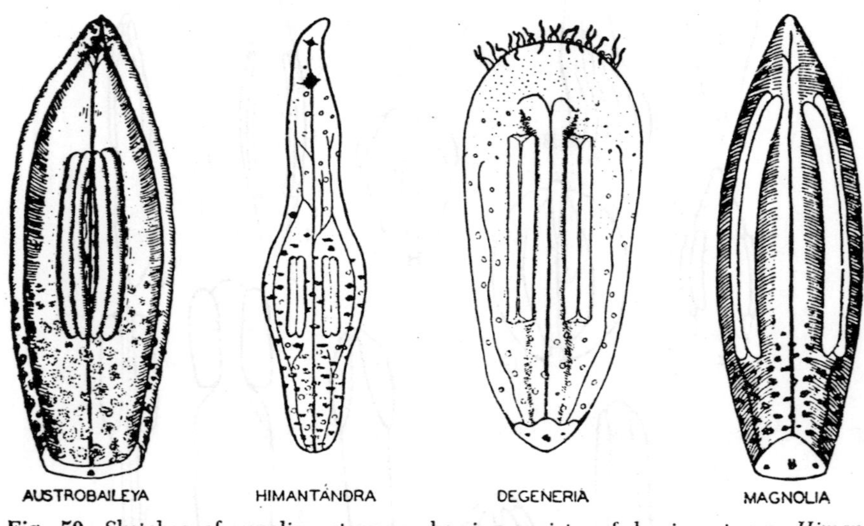

Fig. 50. Sketches of ranalian stamens showing variety of laminar types. *Himantandra* and *Degeneria,* abaxial view; *Austrobaileya* and *Magnolia,* adaxial view. *A. maculatum, H. baccata, D. vitiensis, M. maingayi.* (*After Canright.*)

with the sporangia and the distal part, became the anther. In the anther, the midvein region formed the connective and the sterile, distal part, the appendage.

The Anther

The anther shows great variety in form and in its relation to the filament. Its morphological details are difficult to determine, so great has been the reduction and so elaborate the form of the greatly reduced types. In form, the anther ranges from linear to arrow-shaped, subglobose, and strongly four-angled. Typically, it is two-lobed, deeply so in the more specialized types. The lobes are called *anther lobes* and *anther sacs.* With lateral enlargement of the connective, the anther lobes may be widely separated, and the lobes simulate anthers, as in the Commelinaceae, Melastomaceae, and many Labiatae. The connective may be divided longitudinally, and the separation extend down into the filament, as in *Alnus, Corylus, Ostrya,* and some Malvaceae; each lateral half, with its two sporangia, forms a pseudoanther. Greatly modified forms of anther, frequent in some families, are associated with elaborate methods of pollen distribution.

The form of the connective determines, in large degree, the form of

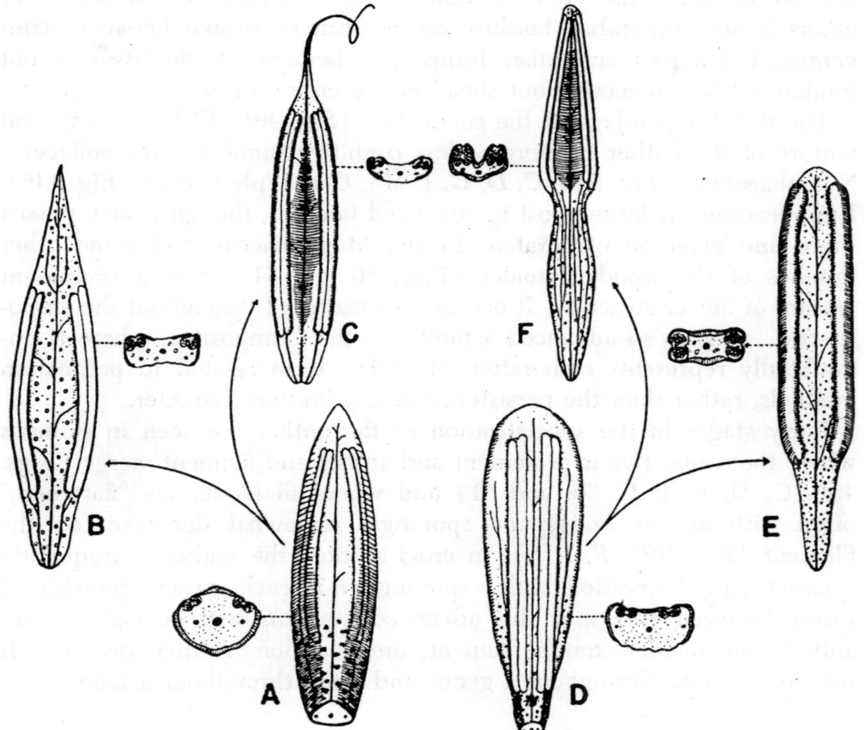

Fig. 51. Diagram representing main trends of specialization in stamens of the Magnoliaceae. Cross sections through fertile parts show position of wall-less sporangia and peripheral endothecial layer. *A, Magnolia maingayi; B, M. nitida; C, M. humori; D, Manglietia forrestii; E, Michelia fuscata; F, Magnolia hypoleuca.* (*After Canright.*)

the anther. In more primitive families, it constitutes a major part of the anther (Fig. 49*B, D* to *F, K, L*), and the sporangia are minor structures; in advanced families, the connective is a slender median axis, sometimes hardly more than a thread or point of attachment for the prominent anther lobes. A series in progressive reduction exists from the massive connective of the Magnoliaceae, *Trochodendron,* and *Tetracentron* to the mere fragment of the Gramineae, where the connective has been described as "absent."

The connective cannot be delimited: proximally, it often merges with the filament (Fig. 49*C, D, F, H, I, L*); laterally, it consists of the reduced and contracted laminar tissues of the ancestral sporophyll which sheathe the sporangia. In the older literature, the connective was called "placental," because of analogy, or supposed homology, with the ovule-bearing structure. The suggestion that "connective" be applied only to "conventional types" of stamens, and "sterile tissue" replace it in others is not acceptable, because no line can be drawn between "conventional" stamens and other forms, and because sterile tissue is not limited to the connective but sheathes the entire anther.

The distal appendage of the connective (Fig. 49*B, F, I, K*), a typical feature of the anther in more or less primitive families—Magnoliaceae, Nymphaeaceae (Fig. 54*B, C, D, G, J, M*), Cercidiphyllaceae (Fig. 146), Eupteleaceae—is largely lost in advanced families, though it may persist there and even be elaborated. In the Magnoliaceae and some other families of the woody Ranales (Figs. 50 and 51), it is a prominent feature of flower structure. It occurs also scattered throughout the angiosperms—even in so advanced a family as the Compositae, where it undoubtedly represents elaboration of anther form related to pollination methods, rather than the persistence of a primitive character.

Early stages in the specialization of the anther are seen in stamens where the connective is prominent and anther and filament merge (Figs. 49*B, C, D, F, I, L*; 50; and 51) and where filaments are "flattened," often with narrow wings and sporangia somewhat decurrent on the filament (Fig. 49*D, F, I, L*). In cross section, the anther is frequently symmetrically four-sided, with a sporangium in each corner. Transitional forms between this and the advanced stamen, with sharply set-off anther and slender, terete filament, are common. Anther structure is usually constant throughout a genus and often throughout a family.

The Filament

The filament ranges in form from broad and winged to terete and threadlike, and from short to long. The broader and shorter types are, in general, the more primitive. Sessile anthers occur occasionally—*Juglans*. Absence of the filament may be the result of reduction in adaptation to special habitats or methods of pollination, as in many aquatic genera—*Najas, Zostera*—or the result of adnation of the filament to the perianth, where the filament is absent superficially, though not lost morphologically, and is often represented by its vascular supply—some of the Proteaceae and Loranthaceae. Some primitive stamens with sporangia near the base and no distinction of filament (Fig. 49*E, F, G, I*) have also been described as sessile—*Himantandra* (Fig. 50).

The terms winged and flattened, as commonly applied descriptively to filaments, imply a modification of a terete form, but winged filaments are usually more primitive than the terete type, rather than more advanced. The broad, short filament in members of some primitive families—Magnoliaceae (Fig. 51), Annonaceae—is hardly more than a narrowed laminar base. The long, slender, terete filament is a highly specialized type.

Though typically ephemeral, the filament is sometimes fleshy and long persistent, or woody and retained even on the fruit—*Kingia.* It may be much enlarged and may have appendages—at the top, *Mahonia;* at the base, *Viola.* The morphological nature of the appendages is difficult to determine and is clearly various. Some appendages are glandular excrescences or modifications of form correlated with pollination methods. The presence, type, and origin of vascular tissue in these appendages is of much value in interpreting their nature. The "hood" and "horn" of the complex stamen of the Asclepiadaceae have been called appendages of the stamen, but only the horn is an appendage, as shown by ontogeny and anatomy; the hood is formed by the filament itself, reflexed upon its base. The petaloid "appendages" of the stamens of *Potamogeton* are adnate bracts. Stipulelike appendages at the base of the filament are occasional and characteristic of some taxa. They apparently vary in morphology; some seem clearly stipular; others represent vestiges of lost stamens of an ancestral fascicle, as in *Parnassia, Sassafras, Benzoin;* still others are merely glandular proliferations. Lateral appendages may be prominent, as in some Amaryllidaceae, where they become petaloid and connate, forming a corona. Anatomical structure and comparison with related taxa are necessary to determine the nature of these appendages.

Anther-filament Relationship. A considerable number of descriptive terms have been applied to types of attachment of anther to filament, but these are conflicting and often obscure in meaning. Morphologically, the anther is not "attached" to the filament, because both are parts of one organ, but the terms are of descriptive and taxonomic value. In the simpler type of stamen, without sharp distinction of anther, where the filament is clearly continuous between the sporangia, the anther is called *innate;* in similar stamens, where the connective is less prominent and differs from the filament tip, the anther is *basifixed.* Where the anther appears to be attached laterally to the distal part of the filament, the anther is called *adnate,* in contrast with *innate.* Adnate is an unfortunate term here and should be dropped, because it implies, incorrectly, a lateral fusion of anther and filament. There is, rarely, a true adnate condition, as in the Melastomaceae and Ericales, where the more or less pendent and inverted anther has become fused to the upper part of the filament. These anthers, now inverted, could well be called adnate but apparently have not been so described. Where the filament is "attached" on the broad side of the anther, the anther is *dorsifixed*. This is a poor distinction, because the point of attachment is difficult to determine when the connective is greatly reduced. Dorsifixed anthers may stand at any angle to the filament. Where the tip of the filament is delicate and the anther is free to turn at the point of attachment, the anther is *versatile*. The place of attachment of the versatile anther varies from the base to the top of the anther. These anthers are dorsifixed and, if attached near the base, may also appear basifixed. Distinguishing between versatile types has caused confusion, as in description of grass flowers. Versatile anthers may be associated with either insect or wind pollination, but are often considered to characterize anemophilous plants. Basifixed is obviously a primitive type; dorsifixed and versatile, advanced types.

The Sporangia

Number of Sporangia. Characteristically, the sporangium number is four, with arrangement in two pairs, one pair on each lateral half of the sporophyll. More than four is uncommon or rare—except in "forked" or "branched" stamens—and represents morphologically complex androecial structure. The larger number is nearly always eight, and this, at least in some taxa, represents connation of two anthers. The morphology of stamens with eight or more sporangia in families with greatly reduced flowers, such as the Balanophoraceae, is uncertain. The apparently multisporangiate condition in anthers, as in the Loranthaceae, Rhizophoraceae, Gentianaceae, represents partition of the sporogenous tissue by sterile plates. Less than four sporangia in the anther is frequent and represents a reduction from the basic four. The lost sporangia are nearly always two; usually both are from one lateral half; rarely, the back members of each half. A sporangium number of two is characteristic of taxa of various rank: of many genera and of families—Epacridaceae, Onagraceae, Labiatae, Lemnaceae; of tribes in some families where other tribes have the normal four—Restionaceae, Stylidiaceae. Where these two sporangia unite at maturity, the anther is termed *unilocular*. (Rarely, four sporangia unite to form a unilocular anther—*Arisaema*, *Callitriche*, some species of *Clusia*.) Reduction to a single sporangium, with some trace of a second, has been reported in the small parasite, *Arceuthobium pusillum*, where the reniform sporangium encircles an erect connective ("columella"). Among the stamens of a flower, sporangium number is variously reduced under zygomorphy; for example, in some of the Proteaceae, one anther may have four sporangia, two others have two each, and the fourth be sterile. The loss of sporangia is often by abortion in early stages, and evidence of the loss may persist in the mature anther.

Taxonomically, the two-sporangiate, unilocular taxa are commonly considered more advanced than the four-sporangiate, bilocular taxa. In the Epacridaceae, the unilocular anthers often show evidence of the fusion of sporangia. The Liliaceae have, typically, four sporangia, but *Smilax* has only two. Sporangium number is usually constant throughout a genus, but, in *Najas* and *Piper*, the stamen is described as having one to four sporangia. Vestiges of the lost sporangia in the form of abortive structures are occasional—Moringaceae, Piperaceae. In highly zygomorphic families with two sporangia, such as the Cannaceae and Marantaceae, the anther may have one lobe normal, with two sporangia, and the other lobe petaloid.

Form of the Sporangium. The sporangium is primitively slender and elongate, as seen in many of the less-advanced families of both dicotyledons and monocotyledons. In advanced stamens, it has been progressively shortened, becoming hemispherical where the anther is globose. Though typically straight, it is frequently crescent-shaped or strongly curved; it may encircle the connective—*Pinguicula*. Some anther lobes are described as "prolonged." The greater length is perhaps related to increase in amount of pollen in anemophilous taxa. The greater length is prominent in versatile types, because the anther is nearly free from the connective.

In the laminar stamen, with two pairs of sporangia, the outer member of each pair is somewhat smaller than the inner. This difference in size is more noticeable in specialized anthers where a cross section shows the well-known "butterfly" form; the two smaller sporangia are, in the laminar stamen, those farthest from the midrib.

Position of Sporangia on the Stamen. In the present century, little attention has been given to the fundamental position of the microsporangia on the angiosperm sporophyll. The sporangia are usually described as "marginal," a position that, superficially, they seem to have, but it is the *pairs* of sporangia, not the individual sporangia, that appear marginal. The typical anther is a greatly reduced and specialized part of the primitive, laminar sporophyll and is of little value in determining the basic position of the sporangia. But the critical studies of woody ranalian families—Degeneriaceae, Himantandraceae—provide descriptions of undoubtedly primitive stamens and a basis for phylogenetic comparisons of sporangium distribution on the sporophyll. In these taxa, the sporangia are all on one side, remote from the margin (Fig. 50). Comparisons in many families show that the "marginal" position was derived by reduction of the lamina, with accompanying "migration" of the sporangia. Evidence that all four sporangia belong on one side of the sporophyll comes also from somewhat less primitive families, the Magnoliaceae and Nymphaeaceae, which have stamens transitional in many characters between laminar and anther-filament types. In these and other families, the relation of the margin to sporangium position is obvious.

An important aspect of sporangium position on the sporophyll seems to have been overlooked. The laminar stamens of some families—Degeneriaceae, Himantandraceae, Calycanthaceae, Annonaceae, Ceratophyllaceae, Lactoridaceae, Lardizabalaceae—have the sporangia on the abaxial side; sporangia are on the adaxial side in other families—Magnoliaceae, Nymphaeaceae, Austrobaileyaceae (Fig. 50). Rarely, on semilaminar stamens, the four sporangia have a marginal or submarginal position—*Drimys*. The position of sporangia in the most primitive stamens—in some taxa, adaxial (ventral); in others, abaxial (dorsal)—is morphologically puzzling. In vascular plants, sporangium position has been generally accepted as an important character of major taxa. The angiosperms appear to have, primitively, sporangia in both positions.

In outline, the evolutionary history of the stamen seems to have been much as follows. The primitive stamen was laminar, with two pairs of sporangia borne on either the adaxial or the abaxial surface. From this simple stamen has been developed the slender, complex organ with marginal pairs of sporangia. Dorsiventral form has been largely lost; the specialized anther of the higher families is more or less four-angled and the filament terete. The marginal position of the pairs of sporangia —one member of each pair apparently adaxial, the other abaxial—suggests that the sporangia belong two on one side, and two on the other, but, in the primitive stamen, all four are on one side (Fig. 50). The marginal position of the pairs of sporangia is clearly derived, but the story of change in location is involved.

Sporangium position in the anther was much discussed between 1820 and 1850, but with little agreement as to its interpretation. The view at that time was that all sporangia belong on the adaxial side, but there was "great doubt whether the anther chambers have arisen where they appear"; "whether two sporangia do not belong on one side and two on the other"; "whether the manifold structure of anthers permits . . . the assumption of one type for all stamens." These questions might be raised today, but an acquaintance with the primitive stamen type, which was unknown to earlier morphologists, now aids greatly in the interpretation of this obscure situation.

In one interpretation of these early days, the stamen was considered inrolled, like the carpel, with its inner, enclosed surface sporogenous, and with the pollen shed through breaks in the sporophyll wall. (That all the sporangia are on one side of the sporophyll is implied.) This theory received little support. At that time, type of dehiscence was considered important in determining sporangium position. Extrorse and introrse anthers (that is, with abaxial and adaxial dehiscence, respectively) were considered derived from sporophylls with abaxial and adaxial sporangia, respectively. The position of the supposed margins of the filament and anther was also used as evidence of sporangium position. Even in the 1820s, it was believed that the margins of stamen and leaf are equivalent. On the basis of similarities in color and vesture, it was argued that the margin of the anther is "in the furrow into which the pollen is discharged." But this interpretation was refuted by the fact that, in some flowers, all the sporangia are on one side. Further, double flowers also demonstrated, in their petaloid stamens, that all four sporangia belong on one side, that the margins are on the "back" of the anther, not in the dehiscence furrows. In 1845, this theory was called "a false interpretation, twenty-five years old." Yet the present-day view that the pairs of sporangia are marginal—the margin between the members of a pair—is continuing that interpretation. But the evidence from the woody Ranales, the Nymphaeaceae, and other taxa that all the sporangia belong on one side must be accepted and the anther interpreted on this basis.

Major modification of sporangium position has occurred as a part of the reduction of the broad, ancestral stamen (Fig. 52). The lamina was reduced in area by progressive lateral narrowing, with the bringing of the sporangia closer to the sporophyll margin (Fig. 52*b, c, d*) and then to the ridges ("corners") of the thickened, four-angled anther, the position in latrorse dehiscence (Fig. 52*e*). From the latrorse position, the two outer sporangia were further moved—"over the edge" to the side opposite the original position (Fig. 52*f*). Associated with this change in sporangium position is the bringing close together of the sporophyll margins (Fig. 52*x x*) on the originally sterile side (Fig. 52*f, g*). These positional changes, brought about by more or less massive, differential growth—contraction and thickening—suggest a partial inrolling of the lamina (Fig. 53), but one basically of tissue rearrangement rather than folding of little modified wings of the lamina, as in carpel closing. This remarkable, evolutionary change in position of sporangia from one side of the sporophyll to the other was, of course, at first ontogenetic; some ontogenetic changes are known—anthers introrse in the bud, extrorse in the flower—in some palms, Polygonaceae.

The Nymphaeaceae and Cabombaceae provide excellent examples of the evolutionary modification of the stamen, showing the change in position of microsporangia from sunken to protuberant and from adaxial to lateral and abaxial. (The sporangia in these families, like those of the Magnoliaceae, are distal on the sporophyll, in contrast with those

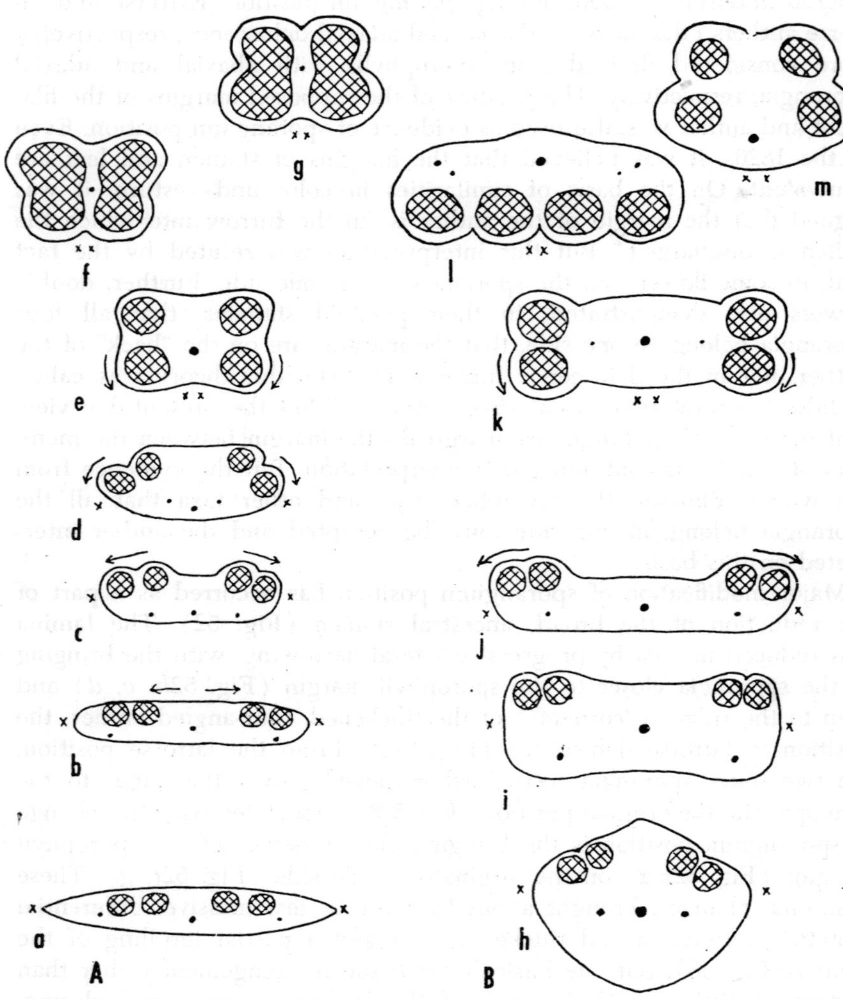

Fig. 52. Diagrams showing evolutionary stages of change in position of sporangia from sunken (*a*) to protuberant (*f, g, l, m*) and of stamen margins *x x*. *A*, *a* to *g*, basic plan showing dehiscence: *a*, introrse; *e*, latrorse; *f*, extrorse; *g*, introrse. *B*, variety of stages present in *Magnolia* and *Liriodendron: h, i, j, Magnolia maingayi, M. nitida, M. grandiflora,* respectively, introrse, sporangia slightly protuberant; *k, M. campbelli,* latrorse, sporangia protuberant; *l, Liriodendron,* extrorse, sporangia protuberant, change of position of outer smallest veins accompanying change of sporangium position; *m, Magnolia acuminata,* latrorse-introrse, sporangia protuberant. (*h, i, adapted from Canright.*)

of *Degeneria* and *Himantandra,* which are median or low in position.)

In form, the most primitive stamen in the families is that of *Victoria* and some species of *Nymphaea* (Fig. 54*J*). This stamen is broad and flat, without division into anther and filament. The four sporangia are elongate, located in the distal part of the sporophyll and deeply embedded. Like the sporangia of laminar stamens in the woody genera of the Ranales, they are wall-less but covered externally by a fibrous layer, commonly called exothecium (Fig. 59). This layer is associated with dehiscence; it is obviously not the sporangium wall, because it extends along the connective beyond the fertile areas (Fig. 55).

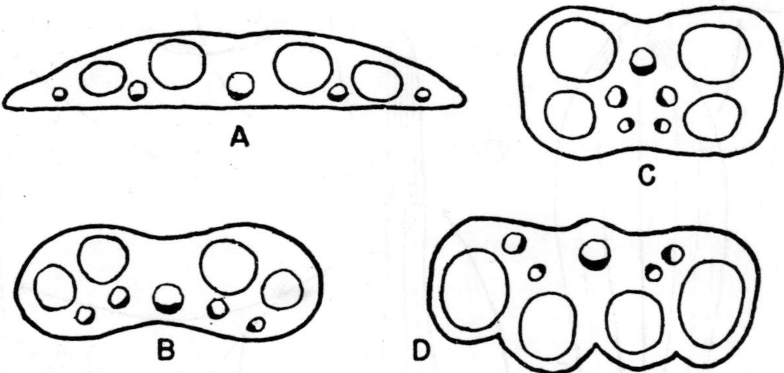

Fig. 53. Diagrams showing progressive changes in vascular structure in the anther accompanying evolutionary migration of sporangia from adaxial to abaxial positions. The change, essentially a down-rolling involving the entire sporophyll, is shown by the inversion of the outer veins in *C* and *D*. Xylem in solid black.

The various taxa in these families show stages in a progressive narrowing and thickening of the lamina throughout its length (Fig. 54). The sterile basal part is greatly reduced in width and thickened, with the formation, in early stages, of a narrow, flat filament; in the last stages, of a terete filament—*Nelumbo, Barclaya,* the Cabombaceae. The fertile section is greatly modified, with the formation of an *anther*. Tissue relations in this part of the lamina are much changed, and the sporangia and even the vascular bundles, in part, occupy new positions. The anther is a complex structure. A thickened median strip, the *connective,* is built up as the marginal areas are reduced in width and retracted. Reduction and rearrangement of the tissues of the laminar wings leave the sporangia surrounded by a wall of laminar tissue, which forms the anther-sac wall. The sporangia appear "free" but are still embedded in laminar tissue.

The sterile, terminal part of the sporophyll is less modified than the fertile part. It is reduced in length and width, becoming a distal ap-

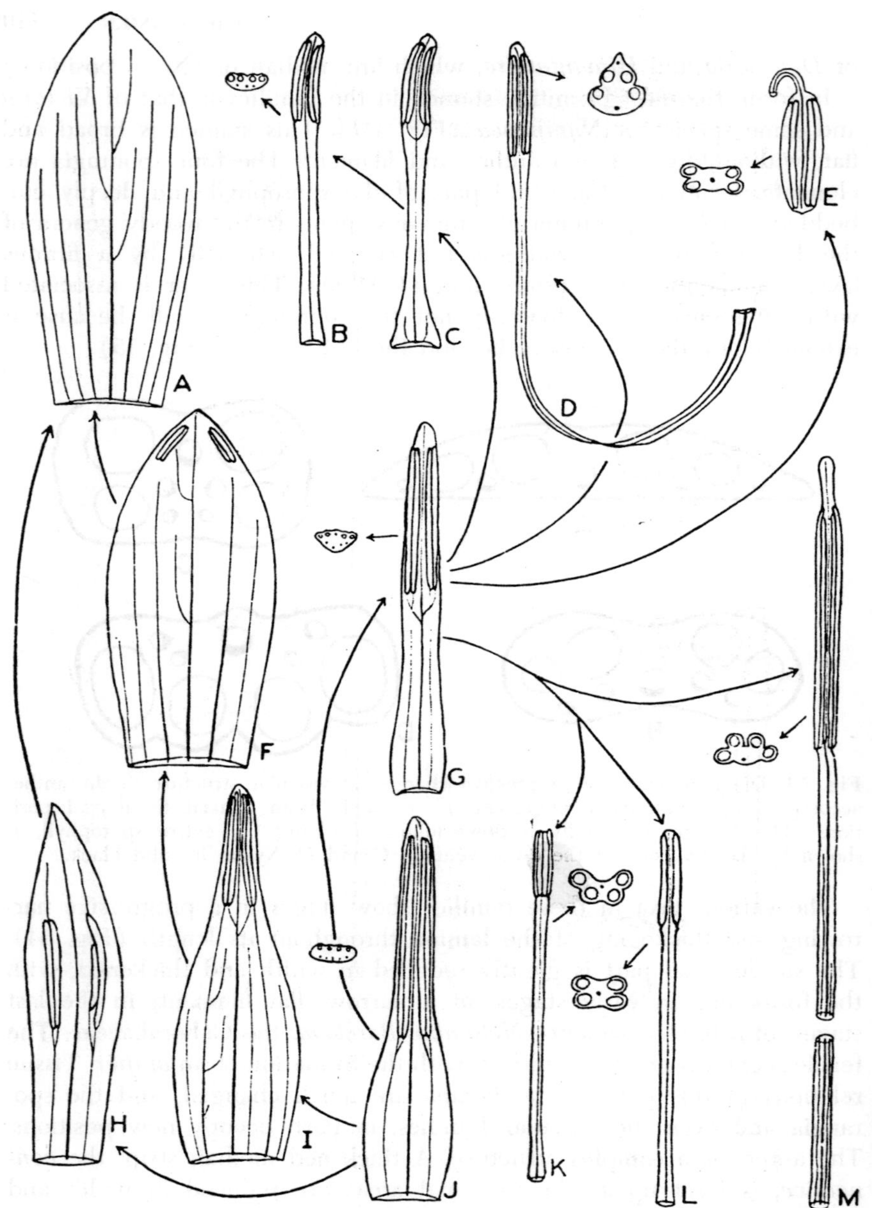

Fig. 54. Diagrams showing the evolution of stamens and petals in Nymphaeaceae. *A*, petal of *Nymphaea heudelotti; B, C,* inner stamens of *N. odorata; D,* stamen of *N. gigantea; E,* stamen of *Barclaya mottleyi; F,* outer stamen of *Nymphaea odorata; H,* outer stamen of *N. heudelotti; I, J,* outer and central stamen of *N. odorata,* respectively; *K,* stamen of *Cabomba aquatica; L,* stamen of *Brasenia schreberi; M,* stamen of *Nelumbo nucifera.* (*After Moseley.*)

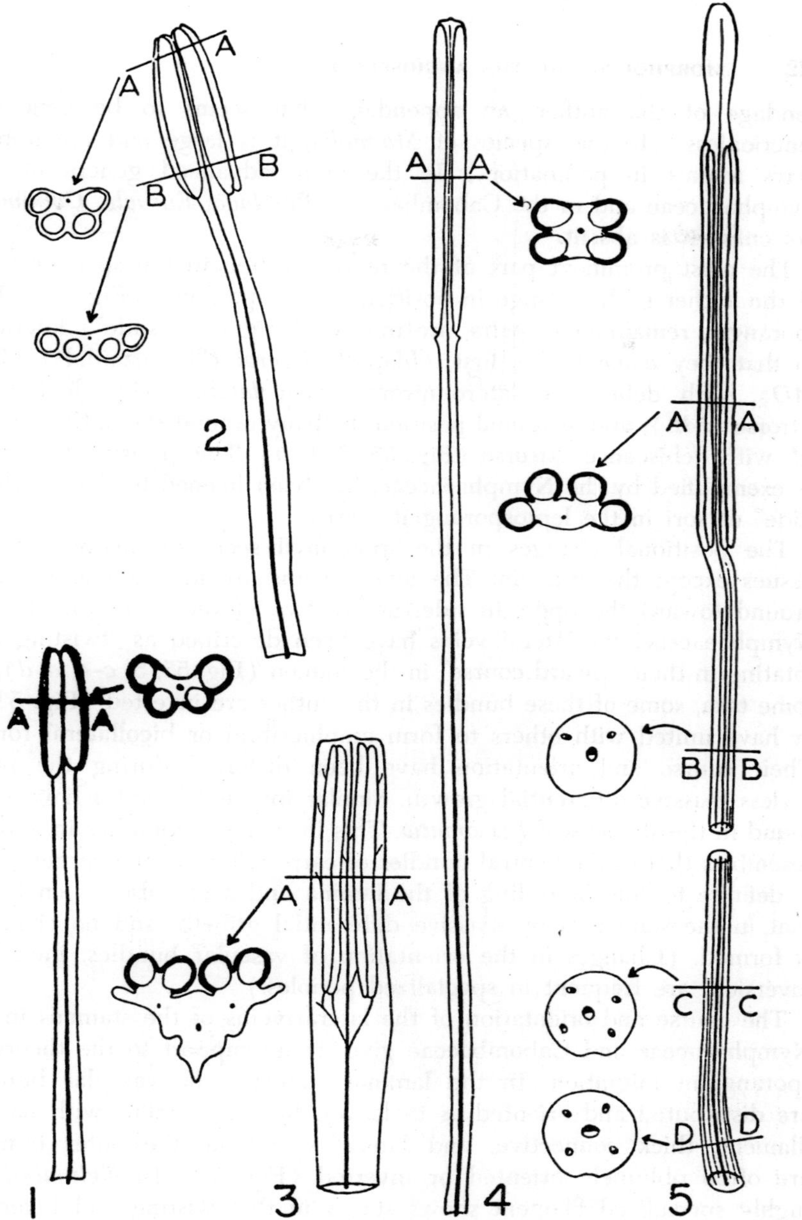

Fig. 55. Specialized stamens of the Cabombaceae and Nymphaeaceae showing vascular anatomy, position, and dehiscence of the sporangia. The sporangia are united in pairs by endothecial tissue. 1, 2, abaxial views and cross sections of *Cabomba aquatica* and *C. caroliniana*, respectively showing anatomy simple, dehiscence latrorse-extrorse in 1, *A–A*, and varying from extrorse, *B–B*, to latrorse-extrorse, *A–A* in 2; 3, *Nuphar variegatum,* anther sacs protuberant in pairs, connective massive; 4, *Brasenia schreberi,* anther sacs semiprotuberant, dehiscence latrorse; 5, *Nelumbo nucifera,* vascular supply complex, filament strands five, twisting and reorienting in their course (*B–B, C–C, D–D*), evidence of differential growth in development of terete from laminar filament. (*After Moseley.*)

pendage of the anther, an appendage that seems to be generally functionless. (In one species of *Magnolia,* it is large and apparently plays a part in pollination.) In the more advanced genera of the Nymphaeaceae and in the Cabombaceae—*Barclaya, Euryale, Cabomba, Brasenia*—it is absent.

The most prominent part of the reorganization in the specialization of the anther is the change in position of the sporangia (Fig. 53). The sporangia, remaining in pairs, are moved ("migrate," "slide") laterally, so that they come to lie, first, *obliquely lateral* ("trapezoidal") (Fig. 54*D*), with dehiscence latero-introrse; then *lateral,* with dehiscence latrorse (54*L*); and as a final position, halfway around the anther, *abaxial,* with dehiscence extrorse (Fig. 55, 2, *b–b*). This positional change, as exemplified by the Nymphaeaceae, has been likened to the "phyletic slide" of sori in the leptosporangiate ferns.

The positional changes in the sporophyll seem to involve all the tissues except the midvein. The lateral veins are moved laterally and around toward the opposite side, as are two of the sporangia. In the Nymphaeaceae, the lateral veins have been described as "twisting and rotating in their upward course" in the stamen (Fig. 55, 5, *c–c, d–d*). In some taxa, some of these bundles in the anther are inverted (Fig. 53*D*) or have united with others to form amphicribral or bicollateral forms. Their course and orientation have been distorted during the more or less massive differential growth. Similar inverted bundles have been found in the stamens of *Eupomatia.* This inversion of bundles in anthers resembles that of the ventral bundles of carpels, but, in the carpel, there is definite folding or rolling of the lamina and a chamber is enclosed; that in the stamen is by massive differential growth, and no chamber is formed. (Changes in the orientation of vascular bundles, including inversion, are frequent in specialized petioles.)

The course and orientation of the lateral veins of the stamens in the Nymphaeaceae and Cabombaceae give strong support to the theory of sporangium migration. In the laminar stamens, the vascular bundles are distributed and oriented as in leaves, but, in stamens with narrow filament, thick connective, and lateral sporangia, the outer bundles are often obliquely oriented or inverted (Fig. 53). In *Nelumbo,* the highly specialized filament shows stages in the twisting and inversion of the lateral veins. Accompanying this distortion in orientation is loss of the weaker lateral veins (Fig. 55, 5).

Progressive change in sporangium position from adaxial to abaxial has been recognized as a feature of advancing specialization in the anther. This change has apparently occurred independently in several, perhaps many, families; the Magnoliaceae and the Nymphaeaceae are excellent examples, for they show many stages in the change. The small family, Cabombaceae, seems to complete the series in the Nymphaeaceae, with the sporangia abaxial in *Cabomba.* It is apparent that, at least in many families, the primitive position for microsporangia is adaxial. The abaxial position in primitive families with especially primitive stamen form—Degeneriaceae, Himantandraceae, Lactoridaceae, Annonaceae, for example—has already been discussed. The Ranales, in their most primitive stamens, present a major morphological problem in the variety of microsporangial position; there are both adaxial and abaxial types. Among ranalian families, some—Magnoliaceae and Nymphaeaceae —show phyletic migration of microsporangia from adaxial to abaxial positions. (Megasporangia in angiosperms are adaxial, without exception.) There seems no question but that the adaxial position is one primitive position and that from this have been derived lateral and abaxial positions. But it seems unlikely that the abaxial position in the highly primitive stamens of *Degeneria, Himantandra,* and *Lactoris* is secondary—that these stamens are highly specialized in sporangium position. It is possible that both positions are primitive, a retention in the stamen of morphological structure of an ancestral taxon. (In the Pteridospermae, ovules were apparently borne on both surfaces of the leaf.)

The old search (1820–1850) for evidence of basic sporangium position by locating the stamen margins was well planned; it should be continued, with further critical studies of marginal meristems in stamens. Ontogeny should add strong support to the theory of the migration of sporangia. Leaves, petals, carpels, and broad stamens (Fig. 50) show marginal meristems. In slender stamens, these meristems have been lost wholly or in part; terete filaments apparently have no marginal growth, but weak marginal growth is found in "flattened" filaments and in the broader connectives. The somewhat flattened filaments of *Liriodendron* have weak marginal meristems; the broader filaments of *Magnolia* have well-marked marginal meristems. (*Liriodendron* seems to have been derived from ancient magnolian stock—its extrorse dehiscence derived from introrse.) Much more information about the development of the sterile tissues of the stamen is needed, especially of semilaminar forms. If the margin of the filament can be distinguished by linear meristems and these meristems followed upward into the connective, the position of the sporangia can hardly be questioned.

Evidence of another type that all four sporangia belong on one side of the sporophyll comes from the close histological association of each lateral pair. The sporangia on each side often fuse, in late stages, and the pollen grains mingle in a common chamber at dehiscence. Still further support for the theory that sporangia migrate is present in stamens where sporangium number is reduced to two. In these stamens, the missing sporangia are the outer members of the pairs of the ancestral lamina, often smaller and nearer the edges of the lamina wings than the others. Latrorse anthers also provide in their form—intermediate between laminar and extrorse or introrse—evidence to support the extreme change in position of the two outer, smaller sporangia. Latrorse dehiscence has commonly been recognized as more primitive than

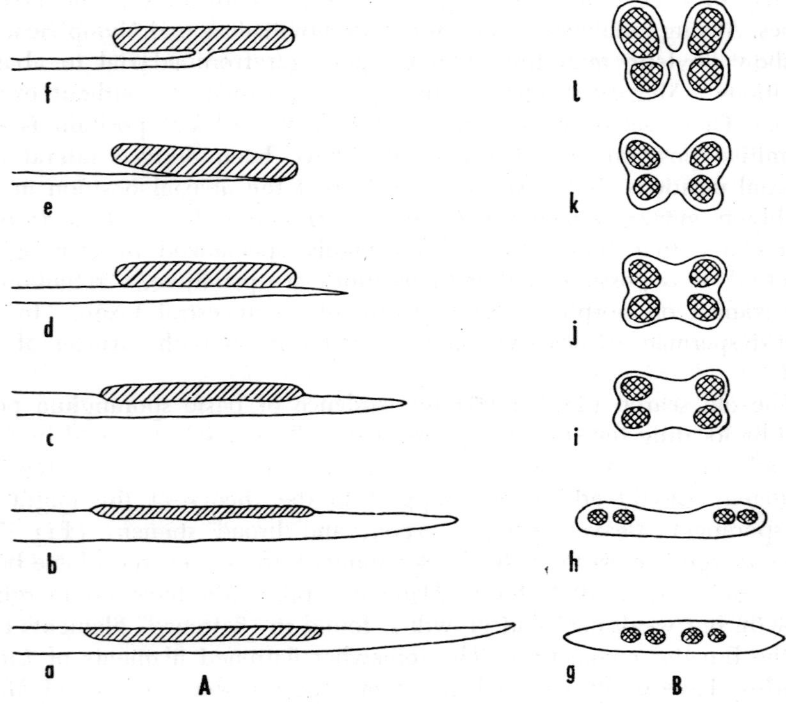

Fig. 56. Diagrams of stamen structure. *A*, longitudinal sections of stamens showing stages in freeing of the anther sacs from the lamina; *B*, cross sections of anthers showing stages in the development of anther sacs by the reduction of the lateral and connective tissues.

extrorse and introrse. The morphologists of the nineteenth century called it "normal"; the other types, specialized. In position on the sporophyll, latrorse anthers provide the step between laminar and extrorse or introrse; they are characteristics of many primitive families—Ranunculaceae, Rosaceae, Butomaceae, Crassulaceae, Monimiaceae, Trochodendraceae.

In position on the sporophyll, the microsporangia of the typical anther are submarginal; in the primitive stamens, they are laminar. Distribution of sporangia is much the same in micro- and megasporophylls, though the similarity is obscured. Primitively, both are laminar; in advanced forms, submarginal; in specialization, the laminar form is lost in microsporophylls, greatly reduced in megasporophylls; the megasporangia are primitively many; the microsporangia always four (higher numbers represent divided sporangia or united sporophylls). Reduction to two or one sporangia is uncommon in stamens, common in carpels. Parallelism in the two types of sporophylls is to be expected. Number and distribution of sporangia are correlated with form of sporophylls.

Primitively, the sporangia were deeply embedded in the tissues of the sporophyll—*Degeneria, Michelia, Himantandra, Ceratophyllum* (Figs. 67, 51, and 49). In the specialization of the anther, progressive reduction and reorganization of the sterile tissues have freed the sporangia from their burial in the lamina, by the formation of anther lobes (Fig. 56). But some of the sterile tissues remained as protective layers about the sporangia and formed the heavy anther-lobe walls. The sporangia within the anther lobes are wall-less, as in the laminar sporophyll. Each lobe of the anther contains the pair of sporangia of one side of the ancestral stamen. As the anther sacs became strongly protuberant, they were gradually freed from the connective (Fig. 56*d, e, f*), until ultimately nearly free [versatile anthers (Fig. 56*f*)]. The separation began at either the distal or the proximal end, or at both ends simultaneously, with resulting variations in point of attachment. Where the anther lobes are nearly "free," they are attached to a small median remnant of the connective, the wall is thin, and the anther consists largely of sporangia, two in each half (Fig. 56*l*). An anther of this type, freed at both ends, is typically X-shaped, with the ends of the halves often spreading—many grasses. Even where the anther lobes are nearly free and the form and position of the sporangia are prominent, the sporangia themselves are not free—in the sense of superficial—for they remain sheathed by a few layers of sporophyll tissue.

Anther-sac Wall and "Sporangium Wall"

The wall of the anther sac is usually described as made up of an epidermis, a hypodermal or *fibrous layer,* and one to several parenchymatous layers. The term *exothecium* has been used for the two outer wall layers—the epidermis plus the fibrous layer—and *endothecium* for layers between the fibrous layer and the sporogenous tissues. Also, the epidermis alone has been called the exothecium, and all layers within it, the endothecium. There has been no consistency in the use of these terms; they cause constant confusion in interpretation and have no morphological value. In unspecialized anthers, the epidermis resembles that of the filament; in highly specialized anthers, with strongly protuberant sporangial areas, the intermediate parenchymatous layers may be absent and the wall two-layered, consisting only of epidermis and fibrous layer, or of the fibrous layer alone, when the epidermis has been lost ontogenetically.

The fibrous layer is associated, mechanically, with dehiscence and varies in position, extent, and structural details with type of dehiscence. In the primitive types, the fibrous layer covers the sporogenous tissue and extends laterally over adjacent sterile tissue somewhat beyond the fertile tissue, as in *Degeneria* and *Himantandra* (Figs. 67 and 51); in advanced anthers, the fibrous layer may be more extensive, covering the protuberant areas and, in some high types, the entire anther. In anthers with apical dehiscence, the fibrous layer is usually absent, except in the region of the pores. In the elaboration of mechanical dehiscence, a series apparently exists in increase in extent of the fibrous layer and in simplicity of wall structure, accompanying increase in protrusion and freeing of the anther sacs. The anther wall becomes simpler by reduction of cell layers; the layers between the tapetum and the fibrous layer are reduced to one layer, and this also may be lost in the highest types. Even the epidermis may be lost—in part or wholly—and the fibrous layer take its place as the outermost layer, its cells remaining fibrous or becoming epidermislike. The loss of layers may be ontogenetic—occurring early or late in the development of the anther —or phylogenetic. Ontogenetic breakdown of the inner layers occurs at about the time of sporocyte formation. In delicate anthers, like those of the grasses, the wall of the anther sacs may have lost all but the outer two layers, the epidermis and a one-layered endothecium. Rarely, the epidermis may degenerate early or be lost just before dehiscence and the anther wall described as one-layered. This one-layered wall consists, at least in part, even in highly specialized anthers, of connective or laminar cells. There is, morphologically, no microsporangium wall in angiosperms—an important character of these plants. (Anther-wall structure is further discussed under Dehiscence of the Anther.) The anther lobe or sac is sometimes called a *synangium,* but, morphologically, this is inaccurate, especially when a "significant" resemblance is seen to the synangium of the Bennettitales. The cycadophyte synangium consists of two connate sporangia; the angiosperm anther lobe consists of two sporangia individually encased in sporophyll tissues, sometimes widely separated by them.

In the massive reorganization of the tissues of the fertile part of the laminar stamen to form the anther, the areas in which the pairs of sporangia lie become protuberant—progressively more and more markedly so, with increased specialization in sporophyll form. The sporangia of the more primitive stamens, sunken in the mesophyll, are wall-less (Fig. 67). With the reduction of the lamina, the sterile tissues around the sporogenous cells persist and, where the protuberances are large and unusually thick-walled, are perhaps added to from nearby regions. The anther-sac wall, thus formed, is laminar in nature, not a true sporangium wall. The entire anther-sac wall or one or more layers of it have frequently been described as the sporangium wall, but the stamens of the Ranales show clearly the naked character of the sporogenous tissue. The protruding sporangia, though histologically and functionally walled, are "morphologically naked"; even in the most thin-walled and free anther lobes, the sporangium is "sunken." In anther sacs, the sporangia are in a position more favorable for pollen distribution than in the center of a laminar stamen. (The megasporangium is probably also naked, though in interpretation of the morphology of the ovule, the nucellus is commonly considered the sporangium wall.)

The importance of the wall-less character of the angiosperm microsporangium seems to have been neglected. The apparent presence of a sporangium wall—the anther-sac wall and the endothecium—has obscured the wall-less condition. Absence of a sporangium wall is probably an important angiosperm character, one that will play a major part in the determination of the ancestry of the angiosperms.

The Pollen-containing Chamber. Several terms have been applied to the pollen-containing chamber: *anther chamber, pollen sac, locule, theca.* Anther lobe, anther sac, and anther half, though obviously covering more than the chamber and its contents, have sometimes been used as synonyms. All these are morphologically loose terms, because they may represent one, two, or four sporangia. The term theca is perhaps the most used and has unfortunately come to be well established, supplanting others because it is less technical. Pollen chamber is an undesirable term, because it may suggest the entirely different pollen-receiving chamber of the gymnosperms. All these terms must remain loose, general terms, but theca should be avoided in technical descriptions because it may represent one, two, or, rarely, more sporangia. It was called obsolete in morphological literature many years ago.

Anthers with two lobes are termed *dithecous;* with one lobe, *monothecous.* (Both these terms are morphologically unsatisfactory, because they involve the loose term theca.) The dithecous anther is a typical complete anther with four sporangia; the monothecous anther is usually a half-anther with two sporangia, the sporangia of a lateral half of a typical anther. But the monothecous anther may have the sporangia of one lateral half abortive or absent, as in many Labiatae; it is rarely a transverse half of an anther, with one sporangium of each lobe abortive. An anther described as monothecous may have one, two, or four sporangia, with their contents united in a common chamber. The anther sac usually contains the pollen of two sporangia, but, in latrorse anthers, where the sporangia are isolated in the four corners, there is usually no fusion of sporangia at maturity and the "theca" represents one sporangium (the anther is "tetrathecous"). In some highly specialized anthers, all four sporangia unite to form one chamber, and the "theca" represents four sporangia—*Callitriche,* some Cucurbitaceae.

Anther-dehiscence Types. On the basis of position or direction of dehiscence, stamen types—better called dehiscence or anther types—are distinguished in taxonomic use as *introrse,* when pollen is freed from the anther directly or obliquely inward in the flower (adaxially); *extrorse,* when directly or obliquely outward (abaxially). (In some genera, anthers that appear introrse in the bud appear extrorse in the flower—and those extrorse in the bud, introse in the flower—because of changes in form or structure of the filament or of movement of the anther on the filament apex.) When pollen is shed laterally, dehiscence is termed *latrorse;* in older usage, "normal." Introrse anthers characterize the majority of angiosperms; extrorse anthers characterize a rather small group of families—Calycanthaceae, Myricaceae, Cucurbitaceae, Fumariaceae, Lardizabalaceae, Aristolochiaceae, Iridaceae, Juncaginaceae, Potamogetonaceae, and a few others. Some families—Liliaceae, Alismataceae—have both types; and, in the Magnoliaceae, *Liriodendron* has extrorse anthers, the other genera, introrse; in *Fagopyrum* and *Persea,* one whorl of stamens is extrorse, the other introrse; in *Cinnamomum,* two whorls are introrse, one extrorse; in *Commelina,* two of the three stamens are extrorse, the third, introrse. Anthers introrse in the bud may become extrorse at flowering—Geraniaceae, Caryophyllaceae, some palms. The latrorse position is intermediate between the primitive laminar and the advanced extrorse and introrse—Ranunculaceae, Butomaceae, Menispermaceae. Figure 52 shows a series of theoretical steps in the modification of the ancestral lamina of the microsporophyll to form the latrorse and, ultimately, extrorse or introrse anther. Even in anthers that, in form, are extrorse or introrse, discharge of pollen may be in a lateral direction—*Solanum, Lonicera, Valeriana, Primula.* Such dehiscence as this characterizes genera but not larger taxa. All anther types could have been derived from sporophylls with, primitively, either adaxial or abaxial sporangia. Orientation of sporangia as indicating extrorse or introrse is commonly apparent early in the ontogeny of the stamen, but often, in early stages, the sporangia are symmetrically placed near the four angles of the primordium and the final position is determined by differential growth. This growth may be strong and even suggest cambiumlike activity (Fig. 57*C, D*).

The terms applied to dehiscence have much descriptive and taxonomic value and are morphologically significant, in that an evolutionary sequence is obvious from the simple extrorse and introrse dehiscence of the primitive laminar stamen to latrorse and to the elaborated extrorse and introrse of advanced anthers. The presence in the same taxon and even in the same flower of extrorse and introrse dehiscence is evidence that these types represent ecological adaptations. Dehiscence in laminar stamens is necessarily simple and determined directly by sporangium position. In advanced anthers, sporangium position is not simple; phylogenetic shifting has brought the outer members of each pair to the side opposite that of their original position, and the resulting obliquely inward or outward dehiscence is derived, secondarily extrorse or introrse. Both extrorse and introrse dehiscence have apparently

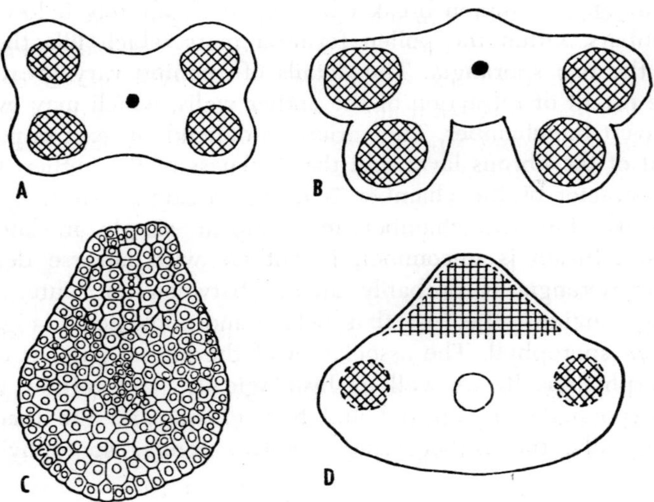

Fig. 57. Diagrams and a sketch of cross sections of anthers showing modification in form of connective at time of flowering. *A, B, Iris pumila,* by differential growth: *A,* young, *B,* mature; *C, D, Nymphaea colorata,* by cambium-like growth, development of ridge on connective: *C,* young, *D,* mature. (*A, B, after Engler; C, D, after Kaussmann.*)

been developed in some taxa from ancestrally simple extrorse—Lauraceae—and from ancestrally simple introrse—Magnoliaceae except for *Liriodendron.* Latrorse dehiscence similarly has two origins. It is clear that extrorse and introrse dehiscence do not necessarily indicate derivation from ancestral forms with abaxial and adaxial dehiscence, respectively. The terms must be considered as chiefly of descriptive value.

Union of Sporangia. In the primitive stamen, the sporangia lie close together in lateral pairs; in advanced stamens, they remain so or have been broguht even closer together by reduction of intermediate laminar tissues. In many families, they unite laterally by breakdown of the separating wall, usually late in ontogeny, as the pollen matures. The pollen grains then lie in a common chamber, from which they are shed through a single opening. Early breakdown of the wall occurs occasionally—*Berberis, Nuphar,* and some orchids. The two pollen masses may mingle or lie side by side without separation by sterile cells—*Cassia,* and some poricidal anthers. In the union of the two sporangia, the tissues of the entire separating wall or of some part of it disintegrate and are largely absorbed. The adjacent outer anther wall, above and beside the separating wall, is commonly also involved in the disintegration; the outer anther-sac walls are retracted, and the intermingling pollen is shed through a large opening. In some taxa, the separating wall persists almost to the outside of the anther, and the chambers discharge through a common break which opens separately below into the two chambers. Often the pollen forms a mass which fills the furrow between the two sporangia. The details of opening vary greatly, especially the extent of retraction of the anther walls, which may even completely invert the chamber. The variations depend, at least in part, upon the extent of the fibrous layer and the thickness of the anther wall. The double character of the chamber is usually evident externally as well as internally, but the chamber may appear simple in late stages. Sporangium fusion is uncommon in anthers with latrorse dehiscence. Union of sporangia is probably always between an "outer" and an "inner" sporangium—the two that belong morphologically together on the laminar sporophyll. The association of the lateral pairs is obviously close, morphologically as well as histologically, and refutes the view that two sporangia are abaxial and two adaxial and the theory that, morphologically, the anther represents two conifer sporophylls "back to back."

Unusual Forms and Arrangements of Stamens

Unusual forms and arrangements of stamens that suggest splitting, forking, and branching received much attention from 1850 to 1900 and are still occasionally discussed. *Chorisis, dédoublement,* and *multiplication* of anthers were considered methods of origin for supposedly branched stamens. Today these terms are little used, even in taxonomic descriptions, and have long had morphological implications that are, at least in part, false.

The stamen is fundamentally a simple, unbranched organ, as shown by the primitive genera *Degeneria, Eupomatia, Austrobaileya.* Many of the so-called branching forms can now, with the aid of anatomy and comparison with the stamens of related taxa, be interpreted more accurately than formerly. Longitudinal division ("splitting") of the stamen occurs in some taxa, as in the Malvales, but, in other taxa, the apparent division represents partial connation of two or more stamens.

"Dichotomous" stamens have been seen as occurring in *Salix* in the interpretations of the "New Morphology," as examples of retention of primitive structure. The "dichotomous" stamens of some species of this genus are pairs of stamens partly united, not forked organs. Two stamens borne on opposite sides of the flower and connate by the bases of their filaments may suggest dichotomy, but anatomy of the flower shows two independent vascular supplies, derived on opposite sides of the receptacle. Taxonomists have correctly interpreted this "doubled" stamen as an end product of reduction in the androecium.

Dédoublement, a doubling—if repeated, a multiplication—of stamens has, since the middle of the nineteenth century, been used in both taxonomy and morphology to explain the presence of two—sometimes more than two—stamens, where one would be expected in a floral pattern (Fig. 48). Among the best known examples of supposed dédoublement are the pairs of longer stamens in the Cruciferae and the pairs of stamens opposite each petal in *Alisma*. In the crucifer flower, two pairs of stamens would fit into the dimerous floral diagram of the family, with two pairs (whorls) of sepals, two of petals, and two of carpels. But six stamens cannot be placed in alternating whorls of two with the other organs. Dédoublement of two of an original four (two whorls) has been used to explain this lack of conformity in the androecium. Neither anatomy nor ontogeny supports the doubling theory; the vascular supply and the primordia of all the stamens arise independently. The members of each pair of stamens in *Alisma* represent independent organs with their vascular supplies derived well apart on the floral stele.

Many objections were raised at the end of the nineteenth century to the acceptance of dédoublement as an explanation of the presence of pairs and clusters of stamens where a single organ is to be expected. Fifty years later, dédoublement, in the sense of a division, seems to be a sound explanation of the origin of those pairs or clusters of "half-anthers" where it is obvious, on anatomical and ontogenetic evidence, that the anthers have been divided longitudinally. In these stamens, the division may extend downward in the filament even to the base—some of the Malvales. But its use as an explanation of pairs of complete stamens, with independent vascular supplies is morphologically incorrect.

Forked stamens, apparently the result of longitudinal division or "splitting," are occasional, as in *Salix*, *Corylus*, and *Ostrya*, but these stamens represent pairs of stamens connate throughout part of their length, rather than forked individual organs. Dédoublement, as a true splitting, a dividing of the entire stamen, has been claimed for *Adoxa*, where pairs of stamens alternate with the petals, with separation occurring in the primordia. (It is doubtful whether this is true dédoublement, and it should be studied further.) The extent to which forking or doubling represents true division, rather than fusion, can be determined only by comparison with related taxa and by anatomical study. (Double flowers have been said to be formed by the ontogenetic splitting and transformation of stamens and carpels.)

The stamen of *Ricinus,* massive, with many branches, each bearing a terminal "anther," has been used as a branched type (Fig. 44). But this interpretation is doubtless incorrect. The nature of this stamen can probably best be determined by comparative studies of inflorescences and flowers throughout the Euphorbiaceae, a family with many "flowers" of difficult interpretation. The staminate flower of *Euphorbia* has been reduced to a single stamen (Fig. 36*A*). The stamen of *Ricinus* is perhaps an inflorescence.

Support for the theory of telomic organization of the sporophylls of angiosperms has been seen in what seems to be dichotomy of the branches of the stamen of *Ricinus.* The tips of the branchlets show an apparent dichotomous forking, with each branch terminated by a sporangium. But the several varieties of the castor bean show that each ultimate dichotomy represents an anther with the two anther sacs nearly free from the connective, which is scalelike and often deciduous (Fig. 44).

The nature of fasciculate or clustered stamens was early discussed by morphologists and taxonomists, who considered the clusters branched or compound stamens, or "stamen systems." (Fasciculate stamens are discussed under Androecium.)

Anatomy of the Stamen

Stamens with well-differentiated anther and filament usually have a single trace which continues as a simple vascular bundle through the filament to the anther and sometimes into the distal prolongation. Branching in the connective is occasional, but the branches bear no relation to sporangium position. In delicate stamens, the vascular supply may be vestigial and discontinuous or even absent; where absent within the stamen itself, it may be present as a vestigial "stub" in the receptacle—some Scrophulariaceae.

A two-trace supply to the stamen is rare: *Austrobaileya, Sarcandra* (a vesselless genus of the primitive Chloranthaceae), some of the Proteaceae, *Victoria* and some other genera in the Nymphaeaceae, *Casuarina,* and some of the Betulaceae. The filaments of *Nuphar, Cyrtandra, Eranthemum,* and *Peristrophe* were long reported to have a pair of bundles, and that of *Doryanthes,* four bundles. In *Casuarina,* two traces enter the filament and soon fuse to form a single concentric bundle. In the Betulaceae and Fagaceae, there is evidence of the double nature of the stamen bundle; in several genera, there are, at the base of the filament, either two separate bundles or one bundle with two xylem strands or two xylem poles; in *Corylus*, the single bundle arises from two traces.

The phylogenetic history of the trace supply of the stamen has been clarified by the better interpretation of nodal structure in the shoot. Changes in number and relation of traces have been the same as those of the leaf. From the basic two-trace supply—characteristic of all appendages—have been derived the three- and the several-trace types by addition of lateral traces to a median double trace (representing the basic two, fused); from the basic two (by fusion) and from more than two (by loss of laterals) has been derived the single trace.

Most laminar and semilaminar stamens have three traces and three major veins. Three traces characterize some of the primitive families: Degeneriaceae, Himantandraceae, Annonaceae, Lardizabalaceae, many of the Magnoliaceae, and Nymphaeaceae. *Eupomatia* has five or seven. Three traces, often with additional laterals, are present in most monocotyledons—Musaceae, Zingiberaceae, Marantaceae. The common one-trace condition has obviously arisen by reduction from a higher number in association with the narrowing of the filament.

Evidence of reduction in trace number from three or more to one by loss of the lateral traces—as also occurs commonly in leaves and carpels—is well shown in the Magnoliaceae. In *Magnolia*, the lateral bundles may be weak and discontinuous; in *Michelia*, some species have three traces, others one, with vestiges of the lost bundles. In these genera, distortion and rupture of lateral bundles are associated with reduction in width of the stamen, accompanied by the development of an anther and the migration of the sporangia. As early as 1824, there was a surprising understanding of stamen anatomy; it was stated that, in slender stamens, the lateral veins vanish and the median persists. The present-day view that the single-bundle supply was derived by reduction from a three-trace supply is very old, not recent, as generally believed.

Branching of the vascular supply is uncommon or rare in stamens with one trace, but is frequent in stamens with more than one trace. Minor lateral branches may connect with major veins or anastomose to some extent with one another, but only rarely does the branching and fusion suggest the reticulate structure of most carpels.

Appendages of the stamen usually have no vascular tissue, but, in the Melastomaceae, the anthers have prominent adaxial and abaxial projections or lobes that are completely vascularized. These appendages have been considered possibly remnants of an ancestral telome system, but they can, at least in large part, be interpreted as projections of the connective, often reflexed and adnate to the body of the anther. The Melastomaceae are an advanced family, and the projections suggest modification of anther form related to uncommon methods of pollination, perhaps ornithophily.

The "trunk" vascular supply to stamen fascicles is a compound structure. That it consists of the connate traces of several to many stamens is apparent from the presence in several primitive families—Paeoniaceae, Dilleniaceae, Monimiaceae—of stages in the union of a group of stamens (Fig. 46). "Branched stamens" are formed in this way. Fascicled stamens have been considered branched stamens that are perhaps homologous with the bennettitalian stamen, but the branched microsporophyll of the Bennettitales has a single, simple trace.

Ontogeny of the Stamen

Ontogeny of the stamen is basically like that of sepal, petal, carpel, and leaf; there is growth in length, width, and thickness. Early development is by apical and marginal meristems (Fig. 58). Marginal meristems are active for periods varying with the form of the filament and anther. In winged filaments, they are well defined and long persistent; in narrower filaments, they cease activity after periods varying with the width of the filament; in terete filaments, they are absent. Weak marginal meristems may be present in thick, subterete filaments. In the connective, marginal growth is difficult to determine, because it is weak and obscured by growth in thickness. Sufficient attention has not, however, been given to details of the ontogeny of the connective and filament. Critical ontogenetic studies should be made of anthers with broad connectives and subterete filaments to obtain evidence in support of the view that both margins are on one broad side of the anther.

The stamen primordium arises on the floral receptacle as a crescent-shaped projection, where the filament is broad, or as a rounded projection, where the filament is terete or nearly so. The primordium elongates rapidly and soon assumes roughly the form of the anther, with some indication of its orientation and relation to the filament. At this stage, the filament is represented merely by the base of the primordium. With increase in size of the primordium, bulging areas indicate the position of the sporangia. These regions develop rapidly and the rest of the anther grows more slowly. Differential growth in the connective and anther wall complete the development of mature anther form. At this stage, differential growth is an important feature in enlargement, because it is largely responsible for the orientation that determines direction of dehiscence. Increase in cell number on one side of the connective—sometimes by almost cambiumlike growth (Fig. 57)—displaces the sporogenous areas toward the other side—toward the adaxial side where dehiscence is introrse, and toward the abaxial side where dehiscence is extrorse. Latrorse dehiscence develops where a subcylindrical primordium becomes more or less four-sided and there is little or no differential growth in the connective. The

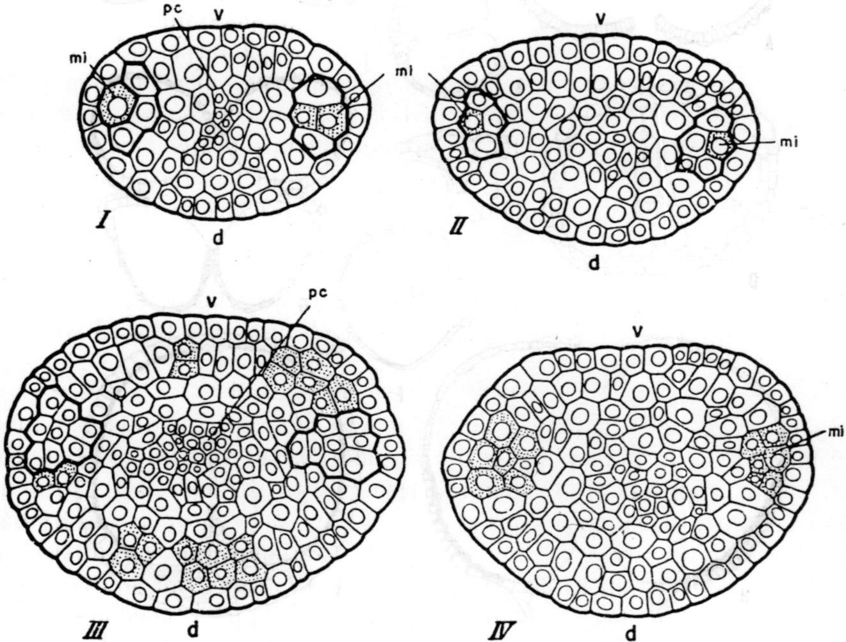

Fig. 58. Cross sections of filament of stamen of *Cleome gigantea* showing marginal development. I to IV, progressive stages in ontogeny. *d, v,* dorsal and ventral sides; *mi,* initials of the meristems; *pc,* procambium. (*After Kaussmann.*)

sporogenous areas remain in the corners. The strongly marked differences of orientation in extrorse and introrse anthers are the result of differential growth in the connective—plus, in some taxa, a twisting at the filament tip, as in some palms—and do not indicate an important morphological difference.

With evolutionary reduction of the sterile tissues of the sporophyll, the protuberance of the sporangial regions increases, but the sporangia remain enclosed by the several-layered anther-sac wall; they do not become "free." The wall layers are progressively reduced, with increasing elaboration of the anther lobes, until only two layers remain, an epidermis and one hypodermal layer (Fig. 59). "Uniseriate" anther walls—so-called "sporangium walls"—are probably misinterpretations of biseriate walls where the epidermis has been lost ontogenetically (Fig. 59*C*, *G*, *H*). Even in aquatic plants with extreme tissue reduction

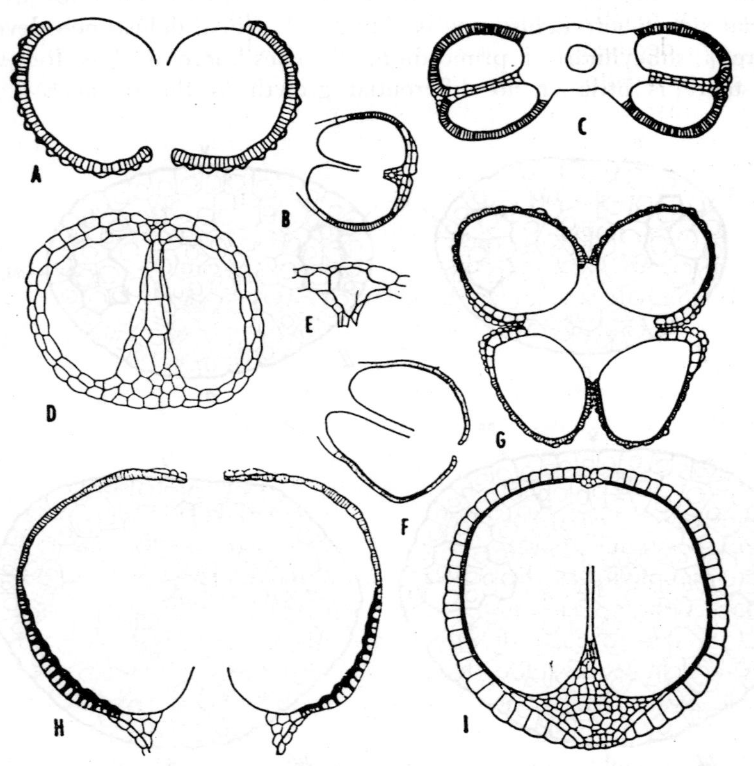

Fig. 59. Cross sections of anther-lobe walls showing reduction at maturity to epidermis and one hypodermal layer in *A*, *D*, *I*, or to the hypodermal layer with fragments of epidermis persisting in the other diagrams. *E*, swollen cells over septum which ruptures the wall at dehiscence. *A*, *Humulus; B*, *Broussonetia; C*, *Dorstenia; D*, *E*, *Zannichellia; F*, *Trema; G*, *Casuarina; H*, *Ricinus; I*, *Zostera.* (*After Staedtler.*)

throughout the flower, the anther wall is two-layered. In the angiosperms, there is no sporangium wall distinct from enclosing tissues of the sporophyll. The description of the microsporangium wall as "many layered" is based on the interpretation of the anther-sac wall as the sporangium wall.

The filament commonly remains very short—sometimes a mere base for the developing anther—until late stages in flower development. Its development may be rapid just before the flower opens—largely as the result of cell elongation—and may continue after the flower is open.

Ontogeny of the Sporogenous Regions. Study of the development of the microsporangium in angiosperms began about the middle of the nineteenth century; it was a popular field of morphological study from about 1870 to 1900, when many taxa were investigated. In the large amount of detailed information available, much confusion in interpretation and terminology exists. This applies particularly to the description of the anther wall and its ontogeny, where topographical terms have been applied loosely. This is not surprising. Most of the early studies were purely descriptive and were rarely compared with one another. Variation in structure of the anther wall (which has been generally assumed to be the sporangium wall) is great. Many of the descriptions were based on the theory that the stamen was cauline in nature, and special attention was given to origin of specific layers from "dermatogen" or "periblem." Deeply sunken sporangia were unknown; the sporangium was considered superficial. Anther wall and sporangium wall were often considered synonymous terms; commonly, the wall of the pollen sac was called the sporangium wall. The possibility that sterile tissues of the sporophyll may form part of the sporangium wall did not enter the picture. The loss of layers in the wall during ontogeny—even of the epidermis—was usually overlooked. Only with recognition of sporangium position as primitively sunken in the sterile tissues of the sporophyll has there been a basis for interpretation of the morphology of the anther wall.

Part of the difficulty of interpreting the nature of the several wall layers outside of the sporogenous tissues lies in the variation in number of these layers. The stamen has, unfortunately, been commonly looked upon as a simple organ, readily interpreted morphologically; actually, it is the most highly specialized of floral organs, resembling an ancestral laminar sporophyll far less than the carpel, and showing much greater variety in form. Adaptation to various types of pollen distribution has brought about not only great range in external form but in the histological structure of the anther. The sporangium, originally buried in sterile tissue, seems to have become "superficial" and, in the highest types, is largely free from the surrounding sterile tissues. Variety of wall structure, seen in the many intermediate stages in this series, with phylogenetic and sometimes ontogenetic loss of cell layers, obscures the story of modification.

Early in the ontogeny of a typical anther, when the beginning of anther form is evident in the primordium, groups of cells in the center of each of the four corners can be distinguished from the cells around them by their large size, denser cytoplasm, and larger nuclei. These cells, in elongate, often crescent-shaped or platelike rows, form the *archesporium* (Fig. 67). The term archesporium, like so many other terms used in flower ontogeny, has had several meanings; it has been applied not only to these rather readily recognized cell clusters but to the central tissues at an earlier stage, from which arise not only the sporogenous tissue and the tapetum but inner wall layers. The archesporial cells—also called the primary sporogenous cells—enlarge radially and divide periclinally, cutting off outer cells which form the *tapetum*. The inner cells, after this division, form the *primary sporogenous cells*. From the tissues outside the archesporium arise the epidermis and wall layers. The terms applied to these cell layers as they develop—*primary parietal layers, parietal layers, primary tapetal layers*—have been so variously and inconsistently used that they are valueless. Doubtless, there is much variation in the behavior of the meristematic layers, as in other meristems, and it is impossible here to ascribe more than general origin and function to specific layers. The subepidermal (fibrous) layer and one or more layers immediately below this are clearly derived from the hypodermis of the young anther; deeper-lying layers, even the layer that becomes the tapetum, apparently may have different origins.

The terms *exothecium* and *endothecium* have been commonly applied to outer and inner layers, respectively, of the mature anther-sac wall, but are used as loosely as they were in reference to layers in the developing wall. Exothecium has been applied to the epidermis alone—the common usage, to the epidermis plus the fibrous layer, and to the outermost layer, regardless of its morphological nature. Endothecium has been applied to the fibrous layer alone, to the wall layer or layers below the fibrous layer, and to all of these together. The confusion in use lies, in part, in the history of the terms, which are old and were at first considered descriptive only. Early students of anther structure overlooked the ephemeral layer below the fibrous layer, as well as the occasional loss of the epidermis. The identity of layers in highly specialized anthers can probably be determined only by ontogeny and comparative studies of related taxa. Exothecium and endothecium have only ecologic and topographic value; they are not of morphological value.

The anther wall consists, typically, of three layers: the *epidermis;* the *fibrous layer*, immediately below the epidermis; and one or more *parietal* or *wall layers* between the fibrous layer and the tapetal cells. But the anther wall may be reduced to only one or two layers (Fig. 59). Fibrous layer is a poor term, since it implies that the layer consists of fiberlike cells. Its cells, though differing greatly from the surrounding cells, are not fiberlike, but the term is an old one and will probably continue in use. Histologically, the fibrous layer may be added to below the sporogenous and tapetal cells, by transformation of parenchyma cells of the connective so that the inner tissues are completely encircled by the fibrous layer. The parietal layers are continuous with sterile cells of the connective.

In further development, the epidermis usually remains simple but its cells may be velvety papillose, as in *Gloxinia* and *Gladiolus,* and the hairs may cover the entire anther, as in species of *Calycanthus.* Stomata —doubtfully functional—are frequent on the connective, rare over the sporangia. Where the fibrous layer is strongly developed, the epidermal cells may be tabular and thin, appearing collapsed and scalelike. They may be early deciduous or lost, as dehiscence approaches, over the entire anther or over restricted areas.

As cells of the fibrous layer mature, lignified or suberized thickenings develop in the walls, contrasting strongly with the thin, cellulosic wall areas between. The thickenings take the form of bars, spirals, or annular bands and often resemble the thickenings of the cell walls of protoxylem, to which they have even been considered morphologically equivalent and, therefore, of phylogenetic significance. But they are doubtless of only ecological importance, representing a structural mechanism involved in dehiscence (Fig. 68). Under drying, the thick and thin wall areas shrink unequally, and the layer, with attached layers, is ruptured. Continued drying of the wall tissues brings about retraction of the borders of the opening; changes in atmospheric humidity cause opening and closing of the rupture. In some genera, as in *Lilium,* the rupture is through a *stomium,* a group of specialized epidermal cells similar to those of the stomium of some ferns. It was long believed, with the existence of the stomium as evidence, that dehiscence was brought about by the epidermis alone. But the epidermis may be lacking at time of dehiscence, especially over the lines of rupture.

Where the sporangia are deeply sunken, the sporogenous cells are covered externally by only two layers, the epidermis and the fibrous layer. The fibrous layer extends a short distance beyond the sporangium, as in *Degeneria* (Fig. 67) and *Himantandra.* With increasing specialization of the anther, the fibrous layer extends farther and farther around the anther lobes, developing internally from parenchyma of the connective where it is often irregular in thickness. The parietal layers are often flattened by enlargement of the tapetal and sporogenous cells and may be crushed and absorbed. This is especially true where there is but one layer. Where the connective is greatly reduced, the fibrous layer may extend all over the anther. In anthers with poricidal or valvular dehiscence, the fibrous layer is usually absent or very weak.

The Thin-walled Anther. It was long ago stated that the microsporangium of the pteridophytes and gymnosperms, with the exception of *Ginkgo,* had only an exothecium (epidermis); that the angiosperms, in contrast, had both an exothecium and an endothecium (cell layers between the epidermis and the sporogenous and tapetal tissues). The validity of this distinction was questioned when it was found that the anther sacs of many angiosperms had walls of only one or two cell layers. It had been pointed out that thin-walled anthers belonged chiefly to taxa then generally considered primitive—Casuarinaceae, Piperaceae, Proteaceae, Urticaceae—and the similarity in sporangium structure to that of the gymnosperms gave support to the view that these taxa were among the most primitive angiosperms. But it was, even at that time, argued that these families were highly specialized in many other characters, that the simple wall structure of the anthers was the result of reduction in specialization. Only with the general recognition that these families are not highly primitive, and that the woody Ranales are the primitive angiosperms, has the phylogenetic significance of the thin-walled anther become apparent.

The deeply sunken sporangia of laminar stamens, without a sporangium wall and with their sporogenous tissues surrounded by sterile cells of the microsporophyll, have important phylogenetic significance. They set the angiosperms well apart in method of microspore-bearing from other vascular plants, except the primitive eusporangiate ferns. The multiseriate anther wall suggests possible relationship of the angiosperms to ancient eusporangiate ferns.

In the 1920s, the uniseriate anther wall was correctly recognized as a reduced structure, but the interpretation received little attention. The evidence presented then can well be outlined here, because it supports the growing opinion that the families in which it occurs are advanced, not primitive.

Reduction of the wall layers comes about by ontogenetic or phylogenetic loss of the innermost or the outermost layers, or of both of these layers. Loss of the epidermis may occur early in ontogeny of the anther or just before dehiscence, or vestiges may remain, especially over the line of fusion of the two sporangia in a sac (Fig. 59*C, G, H*). Scattered remnant epidermal cells may persist over the entire anther sac. Where the epidermis is degenerate or reduced, the fibrous layer takes its place as the outermost layer, retaining its characteristic cell-wall thickenings, as in *Casuarina.*

Variety of structure in thin-walled anthers and the evidence of reduction from thicker-walled types are shown in the following examples. All stages in the reduction can be seen in the Urticiflorae. *Ficus*—perhaps alone in the order—shows a typical anther wall, with well-developed fibrous layer. Genera of the Urticaceae, with a uniseriate wall and no fibrous layer, show the extreme reduction structure. Transition forms occur in some Moraceae, the Ulmaceae, and Cannabinaceae. In *Dorstenia* and *Broussonetia* (Fig. 59*B, C*), a few epidermal cells are present over the region of wall fusion between the two sporangia of the anther sac. *Humulus* (Fig. 59*A*), *Ulmus,* and *Cannabis* show epidermal cells scattered over the fibrous layer. The isolated epidermal cells (Fig. 59*G, H*) are survivors of the typical epidermis of earlier stages of anther development. The simple wall of the Urticiflorae is the result of reduction. Evidence is the presence of isolated epidermal cells, survivors of a typical epidermis of earlier ontogenetic stages, and the absence of the layer of wall cells below the fibrous layer, also often suppressed in ontogeny.

Explosive dehiscence characterizes many of the genera of the Urticiflorae, especially those of the Urticaceae. The pollen grains are hurled into the air by sudden bursting of the anther-sac wall. A remarkable feature of this type of dehiscence is the simultaneous explosion in some tropical genera of great numbers of anther sacs, forming smokelike puffs of pollen grains. The mechanism of explosive dehiscence is not understood. It has been reported to be the result of uniform shrinkage of a uniseriate wall, of internal turgor, and of pressure of maturing pollen grains upon a delicate enclosing wall. A structural feature of the anther sac, unknown in other types, may be a part of the mechanism: there is a short, longitudinal slit in the wall of the young anther. Explosive dehiscence is a highly specialized type developed in association with anemophily.

In *Casuarina,* the simple anther wall suggests superficially the sporangium wall of gymnosperms (Fig. 59*G*), but its outer wall is a typical fibrous layer and the epidermis persists, as a few isolated cells, over the partitions between the sporangia. The anther is not explosive but dehisces normally. In the Proteaceae, the anthers of some taxa—*Leucadendron*—have lost the epidermis completely, and others—species of *Grevillea*—have only scattered surviving cells. In the Piperaceae, *Peperomia* also has persisting remnants of the epidermis.

In submersed aquatic plants, the entire plant is structurally reduced in adaptation to its habitat. The anther sacs of these plants—with the exception of *Zostera* and *Zannichellia*—have uniseriate walls. *Zostera,* a transitional form, has a biseriate anther wall. The outer wall layer is of large parenchyma cells—a true epidermis; the inner layer is of thin cells with weak wall thickenings, which do not function in dehiscence. The pollen is freed by local swellings that break open the sac. *Zannichellia* has a simple, biseriate wall (Fig. 59*D*). The swelling of a small cluster of cells over the septum ruptures the wall (Fig. 59*E*). In

Najas, the simple-walled anther is enclosed by two sheathing structures. Freeing of the pollen is by a forcing apart of the enclosing sheaths, the outer one by growth of the receptacle and peduncle below the flower. Whether the single wall layer of *Najas* represents the epidermis, the fibrous layer, or some sheathing organ is not known.

The Tapetum. The primary sporogenous cells enlarge radially and divide periclinally. The outer cells so formed constitute the *tapetum,* a layer that encases the sporogenous cells and serves as a nutritive tissue for the pollen mother cells and the microspores (Fig. 67). The layer is typically uniseriate, rarely biseriate, but often irregular in outline and thickness, especially on the inner side of the sporogenous tissue. Tapetal cells may be formed also by transformation of outer sporogenous cells and, occasionally, from isolated sporogenous cells. The tapetum is described as derived, in some taxa, from the innermost layer of primary wall cells; in other taxa, partly from this layer and partly from the sporogenous cells. On the inner side of the sporogenous cell mass, it may arise from parenchyma of the connective and here complete the enclosure of the fertile cells. The tapetal cells enlarge rapidly, and their cytoplasm becomes denser. The tapetum is commonly uninucleate but may become bi- or multinucleate. The presence of many nuclei has suggested amitotic division, but an appearance of amitosis is perhaps given by incomplete mitosis. The mature tapetal cells may be uniformly uni-, bi-, or multinucleate or may vary in number of nuclei. At the time of spore formation and immediately afterward, the tapetal cells break down, and their cytoplasm unites to form a *tapetal plasmodium.* The cytoplasm is freed in different ways—by disintegration of the walls or by extrusion through collapsing walls. Cells of the connective abutting on the tapetum may also break down, but it is uncertain whether their contents are added to the plasmodium. The tapetal plasmodium extends among the spore tetrads or the free spores, isolating them singly or in clusters. As the spores mature, the plasmodium nourishes them and builds their walls.

Two types of behavior of the tapetal cells in the development of the pollen grains are distinguished: the *amoeboid* and the *secretory.* In the amoeboid type, the protoplasts of the tapetal cells enlarge and intrude among the spore mother cells, spores, and young pollen grains, before merging as a periplasmodium; in the secretory type, the tapetal cells, without collapsing, become rich in content and "secrete" a periplasmodium. In both types, the periplasmodium ultimately surrounds the maturing spores or pollen grains.

Four types of amoeboid formation of the periplasmodium have been distinguished, but these differ perhaps too little to be important. In the *Sagittaria type,* the plasmodium develops rather late—about the time the microspores are freed from the mother-cell wall—by the intrusion of "tongues" of cytoplasm from the individual tapetal cells among the spores. The nuclei of the tapetal cells migrate into the tongues later, as the processes develop among the pollen grains. Ultimately, the tongues merge to form the periplasmodium, in which the embedded pollen grains enlarge greatly. Vacuoles appear in the periplasmodium, and the nuclei and remnants of protoplasm disappear as the exine develops on the grains. The *Butomus type* of plasmodium is similar to the *Sagittaria type* but appears earlier, while the spores are still in tetrads. The enlarging tapetal-cell protoplasts press among the spores and unite, embedding them. In the *Sparganium type*, the tapetal cells—4-nucleate and vacuolate—enlarge and free their protoplasts at the time of the first reduction division of the spore mother cell. Tetrad formation occurs in the nucleus-rich plasmodium. In the *Triglochin type*, the tapetal cells disintegrate where they lie, and the cytoplasm and later, during the reduction divisions, the nuclei push between the pollen mother cells.

The secretory type of tapetum, like the amoeboid type, varies in behavior; the cell walls disappear or collapse as their cell contents merge in a plasmodium on the periphery of the fertile tissue. Forms of taxonomic significance can perhaps be distinguished, but too little detailed information is yet available. This type occurs in the higher monocotyledons and in many of the dicotyledons. It has been called the primitive type, but the amoeboid type, which characterizes the lower monocotyledons generally and is found in some of the lower dicotyledons, is probably primitive. The resemblance of this amoeboid periplasmodium to that of some of the lower vascular plants supports this view. The formation of a peripheral layer *in situ*—the *Triglochin type*—is the simpler and apparently the most advanced method.

Placentoids. Projections of sterile tissue into the sporogenous tissue have been called placentoids. These projections may be large lobes of connective parenchyma on the inner side of the sporangium—much resembling the placentae of carpels, with which they were once considered homologous—or plates of laminar tissue extending into or through the sporangia, dividing it more or less completely. In the Gentianaceae and Menyanthaceae, sterile cells break up the sporogenous tissue; in *Limnanthemum* and *Menyanthes,* they surround clusters of fertile cells and, in *Gentiana,* individual fertile cells. In the Onagraceae, two to six spore clusters are set apart in a sporangium; in some Orchidaceae (*Phajus*), Rhizophoraceae, Loranthaceae, and Mimoseae, there are several to many well-separated clusters. Transverse partitions give the appearance of multisporangiate anthers—*Butomus.* The partitions may become tapetal wholly or in part, and the groups of sporogenous cells formed in this way may persist together, even to the mature pollen stage. Large, so-called placentoids are merely lobes of connective tissue, not division walls. The sporangial position where all four sporangia lie in one plane, as in *Sassafras* and other lauraceous genera, suggests placentoid division of two sporangia but is the result of major displacement of sporangia by differential growth in the developing anther. Placentoids occur in rather few taxa, chiefly in Labiatae, Acanthaceae, Bignoniaceae, Solanaceae, Scrophulariaceae.

Ontogeny of Spore Mother Cells. The primary sporogenous cells may differentiate directly as spore mother cells—many Compositae, Labiatae, Malvaceae, Portulacaceae—or may divide, usually sparingly, and their daughter cells become spore mother cells. Where division does occur, the adjacent tapetal cells may also divide, maintaining the tapetal sheath about the enlarging cell mass. The spore mother cells (here "pollen mother cells") rapidly increase in size. The nuclei enlarge, the cytoplasm becomes dense, and the walls thicken. Wall thickening is uneven in cells which form pollen grains that remain in tetrads or larger clusters; the outer walls of the cluster are thicker than the internal ones.

Microspore Formation. Microspores are formed from the spore mother cells in tetrads by two closely successive divisions, the *meiotic* or *reduction divisions,* during which the chromosome number is reduced from $2n$ to n. Meiosis is nearly simultaneous throughout a sporangium or anther sac, except in very long anthers, as in *Liriodendron,* where there is an acropetal succession in stages of pollen development. Meiosis follows two somewhat different methods: the *simultaneous method,* where no wall is formed after the first division, the two divisions being almost simultaneous, with four free nuclei in the mother-cell protoplast; and the *successive method,* where a wall is formed after the first division, and each of the cells so separated is then divided. The simultaneous method is characteristic of dicotyledons and the successive, of monocotyledons, but there are many exceptions. There is probably little or no phylogenetic significance in these differences. Cell division in spore formation also takes place by cell-plate formation or by *furrowing,* sometimes called constriction, which occurs by means of the division of the protoplast by the extension inward from the mother-cell wall of wedges of new wall, as the protoplast contracts after nuclear division.

There seems to be much variety in relationship of mother-cell wall and the walls of the four spores. The mother-cell wall may remain intact, even until the pollen grains are mature, enclosing them when they are shed as clusters, or may disappear early or late in pollen maturation. Where the spores remain together as *compound pollen grains* and as *pollinia*, the abutting walls are said to remain uncutinized. The extent to which the mother-cell wall takes part in the formation of the pollen-grain wall by secretion or by modification and adnation to the spore wall probably varies greatly. The wedge-shaped projections that furrow the cytoplasm have been claimed to be formed within a gelatinous inner layer of the mother-cell wall.

The spores are variously placed in the tetrad (Fig. 64); those formed by the simultaneous method are usually in tetrahedral positions; those formed by the successive method, in isobilateral positions. The decussate arrangement is occasional; the linear, rare. The T position is also rare and is probably abnormal. More than one type of arrangement may occur in a species.

The spore mother cell usually forms four spores but may form more or less than four—two, when the second division fails to take place; three, when one of the first pair of cells fails to divide; more than four (*polyspory*), occurring chiefly in hybrids, where divisions are irregular or incomplete. Of four formed, one only may mature; the other three degenerate, often remaining as three vestigial nuclei beside the surviving spore inside the mother-cell wall—Cyperaceae (Fig. 60). Polyploid nuclei and other complex nuclear conditions may result from failure of wall formation in meiosis and from fusion of spore mother cells.

The Microspore Wall. The microspores, at first enclosed by a delicate wall, which may be deposited upon the inside of the mother-cell wall wherever the spore abuts upon it, lie within the mother-cell wall, usually completely filling it. The mother-cell wall may hold the tetrad together for a time, even permanently when the pollen grains remain in tetrads; or it may gelatinize and merge with the spore wall to form a part of the pollen-grain wall. Sometimes, it is apparently absorbed. The walls of the microspores enlarge, expanding and thickening by intussusception. The *exine*—outer layer—is thick and its function is protective, both mechanically and physiologically. It is adapted to changes in volume of the grain—expansion and contraction with changes in humidity. The *intine*—inner layer—is thin and delicate and also readily adapts itself to changes in size of the pollen grain. The wall may become very thick and complex in structure, and the outer layer become sculptured with projecting ridges, spines, and granules ("ornamentations"). Inner and outer layers, usually increased in wall thickness as the spore enlarges, are distinct; the inner is cellulosic, the outer cutinized. In the Epacridaceae, however, the wall of the young spore is thick and becomes thinner as it enlarges. The tapetal plasmodium aids, in large measure, in the increase of size of the spores and, especially, in the thickening of the wall and the building of the projections. When this thickening and ornamentation is completed, the spore is mature. The spore, after germination, when the first cells of the male gametophyte appear within the wall, becomes the pollen grain.

A so-called "third layer" of the anther wall—the layer adjacent to the tapetum—has been called the sporangium wall and persists in only some taxa. But this layer is prominent and persists only in thin-walled, highly specialized anthers, not in primitive stamens, where it would be expected to persist if it were a vestige of a true sporangium wall.

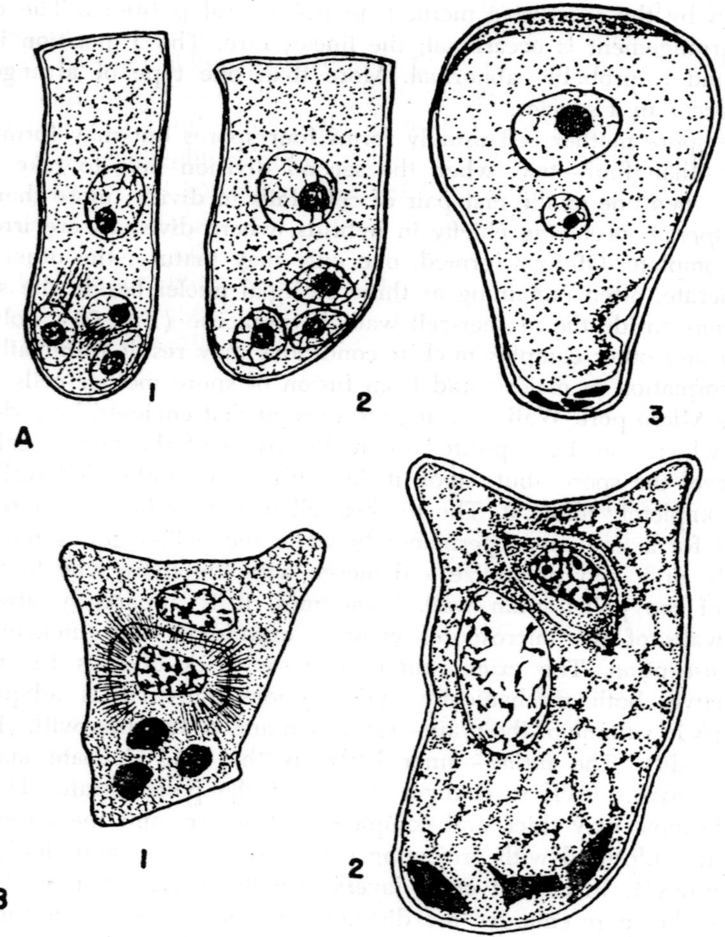

Fig. 60. *Scirpus,* germination of microspore mother cell, three of the daughter cells degenerate. *A, S. lacustris:* 1, the four daughter cells within the mother-cell wall; 2, one nucleus enlarging, three beginning to degenerate; 3, the surviving cell has divided, the other three reduced to remnants. *B, S. uniglumis:* 1, surviving daughter cell has divided (surrounded by phragmoplast), the other three spores aborted; 2, later stage of *B*1, vegetative and generative nuclei maturing. (*After Piech.*)

Morphological Nature of the Stamen

The basic nature of the sporophylls is discussed in Chap. 3 but is further treated here in the light of the comparisons of form and structure discussed in this chapter. The primitive form of the stamen is obviously laminar, like that of the carpel. (Under the New Morphology, the terete-filament type has been called primitive; the broad filament is interpreted as specialized.) In ontogeny, the broader types have marginal meristems like those of carpels and leaves. In specialization, the stamen has been modified more extensively than the carpel and the laminar form largely lost. Sporangium position has also been greatly modified, in such a way that two of the four sporangia appear to be morphologically on the side of the sporophyll opposite its original position, and all four, from an originally sunken position, appear to have become superficial. But, morphologically, all the sporangia are on one side of the sporophyll and are embedded in sporophyll tissues, even though only lightly so in highly specialized anthers. The change in the microsporangium position is largely phylogenetic, but, in some taxa, is still in part ontogenetic (Fig. 57).

The arrangement of sporangia in two pairs has been looked upon as persisting evidence of ancient telomic structure, but the laminar form, clearly primitive in living forms, does not support the view that the pairs of sporangia represent fused terminal telomes. The vascular structure of the laminar stamen shows no evidence of a basic telomic structure. The sporangia lie in pairs between the midrib and the lateral veins and have no connection with the vascular meshwork; if the sporophyll consists of telomes, the sporangia should terminate vascular bundles.

Relationship of the simple and the fascicled stamen is discussed under androecium.

The Peltate Stamen. In the older, descriptive literature, stamens in which more or less free, dorsifixed, or versatile anthers stand somewhat oblique to the filament—a common condition in highly specialized anthers—were called shield-shaped or peltate, but this term did not continue to be used in so simple a sense. Its revival in the early decades of the twentieth century came with its use in extended and modified senses. In one treatment, stamens fall in two major classes: *peltate* (shield-shaped) and *impeltate,* or *epeltate* (not shield-shaped). In impeltate stamens, the connective is the direct continuation of the filament; in peltate stamens, the filament is attached laterally to the anther, which usually stands at an angle to the filament. The impeltate stamen is *arrow-shaped* if the anther sacs extend beyond the attachment of the filament, and *not arrow-shaped* if the sacs do not so project;

it is *unseparated* if there is no clear limit of anther and filament. (Apparently, no term is proposed for the stamen where the sacs project at both ends of the connective.) Peltate stamens are *epipeltate* if filament attachment is on the ventral side; *hypopeltate,* if on the dorsal side.

In the second, more elaborate and morphologically more complex interpretation of "peltate" form in the stamen, the stamen is considered not merely shield-shaped but, like the carpel, a hollowed organ, ascidiform or utriculate. (For a general discussion of the peltate theory as applied to sporophylls, see Chap. 5.) In the stamen, typical hollowed forms do not exist, but examples have been incorrectly described as evidence that they do.

Evidence of peltate form in the stamen is seen, under this interpretation, (1) in ontogeny, where the early form is called peltate; (2) in the so-called three-dimensional (as contrasted with so-called two-dimensional, dorsiventral) form; (3) in organs transitional from stamen to petal; and (4) in petals. Ontogenetic evidence is seen, under this interpretation, in the early ontogeny where the anther is prominent and may stand somewhat obliquely on a basal plate, which, much later, becomes the filament. The oblique position suggests the form of a simple, peltate leaf. (This obliqueness is, actually, merely the early expression of later extrorse or introrse dehiscence.) A slight median furrow, often present at the apex, is considered, under the peltate theory, a vestige of the hollow of an ancestral utriculate stamen (Fig. 61*a*). If the bulging anther base overhangs the filament primordium, as in arrow-shaped anthers (Fig. 61*e*), the enclosed space is said to represent a vestigial cavity. (These minor indentations are merely beginnings in the elaboration of anther form, not vestiges of ancestral cavities; the distal hollow is the beginning of the transverse furrow separating the anther sacs; the lower hollow is formed by differences in diameter of anther and filament, which are evident at an early stage.) The stamen is looked upon, under the peltate theory, as a "three-dimensional" organ, because of the thickness of the anther and the arrangement of the four sporangia in four corners. But this structure is the result of the specialization of the anther, the bringing of two of the four sporangia, originally in a single plane, into a second plane. In some ranalian genera, there are distal hollows in organs transitional between stamens and petals, as in the petals of *Coptis, Helleborus, Aquilegia, Delphinium, Aconitum.* These hollows are secretory areas and definitely glands in petals or staminodes (sterile stamens); they are not vestiges of the cavity of a tubular, ancestral stamen, as suggested under the peltate theory (see Chap. 6). The best evidence that the stamen is not fundamentally a peltate, three-dimensional organ is the type of primitive stamen clearly shown by the Ranales—a simple, dorsiventral, laminar organ.

The theory that peltate form is fundamental in angiosperm sporophylls closely parallels the obsolete theory of carpel polymorphism in its lack of sound morphological basis, lack of convincing evidence, and continuing extension to include all appendages.

The "Diplophyllous" Stamen. Closely related to the theory that the stamen is basically so-called three-dimensional (tubular or four-angled) is the theory that it is fundamentally double in nature, consisting of two

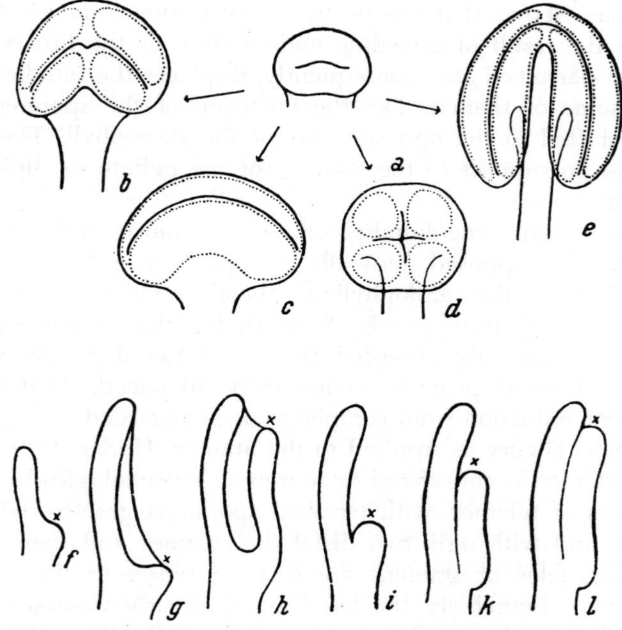

Fig. 61. Diagrams to show, under the peltate theory, the origin of stamen from young peltate leaf. *a*, peltate primordium from which develop the types *b, c, d, e*, each with a median furrow; *f* to *h*, *i* to *l*, median longitudinal sections of a carpel or stamen primordium to show two series of stages in the development of the utriculate form by growth of the cross-zone meristem *x*. Dotted lines outline sporangia. (*After Baum*, 1949.)

united laminae. This theory is suggestive of the old idea that the angiosperm stamen represents two conifer microsporophylls, connate back to back. Evidence in support of this later theory is seen largely in organs transitional from stamen to petal and in so-called diplophyllous ("four-winged") petals, such as occur in teratological forms in the Ranunculaceae, Rosaceae, Saxifragaceae, and Myrtaceae. The so-called double organ is four-winged, suggesting two laminae united by their midribs, and having a common median vascular vein. The abaxial half of the double organ is seen as the true leaf blade; the adaxial half, a prolongation of the "cross zone." The organ is considered fundamentally tubular (peltate), even though it has a vascular core rather than a central cavity.

The distribution, orientation, and type of the lateral veins in the stamens of some taxa have been considered proof that stamens are basically diplophyllous in nature; the filaments in the Nymphaeaceae show some vascular bundles that are inverted and amphicribral or bicollateral (Fig. 55). But study of the complete vascular system of the stamens in the Nymphaeaceae shows that the position, orientation, and type of these bundles are the result of crowding and twisting as the broad ancestral sporophyll is narrowed and, consequently, displaces the bundles. The inversion of some of them is like the inversion of the sporangia which have been shifted to the opposite side of the sporophyll. The inverted bundles give no support to the theories of the peltate or diplophyllous nature of the stamen.

The presence, in some families, of organs transitional from stamen to petal and the supposed, basically diplophyllous nature of both are considered, under the diplophyllous theory, evidence that stamens have been derived from petals. Support for this interpretation—the reverse of the generally accepted theory—is found in the view that simplicity of form is primitive, complexity advanced. That simplicity may represent reduction from complexity is disregarded.

The Telome Theory as Applied to the Stamen. Under this hypothesis, the typical stamen is considered to represent, morphologically, a dichotomous system of telomes, with terminal sporangia greatly reduced and compressed, and with only two distal dichotomies and their sporangia surviving. Fascicles of stamens are seen to represent the survival of several or many branchlets; the basal axis is greatly shortened, and the distal tips are brought together in pairs. Support for this view is seen in the side-by-side pairs of sporangia of the typical anther and in the presence, in some fairly primitive orders—Dilleniales, Parietales, Malvales—of a strong "trunk" vascular bundle which constitutes the central axis of the telome system and supplies several or many stamens.

The pairs of sporangia, characteristic of stamens, naturally suggest dichotomies, and the pairing is perhaps even more prominent in the laminar stamens where all the sporangia lie in one plane. (The two-plane position of the sporangia in the typical anther has clearly been derived from the one-plane position.) The pairing of the microsporangia, prominent in all stamens and especially in the primitive laminar types, contrasts strongly with the basic, scattered position of the megasporangia. If the dichotomy of telomic structure persists in the sporophyll, some similarity should be found in the megasporophyll. And it is perhaps significant also that the outer member of each pair of sporangia is smaller than the inner, a character that is so deep-seated that it is present in most highly specialized anthers.

The vascular structure of stamens, as a whole, gives no support to the hypothesis of telome structure in the stamen; no vascular tips lead to or directly toward the individual sporangia. (Vascular branches are frequent in the connective of some taxa but rarely penetrate the wall between members of the pairs.)

The stamen of *Ricinus* has often been cited as an example of solitary sporangia terminal on dichotomizing branches of the stamen, but the position of these sporangia is now better understood. The anther in this genus has remarkable form; the connective is scalelike, subtending the stalked unisporangiate anther lobes (Fig. 44).

The telomic theory has been considered to aid in the interpretation of stamen fascicles with their complex structure; under this theory, the fascicle can be interpreted as a compacted system of telomes. But, in the vascular system of a fascicle, there is no evidence of dichotomy. And comparative anatomical study of fascicles in several families, especially in the Dilleniaceae and Paeoniaceae, shows that the fascicle is an aggregation of simple stamens (Fig. 46).

Staminodia

Staminodia are sterile floral organs that apparently represent stamens morphologically, resembling them more or less in form and closely in relation to the receptacle and to other floral organs. They have great variety of form. In most angiosperms, the petals are, morphologically, elaborated, sterile stamens and could well be called staminodia. Some taxa have staminodia which are transitional in form and loss of fertility from stamen to petal, and no line can be drawn between stamen and staminodium and between staminodium and petal—in genera of the Nymphaeaceae, Calycanthaceae, Magnoliaceae. Some staminodia are secretory organs, stamens or petals that have become primarily nectaries—*Coptis*. Vestigial stamens, stamens in process of reduction and disappearance, are commonly called staminodia. The remnants of stamens, often mere stubs, in pistillate flowers are good examples of this type.

Staminodia, typically, have the vascular structure of stamens in more or less reduced form. Where the stamen has lateral traces, these may be weak or absent. Staminodia that are mere remnants of reduced stamens may have no vascular supply or a mere basal stub. Vascular stubs of the traces of stamens entirely lost externally may be present in the floral receptacle, as in the flowers of the Scrophulariaceae and in the pistillate flowers of some unisexual genera.

Nectaries

Nectaries are surface areas, emergences, or organs where nectar is secreted. *Floral nectaries* are, morphologically, of two types: localized areas where nectar is secreted, or organs transformed from their original form and function. Nectaries that are modified organs may have restricted secretory areas or be secretory over most of their surface. Secretory areas, such as those on the receptacle, may be superficial only or may be proliferated, forming emergences that suggest organs. Reduced organs, most commonly stamens or petals, are often represented by nectaries. These nectaries may have extensive secretory surfaces—some of the Proteaceae—or only minor nectariferous areas—*Coptis* and *Salix*. Superficial nectaries may be minor or major areas on organs otherwise largely normal—*Aquilegia, Ranunculus, Viola*. *Septal glands*, nectaries on the abutting and partly connate walls of carpels in syncarpous gynoecia, are perhaps the most elaborate of superficial nectaries. They are characteristic structures of many Liliaceae, Amaryllidaceae, Palmae, Bromeliaceae. Where the carpels are not fused by their external edges, the nectar is excreted along a slitlike lateral or distal opening. Where the marginal areas are fused, the nectar flows out through a tubelike passage to a small aperture on the top of the ovary. Direction of dehiscence of anthers is biologically correlated in insect pollination with position of nectaries: extrorse where the glands are below the stamens, introrse where they are above the stamens.

Histologically, the secretory tissue of a nectary consists of subepidermal, small, closely packed cells, with rich cytoplasmic content. The epidermis overlying this tissue varies in structure; rarely, its cells also are secretory. A cuticle, usually thin, is commonly present, although nectaries without a cuticle have been reported. Stomata, usually with enlarged openings, through which the nectar exudes, are present in many nectaries. In other nectaries, the nectar diffuses through the epidermis and cuticle. Rarely, a nectary may consist of a secretory area in the epidermis alone, where the cells are papillose or replaced by multicellular hairs.

The morphological nature of nectaries can best be determined by comparisons with the flower structure of related taxa and by anatomical structure. The number, origin, and type of vascular bundles supplying them give important and usually definite evidence of their nature. The traces of nectaries that represent organs are few and well defined; the vascular supply of superficial nectaries is diffuse, with many minor bundles from several or many points of origin, which may be on more than one floral organ; for instance, on receptacle and ovary, on stamen and petal. The nature of nectariferous discs, especially ring-shaped discs, can often be determined only by the number and position of the major vascular bundles.

Nectaries are present in many places in the flower, on all the floral appendages and, frequently, on the receptacle. No structural nectary occurs in the primitive flowers of many woody Ranales. Their absence in "simple" flowers, such as those of the Amentiferae, has been considered support for the view that anemophily is the primitive method of pollination. But these flowers are obviously greatly reduced, and lack of nectaries may represent loss of these structures. The general story of development of nectaries can be seen in living forms. Where pollination is by beetles exclusively, or nearly so—*Eupomatia, Calycanthus*—there are no nectaries; the insects feed on nonsecretory "food bodies" (Fig. 69). (Superficially and in cell content, these food bodies may resemble nectaries.) In the Magnoliaceae (*Magnolia, Talauma*), nectar is secreted, but not in localized areas; it is described as diffused through petal cuticles and excreted through enlarged stomata on petal bases and carpel surfaces. The entire flower of *Magnolia* has been called a nectary. In some of the Nymphaeaceae, a primitive family, nectar is reported secreted by the perianth. Since a small amount of nectar has been reported in one species of *Magnolia* and one species of *Calycanthus*, genera which have food bodies and beetle pollination, nectar secretion apparently originated while the food-body method existed.

It has been suggested that, in the evolution of the flower, a progressive change has occurred in nectary position, from the base of the flower and the receptacle inward and upward on the appendages. In primitive dicotyledons, nectaries occur on the perianth and on the outer parts of the receptacle; in the higher families, they occur chiefly on the sporophylls or on the "receptacle" (fused perianth and sporophylls) around and above the ovary. In the monocotyledons, the nectaries are on the perianth, stamens, and especially the carpels (septal nectaries). Differences in the morphological nature of the nectaries seem not to have been considered in this view of a phylogenetic change in nectary position, but a general change is evident. Change in nectary position necessarily accompanies fusion of appendages, and, in advanced floral types, nectaries are present higher up in the flower.

The location of nectaries that represent reduced floral appendages may be of much importance in the determination of phylogenetic relationships. For example, the nectariferous disc of some genera of the Proteaceae (four separate organs in other genera) is, from the evidence of comparative form and anatomy, a vestigial corolla, and this family can no longer be placed in the supposedly primitive Apetalae. The nectaries of the Salicaceae, which are reduced perianth parts, demonstrate that the flowers of this family are not primitively simple.

The variety in form, position, and morphological nature of nectaries indicates that these structures have arisen independently in many angiosperm lines. Position and type of nectary are surely important in phylogenetic studies. Flowers now generally accepted as highly primitive—those of the Eupomatiaceae, Himantandraceae, Magnoliaceae, Calycanthaceae, Nymphaeaceae—are chiefly or entirely pollinated by beetles and have no nectaries. Nectar glands are characteristic of angiosperms, except those that are pollinated by beetles and wind.

BIBLIOGRAPHY

The Stamen and Androecium

Aboy, H. E.: A study of the anatomy and morphology of *Ceratophyllum demersum*, Unpublished thesis, Cornell University, 1916.

Arber, A. On the structure of the androecium in *Parnassia* and its bearing on the affinities of the genus, *Ann. Bot.*, **27**: 491–510, 1913.

Bailey, I. W., and C. G. Nast: The comparative morphology of the Winteraceae. I. Pollen and Stamens, *Jour. Arnold Arb.*, **24**: 340–346, 1943.

Baillon, H.: Recherches organogéniques sur les *Eupomatia*, *Adansonia*, **9**: 22–28, 1868–1870.

Baum, H.: Beiträge zur Kenntnis der Schildform bei den Staubblättern, *Oesterr. Bot. Zeitschr.*, **96**: 453–466, 1949.

———: Die Unabhängigkeit der diplophyllous Gestalt der Staubblattspreite von ihrer Funktion als Träger der Pollensäcke, *Oesterr. Bot. Zeitschr.*, **100**: 265–269, 1953.

Brouland, M.: Recherches sur l'anatomie florale des Renonculacées, *Le Botaniste*, **27**: 1–278, 1935.

Canright, J. E.: The comparative morphology and relationships of the Magnoliaceae. I. Trends of specialization in the stamen, *Am. Jour. Bot.*, **39**: 484–497, 1952.

Celakovsky, L.: Über den "eingeschalteten" epipetalen Staubblattkreis, *Flora*, **58**: 481–489, 497–504, 513–524, 1875.

———: Teratologische Beiträge zur morphologischen Deutung des Stamengefässes, *Jahr. Wiss. Bot.*, **11**: 124–174, 1878.

Challenger Expedition: Rpt. Sci. Results Bot., III. Rpt. Bot. Juan Fernandez and Masafuera, 1884.

Chatin, A.: "De l'Anthère." Paris, 1870.

Chifflot, J. B. J.: Contributions à l'étude de la classe des Nymphéinées, *Ann. Lyons Univ.*, n.s., I (10): 1–294, 1902.

Clausen, P.: Über das Verhalten des Antheren-Tapetum bei einiger Monocotylen und Ranales, *Bot. Archiv*, **18**: 1–27, 1927.

Cooper, D. C.: Nuclear divisions in the tapetal cells of certain angiosperms, *Am. Jour. Bot.*, **20**: 358–364, 1933.

Corner, E. J. H.: Centrifugal stamens, *Jour. Arnold Arb.*, **27**: 423–437, 1946.

Daumann, E.: Zur morphologischen Wertigkeit der Blütennektarien von *Laurus*, *Beih. Bot. Centralbl.*, **48** (I): 209–213, 1931.

Diels, L.: Über die Gattung *Himantandra*, ihre Verbreitung und ihre systematische Stellung, *Bot. Jahrb.*, **55**: 126–134, 1917.

Eames, A. J.: Floral anatomy as an aid in generic limitation, *Chron. Bot.*, **14**: 126–132, 1953.

Eckardt, T.: Untersuchungen über Entwicklungsgeschichte und systematische Bedeutung des pseudomonomerous Gynoecium, *Nova Acta Leopold.*, N.F., **5**: 1–112, 1937.

Engler, A.: Beiträge zur Kenntnis der Antherbildung der Metaspermen, *Pringsheim's Jahr. Wiss. Bot.*, **10**: 275–316, 1876.

Frye, T. C.: A morphological study of certain Asclepiadaceae, *Bot. Gaz.*, **34**: 389–412, 1902.

Guérin, P.: Le développement de l'anthère chez les Gentianacées, *Bull. Soc. Bot. France,* 2 sér., **72**: 5–18, 1926.

Hannig, E.: Kritische Untersuchungen über das Vorkommen und die Bedeutung von Tapeten und Periplasmodien, *Flora,* N.F., **2**: 335–382, 1911.

———: Uber Bedeutung der Periplasmodien, *Flora,* N.F., **2**: 209–278, 1911.

Hirmer, M.: Beiträge zur Morphologie der polyandrischen Blüten, *Flora,* **110**: 140–192, 1918.

Hjelmquist, H.: Studies in the floral morphology and phylogeny of the Amentiferae, *Bot. Not. Suppl.* **2**: 1–171, 1948.

Howard, R. A.: The morphology and systematics of the West Indian Magnoliaceae, *Bull. Torrey Bot. Club,* **75**: 335–357, 1948.

Ivancich, A.: Der Bau der Filamente der Amentaceae, *Oesterr. Bot. Zeitschr.*, **56**: 305–309, 385–394, 1906.

Janchen, E.: Die sogenannte Schildform der jungen Staubgefässe, *Phyton.* (*Ann. Rei Bot.*); **2**: 267–270, 1950.

Kasapligil, B.: Morphological and ontogenetic studies of *Umbellularia californica* Nutt. and *Laurus nobilis* L., *Univ. Calif. Publ. Bot.*, **25**: 115–240, 1951.

Kaussmann, B.: Vergleichende Untersuchungen über die Blattnatur der Kelch—, Blumen—, und Staubblätter, *Bot. Archiv,* **42**: 503–572, 1941.

Leclerc du Sablon: Recherches sur la structure et la déhiscence des anthères, *Ann. Sci. Nat. Bot.*, 7 sér., **1**: 97–134, 1885.

Leinfellner, W.: Die petaloiden Staubblätter und ihre Beziehungen zu den Kronblättern, *Oesterr. Bot. Zeitschr.*, **101**: 373–406, 1954.

———: Die blattartig flachen Staubblätter und ihre gestaltlichen Beziehung zum Bautypus des Angiospermen-Staubblattes, *Oesterr. Bot. Zeitschr.*, **103**: 247–290, 1956.

———: Inwieweit kommt der peltat-diplophylle Bau des Angiospermem-Staubblattes in dessen Leitbündelanordnung zum Ausdruck? *Oesterr. Bot. Zeitschr.*, **103**: 381–399, 1956.

Matthews, J. R., and E. M. Knox: The comparative morphology of the stamen in the Ericaceae, *Trans. Bot. Soc. Edinburgh,* **29**: 243–281, 1929.

Money, L. L., I. W. Bailey, and B. G. L. Swamy: The morphology and relationships of the Monimiaceae, *Jour. Arnold Arb.*, **31**: 372–404, 1950.

Moseley, M. F., Jr.: Morphological studies in the Nympheaceae. I. The nature of the stamens, *Phytomorph.*, **8**: 1–29, 1958.

Neumann, R.: Über Antherae anticae und posticae und deren Übergänge in einander, *Bot. Zeit.*, **11**: 353–363, 371–383, 399–403, 1854.

Ozenda, P.: Remarques sur quelques interprétations de l'étamine, *Phytomorph.*, **2**: 225–231, 1952.

Parkin, J.: The protrusion of the connective beyond the anther and its bearing on the evolution of the stamen, *Phytomorph.*, **1**: 1–8, 1951.

Piech, K.: Über die Entstehung der generativen Zelle bei *Scirpus uniglumis* Link durch "freie Zellbildung," *Planta,* **6**: 96–117, 1928.

Pijl, L. van der: The stamens of Ricinus, *Phytomorph.*, **2**: 130–132, 1952.

Salisbury, E. J.: Floral construction in the Helobiales, *Ann. Bot.*, **40**: 419–445, 1926.

Schaeppi, H.: Vergleichend-morphologische Untersuchungen an den Staubblättern der Monocotyledonen, *Nova Acta Leopold.*, N.F., **6**: 389–447, 1939.

Schumann, K.: Blüthenmorphologische Studien: Die obdiplostemonen Blüthen, *Jahr. Wiss. Bot.*, **20**: 349–426, 1889.

Schwarze, C.: Vergleichende entwicklungsgeschichtliche und histologische Untersuchungen reduzierter Staubblätter, *Jahr. Wiss. Bot.*, **54**: 183–243, 1914.

Shoemaker, D. N.: On the development of *Hamamelis virginiana*, *Bot. Gaz.*, **39**: 248–266, 1905.

Smith, A. C.: Studies of Papuasian plants. V. *Jour. Arnold Arb.*, **23**: 417–443, 1942.

Städtler, G.: Über Reduktionserscheinungen im Bau der Antherenwand von Angiospermen-Bluten, *Flora*, N.F., **116**: 85–108, 1925.

Swamy, B. G. L.: A contribution to the life history of *Casuarina*, *Proc. Am. Acad. Arts and Sci.*, **77**: 1–32, 1948.

——— and I. W. Bailey: *Sarcandra*, a vesselless genus of the Chloranthaceae, *Jour. Arnold Arb.*, **31**: 117–129, 1950.

Thoday, D., and E. T. Johnson: On *Arceuthobium pusillum* Peck. II. Flowers and fruit, *Ann. Bot.*, **44**: 813–824, 1930.

Thomson, B. F.: The floral morphology of the Caryophyllaceae, *Am. Jour. Bot.*, **29**: 333–349, 1942.

Troll, W.: Über Diplophyllie und verwandte Erscheiningen in der Blattbildung, *Planta*, **15**: 355–406, 1931.

Van Tieghem, P.: Structure de l'étamine chez les Scrophulariacées, *Ann. Sci. Nat. Bot.*, 8 sér., **17**: 363–371, 1903.

———: Sur les anthères symmetriquement hétérogènes, *Ann. Sci. Nat. Bot.*, 9 sér., **5**: 364–370, 1907.

Von Mohl, H.: "Beobachtungen über die Umwandlung von Antheren in Carpelle." In "Vermischte Schriften," pp. 28–44, Tübingen, 1845.

Warming, E.: Untersuchungen über pollenbildende Phyllome und Caulome, *Bot. Abhandl.*, **2**: 1–90, 1873.

Wilson, C. L. The phylogeny of the stamen, *Am. Jour. Bot.*, **24**: 686–699, 1937.

———: The telome theory and the origin of the stamen, *Am. Jour. Bot.*, **29**: 759–764, 1942.

———: Vascularization of the stamen in the Melastomaceae, with some phyletic implications, *Am. Jour. Bot.*, **37**: 431–444, 1950.

Nectaries

Anderson, C. E.: Some studies on the floral anatomy of the Liliales, Thesis, Cornell University, 1940.

Behrens, W. J.: Die Nectarien der Blüten, *Flora*, **62**: 2–11, 17–27, 49–64, 81–90, 113–123, 145–153, 233–247, 304–314, 369–375, 443–457, 1879.

Bonnier, G.: Les nectaires, *Ann. Sci. Nat. Bot.*, 6 sér., **8**: 1–212, 1879.

Brongniart, A.: Mémoire sur les glands nectarifères de l'ovaire, *Ann. Sci. Nat. Bot.*, 4 sér., **2**: 5–23, 1885.

Brown, W. H.: The bearing of nectaries on the phylogeny of flowering plants, *Proc. Am. Phil. Soc.*, **79**: 549–595, 1938.

Daumann, E.: See reference under Pollination.

———: Zur Phylogenie der Diskusbildungen: Beiträge zur Kenntnis der Nectarien, *Beih. Bot. Centralbl.*, **48**: 183–208, 1931.

Diels, L.: See reference under Pollination.

Fahn, A.: On the structure of floral nectaries, *Bot. Gaz.*, **113**: 464–470, 1952.

———: The topography of the nectary in the flower and its phylogenetic trend, *Phytomorph.*, **3**: 424–426, 1953.

Grassmann, P.: Die Septaldrüsen, *Flora*, **67**: 113–144, 1884.

Jaeger, P.: Nectaires floraux et phylogénèse, *Année. Biol.*, (III) **54**: 111–117, 1950.

Jordan, K. F.: Die Stellung der Honigbehälter und der Befruchtungs-werkzeuge in den Blumen, *Flora*, **69**: 195–226, 243–252, 259–274, 1886.

Porsch, O.: Die Abstammung der Monocotylen und die Blütennektarien, *Ber. Deutsch. Bot. Ges.*, **31**: 580–590, 1914.

Schniewind-Thies, J.: "Beiträge zur Kenntnis der Septalnektarien," Jena, 1897.

Werth, E.: See reference under Pollen and Male Gametophyte.

Wolff, G.: Zur vergleichenden Entwicklungsgeschichte und biologischen Bedeutung der Blütennektarien, *Bot. Archiv*, **8**: 305–344, 1924.

Chapter 5

POLLEN

In most angiosperms, germination of the microspore begins when the flower bud is small, with the enlargement of the spore and the first cell division within the spore. After one or two divisions within the spore wall, a resting period usually occurs. The spore wall, together with its contents of two or three cells in a dormant state, (an early stage of the male gametophyte), is the *pollen grain* (Fig. 62*F*). The microspore is mature long before the dehiscence of the anther, often while the anther is still essentially sessile and the filament undeveloped. In plants that flower in early spring, the anthers usually pass the winter with the sporogenous tissue in the mother-cell stage or, less commonly, in the microspore stage; rarely, the microspore has germinated. In these plants, development may continue in the warmer periods during the winter. Collectively, masses of pollen grains—individual grains or clusters of various sizes—constitute the *pollen.* Where the grains formed from the spores of a tetrad remain permanently together, the grains have been called "compound pollen grains" but are better described as "pollen grains in tetrads."

Pollen grains show great variety of form, size, and sculpturing of the wall (Fig. 63). In form, they are usually globose, ellipsoid, or fusiform, but they may be lobed or angular. Their shape depends, in part, on their moisture content. Drying may greatly change their form, but, when moist, the grains will return to their original shape. Pollen may be shed in either the moist or dry state: that of many Rosaceae is described as shed in the moist state; that of the Compositae, in the dry state. The most extreme shape is confervoid—as in many submersed aquatic genera, *Zostera, Thalassia, Posidonia, Ruppia* (crescent-shaped). Unusually large grains characterize the Cucurbitaceae and very small ones some of the Boraginaceae.

Arrangement of Pollen Grains in the Tetrad. The pollen grains formed by a mother cell are associated in several ways (Fig. 64). The most common arrangements are the tetrahedral—the grains lying at the four corners of a tetrahedron—and the tetragonal—the grains lying at the corners of a square or rhombus. The tetrahedral arrangement is common in the higher dicotyledons. In the formation of the spores, the tetrahedral arrangement usually results from simultaneous divisions;

Drimys is a prominent exception. The tetragonal arrangement results from successive divisions; this arrangement is frequent in monocotyledons, lower dicotyledons, and gymnosperms.

Pollen grains are commonly shed individually or in loose clusters but may remain in tetrads—*Typha,* Droseraceae, Winteraceae, Lactorida-

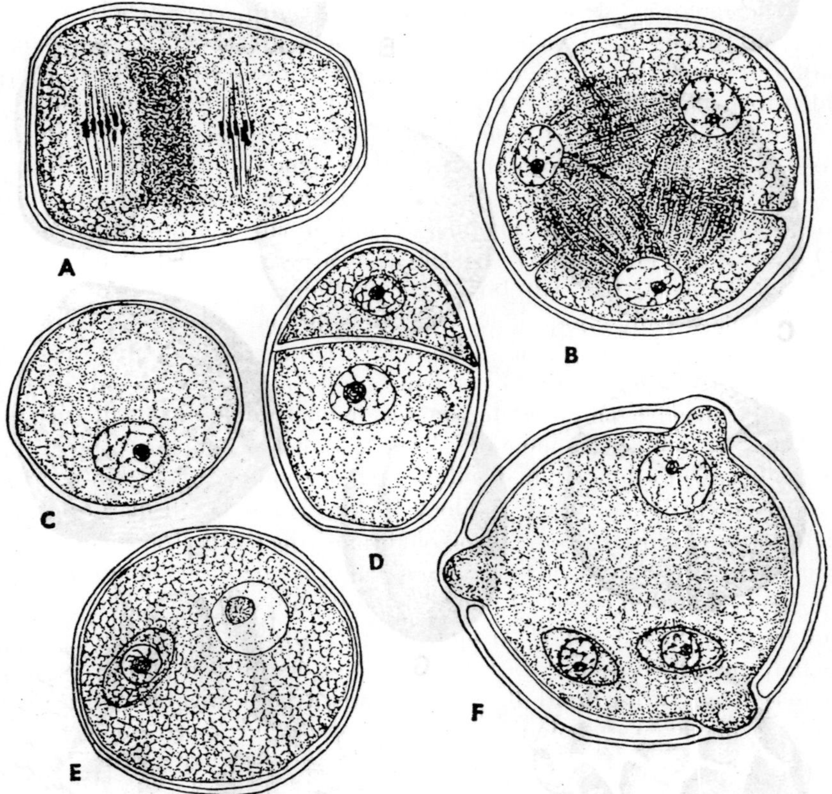

Fig. 62. Microsporogenesis in *Lobelia cardinalis. A,* second meiotic division within the mother-cell wall; *B,* microspores nearly formed, partition walls developing; *C,* mature microspore; *D,* spore enlarged, first cell division completed; *E,* vegetative and reproductive nuclei mature; *F,* mature pollen grain with sperm nuclei. (*After G. O. Cooper.*)

ceae, Juncaceae, several sympetalous families. They are held together sometimes by the persistent mother-cell wall, sometimes by close appression or by adhesive surfaces. The grains may remain in tetrads even in pollinia. Rarely, they remain in pairs called "dyads"—*Scheuchzeria.* They may form *pollinia* or massulae, the compacted contents of individual sporangia, anther sacs, or fused anther sacs—Asclepiadaceae,

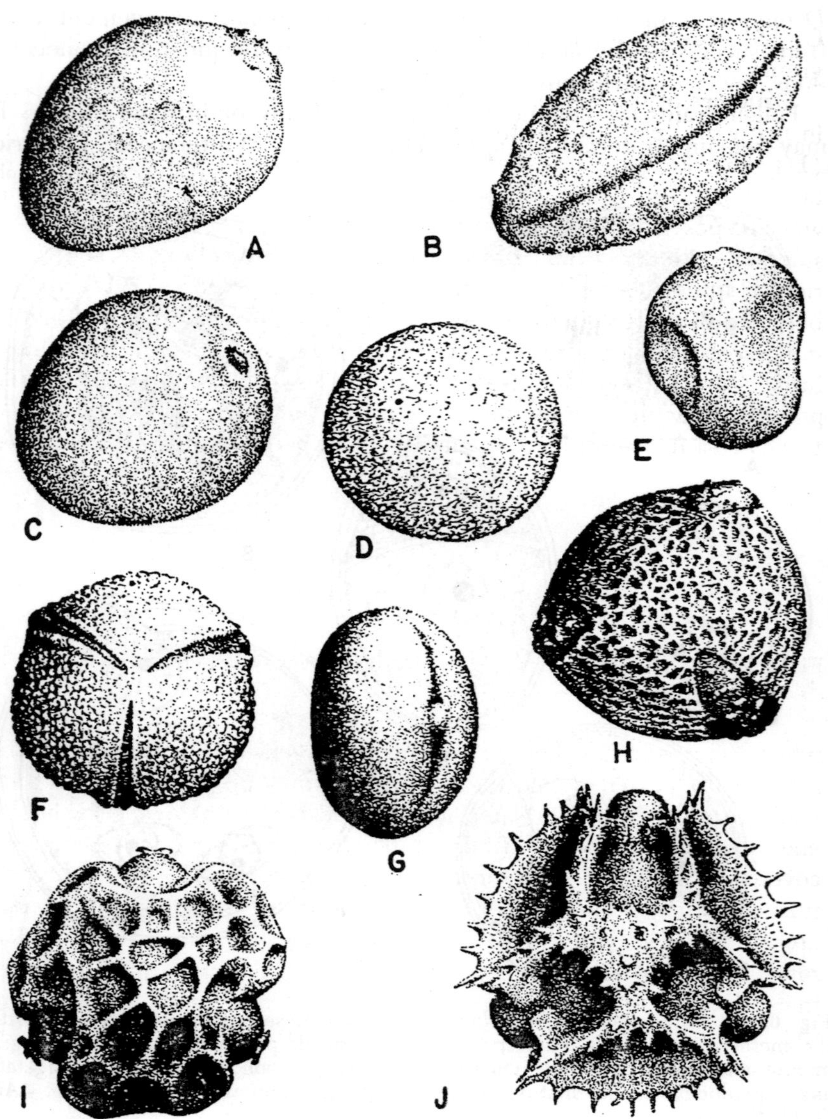

Fig. 63. Pollen grains of angiosperms showing variety in form and sculpturing. *A, Carex stricta,* side view; *B, Liriodendron tulipifera,* ventral view; *C, Festuca elatior,* side view; *D, Populus sargentii; E, Agrostis palustris,* ventral view, contracted; *F, Fagus grandifolia,* polar view, expanded; *G, Castanea dentata,* side view, fully expanded; *H, Salix fragilis,* polar view, expanded; *I, Barnadesia berberoides,* polar view; *J, Taraxacum officinale,* polar view. (*After Wodehouse,* 1935.)

Orchidaceae. In the Mimoseae, pollen is often shed in masses of 8 to 64 grains.

Structure of the Exine. The exine of the pollen grain is often sculptured in many and elaborate patterns, which characterize major or minor taxa (Fig. 63). Pollen grains that remain in tetrads have the surfaces of contact with other spores of the tetrad smooth or only lightly sculptured when exposed by enlargement at germination. The patterns have considerable taxonomic and phylogenetic value. Projections in the form of ridges, spines, and granules are the prominent features. The thin areas between the projections are of two types: *germinal furrows* and *germinal apertures,* or *pores.* The furrows, various in form and distinctness of limit, represent elongate areas where the wall is thin and elastic. They provide structural adaptation to changes in volume of the grain due to changes in atmospheric humidity. The apertures are small thin areas,

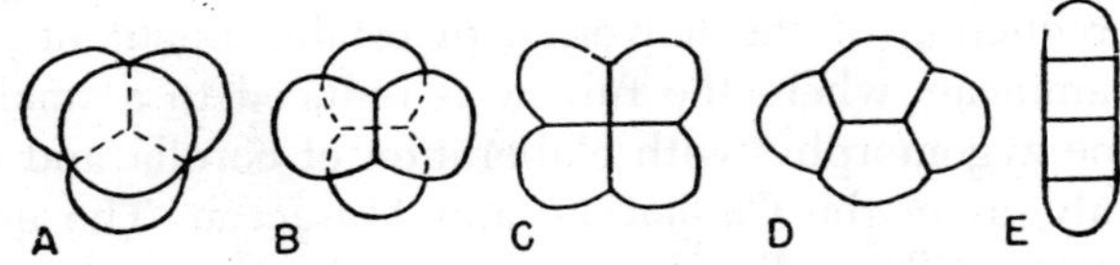

Fig. 64. Diagrams to show the types of arrangement of pollen grains in the tetrad. *A,* tetrahedral; *B,* cross; *C,* square; *D,* rhomboidal; *E,* linear. (*After Erdtman,* 1945.)

which represent positions of possible emergence of pollen tubes. They are usually located within the furrows but may be present in the thicker wall areas. Apertures in some taxa are probably phylogenetically shortened furrows. In walls of elaborate structure, the apertures may be covered with "caps" of thicker wall, which are readily lifted by pressure of the developing pollen tube—Cucurbitaceae. Where no apertures are present, the pollen tubes grow out from the furrows. Germinal furrows are absent in some families, and rounded, thin areas, pores, take their place functionally. These areas may be numerous—about thirty in the Polygonaceae. Specialization of these basic types within the angiosperms has apparently followed several lines, with greatly modified types resulting—both more elaborate and simpler in form. Similarity of type has evolved by convergent changes in lines not closely related.

Pollen-grain Types. There are apparently two major types of pollen grains: *monocolpate* (Fig. 63*B, G*) and *tricolpate* (Fig. 63*F, H*), with one and three germinal furrows, respectively. The monocolpate type is elongate or rounded, with the furrow on the side not in contact with the other grains in the tetrad. This seems to be the primitive form. It is characteristic of the monocotyledons and most of the Ranales, especially the woody families—Winteraceae, Degeneriaceae, Magnoliaceae,

Eupomatiaceae, Calycanthaceae, Lauraceae, and a few other families in other orders (Piperaceae, Saururaceae, Chloranthaceae). Pollen grains of the monocolpate type occur also in the Cycadales, Bennettitales, and pteridosperms. The tricolpate grain, with three meridional furrows, is apparently the basic type in the dicotyledons. It is not known in other seed plants.

Position, form, and number of furrows and other structural features of the exine are important diagnostic characters in the identification of pollen grains. Larger numbers of furrows and their position on the grain are related in part to the arrangement of the spores in the mother cell. The solitary furrow in a distal position—distal as related to position of the spore in the tetrad—seems to represent a primitive form, from which have been derived, along one line, the proximal monocolpate, dicolpate, polycolpate, and acolpate (without a furrow) types; along another line, primitive tricolpate has led to polycolpate and acolpate. In the monocolpate line, specialization tends to eliminate the furrow—both under reduction of the flower, with establishment of anemophily, as in the Gramineae, where the furrow is reduced to a small pore; and under extreme zygomorphy, with elaboration of corolla and androecium in entomophily, as in the Cannaceae and Musaceae. The end products of specialization in the pollen grain seem to be the acolpate and polycolpate types. These have probably been derived from both monocolpate and tricolpate types. Some doubt arises as to the homology of the monocolpate grain with distal furrow with the similar grain with proximal furrow—Annonaceae. Evidence from pollen morphology of apparent relationship among major taxa can be considered important only together with that from other fields.

The fundamental characters of the pollen grain are probably number and position of furrows, form and position of apertures, and pattern of sculpturing. Shape and size are probably of little or no basic importance, though very large and very small size characterize some taxa. Most large families show a considerable range in pollen characters, but pollen type in some is remarkably uniform, as in the Gramineae, where the grains are spherical or ellipsoid, with one furrow.

The pollen of anemophilous plants is usually small, rounded, smooth, rather thin-walled, "dry" (nonadhesive), and with shallow furrows or none at all; it often becomes angular when dry. The pollen of plants distributed by insects and birds is large, sculptured, and often coated with an adhesive waxy or oily substance. Beetle-pollinated plants have simple, thick-walled pollen—*Eupomatia*. Characteristic smooth, thin-walled pollen is present in the wind-pollinated *Populus*, Gramineae, Cyperaceae, *Platanus*, *Plantago*, Betulaceae, Fagaceae, Juglandaceae, Ambrosieae. The Compositae, as a whole, have perhaps the most elaborately sculptured pollen, but, within the family, they show a series in simplification toward loss of sculpturing in the anemophilous genera (Fig. 65). *Artemisia* has smooth pollen; *Xanthium* has lightly sculptured walls, with weak spines. This series is considered evidence that anemophily has been recently established in the Compositae. Scattered, "vestigial" patches of an adhesive layer on wind-borne pollen have been considered evidence of the derivation of anemophily from entomophily. The sculpturing of the exine is clearly correlated with method of pollination.

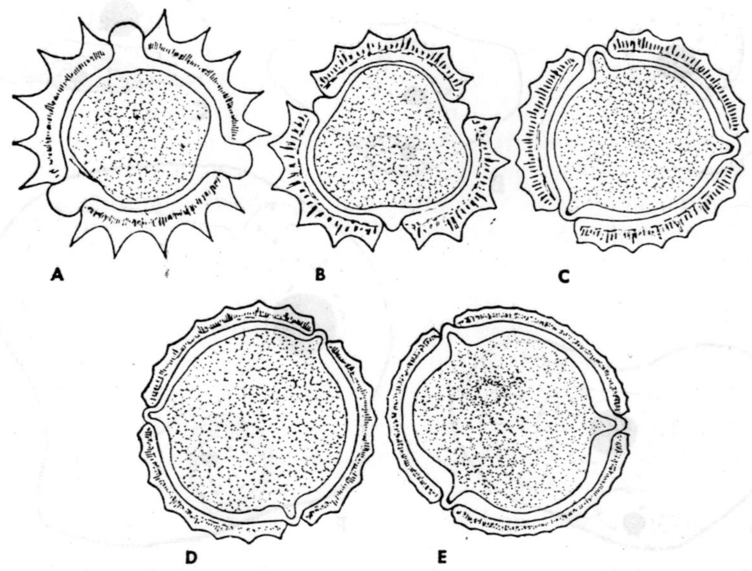

Fig. 65. Pollen grains of Ambrosieae. Diagrammatic equatorial sections showing progressive thinning of exine and reduction in size of spines in a phyletic series from insect to wind pollination. *A, Oxytenia acerosa; B, Chorisiva nevadensis; C, Cyclachaena xanthifolia; D, Ambrosia elatior; E, Xanthium speciosum.* (*After Wodehouse,* 1935.)

Pollen-grain Development. Three types of pollen-grain development have been distinguished: the *normal, Triglochin,* and *Juncus* types. In the normal type, characteristic of most angiosperms, the spore enlarges greatly in volume before dividing to form the generative and tube cells (Figs. 62*C*, *D* and 66*A*, *B*). In the *Triglochin* type (Fig. 66*A* to *D*), cell division occurs in the spore before there has been much increase in size; the major increase in size follows the first cell division. The *Juncus* type, restricted to the Juncaceae and Cyperaceae, seems insufficiently distinct from the other two. The *Triglochin* type is present in lower monocotyledons—*Najas, Triglochin, Ruppia, Lilaea, Aponogeton.*

Sequence within the Sporangium. Development of the pollen grains within a sporangium and within an anther sac is commonly simultaneous, or nearly so, but is successive from one part to the other in some taxa, especially the Ranales and Helobiales. The long anthers of *Liriodendron* show, from one end to the other, successive stages in pollen ontogeny.

Palynology. The cutinized wall of spores and pollen grains is one of the most resistant of organic structures; it persists as a fossil in peat and

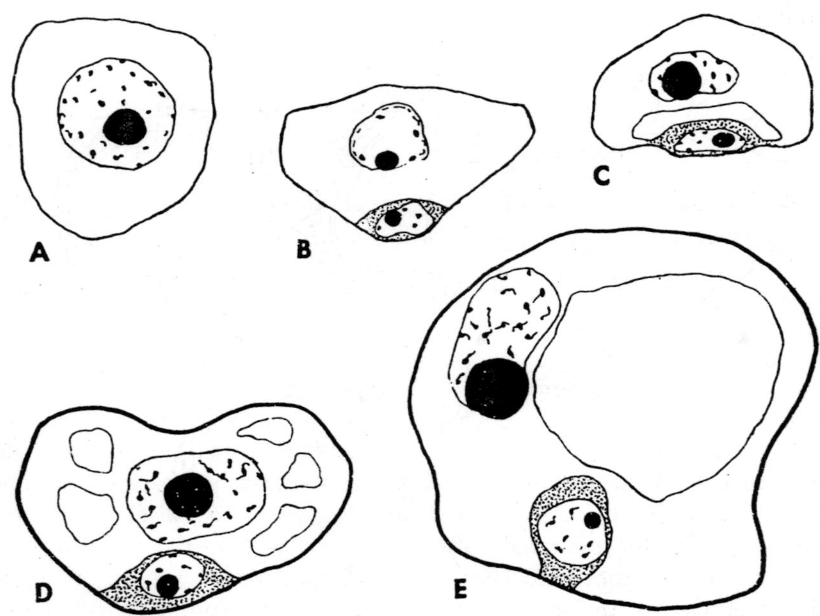

Fig. 66. *Ceratophyllum submersum* L. showing the *Triglochin* type of pollen-grain development. *A*, spore; *B*, *C*, generative cell formation and beginning of vacuolation before enlargement; *D*, *E*, increase in size and vacuolation. (*After Wulff*, 1939.)

in sedimentary rocks of many kinds and all ages—especially in coal and oil shales—when other plant material has been destroyed. Fossil spores and pollen grains are therefore of great importance in the study of the phylogeny of plants. Palynology, the study of spores and pollen grains, has recently been established as a morphological, taxonomic, and phylogenetic section of botanical science and has already contributed information of great value in the determination of relationships among angiosperms. For example, the pollen grains of the Casuarinaceae and of the Amentiferae in general are not of primitive type and support the present generally accepted view that these taxa are advanced, not primitive. Palynology is also of much value in studies of historical distribution and ecology.

The pollen of the monocotyledons appears to be, in general, more primitive than that of the dicotyledons; the monocolpate type is dominant, and the tricolpate type not present. It has been suggested that the prominence of the monocolpate type is perhaps connected with phylogenetic homogeneity, hygrophily, and geophily.

The characteristic monocolpate pollen of the monocotyledons has been called "evidence of their antiquity," because a derivation from the tricolpate Ranunculaceae—as often suggested for the monocotyledons—cannot be considered. The primitive monocolpate pollen of the monocotyledons is evidence at least of their very early origin, whatever their ancestral stock may have been.

DEHISCENCE OF THE ANTHER

Pollen is commonly shed from the anther through longitudinal, slitlike openings in the anther or lamina wall—*longitudinal dehiscence.* Transverse slits occur occasionally—*Alchemilla, Hibiscus, Euphorbia, Chrysosplenium.* Where the sporangia are united in pairs, the longitudinal opening follows the furrow between the sporangia of a pair along the line of sporangial fusion. Dehiscence by small, rounded openings, "pores"—*poricidal* or *porose dehiscence*—is characteristic of a few families—Ericaceae, Epacridaceae, Tremandraceae, Melastomaceae (most genera), Myrsinaceae, some Leguminosae, Ochnaceae, Solanaceae. Pores are usually located at the distal ends of the sacs, rarely at the proximal ends. In the anthers of some Ericaceae, the distal ends are, morphologically, the proximal ends; the anther is inverted in ontogeny.

Arrangement of the sporangia in pairs brings about variety in longitudinal dehiscence. In laminar stamens, the sunken sporangia often dehisce independently; the sporangia have individual endothecial caps. In anthers where the sporangia lie close together in lateral pairs, dehiscence is usually by a common slit in the furrow between the two sporangia. As the pollen matures, the anther-sac wall separating the two sporangia breaks down, and the pollen grains mingle as they escape into the furrow (Fig. 67*F*). Where the four sporangia are isolated in the corners of an angular anther, dehiscence is from each sporangium separately and in a more or less lateral direction.

Dehiscence results generally from hygroscopic shrinkage of the fibrous layer. It may be initiated in regions of existing structural weakness, such as absence of epidermis along the line of dehiscence. Where dehiscence is longitudinal, continued drying brings about retraction of the borders of the openings, and the chambers are opened widely. Changes in atmospheric humidity may cause repeated opening and closing of the anther sacs. The pollen may all be freed at once or may gradually escape. In fusion chambers, the pollen grains probably do not intermingle, except as they escape.

Dehiscence is not always due to changes in the fibrous layer, for this layer is not always present, especially in poricidal anthers. Details of dehiscence in anthers of this type vary considerably, but the pores usually form by disintegration of the anther wall in a small area, with shrinkage of the surrounding tissues. Rarely, a slit occurs in the epi-

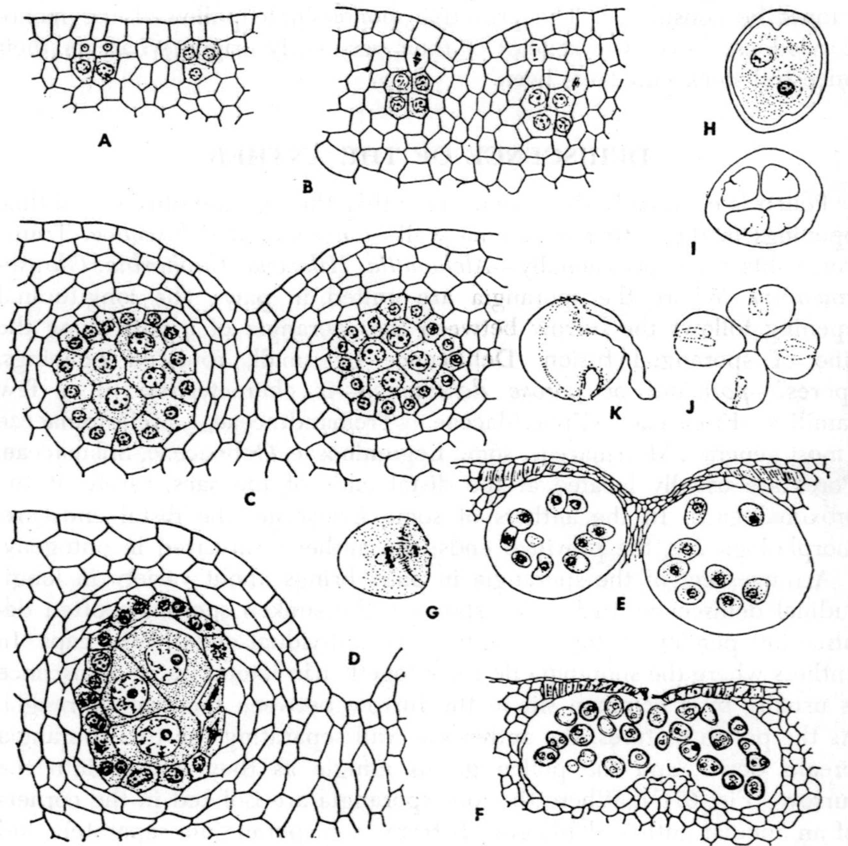

Fig. 67. *Degeneria vitiensis* showing development of microsporangia, spores, and pollen grains. *A, B,* groups of archesporial cells in cross section of the lamina; *C, D,* organization of the tapetum; *E,* sporangia with microspores and the beginning of endothecial caps, tapetum degenerated; *F,* pair of sporangia united by breakdown of intermediate tissue, two-celled pollen grains, and endothecium; *G,* microspore with germination furrow, dividing; *H,* pollen grain at shedding stage, generative cell dark, tube nucleus lighter; *I,* young tetrad in mother-cell wall, showing initiation of furrow on distal face of grains; *J,* pollen grains of a tetrad, just separated, showing furrows; *K,* germinating pollen grain with tube projecting from the furrow. (*After Swamy.*)

dermis where the pore will develop, and the opening is enlarged by shrinkage of the surrounding tissues—some of the Tremandraceae. Where the anther sacs have elongate tips, the sporogenous tissue in these tips may become sterile, forming delicate parenchyma cells, which disintegrate when the pore opens and leave hollow tubes extending to the pore. Cellular breakdown may involve the connective tissue between the tips of the anther sacs, and the two sacs—four sporangia—open through one pore. Poricidal dehiscence has apparently been derived independently in different taxa from longitudinal dehiscence by shortening of the slit. It reaches high specialization where the pollen is freed from four sporangia through a single pore, as in *Cassia.* Forms transitional to longitudinal have some fibrous tissue, restricted usually to areas around the pore—Solanaceae. Association of poricidal dehiscence and a fibrous layer throughout the length of the sac is rare.

Complex form in the anther is often associated with dehiscence. Anthers with poricidal dehiscence may have long, tubular projections ("awns") at the ends of the sacs, through which the pollen is gradually sifted out. In anthers with slitlike openings, the slit may be so shaped as to free a valvelike flap of tissue; this is *valvular dehiscence*—Lauraceae, Berberidaceae. In *Hamamelis,* the valve flap is folded outward and backward and carries the pollen mass out of the chamber as it lifts. In some aquatic genera with submerged flowers, the pollen is freed by disintegration of the anther sac.

Although the general structure of the anther wall has already been discussed, the structure and distribution of the outer layers in their relation to dehiscence need further consideration. The epidermis is characteristically simple but, at anther maturity, may show unusual structure. Its cells may be tabular and greatly flattened, appearing collapsed, as in *Canna, Balsamea,* and many Compositae. It may be lost during ontogeny, either wholly or in part—*Vitis, Grevillea, Asarum,* many Compositae. Along the line of dehiscence, it may be absent—*Aristolochia, Nepenthes.* A narrow band of cells may enlarge greatly when dehiscence approaches—*Crocus, Iris, Bignonia*—and form a specialized opening structure, a *stomium.* The well-developed stomium of *Lilium* is frequently figured. A crest of thickened epidermal cells may develop along both sides of the line of dehiscence—*Passiflora, Lycopersicum.* Epidermal hairs in a fringe along the sides of the slit perhaps aid in the freeing of the pollen grains by hygroscopic movements. Thickening bars like those of the fibrous layer have been reported in the epidermis of *Clandestina.*

The fibrous layer is obviously structurally adapted, by the uneven thickenings of its cell walls (Fig. 68), as a mechanism at least largely responsible for the opening of the anther sac. But there are families and

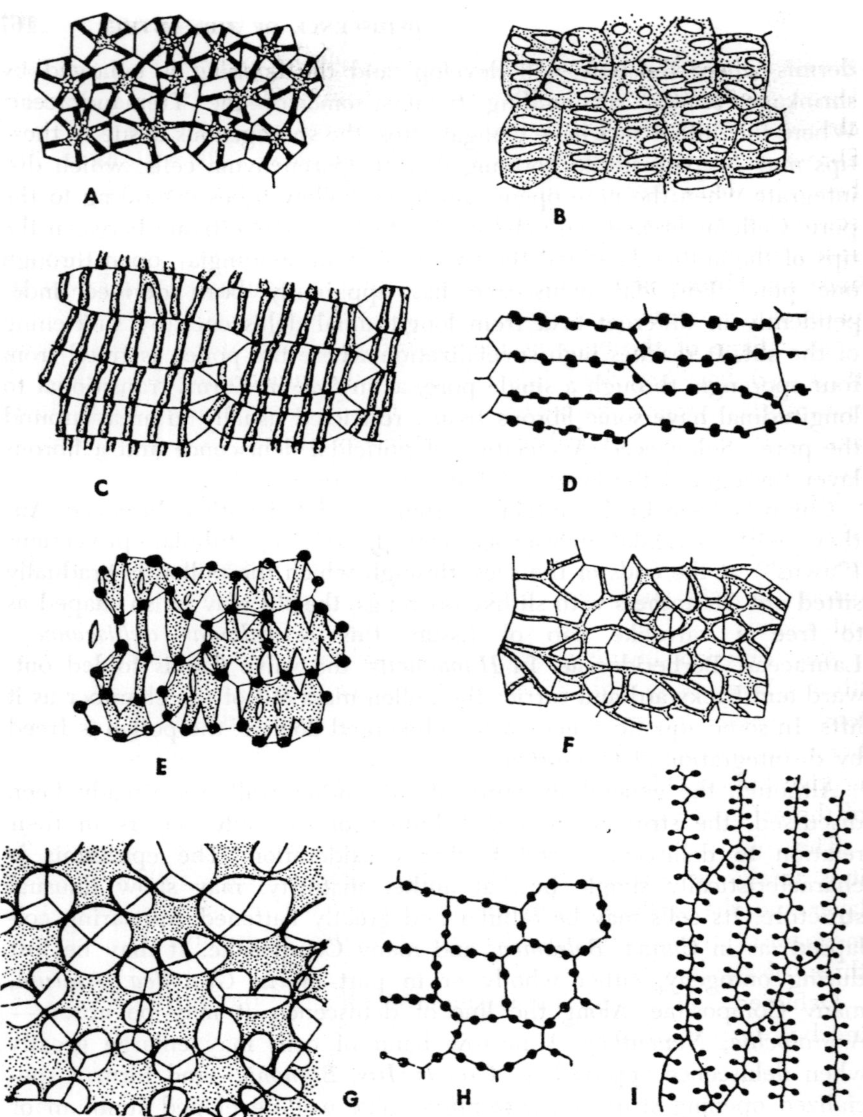

Fig. 68. Fibrous layer of anther-sac wall associated with dehiscence (cells elongated transverse to the anther axis) in face view. *A, Malva sylvestris,* inner tangential wall, bars stellate, no bars on outer wall; *B, Borago officinalis,* inner tangential wall, reticulate, lignified; *C, Lychnis dioica,* inner tangential wall, bars U-shaped, curved on inner wall and arms extending over radial walls; *D, L. dioica,* outer tangential wall without bars, showing bars on radial walls; *E, Delphinium orientale,* inner tangential wall, bars fused scalariformly; *F, Erythraea Centaurium,* inner face of fibrous layer, bars forming a mesh; *G, H, Aquilegia vulgaris,* outer tangential wall, no bars, but showing sections of bars on radial wall; *I, Alopecurus agrestis,* radial wall of epidermis. (*After Leclerc du Sablon.*)

genera in which a typical fibrous layer is absent, and the sacs open by longitudinal breaks—Orchidaceae, Araceae, Asclepiadaceae, Orobanchaceae, *Diospyros.* The anther wall is described as consisting of the fibrous layer only in *Althaea, Vitis, Phytelephas.* (The epidermis has been reported ephemeral in *Vitis.*) The Orchidaceae and Orobanchaceae, considered to lack a typical fibrous layer, have some wall cells with sparse or incomplete bars; these cells probably represent vestiges of an ancestral fibrous layer. Nonfunctioning fibrous cells are also present in parts of the anther sac that do not open—*Berberis.*

The shape of the cells of the fibrous layer and the position of the thickened bars are doubtless related to the functioning of the layer, though there seems to be no consistency in these features. The long axis of the cells is described as parallel to the slit in many genera—*Silene, Erodium, Plantago, Commelina;* and as at right angles to the slit in many other genera—*Pyrus, Tiarella, Lychnis, Geranium.* Shape of the cells appears unimportant, since it differs in related genera—*Silene* and *Lychnis; Erodium* and *Geranium.* The thickening bars usually extend at right angles to the slit; only rarely are they parallel to the slit—*Salvia.*

Any function in dehiscence of the "third layer," that below the fibrous layer, is probably indirect or accessory only. The small, rather loosely packed parenchyma cells of this layer commonly disintegrate just before dehiscence, freeing the outer layers from attachment to the connective below; this disintegration may aid indirectly in the contraction of the external tissues.

The role of the fibrous layer has perhaps been exaggerated, though it is surely the important layer in this function. Apparently both this layer and the epidermis can function alone, but, in most families, the mechanism responsible for dehiscence seems to involve two, or all three, of the layers.

Dehiscence of the anther is similar to that of the carpel in that it involves a sporophyll wall.

POLLINATION

It was long assumed that pollination in the angiosperms was primitively by wind, *anemophily;* that pollination by water, *hydrophily,* and by animals, *entomophily* and *ornithophily,* were derived methods. This conclusion was based largely on the presence of anemophily in the conifers and in simple, and therefore supposedly primitive, flowers, especially those of the Amentiferae. The opinion that the primitive method of pollination was by insects—though argued at the beginning of the twentieth century—received strong support only twenty to forty years later. The views now generally accepted are that the conifers are not on the direct ancestral line of the angiosperms; that the anemophily of the Amentiferae is derived, accompanying high specialization in inflorescence and flower structure; that the primitive flower was bisexual, not unisexual.

Without question, anemophily is advanced in some taxa: in *Fraxinus*, a member of a highly specialized entomophilous family with petaloid flowers; in the Platanaceae, a family of the petaloid Rosales, with greatly reduced flowers and inflorescences; in the grasses and sedges, likewise with greatly specialized and reduced flowers and inflorescences. The recognition of the woody Ranales as the most primitive living angiosperms further emphasizes entomophily as primitive. Pollination in *Eupomatia*—in flower structure, one of the most primitive angiosperms—is by beetles, the ancient and primitive group of insects. The flower structure of this genus is such that pollination by wind is impossible; pollen can reach the carpels only by beetles eating through an enclosing sheath of staminodia (Fig. 142). Pollination is also only, or largely, by beetles in *Calycanthus, Magnolia, Illicium, Paeonia,* and in the herbaceous Nymphaeaceae, especially *Victoria,* now generally recognized as primitive taxa. Little is known of methods of pollination in other genera of the Magnoliaceae and in the primitive families, Winteraceae, Degeneriaceae, Himantandraceae, Trochodendraceae, and Tetracentraceae.

There are strong similarities in features of floral structure related to pollination in *Eupomatia* and *Calycanthus.* In both genera, the stigmas are enclosed in a chamber readily accessible only to chewing insects; food, also available only to chewing insects, is abundant. This food consists of "food bodies," richly protoplasmic, succulent tissue. These food bodies are pads on the surface of the inner staminodia in *Eupomatia laurina* (Fig. 69*F*); apical clusters of cells on inner petals, stamens, and staminodia in *Calycanthus* (Fig. 69*A* to *D*) (a small amount of nectar has been reported in one species) and *E. bennettii;* in *Illicium,* they are "succulent knobs in the center of the flower." There are no nectaries as such in these taxa. In *Eupomatia laurina,* the beetles, trapped inside the floral chamber, rarely escape until the outer floral organs are shed; evidence of their long feeding and imprisonment is the presence of debris and excrement. Pollination by beetles is crude but effective. In higher families, it is replaced by true nectaries, surface areas or organs that supply fluid food over a considerable period.

Beetles are also associated in varying percentages with other pollinizing insects, especially in families with small flowers in dense inflorescences, such as the Cornaceae, Caprifoliaceae, Fagaceae (*Castanea*), Saxifragaceae (Hydrangeae).

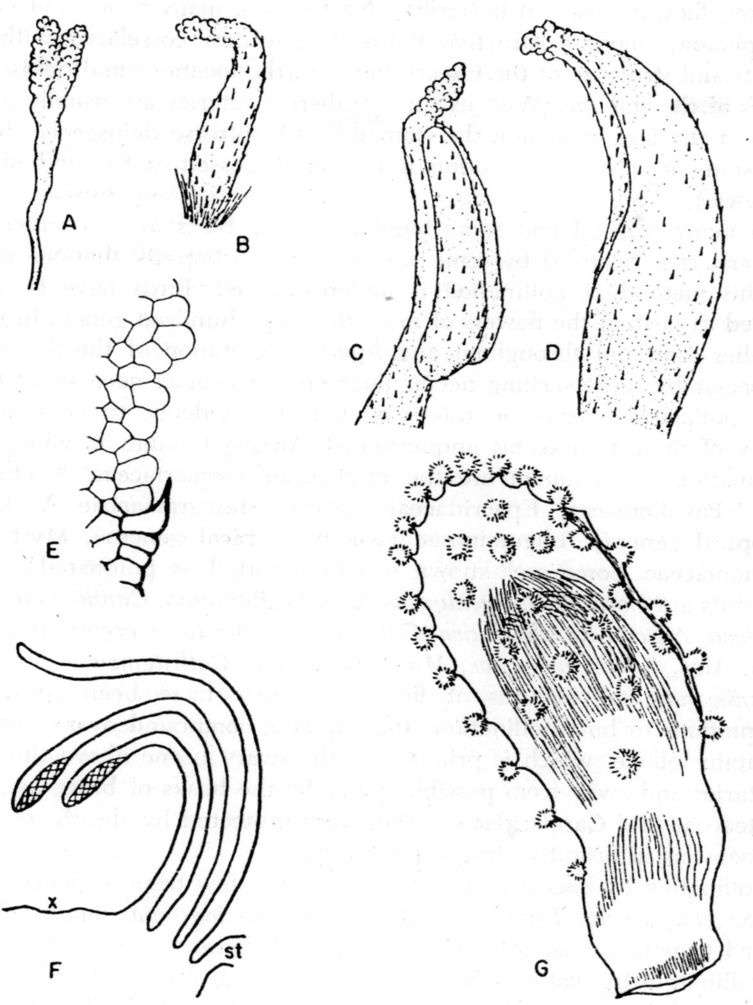

Fig. 69. Food bodies of beetle-pollinated angiosperms. *A, B, C, D,* terminal food bodies on inner floral appendages of *Calycanthus occidentalis: A,* on stigma; *B,* on staminode; *C,* on stamen; *D,* on inner petal. *E,* part of longitudinal section of food body on inner staminodium of *Calycanthus florida* showing thin-walled tissue of food body and, below, cells of connective; *F,* longitudinal section of staminodes of *Eupomatia laurina,* enclosing pollination chamber *x,* the inner two with ventral food bodies; *G,* inner staminode of *Eupomatia bennettii* showing globose food bodies. (*A to D, after Grant; E, after Dauman; G, after Schnizlein.*)

The majority of angiosperms are pollinized by other groups of insects —bees, flies, moths, and butterflies. Nectaries of many forms and varied morphology secrete attractive fluids in positions correlated with the habits and structure of the insects and with the locations and dehiscence types of the stamens. With extrorse anthers, nectaries are usually at the base of the flower outside the stamens; with introrse dehiscence, inside the stamens. (Nectary position is further discussed at the end of this chapter.)

In many tropical and some temperate taxa, birds are the chief pollinizers. It is believed by some students of ornithophily that the extent of this method of pollination is underestimated. Birds have been reported as visiting the flowers of more than four hundred genera in many families scattered throughout angiosperms. Visitation of the flowers of a species by birds seeking nectar does not, of course, necessarily mean that pollination occurs in this way, but the evidence to be found in many of these taxa seems unquestioned. Among families in which bird pollination is prominent are the Proteaceae, Gesneriaceae, Nyctaginaceae, Passifloraceae, Epacridaceae (part), Marcgraviaceae, Violaceae (tropical genera), Leguminosae (woody, tropical genera), Myrtaceae, Bignoniaceae. Some well-known genera reported as pollinated in part by birds are *Loranthus, Eucalyptus, Vanilla, Bauhinia, Canna, Gladiolus, Fuchsia, Acacia, Xanthorrhoea, Cattleya, Pritchardia, Cereus, Anigozanthus, Aloë, Lonicera, Agave, Musa, Jacaranda, Callistemon.*

Some unusual features of flower structure have been considered adaptations to bird pollination: the stipitate, compound ovary (not the stipitate follicle, which is primitive)—the ovary borne above the basal nectaries and away from possible injury by the beaks of birds, as in the Proteaceae and Capparidaceae; the ovary protected by sheaths of fused stamens, as in primitive tropical Malvales.

Pollination by bats is believed to occur in a few tropical genera, such as *Kigelia, Durio, Freycinetia, Erythrina, Barringtonia.* Snails are believed to pollinate some aroids and aquatic plants.

Pollination by water is obviously an acquired method. It may be by contact of flowers on the surface of the water where the staminate flowers break free and float to the pistillate, as in *Elodea* and *Vallisneria;* or by pollen floating submersed, as in *Zostera, Halophila, Cymodocea, Najas,* and some species of *Potamogeton.* (Other aquatic plants have aerial inflorescences, with pollination by wind or insects—many species of *Potamogeton.*) Perhaps the most remarkable method of pollination is that of *Ceratophyllum,* where distribution of pollen is by both air and water. The stamens, freed from submersed flowers, rise to the surface of the water and there discharge the pollen, which sinks slowly through the water to the pistillate flowers.

Anemophily characterizes the Betulaceae, Fagaceae (part), Juglandaceae, Urticaceae, Casuarinaceae, Gramineae, Cyperaceae, Ulmaceae, Restionaceae, Plantaginaceae, Saururaceae, Juncaceae (part), Amaranthaceae (part), isolated genera in other families—*Artemisia, Ambrosia,* and *Xanthium* in the Compositae, *Fraxinus* in the Oleaceae, *Thalictrum* and *Cimicifuga* in the Ranunculaceae, *Sanguisorba* and *Acaena* in the Rosaceae, *Mercurialis* in the Euphorbiaceae, *Rumex* in the Polygonaceae. From this list, it is clear that anemophily is distributed throughout the angiosperm system and in taxa that are highly specialized, not primitive. In these taxa, the flowers are mostly numerous, small, inconspicuous, and odorless, with perianth absent or greatly reduced, not brightly colored. Other features of flower and inflorescence structure that occur frequently in anemophilous taxa are flexuous staminate inflorescences—many Amentiferae; greatly elongate and flexuous pedicels of staminate flowers—species of *Acer;* elongate filaments with versatile anthers—grasses and sedges, many Restionaceae, *Thalictrum, Plantago.* The genus *Acer* shows, in its staminate flowers, a series in loss of perianth, nectaries, and odor, and in elongation of pedicles accompanying the change from entomophily to anemophily. Species like *A. rubrum* and *A. saccharinum* with subsessile flowers, corolla, and odor are entomophilous; *A. saccharum* and *A. Negundo* with long pedicels and naked, odorless flowers are anemophilous; other species show transitional stages. Explosive dehiscence, especially the simultaneous dehiscence in many flowers, as in some Urticaceae, is accessory to wind pollination. Some of the Amentiferae, a chiefly anemophilous group, are apparently in transition from entomophily to anemophily. In the Salicaceae, *Salix* has showy stamens, well-marked vestigial perianth parts, serving as nectar-secreting glands, and pollination is partly by wind and partly by bees; *Populus* is strictly wind-pollinated, without nectar secretion, and has inconspicuous stamens. In the dominantly wind-pollinated Fagaceae, 60 per cent of the insect visitors of *Castanea* (*C. sativa*) are beetles.

In entomophilous plants, transfer of pollen by beetles is apparently primitive; that by bees, butterflies, moths, and flies, is later acquired. Insect pollination may be secondarily acquired from wind pollination, as is apparently the condition in *Ficus, Euphorbia, Hevea, Ricinus,* the Nyctaginaceae, and in some of the Araceae. It is phylogenetically significant that beetles, primitive insects, are associated with the primitive Eupomatiaceae and Calycanthaceae to the exclusion of other insects, and with the also primitive *Magnolia, Illicium, Paeonia,* and the Nymphaeaceae, where they seem to play the major part in pollination. The association strongly supports the now generally accepted view that the woody Ranales are the most primitive living angiosperms. It is evidence that the angiosperms were well established and probably diverse before the higher insects arose. Angiosperms, with well-defined nectaries and stigmas, and higher groups of insects, with hairy bodies, perhaps developed together in Jurassic and Cretaceous times. The closure of the stigma canal and of the carpel itself have also been considered an accompaniment of pollination by the higher insects.

Entomophilous flowers that are visited by bees, moths, and butterflies, have conspicuous perianth, fragrance, and nectar glands. Flies visit chiefly flowers with disagreeable odor and reduced or dully-colored perianth. Flowers pollinated by birds are usually large or closely grouped in large, conspicuous inflorescences, and secrete quantities of nectar. The large, bird-pollinated inflorescences of the Australian waratah, *Telopea speciosissima,* and some species of *Grevillea* drip with glistening nectar, even under dry atmospheric conditions. The highly colored organs in *Telopea* are the bracts and calyces; in *Grevillea,* the calyces. The nectar is secreted by glands that represent vestigial petals in both of these genera.

The pollen grains of wind-pollinated plants are small, light, smooth, and usually nonadhesive, not cohering in clusters; those of entomophilous and ornithophilous plants are strongly adhesive, borne in pollinia, tetrads, or free but cohering in masses. The pollen grains of genera pollinated under water are elongate—filamentous, or confervoid and not adhesive. Vestiges of adhesive material have been found on the wind-borne pollen of taxa with close entomophilous relatives: *Acer* (anemophilous species), *Rumex, Ambrosia, Sanguisorba.* Pollen type, method of pollen distribution, and structure of the gynoecium are, biologically, closely correlated.

THE MALE GAMETOPHYTE

The newly formed microspore has a large, centrally placed nucleus, dense cytoplasm, and a delicate wall. Before germination, it increases greatly in size and volume (Figs. 70*B* and 71*F*). A central vacuole develops, and the nucleus takes a peripheral position. Usually germination occurs only some days or weeks after the spore is mature, though, in tropical plants, it may be immediate. In temperate-climate plants that flower in the spring, the sporangenous tissue may overwinter at the spore-mother-cell stage, the mature microspore stage, or at the two-cell stage of the gametophyte; perhaps more taxa overwinter at the spore-mother-cell stage than at later stages. Germination of the spore begins with cell division and the formation of a small, walled cell, which lies against the spore wall (Figs. 70*D* and 71*G*). This first cell of the gametophyte is rarely formed centrally in the spore by "free cell formation," development without typical wall formation; the phragmoplast forms a delicate cytoplasmic "wall," which delimits the protoplast—

Scirpus (Fig. 60). The formation of this first cell as a wall-less, free protoplast probably marks a forward step in the reduction of the gametophyte. Though definite positions in the spore have been claimed for the first walled cell—and there may be a constant position in a given species—several or many different positions have been reported for angiosperms as a whole. The first-formed cell, walled and peripheral,

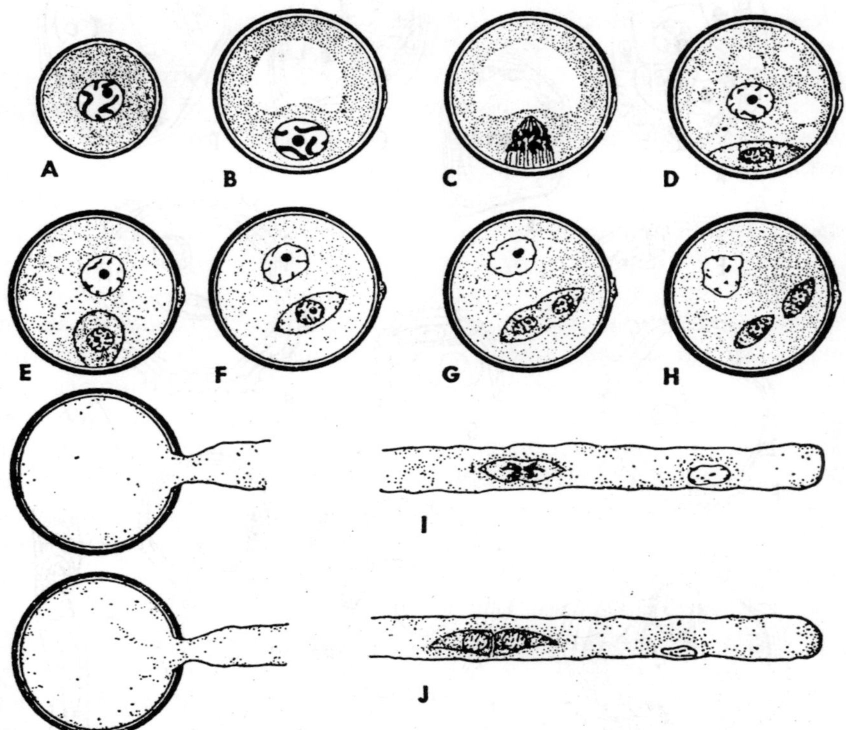

Fig. 70. Diagram showing development of male gametophyte in angiosperms. *A,* microspore; *B,* microspore, enlarged, vacuolate; *C,* spore nucleus dividing; *D,* gametophyte, two-celled, smaller generative cell, larger vegetative (tube) cell; *E,* freeing of generative cell from spore wall; *F,* generative cell embedded in vegetative cell; *G, H,* division of generative cell to form two male gametes within the spore wall; *I, J,* division of generative cell within the pollen tube. (*After Maheshwari.*)

is the *generative cell* (Figs. 70*D* and 71*G*); the larger, naked cell, central and wall-less, filling the remainder of the spore-wall cavity, is the *vegetative* or *tube cell.* The nuclei of these two cells differ not only in size but in structure and staining qualities. The vegetative nucleus has a prominent nucleolus; the generative nucleus, a small nucleolus or none. The generative cell is soon freed from the spore wall and becomes ellipsoid or fusiform. Prominent cytoplasmic differences set it sharply off from the cytoplasm of the vegetative cell in which it is embedded.

The formation of the generative cell takes place during late stages in anther development. In a sporangium, the divisions in the development

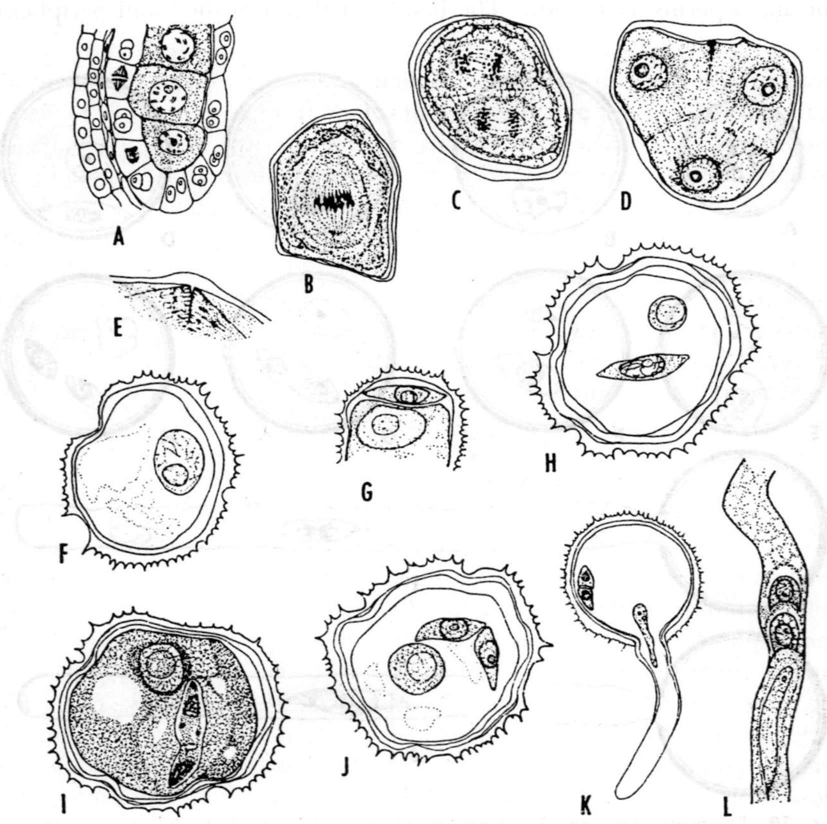

Fig. 71. Microsporogenesis and development of the male gametophyte in *Portulaca oleracea*. *A*, row of spore mother cells in anther sac; *B, C*, meiotic divisions in spore mother cell; *D*, tetrad of spores nearly mature; *E*, detail of spore-wall formation in tetrad; *F*, mature spore, enlarged before germination; *G, H*, generative and tube cells within spore wall; *I, J*, mature pollen grain with generative cell and two sperms; *K*, germinating pollen grain; *L*, part of pollen tube showing tube nucleus and sperms. (*After D. C. Cooper.*)

of the gametophyte are not as nearly simultaneous as are the meiotic divisions, but are not far apart, except in long anthers, such as those of *Liriodendron*, where there is acropetal succession in germination of the spores. In tetrads that remain permanently together, and in pollinia, the divisions may be simultaneous.

The Pollen Tube. Germination of the pollen grain on the stigma occurs after a period of time that varies greatly: from almost immediately—*Saccharum, Sorghum;* five to ten minutes—*Taraxacum, Zea;* two hours; three hours; two days. The grain swells, probably by absorption of the secretions of the stigmatic surfaces; the shape may be greatly modified; furrows are filled out and flat surfaces rounded. The inner wall protrudes at one or more of the thin areas in the outer wall—in the furrows or apertures. In triangular pollen grains, the apertures are at the corners of the grain, and the tube develops from one of these corners—Proteaceae. By rapid apical growth, one of the protrusions becomes the pollen tube. Rate of elongation of the tube varies greatly—with the taxon, and with conditions of temperature and humidity.

In early stages of tube development, the contents of the pollen grain, or most of them, pass into the tube, where further development of the cells may occur. The tube and its contents—together with any part of the contents that may remain within the pollen-grain wall—constitute the male gametophyte. The "pollen tube"—the tube plus its contents—is commonly called the male gametophyte; the contents alone are sometimes described as the gametophyte. Objection has been raised to the inclusion of the tube itself as a part of the gametophyte, but the tube is a development of the microspore wall, and the spore is the first cell of the gametophyte. Both the pollen tube and the embryo-sac wall are enlarged and modified spore walls (walls of the spore mother cell or dyad in tetrasporic and disporic sacs). The contents of the mature pollen grain and of the young pollen tube, commonly called the male gametophyte, represent the immature gametophyte. (The generative nucleus may not yet have formed the male nuclei; the cytoplasm of the tube continues to increase in volume as the tube develops.) The male gametophyte of the angiosperms remains enclosed within the greatly modified spore wall until the male cells are formed. In this enclosure of the male gametophyte within the spore wall, the angiosperms resemble other major heterosporous taxa.

The pollen tube itself, as a male-cell carrier, is characteristic of the highest gymnosperms and angiosperms only; by its aid, the male gametes reach the egg. Accompanying this high specialization of the spore wall has been loss of motility in the gametes. Only in the highest seed plants is fertilization accomplished by nonswimming male cells. The pollen tube represents one of the most difficult adaptations to life on land, the step from fertilization in a fluid by free-swimming sperms—the ancient method—to fertilization by nonswimming sperm cells, carried to the egg by the gametophyte. No animals, and only the highest plants, have made this advance.

The pollen tube is probably an example of parallel development. The coniferophyte gymnosperms—Coniferales and Ephedrales—are so unlike angiosperms as to be unquestionably a stock unrelated in origin to the angiosperms, and their pollen tubes are probably of independent origin. The cycads and *Ginkgo* have tubes functionally different from those of conifers and angiosperms. In these taxa, the tubes, short and haustorial in nature, develop on one side of the pollen grain. The pollen grains become enlarged with increased cytoplasm and break open, discharging the sperms and cytoplasm into a pollen chamber, where the sperms swim in the cytoplasmic fluid to the archegonial necks. In the angiosperms, nonswimming sperms are discharged into the embryo sac.

Germination of pollen grains at a distance from the ovule has been considered an angiosperm character, but, in *Tsuga* and *Araucaria,* the grains germinate far from the ovule and form long tubes extending to the micropyle; the pollen grains of *Agathis* form extensive pollen tubes in the cone axis and ovular tissues before reaching the archegonium. Stages in the phylogenetic development of germination at a distance from the ovules are present in living Ranales. In *Degeneria* and some species of *Drimys,* the pollen grains germinate on a long, ventral stigmatic crest, and their short tubes penetrate between the unsealed "lips" of the carpel directly to ovules close by (Fig. 83). In other species of *Drimys* and other genera of the Winteraceae, progressive acropetal shortening of the crest—the pollen tubes still short—restricts ovule position to the area directly below the crest. With the acquisition of a terminal stigma and a style, the pollen tubes become long, and ovule position is not restricted.

The pollen tube, as it increases in length, usually pushes between the cells of the stigmatic surface and the inner tissues of the stigma and style. In some families, it destroys the adjacent cells, eroding a path to the ovules. Its course may be along the surface of the stylar canal or deep within the stylar tissues. It may pass through vascular tissues (protoxylem) in its course—*Casuarina.* (It resembles in this the course of pollen tubes in the phloem of araucarian conifers.) Where a stylar canal is present, the tube may pass wholly or partly along the wall of the canal. Entering the ovarian chamber, the tube may continue along the wall to the ovule or cross directly to the micropyle. It follows lines of *transmitting tissue* (often wrongly called *conducting tissue*), the extent and distribution of which within the stylar canal and the ovary have not commonly been recognized. The course of the tube, though obviously toward the ovules, has been considered not directed structurally. But, in many taxa, the tube follows direct lines through the ovary, predetermined by the distribution of transmitting tissue (Fig. 77), which is by no means restricted to the style, as often implied. In the more specialized taxa, the tube seems to take the shortest course along the surface of the ovary wall and placenta to the ovule. To reach the micropyle, the tube may pass along the surface of the ovule to the micropyle, but it may enter the ovule and follow the vascular bundle, running through protoxylem cells in funicle and raphe—*Casuarina*. The crossing of a space, as a "short cut" to the micropyle—described in some taxa—has been called an "extraordinary accomplishment," because it involves the continuing growth of the tube tip, without contact of the tip with tissue of the ovary. Where no ovule, as such, is present —some parasites, such as *Viscum*—the tubes pass through carpellary tissues. Branching of the tube is rare except in chalazogamous taxa, where it may take place at various levels in the ovule.

Time of growth of tube from stigma to embryo sac varies greatly and is not controlled by distance from stigma to embryo sac; the tip may grow rapidly or slowly, and many factors enter into its elongation. The tube may grow from pollen grain to the egg over some millimeters in a few hours. But many days, even weeks and months, may pass between pollination and fertilization—in the witch hazel, about six months; in the black oaks, about twelve months; but these periods include winter dormancy.

Entrance of the pollen-tube tip directly through tissues of the ovule, rather than by the micropyle, was first found in *Casuarina* in 1891 and termed *chalazogamy* (Fig. 103), because the tube penetrates the chalaza on the way to the embryo sac. Chalazogamy was soon found in several other amentiferous genera, and, since the Amentiferae were then considered the primitive angiosperms, chalazogamy was considered a primitive character. (A classification of angiosperms as Chalazogams and Porogams was supported by a few botanists.) But chalazogamy is now known to be present in taxa in other parts of the angiosperm system—Anacardiaceae, Cucurbitaceae, Rosaceae, Chloranthaceae—and must be considered a specialized, rather than a primitive, character. Penetration of the tube through the integuments is sometimes termed *mesogamy*, and the term *basigamy* was suggested to replace chalazogamy. "True chalazogamy" is said to occur only when the tube enters the embryo sac at its base. Entrance of the tube by the micropyle has been called *porogamy;* all other methods, *aporogamy*. In a few taxa, there is variation in place of entry of the tube—*Epilobium, Brassica*. Entrance of the tube into the embryo sac at any point other than the micropylar end is probably rare; tubes entering the ovule below the micropyle usually pass along the embryo-sac wall until they reach the position of the egg. Pollen tubes may persist in the nucellus long after fertilization, a condition frequent in the Proteaceae.

The parasitic behavior of the pollen tube is reported to reach an extreme in *Casuarina*, where it is described as destroying germinating megaspores; in this, it resembles the tubes of the araucarian conifers, which destroy the contents of immature archegonia in their course.

The Tube Nucleus. The nucleus of the vegetative cell, the tube nucleus, commonly shows signs of degeneration as the generative cell matures, and may change in shape. It has been generally stated that, as the development of the pollen tube begins by extrusion of the intine of the pollen-grain wall, the tube nucleus passes into the tube and, as the tube grows, maintains a position in the distal part, where it "governs apical growth." But the supposed function of this nucleus has been questioned. The nucleus may remain within the spore wall; it is often degenerate before the beginning of tube growth; it may not enter the tube until the tube is well developed; it often follows, rather than precedes, the generative cell or the gametes (the relative position of the two or three nuclei is perhaps a matter of chance); it may not progress far, and the male cells may pass it. The tube may branch, even repeatedly—*Fagus* and other Amentiferae—and the branches which have no nuclei develop as well as that with the nucleus. The vegetative cell has also been called a vestigial cell, the last survivor of the vegetative tissue of an ancestral, free-living gametophyte. As such, it would be homologous with the one or two prothallial cells of the cycads, *Ginkgo*, and the conifers.

Male Gametes. Division of the generative cell, forming two male gametes or sperms, may follow soon after it is mature or may be delayed until the cell has passed into the pollen tube. Within the tube, division may occur close to the base or far along, even well down in the style—*Lilium*. In some species, it apparently may occur in either place. Division of the generative cell in the ungerminated pollen grain as it lies on the stigma also occurs. The pollen grain is either 2- or 3-nucleate —3-nucleate if the male gametes have been formed before shedding. It is uncertain whether the two-celled or the three-celled pollen grain is more common. The male gametes are definite cells, with limiting cytoplasmic sheaths, not naked nuclei, as often formerly believed. In shape, they are spherical, ellipsoid, lenticular, or vermiform. Their shape may change as they pass through the pollen tube. The two gametes are usually alike in size. Differences in size and staining quality—the gamete that unites with the egg, often smaller and staining less deeply—have been reported in a few species.

The question of possible motility of the male cells has received considerable attention. In the lower gymnosperms, the sperms are ciliate and actively motile; in the higher gymnosperms, the pollen tube is a "male-cell carrier," as in the angiosperms, and independent movement seems to have been lost. It has been claimed that the sperms of angiosperms have some capacity for "active movement" independent of cytoplasmic streaming in the tube. Part of the evidence for motility of the sperms is the frequent fusiform or vermiform shape of the gametes —shapes that suggest the sperms of lower vascular taxa. The generative cell also, when in the tube, has been described as moving "actively and independently." Elongate and vermiform shape of the male-cell nuclei is not, in itself, evidence of motility; nuclei of this type are frequent in elongate cells of many tissues. The older view, generally maintained, was that the gametes are carried by cytoplasmic streaming in the pollen tube. Movement of the gametes along the pollen tube where two-way movement in the cytoplasm occurs has been considered further evidence of independent movement of the gametes and proof that these cells are not carried simply by streaming of the cytoplasm as the tube rapidly elongates. Changes in turgor of the tube have also been considered the major cause of passage of the gametes or generative cell along the tube.

Terminology of the Male Gametophyte. Since the later part of the nineteenth century, there have been many changes in the terms applied to the various structures and stages in the development of the male gametophyte. Attempts to homologize the pollen grain with the antheridium of lower groups have been, in some measure, responsible for the choice of terms. Changes in terminology have also come about by the demonstration, with new techniques, that some stages described as represented by "nuclei" are definite cells, with delimited cytoplasmic sheaths. *Tube nucleus* and *vegetative nucleus*—long used for the nucleus of the vegetative cell—are satisfactory and useful terms. They will continue in use, even though, as generally believed, the tube nucleus probably does not control the development of the pollen tube. The vegetative cell is a naked cell, lying within the microspore wall. The pollen tube is an extension of part of the inner layer of the spore wall; it does not constitute "the male gametophyte," even though it may enclose most of the male gametophyte. The generative cell was early called the *spermatogenous cell,* a term as good, or better, than *generative cell,* which is now used in the broader treatments. Both of these terms have been supplanted by *antheridial cell* in many treatments since 1912. The term antheridial cell brings confusion to the description of the male gametophyte, unless it is used as a part of a comparative study of the antheridium throughout vascular plants. The term *generative nuclei,* as applied to the nuclei of the male gametes, is unfortunate, because it suggests nuclei of the generative cell. The nuclei formed at division of the generative cell are the nuclei of the male gametes. They have been called *male nuclei, generative nuclei,* and *spermatogenous nuclei,* all good, in a descriptive sense, for the nuclei of the gametes themselves. But it has commonly been assumed these nuclei are the gametes, that definite male cells are not usually formed. Improved techniques have demonstrated that each nucleus is surrounded by a definitely limited sheath of cytoplasm. The male gametes are cells.

BIBLIOGRAPHY

Pollen and Male Gametophyte

Bailey, I. W., and C. G. Nast: The comparative morphology of the Winteraceae. I. Pollen and stamens, *Jour. Arnold Arb.*, **24**: 340–346, 1943.

Beer, R.: Development of the pollen grain and anther of some Onagraceae, *Beih. Bot. Centralbl.*, **19** (I): 286–313, 1905.

Canright, J. E.: The comparative morphology and relationships of the Magnoliaceae. II. Significance of the pollen, *Phytomorph.*, **3**: 355–365, 1953.

Cooper, D. C.: Microsporogenesis and the development of the male gametes in *Portulaca oleracea*, *Am. Jour. Bot.*, **22**: 453–459, 1935.

Cooper, G. O.: Microsporogenesis and development of seed in *Lobelia cardinalis*, *Bot. Gaz.*, **104**: 72–81, 1942.

Cranwell, L. M.: New Zealand pollen studies: The monocotyledons, *Bull. Auckland Inst. and Mus.*, **3**: 1953.

Engler, A.: Beiträge zur Kenntnis der Antherenbildung der Metaspermen, *Jahr. Wiss. Bot.*, **10**: 275–316, 1876.

Erdtman, G.: "An Introduction to Pollen Analysis," Waltham, Mass., 1943.

———: Pollen morphology and plant taxonomy. V. On the occurrence of tetrads and dyads, *Svensk. Bot. Tidskr.*, **39**: 286–297, 1945.

Garcide, S.: The developmental morphology of the pollen of the Proteaceae, *Jour. So. Afr. Bot.*, **12**: 27–34, 1946.

Goebel, K.: "Die Entfaltungsbewegungen der Pflanzen," Jena, 1920.

Hanf, M.: Vergleichende und entwicklungsgeschichte Untersuchungen über Morphologie und Anatomie der Griffel und Griffeläste, *Beih. Bot. Centralbl.*, **54A**: 99–141, 1935.

Hannig, E.: See first reference under Stamen and Androecium.

Hotchkiss, A. T.: Pollen and pollination in the Eupomatiaceae, *Proc. Linn. Soc. N.S.W.*, **83**: 86–91, 1958.

Knuth, P., and E. Loew: "Handbuch der Blütenbiologie," vols. I–III, Leipzig, 1904–1905.

Maheshwari, P.: The male gametophyte of angiosperms, *Bot. Rev.*, **15**: 1–75, 1949.

———: "Introduction to Embryology of Angiosperms," New York, 1950.

Martens, P.: La graine et le tube pollinique: Réflexions sur les caractères propres des phanérogames, *Bull. Acad. Roy. Belg. Cl. Sci.*, 5 sér., **33**: 919–943, 1949.

Matthews, J. R., and E. M. Knox: See reference under Stamen and Androecium.

Murbeck, S.: Über das Verhalten des Pollenschlauches bei *Alchemilla arvensis* und das Wegen der Chalazogamie, *Lunds Univ. Arsskr.*, N.F., (Avd. II) **9**: 1–46, 1901.

Newman, I. V.: Studies in the Australian acacias. IV. The life history of *Acacia baileyana* F.V.M. Part 2. Gametophytes, fertilization, seed production and germination, and general conclusion, *Proc. Linn. Soc. N.S.W.*, **59**: 277–313, 1934.

Parmentier, P.: Recherches sur le pollen des dialypetalées, *Jour. Botanique*, **15**: 150–166, 194–204, 218–223, 419–429, 1901.

Piech, K.: Zur Entwicklung der Pollenkörner bei *Scirpus lacustris* L., *Bull. Acad. Polon. Sci. et Lett. Cl. Sci. Math. et Nat.*, sér. B., **1924**: 113–123.

———: Über die Entstehung der generativen Zelle bei *Scirpus uniglumis* Link durch "freie Zellbildung," *Planta,* 6: 96–117, 1928.

Pohl, F.: Der einfaltige Pollen, seine Verbreitung und phylogenetische Bedeutung, *Beih. Bot. Centralbl.,* 45: 59–75, 1928.

———: Die Kittstoffreste auf der Pollenoberfläche windblütiger Pflanzen, *Beih. Bot. Centralbl.,* 46: 286–305, 1929.

Porsch, O.: Geschichtliche Lebenswertung der Kastanienblüte, *Oesterr. Bot. Zeitschr.,* 97: 269–321, 1950.

Schnarf, K.: Studien über den Bau der Pollenkörner der Angiospermen, *Planta,* 27: 450–465, 1938.

Shoemaker, D. N.: See reference under Stamen and Androecium.

Swamy, B. G. L.: See reference (1948) under Stamen and Androecium.

———: Further contributions to the morphology of the Degeneriaceae, *Jour. Arnold Arb.,* 30: 10–38, 1949.

Tammes, P. M. L.: On the origin, number and arrangement of the places of exit on the surface of pollen grains, *Rec. Trav. Bot. Néerl.,* 27: 1–82, 1930.

Werth, E.: Ueber die Bestäubung von *Viscum* und *Loranthus* und die Frage der primitivität der Windblütigkeit vie der Pollenblumen bei den Angiospermen, *Ber. Deutsch. Bot. Ges.,* 41: 151–164, 1923.

Wodehouse, R. P.: The phylogenetic value of pollen-grain characters, *Ann. Bot.,* 42: 891–934, 1928.

———: "Pollen Grains," New York, 1935.

———: Evolution of pollen grains, *Bot. Rev.,* 2: 67–84, 1936.

Wulff, H. D.: Beiträge zur Kenntnis des männlichen Gametophyten der Angiospermen, *Planta,* 21: 12–50, 1933.

———: Die Entwicklung der Pollenkörner von *Triglochin palustris* L. und die verschiedenen Typen der Pollenkörnentwicklung der Angiospermen, *Jahr. Wiss. Bot.,* 88: 141–168, 1939.

——— and P. Maheshwari: The male gametophyte of angiosperms, a critical review, *Jour. Indian Bot. Soc.,* 17: 117–140, 1938.

Dehiscence

Artopoeus, A.: Über den Bau und die Offnungsgeweise der Antheren und die Entwicklung der Samen bei den Ericaceen, *Flora,* 92: 309–345, 1903.

Chatin, A.: Causes de la déhiscence des anthères, *Compt. Rend. Acad. Sci. Paris,* 70: 201–203, 410–413, 644–648, 1870.

Leclerc du Sablon: Recherches sur la structure et de la déhiscence de anthères, *Ann. Sci. Nat. Bot.,* sér. 1: 97–134, 1885.

Matthews, J. R., and C. M. Maclachan: The structure of certain poricidal anthers, *Trans. and Proc. Bot. Soc. Edinburgh,* 30: 104–122, 1930.

Schips, M.: Zur Öffnungsmechanik der Antheren, *Beih. Bot. Centralbl.,* 31: 119–208, 1913.

Van Tieghem, P.: Observations sur la structure et la déhiscence des anthères des Loranthées, suivies de remarques sur la structure et la déhiscence des anthères en général, *Bull. Soc. Bot. France,* 42: 363–368, 1895.

Woysicki, Z.: Recherches sur la déhiscence des anthères et le role du stomium, *Rév. Gén. Bot.,* 36: 196–212, 250–268, 1924.

Pollination

Daumann, E.: Das Blütennecktarium von *Magnolia* und die Fütterkoerper in der Blüte von *Calycanthus, Planta,* 11: 108–116, 1930.

Delpino, F.: "Ulteriori Osservasioni sulla Dicogamia nel Regno Vegetali," Milan, 1868–1875.

Diels, L.: Käferblumen bei den Ranales und ihre Bedeutung für die Phylogenie der Angiospermen, *Ber. Deutsch. Bot. Ges.*, 34: 758–774, 1916.

Grant, V.: Pollination systems as isolating mechanisms in angiosperms, *Evol.*, 3: 82–97, 1949.

———: The protection of the ovules in flowering plants, *Evol.*, 4: 179–201, 1950. (Bird pollination.)

———: The pollination of *Calycanthus occidentalis*, *Am. Jour. Bot.*, 37: 294–297, 1950.

Hotchkiss, A. T.: See reference under Pollen and the Male Gametophyte.

Pijl, L. van der: Fledermäuse und Blumen, *Flora*, 131: 1–40, 1936.

Pohl, F.: Beziehungen zwischen Pollenbeschaffenheit, Bestäubungsart und Fruchknotenbau, *Beih. Bot. Centralbl.*, 46: 247–285, 1929.

———: See second reference under Pollen and the Male Gametophyte.

Porsh, O.: Kritische Quellenstudien über Blumenbesuch durch Vögel, *Biol. Gen.*, 2: 217–240, 1926; 3: 171–206, 475–548, 1927; 5: 157–210, 1929; 6: 133–146, 1930.

———: Vogelblümenstudien, *Jahr. Wiss. Bot.*, 63: 553–706, 1924; 70: 181–277, 1929.

———: Die Vögel als Blumenstäuber, *Biol. Gen.*, 9: 239–252, 1933.

———: Säugertiere als Blumenausbeuter und die Frage der Säugertierblume, *Biol. Gen.*, (I) 10: 657–685, 1934; (II) 11: 171–188, 1935.

———: Windpollen und Blumeninsekt, *Oesterr. Bot. Zeitschr.*, 103: 1–18, 1956.

Sahni, B.: Pollen grains in the stylar canal and in the ovary of an angiosperm, *Curr. Sci.*, 4: 587–589, 1936.

Schnizlein, A.: "Iconographia," vol. III, Bonn, 1843–1870.

Stebbins, G. L., Jr.: "Variation and Evolution in Plants," New York, 1950.

Werth, E.: See reference under Pollen and the Male Gametophyte.

Chapter 6

THE GYNOECIUM

The ovule-bearing organs of the flower, the carpels or megasporophylls, make up the gynoecium. They range in number from many to one and, in arrangement, from spiral to whorled. The carpels may be free from fusion with one another—the gynoecium *apocarpous*—or connate in various degrees—the gynoecium *syncarpous.* In the early decades of the twentieth century, the term *coenocarpous* has been applied by some authors to gynoecia with connate carpels that are several-chambered, with axillary placentae; and the term *paracarpous* to syncarpous gynoecia with a single chamber and parietal, basal, or free central placentation. This distribution has morphological value, but paracarpy has perhaps arisen in more than one way: by phylogenetic union of open carpels, and by modification of coenocarpic gynoecia. A line between apocarpy and syncarpy is difficult to draw, because fusion may be slight and even developed late in ontogeny. Syncarpy is discussed later in this chapter.

The gynoecium perhaps shows more simply than the androecium and the perianth the major changes in the evolution of the flower. Especially prominent are the advances from spiral to whorled arrangement, from free to fused members, and from many to one which is pseudoterminal.

Numerous spirally arranged carpels characterize rather few families —Magnoliaceae, Annonaceae, Eupomatiaceae, many of the Ranunculaceae and Rosaceae; usually apocarpous, they may become syncarpous in fruit—Annonaceae. Gynoecia with few, spirally arranged carpels are few, but, in syncarpous forms, the spiral arrangement may be determinable only anatomically—*Berberis.* The flower of *Scheuchzeria* has been cited as an example of the spiral placing of a few carpels, but this "flower" is perhaps an inflorescence. Spiral arrangement is difficult to determine in gynoecia with only two carpels, but, in these gynoecia, one carpel overtops the other—*Jeffersonia, Epimedium.* Oblique adnation to the receptacle by individual carpels brings about "false coenocarpy," a type of union difficult to determine morphologically, because distinction must be made between the tissues of the receptacle and those of the carpels. False coenocarpy is discussed further under Syncarpy. Reduction of fertile carpels to few and to one is common.

Whorled carpels represent modification of spiral arrangement. This modification is often readily seen in internal structure where there is little superficial evidence.

The Terms Pistil and Locule. The term pistil is unfortunately loosely used and confusing to students in its various morphological implications. It has been defined as the "unit of the gynoecium," but this "unit" may be a single carpel—in apocarpous gynoecia—or a group of carpels—in syncarpous gynoecia. The use of modifying adjectives—*simple* where the gynoecium consists of a single carpel, and *compound* where it consists of more than one carpel—are helpful, but "compound" is sometimes used for both apocarpous and syncarpous gynoecia. "Pistil," as a term applied to gynoecia, should be avoided as far as possible.

The term locule is also loosely used, morphologically. It is applied to the space enclosed within a single carpel, and to that within connate carpels, both the common chamber and that within a ring of closed carpels—*Sempervivum, Phytolacca.*

Discussions of syncarpy and the inferior ovary conclude the discussion of the carpel.

THE CARPEL

The angiosperm megasporophyll, the carpel, is, like the stamen, an elongate appendage, primitively of laminar form. It is like the stamen also in that it is leaflike in its relation to the stem—position, arrangement, vascular connections (traces), anatomy, and ontogeny (apical and marginal meristems). It differs from the microsporophyll in that the megasporangia are always on the adaxial side, whereas the microsporangia may be on either the adaxial or the abaxial side. In the carpel, the sides of the lamina have been folded or rolled adaxially toward the midrib. The folding or rolling encloses the megasporangia (ovules) in a chamber, the *locule.*

Form of the Carpel

In its typical form, the carpel resembles a folded or rolled leaf lamina, with margins usually appressed or fused. Ancestrally, it was undoubtedly a stalked, dorsiventral organ. The folded organ is still incompletely closed at pollination time in some taxa. Genera with the *stipe,* a stalk, are frequent in both monocotyledons and dicotyledons—Helobiales (*Zannichellia, Scheuchzeria, Ruppia, Althenia*); primitive Liliaceae (*Tofieldia, Veratrum*); many woody Ranales (*Degeneria, Drimys, Calycanthus, Bubbia,* some Annonaceae, *Austrobaileya*); many Ranunculaceae (*Coptis, Eranthis, Cimicifuga, Helleborus,* even some achene-bearing genera, such as *Thalictrum*); *Cercidiphyllum; Eupteleα; Paeonia;* manv Dilleniaceae; primitive Rosaceae (*Physocarpus*). The stipe is sometimes confused with the *gynophore,* an elongate part of the receptacle that bears the entire gynoecium. Anatomy shows that the stipe is a part of the carpel, not of the receptacle; the vascular traces of the carpel arise from the receptacular stele and are independent in the stipe, with the ventral traces inverted (*Coptis, Tofieldia*). Stigma, style, and ovary are parts of the lamina of the sporophyll, functionally distinct but often merging. The style, in some taxa, represents a sterile part of the ovary; in other taxa, it seems to be formed by a secondary elongation of the distal part of the carpel.

The carpel varies from simple, without differentiation into fertile and sterile parts, to complex. The complex form is divided into a proximal, ovule-bearing part, the ovary; a distal, pollen-receptive part, the *stigma;* and a median sterile section, the *style.* Usually terminal on the ovary, the style appears to be lateral in some highly specialized taxa—the Labiatae and achene types in the Rosaceae (Potentilleae, Chrysobalanoideae). In these plants, ontogeny and the course of the midrib bundle show that the lateral position is secondary. The style is terminal in early stages, but growth of the ovule and seed on the dorsal side displaces the style. The style appears nearly basal, seeming, in later stages, almost independent of the carpel. But anatomy shows that the dorsal bundle of the carpel is continuous from the base of the carpel over the ovule—along the dorsal side and top of the enlarged ovule-bearing base, down to the base of the style and upward through the style. Obviously, the two structures are parts of one organ. (Under the carpel polymorphism theory, the two parts of the carpel were interpreted as two independent organs—one solid, the style, and one valvular, the ovary.)

The carpels of some of the woody Ranales illustrate the simple type (Fig. 72). They are elongate, usually without distinction of ovary, style, and stigma. The many ovules are borne throughout an elongate chamber. The pollen in many is received on a longitudinal *stigmatic crest* (Fig. 72*C, F, H, I*), consisting of the papillose surfaces of the margins and borders of the lamina, which have come together in the closing of the carpel. The double nature of the crest is evident in its more or less strongly two-lobed form, the margins flaring back somewhat from the line of contact (Fig. 72*F, H*). In some species of *Drimys,* in *Himantandra* and *Degeneria* (Fig. 83*A, B*), the crest extends from the apex to the base of carpel (taxonomic descriptions read "stigma decurrent"); in other taxa, it is restricted to the distal part (Fig. 72*D, E, H, I*). In these primitive taxa, the marginal areas of the lamina are appressed but not fused histologically; closure of the carpel is by interlocking of projecting papillae. The pollen germinates on the stigmatic crest over its full length, and pollen tubes may penetrate directly to the nearest ovules. Elongate stigmas may represent primitive form, as in many

Ranales, or advanced form, an adaptation to wind pollination. In *Eupomatia* (Fig. 72A), connation of the carpels restricts the exposed stigmatic area to the apex of the carpel, and the papillose surface extends downward inside the carpel to and around the ovules, forming an

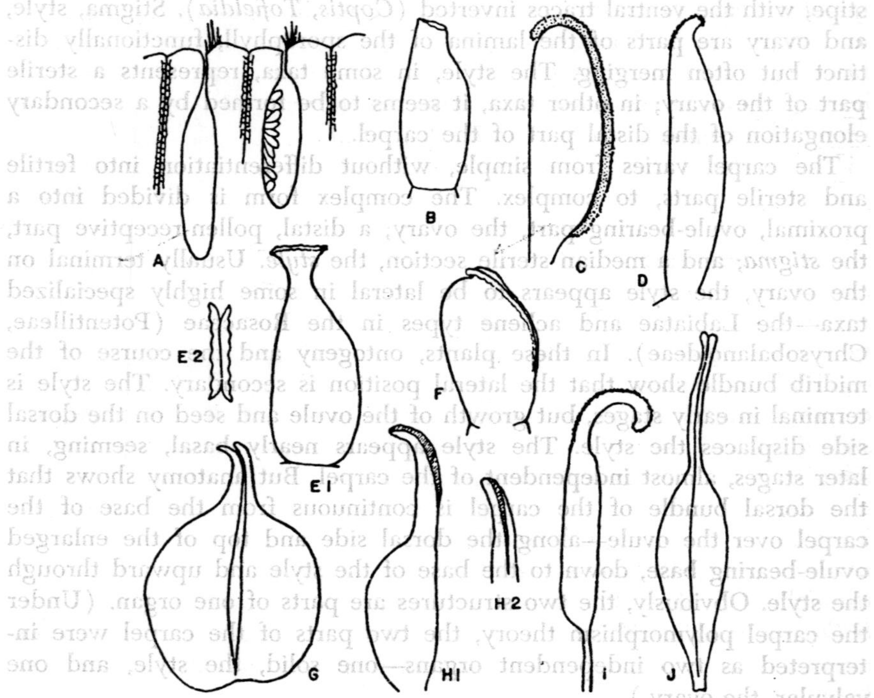

Fig. 72. Some primitive carpel types. *A, Eupomatia laurina,* part of longitudinal section of gynoecium showing connate carpels (spirally arranged) and stigma reduced to a few papillae on gynoecial "floor"; *B, Cimicifuga racemosa,* showing early stage in development of style and no definite stigma; *C, Drimys membranea,* showing decurrent stigma, stigmatic crest; *D, Caltha palustris,* showing slightly decurrent stigma, no style; *E, Hydrastis canadensis: E*1, showing weak style and two-lobed stigma with flaring borders, *E*2, face view of stigma; *F, Degeneria vitiensis,* showing stigmatic crest; *G, Chamaerops humilis,* open carpel with style; *H, Butomus umbellatus: H*1, carpel with style and stylar canal, *H*2, detail of stylar canal; *I, Coptis trifolia,* showing stigma decurrent on style and a long stipe; *J, Physocarpus opulifolius,* showing elongate style and terminal stigma. (*F, after Bailey and Smith; G, after Juhnke and Winkler.*)

"internal stigma." The restriction here of pollen-tube transmitting tissue to the inside of the carpel, except at the apex, resembles the phylogenetic shortening of the stigmatic crest by the closure of the carpel as the typical stigma developed.

The basic form of the carpel is that known, when mature, as the *follicle.* The follicle is the classical carpel, the form long accepted as the primitive type, the type from which have been derived various specialized types, such as the *achene*. (The terms follicle and achene are usually applied tc mature carpels, fruits.) The follicle is primitive in its elongate form, numerous ovules, simple dehiscence, and the frequent absence of a style; it is advanced in the delimitation of ovary and stigma. It is characteristic of many genera of the more primitive families of both dicotyledons and monocotyledons. Carpels of the following genera have many folliclelike characters: *Degeneria, Drimys* (sect. Tasmannia), *Butomus, Trollius, Caltha, Coptis, Physocarpus, Akebia, Telopea, Paeonia, Tofieldia, Scheuchzeria.*

Closure of the Carpel

One of the distinguishing characters of the angiosperms is the enclosure of the ovules by the megasporophyll; the pollen grains do not reach the ovule but germinate at a distance, in contrast with the condition in the gymnosperms, where the ovule is "naked" and the pollen is received and germinates on the ovule. There are exceptions in the conifers to the place of germination of the ovule; in *Araucaria, Agathis,* and *Tsuga,* the pollen germinates at a distance from the ovule. In the angiosperms, there are taxa in which the carpel is not completely closed at pollination time—*Degeneria, Drimys,* Butomaceae, Hydrocharitaceae, *Sparganium, Sassafras, Coptis, Tiarella, Platanus.* The "opening" is very narrow, and pollen grains probably never enter the ovarian locule through it. (Pollen has been reported in the stylar canal in *Butomopsis, Hillebrandia, Reseda,* but probably does not germinate there.) The opening is usually quickly closed but may still be present in the mature fruit.

"Union" of the carpel borders shows all stages from the presence of an open slit to loosely and tightly interlocking papillae and to histological fusion—a union so complete that no evidence remains in tissue structure (Fig. 74*A, B*). In the open carpels of *Degeneria* and *Drimys,* the opening is partially closed by the papillose cells of the stigmatic crest (Fig. 73). In carpels where the uniting margins or borders are closely appressed but without histological union, the epidermal layers may be merely coherent, without elaboration of surface-cell form, but the epidermal layers are often rugose or toothed and interlock as they mature; evidence of the identity of the individual layers may be only in cell form and position. Ontogenetic adjustments among the uniting layers often obscure the line of fusion. All these conditions are found in carpels in which union is ontogenetic; where fusion is congenital, there is no histological evidence of the union. Morphological nature of the carpel may be evident in the maturing and mature fruit, as in the peach and cherry.

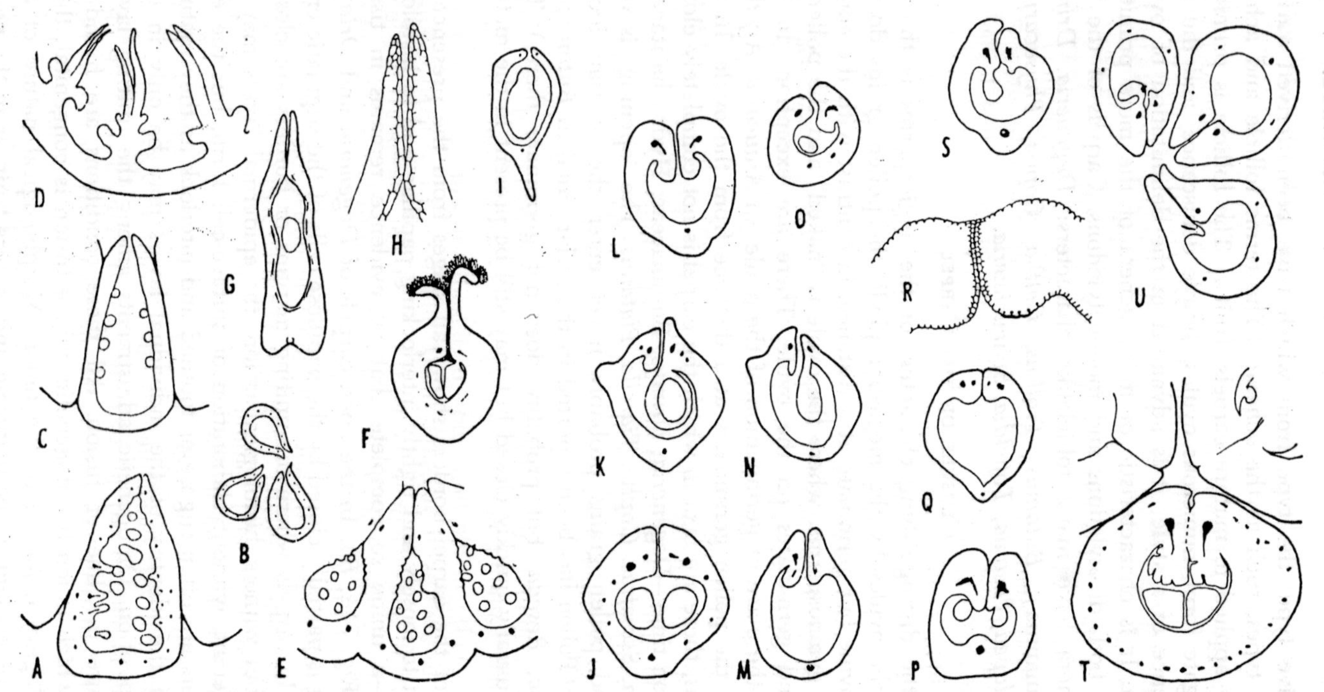

Fig. 73. Cross sections of carpels showing variety in degree of closure at flowering stage, from clearly open to appressed. *A* to *E*, laminar placentation; *F*, *J* to *U*, submarginal; *G* to *I*, basal placentation. *A*, *B*, *Butomus umbellatus;* *C*, *Limnocharis;* *D*, *Stratiotes aloides;* *E*, *Trochodendron;* *F*, *Degeneria vitiensis;* *G*, *H*, *Alisma;* *I*, *Sagittaria heterophylla;* *J*, *Actaea rubra;* *K*, *Trollius laxus;* *L*, *Coptis trifolia;* *M*, *Spiraea salicifolia;* *N*, *Tofieldia racemosa;* *O*, *Schisandra;* *P*, *Cercidiphyllum;* *Q*, *R*, *Physocarpus opulifolius;* *S*, *Euptelea;* *T*, *Paeonia;* *U*, *Sedum ternatum.* (*C*, *after Troll;* *D*, *after Salisbury;* *E*, *after Swamy.*)

Conduplicate and Involute Closure. The closing of the carpel comes about, structurally, by an upturning of the sides of the lamina and the bringing together adaxially of the surfaces or edges, with more or less complete fusion. (Recent description of abaxial folding in *Cercidiphyllum* was based on the error of interpretation of the inflorescence as a flower.) The upturning may be a simple folding on the midrib region a an axis, with the margins lying side by side (Fig. 74*A*), or it may be

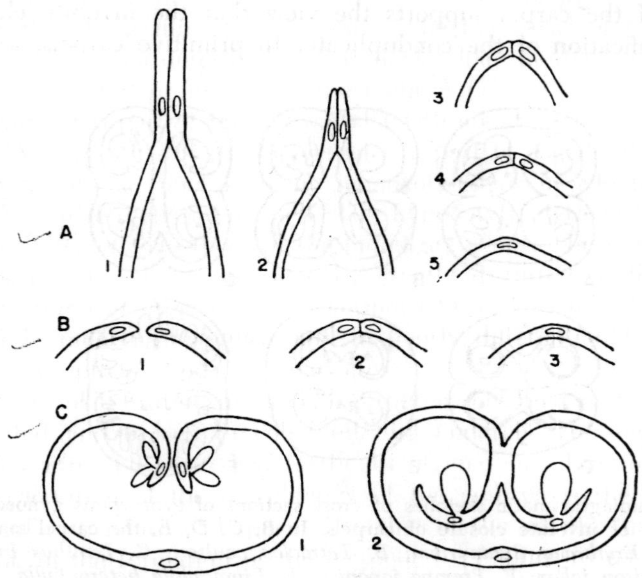

Fig. 74. Semidiagrammatic sketches of cross sections of carpels to show two methods of closure of carpels. *A*, conduplicate method, borders of the lamina meeting face to face by ventral surface, vascular bundles half-inverted; *B*, involute method, margin of lamina meeting margin (marginal initials meet), vascular bundles inverted; *C*, involute method, borders of the lamina meeting by their dorsal surfaces, margins free.

a more or less extensive incurving or inrolling, with contact face to face by the edges (Fig. 74*B*) or, where the inrolling is strong, by parts of the abaxial surface, with the margins free within the locule (Fig. 74*C*). The carpel has apparently closed independently in many lines and in a variety of ways. A primitive method of closing is that of a simple folding so that considerable parts of the adaxial surface come together; a carpel, closed in this way, is called *conduplicate* (Fig. 74*A*). Conduplicate closing has been considered the basic type from which all others have been derived, but a survey of many families suggests that this method is only part of the history of carpel closure. Carpels in which the lamina sides appear rolled upward and inward have been termed *involute*. The rolling appears commonly to have brought the edges of the lamina together (Fig. 74*B*); where the inrolling is greater—an uncommon condition—contact is by strips of the abaxial surface. It seems improbable that all the types of involute carpel have been derived from the conduplicate; this derivation would involve a change from contact by the adaxial surface only to contact by the abaxial surface—a major change, far more complicated and circuitous than usually found in evolutionary derivation. The vascular anatomy of the carpel supports the view that the involute closure is not a modification of the conduplicate. In primitive carpels, a narrow

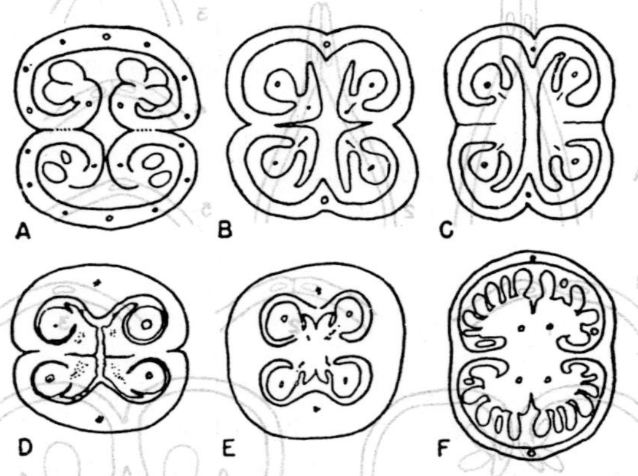

Fig. 75. Semidiagrammatic sketches of cross sections of syncarpous gynoecia showing examples of involute closure of carpels. In *B, C, D, E,* the carpel margins are obvious. *A, Erythraea Centaurium; B, Tetraclea coulteri; C, Isanthus brachiatus; D, Clerodendron fallax; E, Premna japonica; F, Limnophila heterophylla.* (*A, after Baum,* 1949; *B, C, after Stauffer; D, E, after Junell; F, after Hartl.*)

marginal strip of the lamina is sterile. In specialization of the carpel, this sterile band is progressively narrowed (Fig. 74*A*, 2) and, in some taxa, lost (Fig. 75*B* to *E*), and the ovules seem to be borne on the margin itself (Fig. 75*A*, *F*). Where, in involute carpels, the sterile strip is lost and the edges meet directly, the ventral bundles and the placentae are brought close together and often fuse into a common bundle (Fig. 74*B*) and common placenta. Where the sterile margin is narrowed but persists, contact is by the abaxial surface, and the sterile band projects inward as two flaring edges (Fig. 74*C*, 1); contact is in the region of the vascular bundles. Where the inrolling is greater, contact is farther back on the abaxial surface, and the ventral bundles and placentae are well inside the locule (Fig. 74*C*, 2). Extreme inrolling is present in the Labiatae, where the inrolled borders may reach the midvein and sometimes become fused with tissues of that region.

The difficulty of determining whether a carpel of advanced form, with apparently inrolled borders, is involute or conduplicate is increased by the trend throughout carpel specialization toward reduction of the sterile borders. In primitive conduplicate carpels, as in *Degeneria*, these borders are wide; in less primitive types, they are narrower and, in advanced carpels that can still be recognized as conduplicate, they are narrow. In carpels that are probably involute, the sterile borders range from fairly wide to absent, with ovules borne on the very edge. Intermediate stages show vestigial borders. The recognition of these reduced borders as such demonstrates an inrolling that brings together borders of the dorsal surface of the carpel. Evidence supplied by position of the marginal meristems in carpels of the rolled type is probably not on record and is important for the determination of basic types of carpel closing.

Where the sterile margins are greatly reduced or lost, the two placentae are brought close together and unite, in greater or less degree, in a common placenta. The placenta may show evidence of its double nature in bilobed form. The ventral bundles, which underlie the placentae, may lie side by side or be fused as one simple bundle (Figs. 74*A*, 4, 5 and *B*, 2, 3), regardless of external form of the placenta. A deeply two-lobed placenta with the lobes divaricate and the epidermal layers continuous through the line of union is evidence of fusion by the dorsal surface well back from the edges (Fig. 74*C*, 2). Orientation of the ventral bundles may be of aid in determining type of folding. In typical conduplicate carpels, these bundles are half inverted; in carpels where edge meets edge or the edges are rolled in, they are typically completely, or nearly completely, inverted. In carpels apparently of involute type, whether in apo- or syncarpous gynoecia, inversion of the ventral bundles is typical. Variations in degree of inversion occur occasionally in individual carpels, but this accompanies variations in external form.

The assumption that all carpel types are derived from the conduplicate involves complicated changes in orientation of the borders and margins in the phylogenetic development of the more common types. In the conduplicate type, *ventral* surfaces of the carpel borders are in contact, and the margins lie side by side (Fig. 74*A*, 1, 2); in the more common types, the margins may come face to face, with no contact of the borders, or the margins, still side by side, are turned inward, and the *dorsal* surfaces of the borders are in contact (Fig. 74*C*, 1). Where the inrolling is continued beyond the line of contact, the borders flare and may recurve strongly (Fig. 74*C*, 2).

It is difficult to see, in the involute-margin types, modifications of the conduplicate condition, especially in the light of the existence today of open carpels in unrelated taxa and the undoubted occurrence of independent closure in several or many phylogenetic lines. (Outstanding among the contributions of comparative morphology is the discovery of the frequent occurrence in phylogenetic progress of the same structural modification—dioecism, gamopetaly, zygomorphy, the inferior ovary, the solitary ovule, the vessel.) The carpel is closely like the leaf in many ways. In vernation, the leaf is folded or rolled in several ways: conduplicate, involute, revolute. In ecological modification, the leaf blade may, under reduction, be folded or rolled upward or downward, and no type apparently is basic to the others. It would be strange, in the light of the behaviour of the leaf in vernation and in ecological modification, if the carpel, so like the leaf in form, ontogeny, and anatomy, closed in only one way—that all forms of closed carpels have been derived from the conduplicate type.

That the closed carpel is not an ancient angiosperm character is evident not only in the presence of open carpels, even in some fairly advanced families, and the frequent presence of those with borders merely appressed, but also in the presence of carpels closed by adnation to the receptacle. Fusion, by both connation and adnation, clearly preceded closure of the carpel.

Position, number, and size of ovules may, in part, determine method of enclosure. Conduplicate closing perhaps provides better enclosure where there are many ovules scattered over the lamina; involute, better space for ovules massed in longitudinal rows near the margins. Carpels that are clearly conduplicate usually have laminar or sublaminar placentation—an association of primitive characters, but not necessarily evidence that involute carpels also may not be primitive. The majority of carpels appear involute, with the sterile border greatly reduced or absent, and the margins meeting, edge to edge. Though the existence of marginal meristems in carpels is well known, little attention has been given to late stages in ontogeny, where position of the marginal meristems should aid in determination of the basic type of closing. Available information indicates that, in at least some taxa, the marginal meristems meet directly—evidence of involute closure. Closure of the carpel, where ontogenetic, results in the formation of a more or less distinct ventral suture. The suture varies greatly in form and distinctness with the degree of inrolling and of completeness of histological fusion. The openings may extend the full length of the ovary or be restricted to the median or distal parts; in *Platanus*, the opening is very short. Closure of the carpel is clearly still in process of establishment in the angiosperms. The closed carpel is not a fixed, universal character in "angiospermy," though it serves well as one of a group of distinguishing characters.

Closure by Adnation to Receptacle. Carpels that stand obliquely on the receptacle may be closed at the base by adnation of the margins to the receptacle. In these carpels, the locule is enclosed, in part, by tissues of the axis. Apparently, the ontogeny of these carpels has not been described, but the adnation is probably congenital. The crescent-shaped carpel primordia and the primordium of the receptacle develop together; the flanges of the carpel primordium arise united with receptacular tissues. Higher up, the carpel margins unite with each other.

Closure by Connation. In some syncarpous gynoecia, carpels open at the base are united by their lateral walls. Their locules unite, forming a common ovarian chamber—*Phytolacca, Sabal* and other palms. This type of gynoecium is part of the evidence that connation and adnation were present in the gynoecium before the carpel closed; that the closed carpel was not a character of the earliest angiosperms.

Closure by Ontogenetic and Phylogenetic Fusion. The closure of the carpel is commonly ontogenetic; the margins of a primordium, crescent- or horseshoe-shaped in cross section, develop toward one another and unite. But the closure may be congenital; the carpel arises closed from a ring-shaped primordium. A primordium that is at first crescent-shaped may become ring-shaped at the base and, with continuing growth from the base, form a carpel that is closed distally during ontogeny but arises fused proximally. In many advanced families, the carpel borders have become congenitally concrescent; the closed carpel is phylogenetically established. (Failure to recognize the nature of the ring-shaped carpel primordium has been largely responsible for the peltate theory of basic carpel form.) A common comparable condition in the corolla helps to make the complex carpel development clear. The typical gamopetalous corolla arises from a whorl of separate primordia, each representing one of the ancestrally free petals. After elongation has continued for some time, the petal primordia fuse into a ring, which forms the corolla tube. The bases of the petals, evident by their independent vascular supplies, are fused congenitally (or phylogenetically); they arise fused.

The Complex Carpel

The differentiation of ovary, style, and stigma from the simple, primitive carpel was a gradual one, which took place in many lines, with elaboration of style and stigma proceeding at different rates. The stigma is sometimes sessile, the style undeveloped in some of the Winteraceae and *Euptelea;* the style may be well developed, the stigma still a primitive stigmatic crest, decurrent on the style, as in *Cercidiphyllum*. Morphologically, the style is usually the distal part of the primitive carpel, with its ovules lost; several families show gradual transition from ovary to style, with vestigial ovules and traces for lost ovules in the transitional region. Elongation of the style is characteristic of many taxa. The nature of the stigma is evident in the stigmatic crests of primitive carpels, especially those of the woody Ranales, such as *Degeneria* and the Winteraceae. The pollen-receiving surface of the stigma is characteristically papillose and often secretory. Where this area extends the full length of the carpel, forming an undifferentiated stigma, a stigmatic crest, the papillae cover the margins of the carpel and narrow adjacent bands of the lamina surface. Where the marginal areas are merely approximated or appressed, the carpel still open, the papillae fill the slit loosely or compactly. In some conduplicate carpels, the margins of the carpels flare (Fig. 72*C*, *F*, *H*), exposing a narrow band of the adaxial carpel surface and making the stigmatic crest two-lobed (*Degeneria*, *Drimys*). The two-lobed form of crest, conspicuous in the most primitive carpels, is carried over into some advanced types and doubtless explains the two-lobed stigmas of some taxa. (The double lobing of stigmas of other taxa may be morphologically different.) In stigmatic crests, the papillae may occur not only on the carpel margins but on adjacent narrow bands of both dorsal and ventral surfaces (*Degeneria*, Winteraceae). In the Winteraceae, they may extend inward on the ventral surface to the placental ridge, around the ovule bases, and, in *Degeneria*, farther in over much of the carpel surface, forming an internal continuation of the stigmatic area, even into sealed parts of the carpel. A similar "internal stigma" is present in *Eupomatia*, where, in a carpel completely closed by connation, stigmatic papillae extend downward from a distal pore to and around the ovules. In the Annonaceae (*Artabotrys*) also, the papillate tissue extends downward along the carpel margins to the ovule bases, as a *transmitting tissue* for the pollen tubes. Internal stigmatic tissue in a sealed carpel is doubtless a remnant of the more extensive stigmatic area of primitive open carpels.

The carpel, as a simple, folded, or involute laminar appendage, with several to many ovules, varies greatly in form: with or without clear distinction of ovary, style, and stigma; with or without style; stipitate or nonstipitate; completely or incompletely sealed. Evolutionary changes in these characters have progressed unequally, and primitive form in one character may accompany advanced form in another. The carpel of this type is *follicular* and matures into the fruit type termed a follicle.

Modification of the Follicular Carpel

Specialization of the follicular carpel has been primarily in reduction in size and in ovule number, simplification of vascular supply, and loss of dehiscence. In extreme reduction, only one ovule persists. The fruit formed from such a carpel as this, when indehiscent, is an *achene*. All stages in the development of the achene type of carpel are present in the Ranunculaceae and Rosaceae. The change is evident, superficially, in the shortening of the ovary and the presence of abortive or vestigial ovules. That the achene is a reduced follicle is shown in Fig. 42; the number of traces is reduced by loss and by fusion, and traces to lost ovules persist in the ovary wall. *Clematis, Caltha, Calycanthus*, and *Adonis* also show derivation of the achene from the follicle.

Reduction and simplification of the carpel have occurred independently in many families and apparently from follicular types that were at various stages of specialization: from those with well-developed style—the style retained—as well as from those with sessile stigma; from stipitate carpels—the stipe retained—and from sessile carpels. Reduction may proceed beyond functional form, with the last ovule lost and the locule nearly or quite closed—the "solid" carpel.

The Solid Carpel

The term solid, as applied to carpels, became prominent in the 1920s and 1930s when the theory of carpel polymorphism aroused considerable discussion. But the term was used in the older literature, much more soundly, for sterile carpels with locule greatly reduced or absent (Fig. 76), carpels that are vestigial or abortive. In its early use, the term was descriptive, and its nature, a reduction form, correctly interpreted. Sterile carpels are frequent in many taxa and show all stages in structural reduction—compressed and "consolidated" in various ways and degrees; rodlike, free or adnate to normal carpels; short stubs. In syncarpous gynoecia, they may be represented by vascular supplies only. Vestigial carpels are most readily recognized in staminate flowers of monoecious and dioecious genera. In some unisexual flowers that have no external remnants of reduced carpels, stubs of the carpel traces are present in the receptacle. In syncarpous taxa that show series in gynoecial reduction, solid carpels are often apparent as remnant structures. *Triglochin* has obvious solid carpels; within the genus, carpel number is reduced from six to three; species with three normal carpels have three sterile carpels, with little or no locule (Fig. 76*F*). (The solid carpel as a reduction form is discussed further under Syncarpy.)

The Stigma and Transmitting Tissue

In the more primitive carpels, the stigma is not clearly set apart: it merges into the style or ovary; in more specialized forms, it is usually distinct. In such genera as *Degeneria* and *Drimys*, it is represented by the stigmatic crest, consisting of proliferated and papillose marginal bands extending the full length of the lamina (Fig. 72*C*). These bands form a pollen-receptive area, which is doubtless the primitive stigma. Restriction of the crest to its distal part, with reduction and sharper delimitation, has formed the specialized stigma of the higher taxa. Intermediate forms occur—*Butomus, Caltha, Trollius.* Some decurrent stigmas are probably specialized, rather than primitive, types, as in anemophilous taxa—*Cercidiphyllum* (Fig. 146), Juglandaceae, *Thalictrum,* some Restionaceae (*Anarthria*).

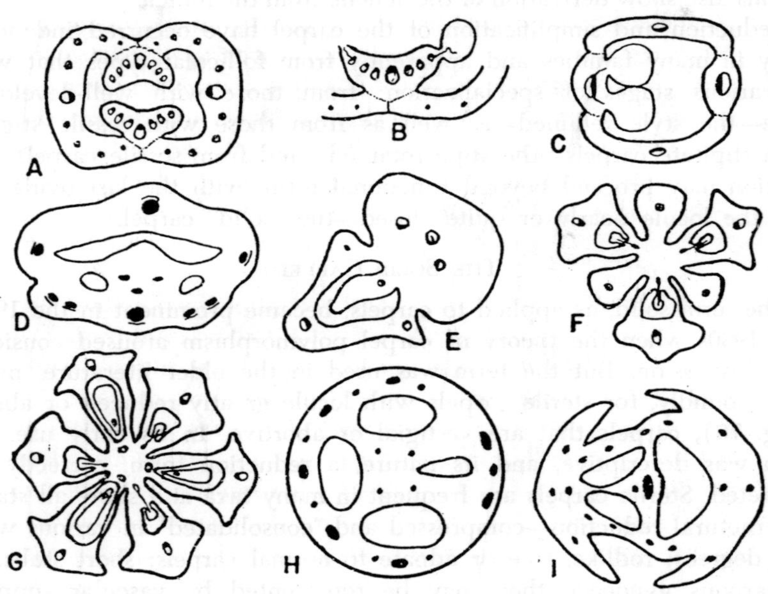

Fig. 76. Semisolid and solid carpels in syncarpous gynoecia. *A, B, Glaucium flavum; C, Chelidonium majus,* one of the two pairs of carpels without loculus, one carpel showing ovules; *D, Valeriana* sp., two solid carpels, each with vestigial loculus and one normal carpel; *E, Achlys* sp., two carpels, one fertile and one solid (without loculus); *F, Triglochin palustris,* three normal, three solid carpels; *G, T. maritima,* six normal carpels; *H, Triosteum perfoliatum,* three normal carpels with large loculi, one solid, sterile, with narrow loculus; *I, Linnaea borealis* var. *americana,* three carpels with two adnate bracts, one normal, two solid. (*A, B, C, after Van Tieghem; D, after Dyal; E, after Chapman; F, G, after Uhl; H, I, after Wilkinson,* 1949.)

The stigma and the *conducting tissue*—much better termed the *transmitting tissue,* because vascular tissue is commonly called the conducting tissue—represent the surviving and specialized parts of the stigmatic crest and the accompanying papillose carpel surface, which, in primitive carpels, extended between the carpel margins, over their internal surface, about the ovules, and even farther in on the laminar surface. In carpels with a stigmatic crest, the pollen tubes pass directly to ovules that are close to the place of germination. Restriction of the pollen-receiving surface involves, for the pollen tubes, a much longer course, but one still largely through the transmitting tissue.

The transmitting tissue is commonly thought of as filling the stylar canal only, but it was shown as early as the 1870s and 1880s to extend,

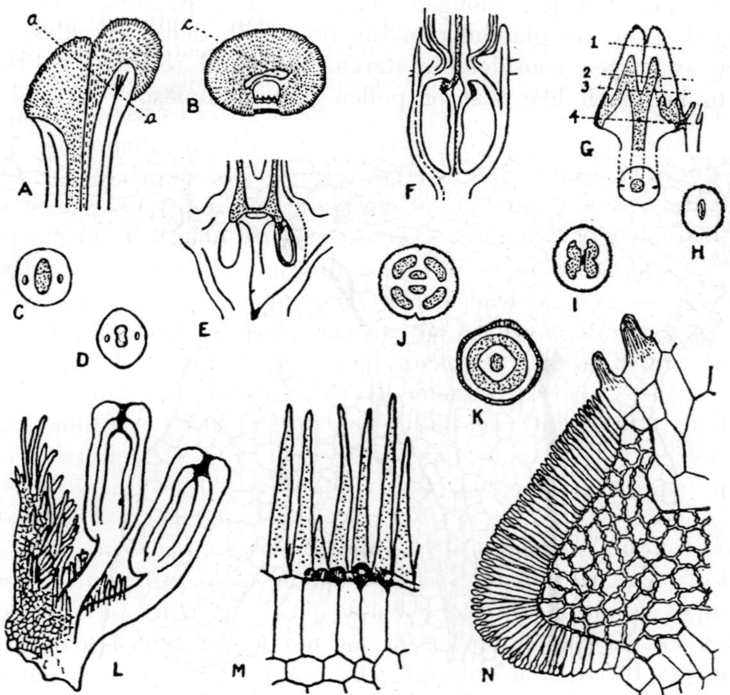

Fig. 77. Transmitting tissue. *A, B, C, D, Verbascum thapsus: A,* longitudinal section of style and stigma; *B,* section at *a–a* in *A, c,* stylar canal; *C, D,* cross sections of middle and base of style, respectively. *E, F,* longitudinal sections of ovaries of *Adoxa moschatellina* and *Fedia cornucopioides,* respectively, showing extent of transmitting tissue (shaded) in style and ovary; *G* to *K,* longitudinal section and cross sections of stigma of *Heliotropium peruvianum: H, I, J, K,* cross sections at levels 1, 2, 3, 4, of *G,* respectively. *L, Philodendron cordatum,* papillose transmitting tissue on receptacle and filaments; *M, Vinca minor,* papillae with drops of oil at base; *N, Heliotropium peruvianum,* vertical section of border of stigma, shown in *G.* (*A to K, N, after Guégen; L, M, after Capus.*)

in the great majority of families, over the placenta and considerable parts of the ovary wall (Fig. 77*E, F*). The presence of a papillose surface over areas of the wall where ovules have been lost is probably evidence of primitive laminar placentation.

The transmitting tissue is formed from superficial carpellary—rarely ovular—tissues, from the epidermis, or from the epidermis and hypodermal tissues (Fig. 78). Cell modification, the development of papillae or hairs, and proliferation of tissues form the stigmatic surface. Cell divisions in the epidermis are chiefly tangential; those of inner layers are also largely tangential but may be in all planes. Papillose cells form the most common stigmatic surface. Cells of the transmitting tissue are delicate, more or less elongate, thin-walled cells, sometimes loosely arranged. They are glandular in function, with dense cytoplasm, large nuclei, and often considerable starch. The fluid secreted by them is said to be much like that of pollen-grain cytoplasm. Oil and other

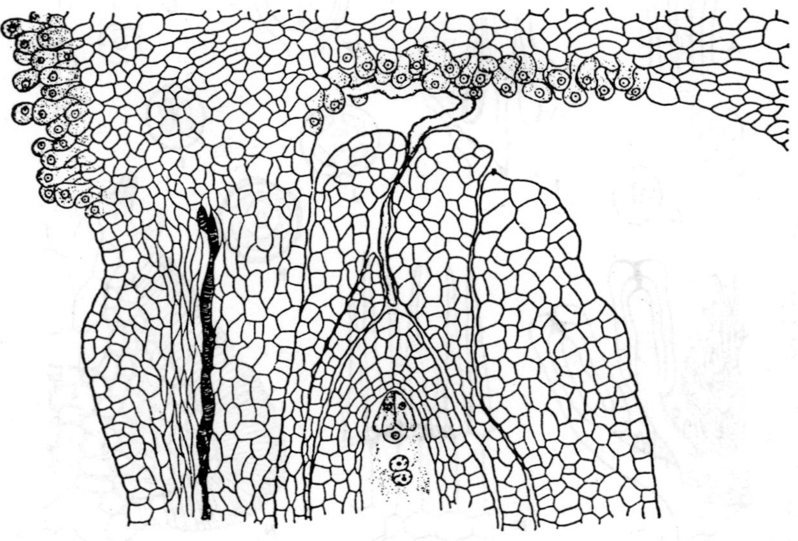

Fig. 78. *Degeneria vitiensis.* Micropylar part of ovule and adjacent carpellary tissue at pollination, showing transmitting tissue and course of pollen tube to micropyle. (*After Swamy.*)

fatty substances are frequently present, mingled with mucilaginous substances formed by cell-wall breakdown. Chlorophyll is common in all parts of the tissue. Some transmitting cells are simple or compound hairs, as in *Zea.* Plumose stigmas, such as those of many grasses and other anemophilous taxa, have numerous delicate branches that consist of papillose cells surrounding a single protoxylem element. Pollen grains reaching the stigmatic tissue are retained by entanglement among the projecting cells or are held on cell surfaces by their rough exines or oily surfaces, or by the mucilaginous or sugary secretions of the stigmatic cells.

Cells of the transmitting tissues, especially those of the stylar canal and of the upper ovary wall, may become separated by the dissolution of the middle lamellae and float free in a mucilaginous fluid (Fig. 79*E*, *F*, *H*). This condition has been considered rare—"found in orchids only"—but was long ago reported for many taxa.

Function of the Transmitting Tissue. The discovery, in 1823, of pollen germinating on the stigma aroused renewed discussion of the mystery of the method of fertilization. Two theories had been held in the pre-

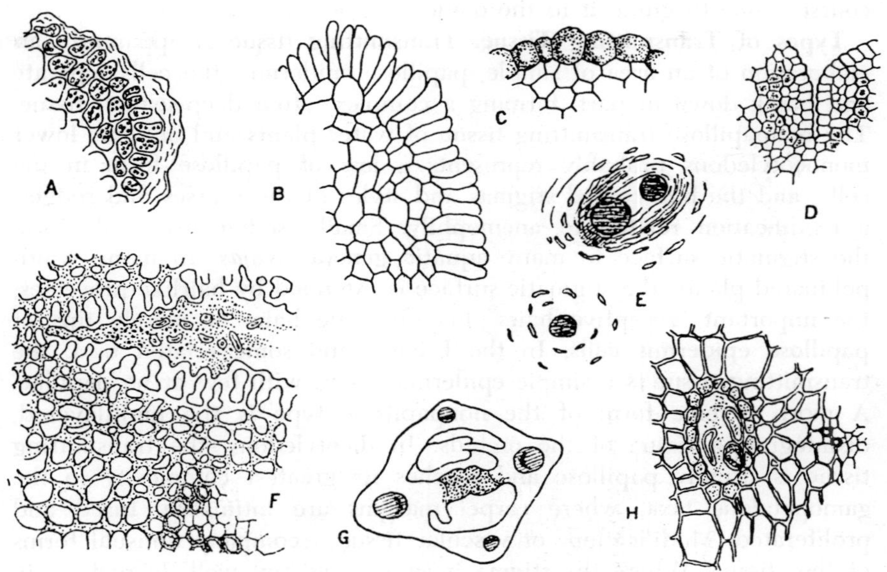

Fig. 79. Transmitting tissue. *A*, *Verbascum vernale*, transverse section of placental tissue, parenchyma cells swollen and filled with starch; *B*, *Pittosporum sinense*, papillose surface of ovary wall near ovules; *C*, *Solanum glaucophyllum*, transverse section of surface of placenta, epidermal cells filled with starch; *D*, *Rubus odoratus*, transmitting tissue of the carpel margins at top of ovary; *E*, cross sections of pollen tubes filled with starch grains in the gelatinous substance formed from the breakdown of the middle lamellae of the transmitting tissue; *F*, *Deherainia smaragdina*, cross section of part of the style showing stylar canal filled with mucilage in which the pollen tubes run; *G*, *Cheiranthus cheiri*, transverse section of ovary, showing central transmitting tissue; *H*, *Fumaria major*, cross section of stylar canal filled with granular mucilage and lined with a conducting epidermis against which the pollen tubes lie. (*After Capus.*)

ceding century: that the "fertilizing dust" (pollen) contained an "*esprit volatil*," which entered the vascular tissue of the stigma, passed down to the placenta, entered the ovules, and fertilized the "embryo"; that the pollen grains themselves "descended" into the ovary through open stylar canals (then believed to occur in all plants) and formed the "basis of the embryo." The discovery that, in many plants, the stylar canals were closed completely, or for part of their length, and that the pollen tube was seen to enter the stigma and its tip to be present at the ovule ended the pollen-grain-descent theory. The distribution of the transmitting tissue probably led to the discovery of its function. It had been noted also that considerable time elapsed between the presence of pollen on the stigma and the presence of an embryo. Finally, the pollen tube was followed from stigma to ovule, and a "double role" was assigned to the transmitting tissue: "to nourish the tube in its long course" and "to guide it to the ovule."

Types of Transmitting Tissue. Transmitting tissue is apparently an elaboration of an area of simple, papillose epidermis; the cells separate and break down in part, forming a complex, often deep-seated, tissue. The nonpapillose transmitting tissue of water plants and of some lower monocotyledons probably represents a loss of papillose form in the cells; and the hair-coated stigmas and styles of the grasses and sedges, a modification related to anemophily. Small, isodiametric cells form the stigmatic surface in many aquatic genera—*Najas*. In many wind-pollinated plants, the stigmatic surface is extended by hairlike processes; the important "receptive hairs" may be true hairs or merely tips of papillose epidermis cells. In the Liliales and some related taxa, the transmitting tissue is a simple epidermal layer, with few or no papillae. A most extreme form of the nonpapillose type is the multilayered, disintegrating tissue of the orchids. In dicotyledons, the transmitting tissue is usually papillose and reaches its greatest complexity in the gamopetalous taxa, where carpel margins are intimately fused and proliferated. Modifications of vascular tissue accompany unusual forms of the stigma. Where the stigma is simple and not well defined, as in *Zea*, the vascular tissue becomes gradually weaker, distally; where the stigma is large or lobed, the vascular tissue branches peripherally into many delicate strands.

Form of the Stigma. Variations in form of the stigma are largely related to the morphological structure of the carpel and gynoecium. A more or less distinctly two-lobed stigma, retaining, in part, the two-lipped form of the primitive ancestral crest, is frequent in less advanced taxa. Anatomically, two-lobed stigmas are of two types: the median vascular bundle dichotomously forked, the lateral bundles shorter; the median bundle short, the laterals prolonged. These vascular variations reflect the fundamental structure of the angiosperm node and appendages—a median trace, double in nature, with a pair of laterals (Fig. 6C) (the "three-trace supply," formerly considered basic for carpels). Two-lobed stigmas seem to be primitive types; the highly specialized stigma is simple—globose or cylindrical—or elaborate in form—plumose or dissected. Greatly increased stigmatic surface is associated with wind pollination. A well-differentiated stigma—in contrast with the stigmatic crest—and a sealed carpel have been called characters associated with insect pollination. The commissural stigma is discussed later in this chapter.

The Stylar Canal. The stylar canal is a space enclosed by the uniting margins of a carpel—in syncarpous gynoecia, of all the carpels. It is usually closed for part or all of its length by transmitting tissue. In highly specialized syncarpous pistils, it may be complexly branched; the branches may extend as strips of transmitting tissue beyond the bases of the styles along diverse courses to the placentae. The transmitting bands may vary in color from the ovary wall and, in the dicotyledons, may be somewhat collenchymatous and easily dissected from the ovary walls. A hollow style is occasional—*Viola, Campanula, Reseda, Lilium, Yucca, Erythronium,* Butomaceae. Pollen grains have been found in the open canal of *Butomopsis* and *Hillebrandia* but have not been seen germinating there. In syncarpous gynoecia, the term stylar canal is often applied to the transmitting tissue of the style only. It has been suggested that it be applied to all parts of the branched transmitting system, but the distal parts may be surface areas; where there is complex extension of the tissue, it could well be termed the *stigmatic-tissue system.*

The Terms Stigma, Transmitting Tissue, Stigmatic Tissue, Stigmatic Crest. The stigma has been variously defined: morphologically, as the distal part of the carpel, modified in form in adaptation to pollen reception; functionally, as the tissues of the carpel, external and internal, concerned with receiving and transmitting the pollen tube. The transmitting tissue within the style and ovary has been called the *internal stigma.* Where carpel borders with papillose epidermal cells are appressed and sealed ontogenetically, the external hairs are lost, and the internal hairs form an internal stigmatic line that extends downward along the placental line and around the ovules. In the primitive carpels of *Eupomatia,* which are without style and with an external stigma of a mere tuft of apical hairs, connation has resulted in the loss of the stigmatic crest, except for the internal band of papillose cells (Fig. 72A).

The genus *Drimys* shows stages in the restriction of the stigmatic area to the distal part of the carpel and the internal surface. In the section Tasmannia, the carpel borders are only appressed, and closure is by glandular hairs; in the section Wintera, the borders are sealed, and the crest reduced to a distal area. If the terms transmitting tissue and conducting tissue were discarded, all the tissues that nourish and control pollen-grain germination and pollen-tube growth could well be called the *stigmatic tissues.* The stigma could then be loosely defined as the specialized distal part of the carpel, covered wholly or in part by stigmatic tissue. The *stigmatic crest* is a primitive form of stigma where the stigmatic tissue is present along the carpel margins, below as well as at the apex (Figs. 72*C*, 83).

It has been claimed that, morphologically, the stylar canal represents the distal parts of modified or vestigial ventral carpellary bundles, but this theory is based on misinterpretation. Vestigial bundles, representing lateral veins, are often present in the style and may lie beside the stylar canal. Both ventral bundles and the canal are present even in one of the examples cited to support the theory. (The Boraginaceae show well the strands of transmitting tissue beside and distinct from the ventral carpellary bundles.)

Placentation

Placentation in Free Carpels. Ovules are borne on the adaxial surface of the carpel lamina. Evidence from comparative studies and from anatomy demonstrates that there are no cauline ovules in the angiosperms; all are borne on carpels, though some appear to be terminal on the receptacle. The ovules may be distributed over most of the surface of the lamina, with only a narrow marginal strip and sometimes the midvein line sterile, or they may be greatly restricted in location in several ways. The pattern of ovule arrangement on the carpel constitutes *placentation.* The area where the ovules are attached, usually more or less enlarged as an emergence of the laminar tissues, is the *placenta.* In primitive taxa, there may be little or no modification of the region where the ovules are borne; in these groups, the placenta is merely a location. But position of ovule attachment is important both morphologically and taxonomically. Absence of an emergent placenta is probably the primitive condition; *Drimys* and *Degeneria* have only slight placental ridges. The ovule-bearing projection may be simple, a mere low cushion or ridge, but it is frequently large and complex in form and structure, and, together with the ovules, may nearly fill or even divide the locule (Fig. 80). It may form a major part of the ripe fruit and even project from the ovary walls. The form of the placenta varies greatly in number of ovules, type of ovule arrangement, and form of locule. When the placenta is close to the margin and where, in specialized carpels, the marginal band is reduced or lost, interpretation of ovule position may be difficult.

Where the fertile area is all or most of the ventral carpel surface, the entire surface is the placenta. Where the fertile area is restricted to submarginal areas and the margins are fused so that the fertile areas are brought together, the placenta is double in nature, consisting of strips from two lateral areas. Two-lobed form (Fig. 80*C*, *D*, *E*) may give evidence of this union, but often the fusion is anatomically complete and the placenta is simple in form and structure. It has been suggested that the term *half placenta* be used for each of the parts of this double structure, but this would involve unnecessary complications in description, especially in syncarpy, where fertile areas from adjacent carpels

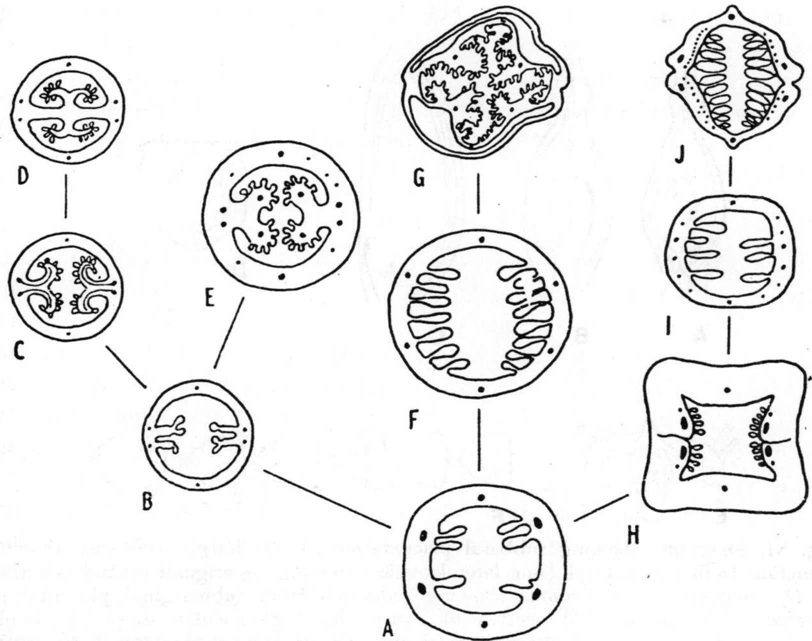

Fig. 80. Chart showing placentation in the Gentianaceae. The variety of types show theoretical modifications of basic form and possible evolutionary relationships of types. Placenta merely a position in *A, F, I, J*; a ridge in *H*; a carpellary flange in *B, C, D, E, G*. *A, Gentiana chinensis; B, Chironia baccifera; C, C. palustris; D, C. densiflora; E, Sabatia; F, Crawfordia fasciculata; G, Bartonia virginica; H, Frasera speciosa; I, Gentiana linearis; J, G. crinita.* (*After Lindsey.*)

unite to form a common placenta. Placentae would have to be described —as were carpels under the polymorphism theory—as "one-half plus one plus one-half" in nature. The term placenta should be used loosely for the fertile area of a carpel or united carpels.

Types of Placentation. In both morphology and taxonomy, the use of terms describing placentation is inconsistent and confused. A term given to placentation in free carpels has been applied to a morphologically different type in connate carpels. And differences in interpretation of the nature of the syncarpous gynoecium—whether it is wholly carpellary or partly cauline in nature—have greatly complicated the terminology of placentation. Morphological consistency requires interpretation of placentation in the simple, free carpel as the basis for that in the fused-carpel ovary.

The follicle type of carpel, with the carpellary margins approximated or united, is generally accepted as primitive. In this carpel, there are two types of placentation. Commonly, the ovules are borne in rows near

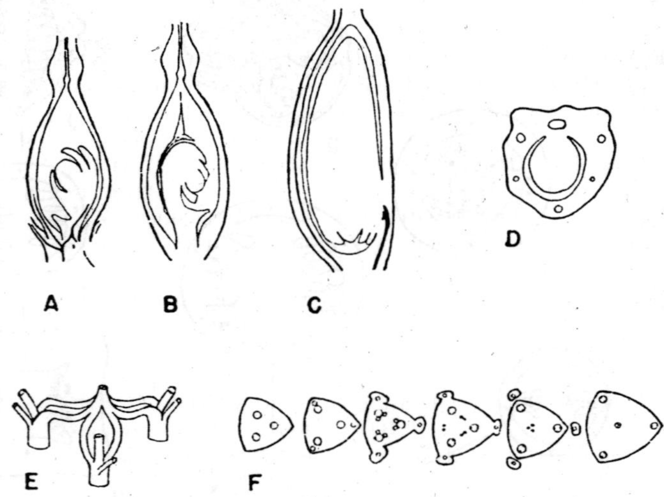

Fig. 81. Diagrams showing subbasal placentation. *A, B, Scirpus robustus,* showing reduction from free central (one lateral ovule surviving on original central column); *C, D, Saccharum* and *Bromus* showing reduction from submarginal placentation: *Saccharum* in longitudinal section of ovary showing vascular supply to ovule, *Bromus* in cross section of ovary at level of ovule attachment showing three carpels (one fertile), three dorsal and three (double) ventral bundles, the ovule supplied by the large ventral; *E,* diagram of typical vascular supply for flower in Cyperaceae; *F,* series of cross sections of flower of *Cyperus dentatus* showing origin and path of traces to carpels and stamens: three carpels, each with dorsal and two ventral traces, the pairs of ventrals uniting in a central column, which supplies the ovule traces. (*A, B, E, F, after Blaser.*)

the margins—*submarginal placentation;* infrequently, they are distributed broadly over the lamina—*laminar placentation* (also termed *diffuse, scattered, superficial, reticulate*). Laminar seems morphologically preferable. Reduction in ovule number in the carpel with submarginal placentation, together with modification of carpel size and form, brings about changes in apparent ovule position; the one, two, or few surviving ovules are usually borne near the base or apex of the locule, and the placentation is termed *basal* (Fig. 81) and *suspended* (apical) (Fig. 82), respectively. Basal placentation, derived directly from laminar

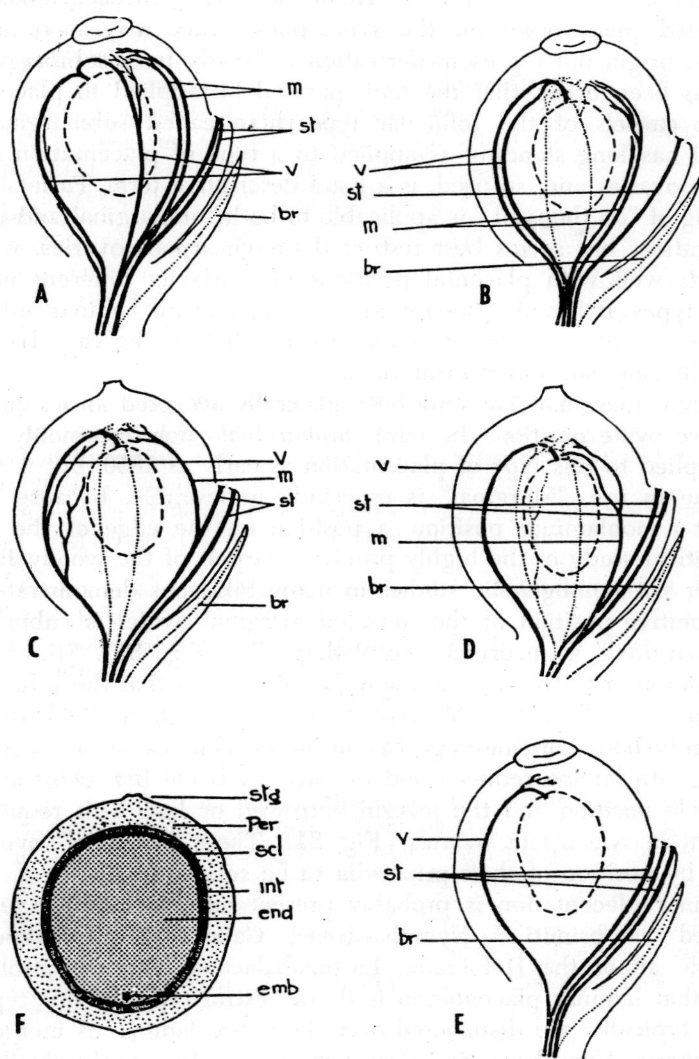

Fig. 82. *Sarcandra* (*Chloranthus*). *A* to *E*, longitudinal diagrams showing vascular structure of ovary with solitary suspended ovule. Vascular supply to ovule from both ventral bundles in *A*, *B*, or from one only in *C*, *D*, *E*; *F*, cross section of fruit. *br*, bract; *emb*, embryo; *end*, endosperm; *int*, inner integument; *m*, dorsal bundle; *per*, pericarp; *scl*, stony layer; *st*, stamen traces; *stg*, stigmatic end of fruit; *v*, ventral bundle. (*After Swamy and Bailey.*)

placentation, is apparently rare—Helobiales, Nymphaeaceae. Basal and suspended placentation in the syncarpous ovary may have a more complex origin but represent derivation indirectly from submarginal.

It has been urged that the term parietal be applied to placentation in free carpels of the follicular type, here called submarginal, but parietal has long standing as applied to a type of placentation in syncarpous ovaries and, so used, is a good descriptive term. Parietal, in its meaning of "on the wall," is applicable to both submarginal and laminar placentation, but seems best restricted to syncarpous ovaries, where it contrasts well with placental positions of markedly different morphological types. It has also been used as a synonym for laminar, especially where the ovules are very few and are isolated on the carpel wall between the median and ventral veins.

Though "marginal" has long been generally accepted as a simple and primitive ovule position, the term *submarginal*—now commonly used—was applied to this type of placentation as early as 1850, but it did not continue in use. "Marginal" is especially undesirable, because it may suggest a nonlaminar position, a position on the edge of the carpel. The critical study of the highly primitive carpels of the woody Ranales, together with ontogenetic studies in many taxa, has demonstrated that the primitive position of the so-called marginal ovules is submarginal, and "marginal" is incorrect, morphologically (Fig. 83). Study of the placentation of free carpels in many families shows that the submarginal position is characteristic of most taxa. In some taxa, the ovules are apparently borne on the edge of the lamina, but these are specialized carpels with lamina reduced and the strip of blade between the ancestral ovule position and the margin narrowed or lost. This reduction is frequent in syncarpous ovaries (Fig. 84). These "marginal" ovules are shown by position of their primordia to be submarginal.

Laminar placentation is probably present only in families generally accepted as primitive—Nymphaeaceae, Cabombaceae, Butomaceae, probably all of the Helobiales, Lardizabalaceae. There is strong evidence that laminar placentation is the ancestral type. In this type, the ovules, typically, are distributed over the entire lamina, as in *Butomus, Hydrocharis, Nymphaea,* and derive their vascular supply chiefly from the smaller meshwork bundles of the lamina, rarely directly from the median or major lateral veins. In reduction in ovule number and restriction to the submarginal position, a marginal strip of the lamina and the median line become sterile, and the ovules derive their vascular supply from two major lateral bundles, one on each side, not from the laminar vascular meshwork. The Degeneriaceae and Winteraceae show evidence of transition from laminar to submarginal placentation. In *Degeneria,* though most of the ovules derive their vascular supply from the ventral veins, the distal ovules are connected with branches from the dorsal bundle, and a few ovules derive their supply from a meshwork of veins connected with both dorsal and ventral bundles (Fig. 83). In *Drimys,* the ovules are in two submarginal rows, with the ovule traces derived, in part, from the ventral bundles and, in part, from anastomosing branchlets from both dorsal and ventrals. These

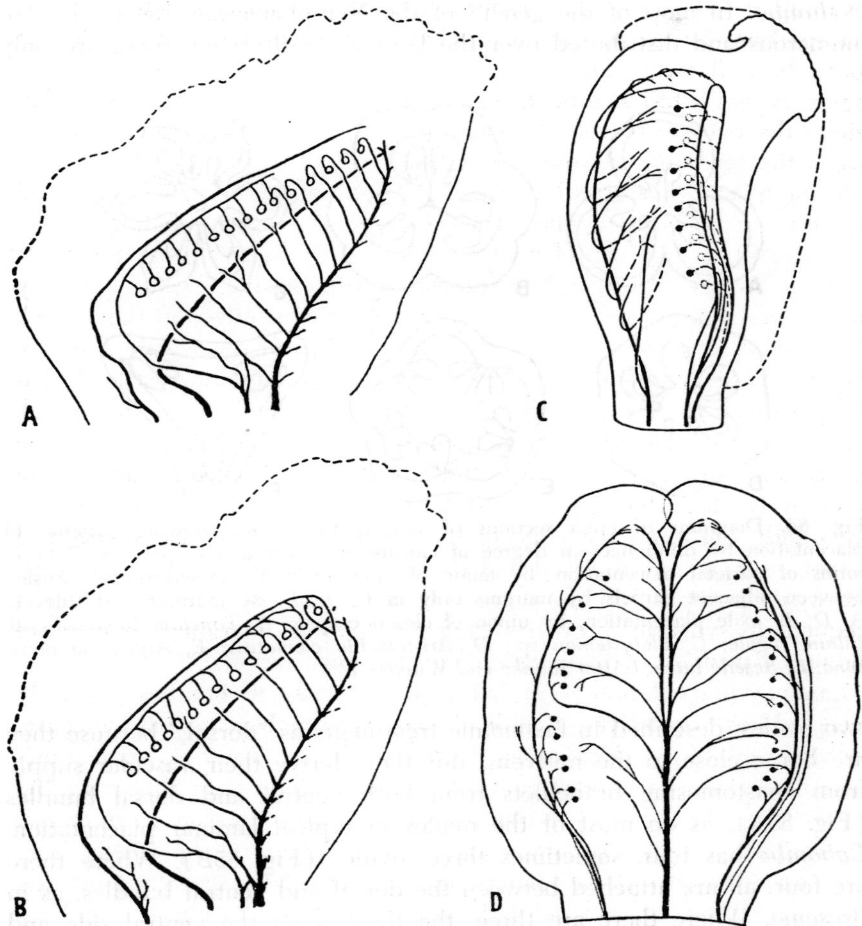

Fig. 83. Diagrams showing placentation transitional from laminar to submarginal as shown by vascular structure. Ovules in loose, submarginal rows with traces derived from dorsal and ventral bundles and from the small, connecting strands. *A, B, Degeneria,* showing variations in ovule supply; *C, D, Drimys* (Tasmannia type): *C,* lateral view of carpel; *D,* carpel split ventrally and spread open. For clarity, ovules are omitted, small circles and black dots indicate position of micropyles; dotted lines indicate position of stigmatic crest. (*A, B, after Swamy; C, D, after Bailey and Nast.*)

two primitive families, with placentation transitional from laminar to submarginal, seem to show that the primitive ovule position was laminar and that the submarginal type has been derived from the laminar by restriction of ovules to the near-marginal areas.

A similar evolutionary step is evident in the Nymphaeaceae, Cabombaceae, and Ceratophyllaceae, where ovule number is reduced in the more specialized genera—*Nuphar, Brasenia, Cabomba, Ceratophyllum, Nelumbo*. In most of the genera of the Nymphaeaceae, the ovules are numerous and distributed over the lamina; in *Brasenia,* there are only

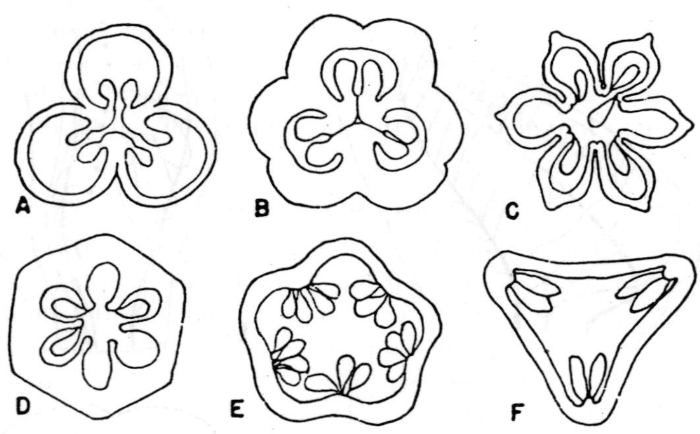

Fig. 84. Diagrams of cross sections of syncarpous ovaries showing varieties of placentation by differences in degree of closure of constituent carpels. *A, C* to *F,* forms of parietal placentation, by union of open or partly closed carpels; fusion between adjacent carpels by margins only in *C, E, F;* by margins and sides in *A, D; B,* axile placentation, by union of closed carpels. *A, Tofieldia calyculata; B, Lilium regale; C, Platystemon* sp.; *D, Aristolochia clematitis; E, Argemone mexicana; F, Reseda lutea.* (*After Juhnke and Winkler.*)

two ovules, described in taxonomic treatments as "dorsal," because they are borne close to the midvein. But they derive their vascular supply from anastomosing branchlets from both ventral and dorsal bundles (Fig. 85*A*), as do most of the ovules in typical laminar placentation. *Cabomba* has four, sometimes three, ovules (Fig. 85*B*). Where there are four, all are attached between the dorsal and ventral bundles, as in *Brasenia.* Where there are three, the third is on the ventral side and above the others, and the vascular traces of all three come from the ventral bundles: those to the lower ovules as laminar branches across the carpel sides, that of the upper ovules, directly. Ovule position and vascular supply in these two genera clearly indicate derivation from laminar placentation, as would be expected in these highly specialized genera of the nymphaeacean line. The carpel of *Ceratophyllum,* with only one ovule, has a reduced vascular supply, consisting of simple, unbranched dorsal and ventral bundles (Fig. 85*C*). The ovule is described as "borne on the dorsal bundle." Though illustrations show no vascular trace, the ovule lies directly over the dorsal bundle and seems to represent a last stage in reduction from laminar placentation. These three genera apparently show derivation of few-ovule placentation directly from the laminar type. In contrast, in carpels with one or two ovules in apocarpous Rosaceae and Ranunculaceae, basal placentation is clearly reduced from submarginal (which, in turn, has been derived from laminar placentation) (Fig. 42).

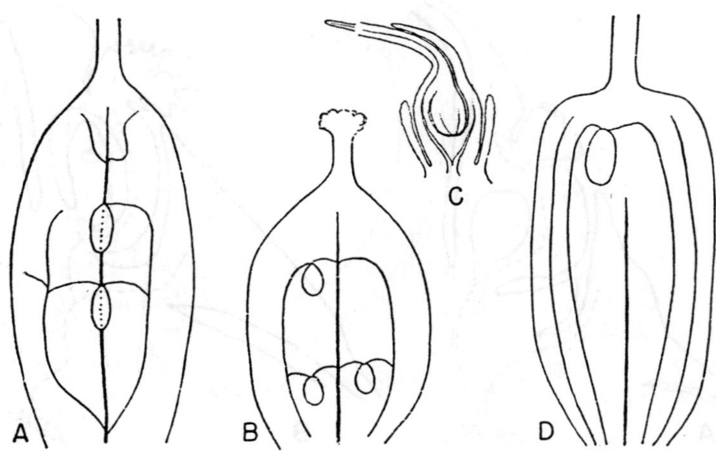

Fig. 85. Diagrams of ovaries showing reduced laminar placentation. *A*, *Brasenia*, two ovules at union of midvein and lateral veins; *B*, *Cabomba*, three to four ovules on transverse veinlets; *C*, *Ceratophyllum*, vascular supply reduced, remnants of dorsal and ventral bundles; *D*, *Nelumbo*, ovule solitary, attached at union of lateral veins. (*A, B, D, adapted from figures and descriptions of Strasburger and Saunders; C, after Eckardt.*)

Further evidence that laminar placentation is primitive is seen in the Butomaceae and Hydrocharitaceae, families in which this type of placentation is associated with open carpels and, in the Butomaceae, with decurrent stigmas. Laminar placentation, though little discussed, has been considered a specialized, rather than the primitive, type, a form derived from "marginal" by increase in the fertile area of the sporophyll. But the presence of the laminar type in primitive families of both dicotyledons and monocotyledons and of transitional forms in the Ranales and Cabombaceae supports the view that placentation in the angiosperms was primitively laminar.

In greatly reduced placentation, interpretation of ovule position may be difficult; evidence from anatomy, ontogeny, and comparison with ovule position in related taxa are all necessary where a carpel has only one or two ovules. These ovules may be survivors of the many ovules of either laminar or submarginal clusters. The terms basal, median, distal, suspended are descriptive, rather than morphological. "Basal" should be interpreted as morphologically *subbasal.* The "basal" ovules of many taxa, such as *Anemone* and *Potentilla,* which have abortive ovules (Fig. 42*N*) and traces to lost ovules along the ventral suture above the normal ovule, show that the functioning ovule is the lowest in the ancestral row. (These genera are achene-bearing taxa, with the achene obviously

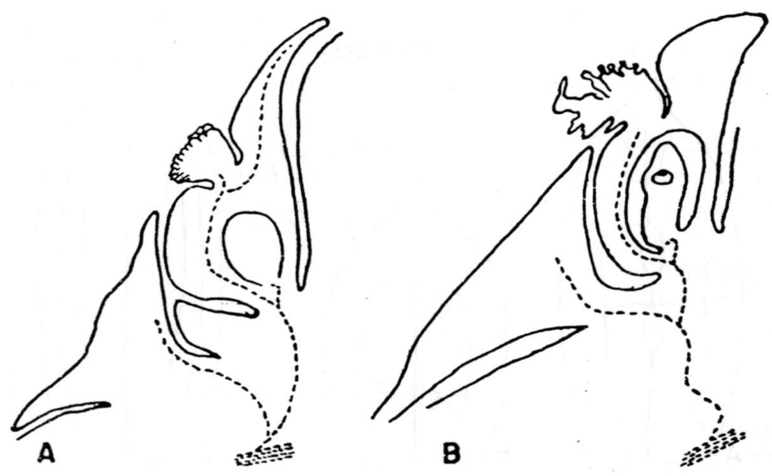

Fig. 86. Longitudinal sections of flower of *Peperomia* showing solitary, "basal" ovule with form of ovule trace, hooked, indicating ovule to be lateral, not basal and cauline. *A, P. argyreia; B, P. cniapas.* (*After Murty.*)

reduced from a follicle.) That the apparently basal position in achenes is morphologically submarginal is evident in many taxa by the course of the ovular trace, which shows that the apparent basal position has been obtained by differential growth, a downward "migration of the ovule"; the trace *descends* to the so-called basal position from an originally more distal position (Fig. 86)—*Ranunculus, Piper.* (Solitary, so-called basal ovules in syncarpous ovaries may show similar evidence of change in position.) Objection has been raised that this roundabout course of the ovule trace has no significance, that it is the result of physiological demand, but vascular bundles, developing in relation to function, as in the fleshy tissues of fruits, do not follow circuitous routes. The term basal is perhaps more often applied to the solitary ovule of a syncarpous ovary, where it is considered to be cauline (Fig. 81). These ovules, as shown by anatomy and comparison with the placentation of closely related groups, are always appendicular. Interpretation of the basal ovule in free carpels as cauline is nonsensical, especially when applied to stipitate achenes; minute projections of the stem must be assumed to pass through the stipe of each carpel.

The solitary surviving ovule of a submarginal row is usually the proximal member but may be the distal member, as in *Leitneria*. Rarely, if ever, is it a median one. The term *median placentation* has been used occasionally—apparently always loosely and in a descriptive sense—for ovules borne "in the middle of the lamina," on the dorsal side, on the ventral side, or between these, as in *Brasenia* and *Cabomba*. *Median* has been suggested as a replacement for submarginal and, in "peltate" carpels, as a term to refer to the condition where the solitary ovule is borne on the "cross zone." Median, as a descriptive term for placentation, is hopelessly loose morphologically and should not be used. It has been combined with an even poorer term, *lateral*—in "median lateral." Both "lateral" and "median" have been applied to ovules borne on the dorsal and ventral edges of the folded carpel and on the lamina between. The vestigial ovules in achenes which are obviously submarginal and ventral have been called "lateral median," and the ovules in achenes of the *Ranunculus* type have been described as "median lateral." These terms are morphologically meaningless and of doubtful descriptive value.

Axial, parietal, and *free central placentation* are discussed under Syncarpy.

The U-type Placenta. As a part of the demonstration of carpel morphology according to the peltate theory, a form of placenta called the *U-type* has been described and interpreted as "the placentation type of the angiosperms." The interpretation has been the result of the extension of the theory of the underlying peltate nature of floral appendages to greatly reduced carpels, with emphasis on the "cross zone" as a morphologically important part of the organ, especially of the carpel. The U-shaped placenta has been given this name because it surrounds the base of the ventral furrow or slit of a so-called peltate carpel. It is described as present in its typical form in achenes of the Rosaceae and Ranunculaceae, where the ovule is borne at the base of the "slit"—and of the "U"—on the "cross zone," with the arms of the placenta extending upward along the united margins of the carpel. The arms of the placenta, sterile in most of these taxa, may be fertile. The placenta, in these achenes, is merely a position, or area, where the ovule is borne.

A consideration of placentation in general and the ontogeny and basic morphology of the carpel shows that the U-type placenta is merely a modified form of the submarginal type, a form resulting from advance in carpel closure from ontogenetic to congenital fusion, together with reduction in ovule number to one. The primordium of the more primitive carpels is crescent-shaped in cross section, and closure is ontogenetic; that of advanced carpels is ring-shaped in c oss section, and the maturing carpel is tubular and its margins are united congenitally. The ontogeny of many carpels combines these two methods. The primordium, at first crescent-shaped, becomes ring-shaped later. The carpel, so formed, is distally open, and closed only in late stages; it is proximally tubular, arising closed. (A tubular corolla with petallike lobes has similar ontogeny.)

The so-called U-type placenta is far from primitive and cannot be considered in any way a basic type. It is a high type of placenta, the result of extreme reduction in carpel length and ovule number. According to the theory that the U-type is the basic kind of placenta, the typical double-row, submarginal placenta is considered derived from the U-shaped type. This is an example of reading a series from the apparently simple to the complex, because the simple was accepted as always primitive. The series obviously runs in the other direction; achenes are surely highly specialized carpels.

In carpels in which the primordium changes during ontogeny from crescent-shaped to ring-shaped, the level of transition is a point or line where the margins meet. This transverse connecting bit of margin has been termed, under the peltate theory, the *cross zone.* It varies in width with the form of the crescent-shaped primordium but is usually very narrow. The peltate theory stresses the cross zone as of much morphological importance, but its existence is incidental to the change in primordium shape, the beginning of congenital fusion of the carpel margins. Recognition of the zone as significant of basal form in carpel morphology has led to far-reaching misinterpretations in structure.

The cross zone has been considered a placenta because, as seen in achenes, the solitary ovule is attached at that level, apparently on the connecting bit of transverse margin. But the ovule is the lowest member in the submarginal placenta of the ancestral follicle, as shown by anatomy and by the presence, in *Clematis* and *Anemone,* of vestigial ovules above the normal ovule. Placentation called U-form has also been called lateral, median, median laminar, median lateral, ventral median, and submarginal lateral. Obviously, it is a reduction form of submarginal. In both apocarpous and syncarpous gynoecia, reduction in ovule number introduces much difficulty in the interpretation of ovule position and the terminology of placentation; comparative, anatomical, and ontogenetic studies are essential in interpretations.

Reduction in Laminar Placentation. The evolutionary history of placentation is clearly one of reduction of ovules from many and indefinite in number to much smaller numbers and, ultimately, to two or one.

The primitive carpel has numerous ovules distributed over the adaxial surface, except perhaps for a narrow marginal band. Ovules are occasionally borne even on the midvein area, though less abundantly than elsewhere. There seems to be no definite pattern of position or of origin of vascular supply. Each ovule is supplied by a slender trace, derived largely from the meshwork of small bundles of the lamina; the traces represent the tips of more or less free veinlets, or shorter strands derived at points of anastomosis in the meshwork and involving small veins only, or small and larger veins, even the midrib. With reduction in number, the ovules are first restricted to rows near the margins, with their traces derived from ventral veins—submarginal placentation. The rows are at first rather broad and loose, with traces of some of the ovules derived from the lamina meshwork (*Degeneria, Drimys;* Fig. 83*D*); with further reduction, the rows become linear, with all traces from the ventral veins; further reduction of ovules in the submarginal lines leaves few, and ultimately two or one. The persisting ovules are commonly the proximal members of the row—often with vestigial, distal ovules above. Persisting distal ovules—two or one—are uncommon or rare (Proteaceae), and median ovules are rare. "Basal," "suspended," and "median" solitary ovules, especially those of syncarpous ovaries, represent various morphological situations and require individual interpretations (Figs. 81 and 82). Solitary ovules may have more than one trace, and the traces may come from the two ventral bundles or even, very rarely, in syncarpous ovaries, from the ventrals of different carpels. Fusion of the major carpellary veins—characteristic of many achenes (Fig. 42)—may suggest the derivation of ovule traces directly from the dorsal vein.

The Nymphaeaceae and Cabombaceae provide excellent examples of reduction to few, two, and one ovule directly from lamina distribution. The vascular structure of the carpels shows that, in reduction in ovule number, intermediate submarginal stages have been omitted. Most of the genera in the Nymphaeaceae have the numerous ovules scattered widely over the carpel surface; *Nelumbo* has only one; in the Cabombaceae, there are three (rarely four); and in *Brasenia,* two (rarely three) (Fig. 85). Ovular position in these genera has long been a morphological puzzle, but their positions on the carpel walls and the derivation of their traces show that they represent scattered members of ancestral laminar distribution.

In taxonomic treatments, ovule position in these taxa has been described as "parietal" and "on the sides of the carpel"—more specifically, in *Cabomba,* "on the lateral walls" and, in *Brasenia,* "along the dorsal suture"; these locations are also called by the morphologically better terms "laminar lateral" and "laminar dorsal," respectively. In

Cabomba, two of the three or four ovules are borne on the side walls, near or below the center of the carpel, the others similarly above (Fig. 85*B*). Horizontal veinlets extend from the midrib and from the laterals across the carpel side just above the ovules, and the ovule traces are derived at the point of anastomosis of each pair. This has been described as "suspended by the sling method." In *Brasenia,* the vascular system is similar, but the ovule trace is derived at the point of union of the midvein and a transverse veinlet; the ovules, therefore, are borne along the midvein. In *Nelumbo,* the ovule is attached along the ventral suture, deriving its trace in the "sling" fashion at the point of union of horizontal veinlets which connect two "posteriolaterals." *Ceratophyllum* has a solitary ovule attached along the dorsal vein. In these four genera, the placentation is clearly modified laminar (Fig. 85). Some of the Magnoliaceae (*Michelia*) also show reduction of laminar placentation —ovule supply from both dorsal and ventral veins.

Morphological Nature of the Carpel

The problem of the fundamental nature of the carpel—axial or appendicular—leads far back in the history of vascular plants, back to the differentiation of the ancient, thalloid plant body. Because of similarity in form and structure, the leaflike nature of the carpel has been recognized since the earliest days of plant study. Detailed comparative studies in ontogeny and anatomy, especially those of nodal structure, have shown that the similarity is much closer than had been believed; that many supposed differences in position, origin, and development of these organs are based on misinterpretations. Discussion of the possible homology of carpel and leaf has been prominent in botanical history. The carpel has been called a fertile leaf, and the leaf, a sterile sporophyll. Leaf and sporophyll have been called organs *sui generis.* The theory of phyllospory and stachyospory has added the new term *stegophylls* for carpels that "surround" and "protect" stachyosporous ovules. Interpretation under this theory implies that the appendages commonly called carpels are of two basically different types —foliar and cauline—an interpretation that is unacceptable by morphologists. In 1956, the interpretation of the carpel as a leaflike appendage was challenged again; it was considered axial, a part of the receptacle, on which are borne "buds," the ovules. In support of this view, the phyllospory-stachyospory theory maintains:

1. The inversion of bundles has no significance. The inversion is merely the "aberrant result" of the "moving inward" of a vascular bundle to a part radially inward from its point of origin.

2. Pairs of normally oriented bundles in the ovary wall—as in parietal placentation—are merely "two adjacent bundles."

3. The vascular systems supplying ovary wall and placenta are independent.

4. The recent recognition of the two-trace nature of the primitive angiosperm leaf breaks down the homology of carpel and leaf, for the carpel is a three-trace organ.

5. The sessile nature of carpels is evidence that the carpel is axial, a part of the receptacle.

6. The placenta is an independent structure, also of axial nature, bearing "terminal buds" (ovules).

In support of the appendicular theory, are the following facts:

1. The inversion of the lateral bundles is clearly the result of upturning and inrolling of the sides of laminar organs; they do not arise inverted but are inverted during their course to the carpel or within the carpel base. Similar inverted bundles are common in petioles, where there has been a rolling of the margins and a flattening. Inverted bundles are also found in many veins in cylindrical and rolled leaves of xerophytes.

2. The pairing of normally oriented bundles in a syncarpous ovary wall is not a mere "happenstance." The pairs are the ventral bundles of adjacent carpels, fused margin to margin, as open and partly upfolded laminar carpels. The bringing together is ontogenetic in some taxa; congenital in others.

3. Independence of vascular systems of ovary wall and placenta is normal; the traces which supply these parts arise at different levels on the stem and wholly independently. (They may be freely connected by small veins.) If carpels are to be interpreted as axillary on this basis, so are leaves and all other floral organs in which there are several traces. In laminar placentation, there is no distinction of vascular "systems."

4. The dorsal trace of the carpel is frequently double; the stamen, in some families, has two traces; the cotyledon, in most families, has two or four traces and shows the origin of the three-trace condition by fusion of the median two (Chap. 9). Anatomically, carpel and leaf are now brought even closer together. (Recognition of the two-trace node as primitive has replaced the long-held theory that the three-trace node was primitive.)

5. The carpel in primitive families, both monocotyledonous and dicotyledonous, is stipitate; the sessile carpel is specialized.

6. The placenta is merely a location on the carpel where ovules are borne and where, often, the fertile area becomes enlarged. It appears to be an independent structural part only in highly specialized ovaries where greatly enlarged or where lateral-wall connections have been lost, as in free central and basal placentation. Where enlarged, it has no uniform pattern of vascular supply, as an independent axis would

have. If a carpel be considered an axis, it would be a hollow structure containing other axes, the placenta and its branches, the ovules.

The claim of homology of carpel and axis fails to explain the existence of open carpels and the ontogenetic closing of carpel primordia. It completely disregards anatomical structure and ontogeny, and is valueless in morphological interpretations.

When evidence from all fields is considered, none of the twentieth-century concepts of the nature of the carpel can displace the classical view that the carpel is a fertile lateral appendage of a determinate stem tip. In details of position and origin on the stem, and of ontogeny and anatomy, it is like the leaf and is clearly of leaf rank. Comparative studies of the stamens of primitive dicotyledons show that this sporophyll, although much more modified than the carpel, is, like the carpel, basically leaflike in form, structure, and ontogeny. The similarity between the two types of sporophylls strengthens the classical view.

Axial Nature of the Carpel. Interpretations of the sporophylls as branches of the axis, not appendages, take various forms. The simplest of these considers the flower a branching stem. The sporophylls are seen as lateral branches. These are subtended by bracts, the sepals and petals. In very few flowers can position of the floral organs—sporophylls radially internal to outer organs—be seen to suggest this relationship. In anatomical structure, all the organs are leaflike, and their vascular traces are like those of leaves, not branches.

The concept that the carpel is an aggregate of telomes—basic elements of an ancient, ancestral thalloid body—can perhaps be considered an "axial" theory. The carpel, so interpreted, is made up of a determinate, much-branched lateral part of the body; the distal branchlets, fused, form a laminar structure. If the plant body of the angiosperms is interpreted in terms of telomes, the carpel can well be so described. Consideration of the primitive plant body as made up of basic units, telomes, is doubtless of value for the understanding of the more primitive taxa, but its value in the interpretation of the higher taxa, where axis and appendages have become established as morphological units, is doubtful.

The Sui Generis Theory. Akin to the theory that carpels and stamens do not exist as morphological units is the concept that the sporophylls are "organs *sui generis,*" in other words, organs neither homologues of leaves nor leaflike in nature. The basis for this interpretation was found in supposedly fundamental differences in the structure and behaviour of the apical meristems of reproductive and vegetative stems. The floral apex is considered limited by the development of the carpel, which is described as terminal; proliferation of the apex does not occur. The vegetative axis is unlimited in development, and the leaf is a lateral organ. The leaf arises as a flattened organ; the carpel as a tubular organ; if leaflike, it would be laminar in early stages The floral apex is broader and flatter than the vegetative apex.

The structure of many flowers shows that these differences do not exist. The broader apex of the floral stem tip is the result of the close arrangement of the appendages and the cessation of growth in length, and is seen in determinate vegetative stems and in short-shoots and thorns. The carpel is not terminal on the floral apex but lateral, as shown by ontogeny and by vascular anatomy. The axis continues beyond the uppermost carpels in many taxa, as is evident in external form and in vascular structure. The carpel develops like a leaf in every detail, from its origin in the apical meristem to apical and marginal growth of the lamina. The primitive carpel is, in early stages, laminar in form (it closes by the ontogenetic fusion of the margins); only the highly specialized carpel arises as a tubular structure (fusion of the margins is congenital). The *sui generis* theory is without support from ontogeny or vascular anatomy.

Theories of "Acarpy." The theory that floral appendages do not exist as morphological units received some attention in the early twentieth century. It has been expressed in two forms, with supporting evidence derived chiefly from ontogeny and physiology. The appendages have been considered mere emergences, determined in position, number, and nature by nutritional and spatial relationships in early stages of stem-apex ontogeny. Vascular structure, under these theories, has no significance; its form is dependent upon physiological demand; it is not conservative of ancestral form and has no phyletic significance.

Under one interpretation, the flower is considered to consist of a determinate stem tip, divided into lower, sterile, and upper, fertile areas. Emergences arising on the lower part are sterile and form bracts and sepals; those on the upper part bear sporogenous tissue and constitute the stamens, ovules, placentae, and carpel walls. The lower fertile emergences may become sterile and form petals and staminodia. Above the fertile emergences, other sterile structures, styles and stigmas, appear. The carpel is looked upon as a phyllocladlike, ovule-bearing emergence; the ovules as enclosed by the inrolling of the margins.

The basis for this understanding of flower structure is that each floral type is an individual expression of flowering; the comparison of the achene with other fertile emergences need not imply evolutionary development from ancestral carpels of any type. These statements imply that there is, for the carpel (as for all appendages), no basic plan or form, no natural relationship among types; follicle and achene, superior and inferior ovaries are autonomous structures and bear no phyletic relation to one another.

Another interpretation of the appendages of the floral meristem is based on the pattern of origin of these organs on the stem apex. Leaves are considered to be produced in helices; the sepals, and sometimes the petals, stand in the same series at the end of the spirals, but the arrangement of stamens and carpels is said not to continue the helices. (No mention is made of the taxa—Nymphaeaceae, Paeoniaceae—in which the sporophylls do continue the spiral series of the perianth.) Leaves are said to be formed by an "annular meristem," a part of the tunica, but, near the apex, the tunica is used up in the formation of the lower sterile appendages, and the corpus alone is involved in the formation of the fertile appendages. The interpretation of the histological behavior of the tunica and corpus when the floral organs are being formed is involved and obscure. It appears that, since the sporophylls have—in the opinion of the proponent of this theory—an origin entirely different from that of leaves, they are not of leaf rank; apparently they are considered as of less than organ rank. By another morphologist, they have been called "lobes of the receptacle."

The view that floral organs are mere excrescences of the axis and not independent organs can be held only if anatomical structure and early ontogeny are meaningless, and if morphology has no significance. The theories of acarpy "hardly come within the province of morphology."

The Caytonialian Theory. A Mesozoic group of fossil seed plants, the Caytoniales, provide the basis for another theory of the origin and nature of the carpel. Although the relationships of these plants are uncertain, they have received considerable attention as possible ancestors of the angiosperms, because their ovule-bearing organs have been interpreted as representing an early stage in the evolution of the carpel. The Caytonialian megasporophyll is a frondlike, pinnate, dorsiventral organ, consisting of an axis, which bears two lateral rows of globose "cupules," or "ovaries." Each cupule has a median longitudinal vein, along which ovules are borne. A folliclelike carpel is considered derived by the reduction and condensation of this fertile organ; the axis is greatly shortened and the cupules reduced to the lowest member of each row, and these fused to the axis and to each other. In the complex organ so formed, the axis of the ancestral sporophyll would form the midvein of the follicle; the vascular bundles of the two cupules would form the ventral bundles along which the ovules are borne; the ovules of each cupule form one of the two rows. This seems a farfetched and complex derivation for so simple a structure as the follicle.

Structurally, the carpel and the Caytonialian sporophyll are wholly unlike; the carpel is palmately veined, the Caytonialian organ strongly pinnate. The ovule of the angiosperms has two integuments; that of the Caytoniales, one.

It is claimed that this theory of the origin of the carpel was not originally considered to suggest the origin of the angiosperm carpel itself from the Caytonialian megasporophyll, but as a possible method of origin of the carpel from a pteridosperm sporophyll. But, in recent years, it has been suggested many times that Mesozoic pteridosperms are possible ancestors of the angiosperms, and the theory has been considered to imply this origin.

Homologies of the Carpel with Gymnosperm Structures. Comparison of reproductive structures of some living gymnosperms with simple angiosperm flowers has been believed to demonstrate homology between the integument of the gymnosperm ovule and the angiosperm carpel. In *Ephedra,* the pair of bractlets that enclose the naked, apparently terminal, ovule are considered homologous with the carpels of a two-carpellate flower of *Peperomia,* demonstrating the integumentary nature of the carpel. As a part of the search to find in the gymnosperms the origin of the carpel, the cone of *Juniperus* has been claimed to give important evidence. The ovules of this conifer are "basal" on the fertile scale and interpreted as cauline ovules surrounded by sterile scales, the homologues of angiosperm carpels. *Ephedra* and *Juniperus* are considered to show that the carpel is a sterile appendage which surrounds cauline ovules. But the plants used to demonstrate these homologies are highly specialized: *Ephedra,* a surviving remnant of an ancient stock; *Juniperus,* a conifer with greatly reduced cones; the Piperaceae, angiosperms with greatly reduced flowers. The simplicity of great reduction suggests a similarity in morphological structure that does not exist.

Carpel Polymorphism. The theory that the gynoecium consists of carpels of more than one basic type was proposed by E. R. Saunders in 1923, and, for fifteen years following, many descriptive articles and, ultimately, a two-volume book, "Carpel Polymorphism," were published. This theory presented an entirely new interpretation of the gynoecium, one wholly at variance with the classical view. Two basic types of carpel, the *valve* and the *solid,* were described and, later, a third, the *semisolid,* was added. Recognition of these types multiplied twice to several times the number of carpels in the gynoecium of most families and commonly doubled the number of whorls. For example, the gynoecium of most legumes consists, according to this theory, of two carpels, one valve and one solid, that of the Liliaceae of six, rather than three, in two whorls, one valve, one solid. Large numbers of carpels were described in gynoecia commonly considered to have one or two: 10 to 14 in the peanut, *Arachis,* and *Scorpiurus;* 16 in *Brassica;* 40 to 50 in *Rapistrum.*

Definitions of these valve, solid, and semisolid types are difficult to make, because the basis for each type was continually changed, as the theory was extended, and the types are not distinct. The valve carpel is described as open, commonly sterile—if fertile, with one or more ovules on the lateral parts; the solid carpel, as "consolidated," columnar, narrowed, usually fertile, with one or many ovules on the flanks; the semisolid or "pseudovalve" as resembling the solid type but with lateral wings, approaching in form the valve type. The interpretation of structure of the simple, follicular carpel well illustrates the two basic types: (1) a solid carpel, the dorsal part, with the midrib vascular bundle and a narrow band of surrounding tissues—broad and winglike in its semisolid variation; and (2) a valve carpel, the larger, ventral part, with the ventral vascular supply and the sides of the carpel. No line of demarcation between the carpels was described.

The basis for the establishment of these types was anatomical. Each vascular strand supplying the gynoecium or the individual carpel was considered evidence of the presence of a carpel. The explanation of the pair of bundles or a double bundle on the ventral side of many carpels was that the bundle is split longitudinally, in preparation for dehiscence along the suture. The number of vascular traces supplying a gynoecium, apocarpous or syncarpous, was seen to indicate the number of carpels present, and the origin of these traces at two levels indicated two whorls of carpels. That the members of the "double trace" arose in the receptacle at two points, often widely separated, was not considered.

The polymorphism theory received very little support at any time and was refuted from several viewpoints. It became prominent because of the large amount of descriptive material published by its promoter over a long period, and the ingenuity shown in the interpretation of the gynoecia of many families under a theory that appeared fantastic to many morphologists. It is valueless because, though its basis was anatomical, its interpretations of vascular structure were made without regard to the basic vascular structure of the plant body. The fundamental error lay in the misinterpretation of stelar and trace structure in the floral receptacle and of the vascularization of the carpels themselves. Axis and appendages have a constant vascular relation at the node. If the leaf were interpreted on the same basis as the carpel under the polymorphism theory, each leaf with three-trace vascular supply would consist of two basic leaves, one a median "solid" leaf, and two lateral wings, which, together, form a "valve" leaf. Aside from the anatomical error, many inconsistencies rendered the morphological interpretations unacceptable.

The detailed descriptions of flower anatomy published as a basis for this theory are apparently generally accurate and provide much valuable factual material for the interpretation of flower morphology. The publication of the theory appeared, for a time, to throw discredit on the value of evidence from vascular anatomy but stimulated discussion of the basic anatomy of floral organs and contributed in this way to a knowledge of flower morphology.

The Peltate Carpel. The theory that the carpel is basically a peltate organ arose from similarities seen between the carpel and the peltate leaf and has received much attention in continental Europe since 1925. Under this theory, the carpels and gynoecia of many taxa have been reinterpreted, with support from details of ontogeny and anatomy. Peltate, in the sense of shield-shaped, as commonly used in leaf description, is an unfortunate term for use in this theory; the adjectives ascidiform, tubular, and utricular, which have been used less frequently, much better describe the form of the carpel. The peltate theory presents a morphological origin for the carpel more complex than the simple folding or inrolling of the margins of the lamina. Peltate form in the carpel is assumed to have arisen by the turning upward (ventrally) of the basal lobes of the lamina and their fusion, margin to margin, as in the formation of peltate leaves. Where the two marginal meristems meet, they unite, forming a transverse meristem, the *cross zone*. As the primordium elongates, the cross zone, continuous with the marginal meristems, is said to build up a ventral strip of carpel wall, which, united with the lateral walls, forms a tubular organ. Under this theory, the ovules are borne on the wall formed by the cross zone. The cross zone is considered a region of much morphological and functional significance. Its presence or absence, its position, and its degree of fertility—number of ovules borne—are seen to give much information concerning the evolution of the carpel.

Carpels are described as *peltate* or *epeltate* (not peltate). Peltate carpels have great variety of form, variety that is said to depend chiefly on the extent of the activity of the cross zone, as seen in the length of the carpel stipe and of the part of the carpel below the cross zone.

Two types of peltate carpels are described: *manifest peltate,* with well-defined stalks and a tubular lamina that has been formed by a well-marked cross zone; *latent peltate,* intermediate between manifest peltate and epeltate, with a very short tubular base and the cross zone only "suggested," or present only in early ontogeny. Achene types, such as the carpels of *Thalictrum,* are cited as the best examples of the manifest peltate type. Forms intermediate between achenes and follicles, such as those of *Calycanthus* and many genera of the Rosaceae and Ranunculaceae, are given as examples of latent peltate carpels. The "primitive achene" of *Ranunculus,* described as having a solitary ovule on a "latent" cross zone, is considered to represent a "first stage" in the building of the tubular follicle. In the follicles of the Ranunculaceae, an elongated fertile strip is claimed to be built up by a cross zone more active than that of *Ranunculus*. The carpel of *Calycanthus* is cited as showing steps in the transition from the primitive achene to the follicle: its short stipe and ovary represent a first stage in the evolution of a longer stipe and a longer tubular ovary; the two ovules, a step from the one of the achene to the many of the follicle. The follicle, with its many ovules on this long, fertile strip built by the cross-zone meristem, is interpreted as a higher type of carpel. (This interpretation of follicle-achene relationship is the reverse of that based on much generally accepted evidence from comparative form, vestigial structure, and vascular supply.)

Epeltate carpels "show no evidence of peltation" at any stage; they have no cross zone. The carpel primordium arises crescent- or horseshoe-shaped in cross section and remains so; the closure of the carpel is wholly ontogenetic, and there is, therefore, "no strictly tubular part"—that is, a part formed by a cross zone. Few taxa with carpels of this type are listed: Butomaceae, Hydrocharitaceae, Aponogetonaceae, Nymphaeaceae, Alismataceae, Scheuchzeriaceae. (Both follicles and achenes are present in this group of families.) The absence of a cross zone in these plants is considered remarkable and explained only as the result of reduction. Whenever a stipe is present in epeltate forms, this theory maintains that there must have been originally a cross zone.

Lines are not clearly drawn between the types of peltate carpels, and opinions as to the most primitive form seem to differ. The epeltate forms are considered reduced and derived from the peltate by loss of the cross zone. The latent peltate (achene) form is considered less advanced than the manifest peltate. The manifest peltate form is called the "basic form from which the others are derived."

Abundant evidence from anatomy, ontogeny, and comparative morphology of the families listed as having epeltate carpels, supports the view that the peltate theory reads the series in carpel form in the wrong direction; simplicity, here, is advanced, not primitive. The theory that the peltate form is basic for the carpel is not supported by comparisons throughout the angiosperms. If leaf form were interpreted similarly, the shield-shaped and pitcher-shaped leaves would be considered basic types. The taxa considered epeltate are phylogenetically primitive, according to the evidence of ontogeny, anatomy, and flower structure, and, in them, the peltate form should be seen in its simplest expression, if it is basic. Carpels of these taxa are interpreted as reduced from peltate types. The peltate theory assumes as primitive a carpel of the achene type, with a single ovule borne on the cross zone, a secondarily developed meristem. But the solitary ovule is unquestionably the highest type of placentation and the achene a reduced form of the follicle.

Ontogeny shows the nature of the peltate carpel. In early development, the typical carpel is like the leaf. The primordium arises as a moundlike projection, which soon becomes crescent- or horseshoe-shaped in cross section. If the primordium remains crescent- or horseshoe-shaped, the carpel formed is, under the peltate theory, "epeltate"; if the basal meristem of the primordium extends laterally on the receptacle and becomes ring-shaped, the part of the carpel formed thereafter is tubular and the carpel is peltate.

Epeltate carpels are closed ontogenetically; peltate carpels arise with congenitally fused margins. From the standpoint of ontogeny, many carpels are epeltate in early stages and become peltate in later stages. Where the primordium becomes ring-shaped very early—fusion of the margins is complete except at the carpel tip—the carpel is latent peltate, with a cross zone "only suggested." The peltate carpel, tubular wholly or in large part, developed from a ring-shaped primordium, is an advanced, not a primitive, type, in which fusion is phylogenetically established. (Similar congenital fusion is present in many parts of the plant.)

The various types of carpels described under the peltate theory are shown by ontogeny to represent merely stages in the evolutionary modification of the carpel. The epeltate carpel is a primitive carpel, with closure ontogenetic; the peltate types are advanced, with closure, at least in part, congenital; a ring-shaped meristem forms the tubular part. Carpel closure may be wholly ontogenetic, wholly congenital, or partly ontogenetic and partly congenital, but the peltate theory fails to recognize congenital fusion. The "cross zone" represents the level in the developing carpel between the distal part of the carpel, where closure is ontogenetic, and the proximal part, where it is congenital. A ventral longitudinal "strip" of the carpel wall is indeed formed by the section of the basal meristematic ring between the edges of the originally crescent-shaped primordium. But this strip is not delimited structurally; it is initiated by basal growth, as is the rest of the carpel, not by a cross-zone meristem, as implied by the peltate theory. This strip consists of the marginal strips of the lamina, the fertile areas, but it is not a distinct, morphologically significant part of the carpel. The "cross zone" appears important in achenes where the carpel is shortened and the ovules reduced to one, because the surviving ovule is attached at this point, the remnant of the fertile, ventral border of the closed carpel.

Ontogeny is claimed to support the peltate theory, in that the carpel primordium "assumes very early the peltate form with the formation of a cross zone." But this condition is demonstrated in achenes, where carpel length is greatly reduced and ovule number is reduced to one. The cross zone on which the solitary ovule is borne is considered "latent," a supposedly primitive stage in the development of the peltate carpel; it is, rather, a vestigial condition. The "unifacial" character of the carpel is considered important evidence of the peltate nature of the carpel, and the presence of a unifacial "stalk" (stipe) is considered further proof of peltate form; the stalk "must have been formed by a cross zone." But the stipitate carpel is characteristic of many primitive taxa—some of the woody Ranales, Ranunculaceae, Rosaceae, Helobiales—and is clearly a basic feature of the primitive carpel, rather than an advanced character. The unifacial form of the carpel and the stipe is not evidence that the carpel is basically a peltate organ. Simple folding or inrolling of the carpel forms a unifacial structure. Anatomy supports the involute, not the peltate theory, of carpel nature by the presence of the strong dorsal bundle and the pair of smaller, inverted ventral bundles on the side opposite the dorsal. Support for the peltate theory is seen in the arrangement of the bundles—"in a cylinder"—but size, number, and arrangement are those of a folded, bifacial organ. "In a cylinder" is morphologically an incomplete description.

The peltate theory must explain the presence of the prominent ventral suture of the carpel which, under this theory, is built up by a simple transverse (cross-zone) meristem. The explanation given is that the suture is the result of earlier and stronger growth on the dorsal side. But this does not explain the double epidermal layers connecting dorsal and ventral epidermises, which are present in many carpels. The typical mature carpel is, of course, ascidiform, in that it is tubular, but this form is attained, morphologically, by simple folding or inrolling and fusion, not by the complicated process of union of basal lobes and development of a new "meristem."

The peltate theory has been extended far beyond its original basis; it has been applied also to stamens, petals, and cotyledons. The course of development and the elaboration of the theory parallels that of the theory of carpel polymorphism. From an unsound base, the theory has been extended broadly to other floral organs, with a consequent amazing entanglement of superfluous interpretations. Like the polymorphism theory, it is unsound in basis and in detail.

The Solitary Carpel. Gynoecia that appear to consist of a single carpel are of two types: those that are, morphologically, single carpels and those that consist of more than one carpel but resemble a single carpel and are well described as "pseudomonomerous." The single carpel is a surviving member of a multicarpellate gynoecium, as shown by comparative studies in several families, especially the Ranunculaceae, and from the evidence of vascular anatomy, which shows vestigial vascular supplies of the lost carpels. *Cimicifuga* and *Delphinium* show well the origin of solitary carpels.

The position of the solitary carpel on the receptacle has been discussed in relation to the nature of the carpel. Evidence that the carpel is a leaflike, lateral organ is seen in the more or less clearly lateral position in many taxa, but, in other taxa, the carpel appears to be median and terminal. A median position has been considered evidence that the carpel is cauline, not appendicular, in nature, that it is a continuation of the stem. But ontogeny shows that the solitary carpel in the Leguminosae is a lateral structure; the carpel primordium arises on the side of the floral apical meristem and is moved to a "terminal" position by differential growth. Vascular traces to carpels are derived laterally from the receptacular stele and give evidence of the morphologically lateral position of carpels that appear terminal, especially in flowers in which bundles of the receptacular stele continue beyond the levels of departure of the traces. In unisexual flowers, a single carpel may be the only floral appendage and may seem to be a continuation of the pedicel—*Euphorbia, Heliconia* (Fig. 36), *Najas*—and, therefore, cauline, but a typical carpellary vascular supply and dorsiventral form show its appendicular nature.

Ontogeny of the Carpel

The primitive carpel arises on the floral meristem as a more or less crescent-shaped primordium, which soon becomes broadened by lateral extension. Apical and marginal meristems appear early and increase its length and width (Fig. 87). The presence of marginal meristems in the carpel was recognized only in the early twentieth century. The apparent absence of marginal meristems had earlier been considered evidence that the carpel was unlike the leaf. Differential growth brings about an upturning and bringing together of the lateral "wings," with, ultimately, their appression and more or less complete histological union. Comparative studies of the general method of closure of the carpel seem not to have been made. The bringing together of the laminar wings has been loosely described both as an upfolding and as an inrolling. The carpel closed by a simple upfolding is termed *conduplicate;* that by inrolling, *involute*. Union of the contacting margins has been commonly described as "by the margins." But "margin" is a poor descriptive term here, because it may mean either the edges of the blade or marginal strips (of either surface) near the edges.

Earlier descriptions of carpel closure as an upfolding or inrolling appear to be generally accurate, though little attention has been given to details of contact from the standpoint of position of the marginal meristems. Histological studies of the development of some primitive carpels—especially *Degeneria*—show a simple closing by folding; the conduplicate carpel so formed has the ventral surfaces of the lamina halves in contact, with the edges side by side. The frequent occurrence of the conduplicate type in primitive taxa and its simplicity have been

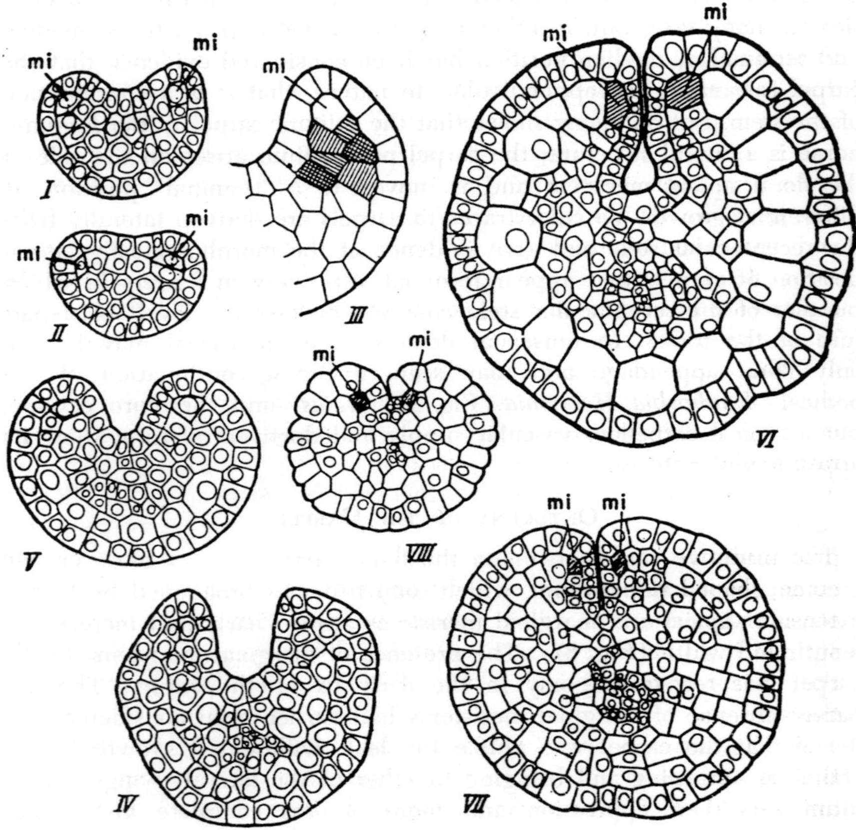

Fig. 87. Cross sections of carpel at various stages of development to show marginal meristems. I to III, *Anemone silvestris:* I, II, sections of very young carpel, proximal and distal, respectively; III, part of I, showing by shading the cells derived from the marginal initial. IV to VIII, *A. Pulsatilla:* IV, V, sections of young carpel, proximal and distal, respectively; VI to VIII, sections of carpel with marginal growth completed. *mi,* marginal initial. (*After Sprotte.*)

considered evidence that all other types have been derived from this by reduction of the marginal strips and a withdrawal of the edges, together with the ventral, half-inverted, placental bundles. But carpels with edges and margins deeply inrolled are complex and seem not to have been derived from the conduplicate type. The carpels of some taxa seem to close edge to edge; in others, the margins are inrolled and, in contrast with closure in the conduplicate carpel, the *dorsal* surfaces meet and fuse. In some taxa, the marginal strips of the lamina are deeply inrolled and recurved, so that the placentae are turned laterally or even inverted (Fig. 74).

The problem of methods of the phylogenetic closure of the carpel needs critical study; is there only one basic method, by upfolding, or is there another, by inrolling? There is strong evidence that the conduplicate carpel represents only one type. In leaf vernation, there are several types of folding, including involute and conduplicate types; under xerophytic conditions, involute leaves are common. If involute types have been derived from conduplicate, the change in areas of contact from the ventral to the dorsal surface is a major step. The position and orientation of the ventral (submarginal) bundles in relation to the areas of union in closing is important in determining whether involute closure has been derived from conduplicate or is of independent origin. These bundles point out placental position in submarginal placentation and the extent of insertion of marginal bands of the lamina within the closed carpel. The derivation of carpel type with broad bands of lamina enclosed within the carpel from the conduplicate carpel requires elaborate evolutionary modification: the opening up of a conduplicate carpel with appressed ventral surfaces, followed by an inrolling of marginal strips, with union of dorsal surfaces. The involute carpel is, in all probability, an independent type. Closing of the carpel has doubtless come about independently several or many times, as have advances in other characters; it is still in process in many living taxa. It would be strange if all closures were structurally alike, especially in the light of the variety of folding and rolling of leaves in buds and cotyledons in the seed.

In all studies of carpel structure and floral anatomy, it is important that the carpel be recognized as an immature organ at flowering time and should not be compared with fully developed stamen or petal. The vascular tissues, especially the connections of the traces of the carpel to the vascular receptacular stele, are frequently immature at flowering time; the downward extension of the traces may be incomplete. In the Capparidaceae-Cruciferae line, this late development is marked; connections are not completed until the fruit is partly developed. Interpretations of the gynoecia of these taxa based on the statement that the carpellary traces end blindly in the pith have no value.

The advanced carpel is congenitally closed; it arises from a ring-shaped primordium as a tubular organ. The carpels of many taxa are closed in part ontogenetically and in part congenitally; in early ontogeny, the primordium is crescent-shaped; later, it becomes ring-shaped. Failure to recognize the existence of this combination of types of carpel building has been the basis for many of the errors of interpretation that have formed a basis for the peltate theory, with its obscure "meristem," the "cross zone."

Anatomy of the Carpel

In vascular structure, carpels are closely like leaves. A three-trace, three-gap system was generally accepted as basic for the carpel through most of the first half of the twentieth century, but studies of the woody Ranales during the later part of this period have shown that a two-trace, one-gap is probably primitive for all appendages of the stem (Fig. 6). The three-trace system, doubtless an early modification of the two-trace system, is common in carpels. Phylogenetically, it has been formed by the union of the two traces of the primitive organ to form a morphologically double median trace, with two additional lateral traces, one on each side, with separate gaps. The several-trace and the single-trace types have been developed from the three-trace, three-gap, as in leaves.

A two-bundle, median carpel trace, consisting of two independent bundles, is rare in carpellary modes (although common in cotyledonary nodes), but double and two-forked midveins are frequent. Some two-lobed stigmas are supplied by the two midveins or their branches. Carpels with more than three traces are frequent. Carpels that are structurally reduced have one, or three in various stages of reduction to one. Reduction in vascular supply—traces and laminar bundles—parallels reduction in ovules. Both median and lateral supplies are reduced by distal shortening and by lateral union. That loss of lateral (placental) bundles is by progressive distal shortening is apparent from comparisons in achenes of related genera; that the reduction may extend to elimination of the traces and their gaps is seen in some species where individual carpels have from one gap a single trace representing traces fused before entering the carpel. (Lateral fusion is a stage in the loss of the lateral traces.) Reduced follicles and achenes show stages in the simplification by reduction of the vascular supply—traces and laminar strands (Fig. 42). Shortening of the ventral (placental) bundles ultimately brings the ovule to a position where it appears to be borne on the dorsal (midrib) bundle, but anatomical comparison with other achenes shows that the ovule is morphologically borne on a fusion bundle that represents the united dorsal and ventral bundles. An ovule in this position has been called "basal" (a satisfactory term for simple description, but incorrect morphologically, because the ovule belongs on a submarginal placenta). An ovule in this position has also been called "cauline," because it appears to be strictly basal (if anatomy is disregarded) and merely surrounded by the carpel.

SYNCARPY

Connation of carpels is a prominent feature in the specialization of the gynoecium and is found in great variety of position and extent of fusion. Degree and type of closure of the individual carpels that enter into the compound structure increase the complexity of syncarpous gynoecia. These gynoecia are often difficult to interpret when the fusion is intimate, involving the vascular tissue, and reduction has occurred. But all gynoecia are interpretable in terms of the simple, free carpel when ontogeny, vascular structure, comparison with gynoecia of related taxa, and the modifications that commonly accompany reduction are considered. Syncarpy has arisen independently in many taxa, and different degrees of fusion may occur within a genus.

Connation may occur among any number of carpels and among those arranged in spirals, as well as in whorls. It is infrequent among spirally placed carpels—the Annonaceae, Eupomatiaceae, *Zygogynum,* some Berberidaceae. In spiral arrangement, adjacent carpels meet at different levels; the base of the ovary of one carpel may be fused with the top of the ovary of another. Fusion among carpels is primarily by the sides (dorsal surface) of the folded or inrolled laminae; infrequently, it is by the ventral margin only. It is seen in all degrees of histological union and longitudinal extent. The union may be throughout carpel length, by the base or apex only, by the middle only (rare), or by both apex and base, with the carpels free in the central region; it may involve the entire lateral walls, or only parts of them—the margins, median or distal strips. The form of the individual carpel and its primordium—crescent-shaped or tubular—controls, in part, the area of contact between them. Ovaries, styles, and stigmas are involved where fusion is throughout the length of the carpels; where it is incomplete distally, styles and stigmas are free. (The number of styles and stigmas does not necessarily indicate the number of carpels involved in the fusion.) When fusion is at the base only and is of slight vertical extent, the gynoecium may appear apocarpous—*Butomus, Scheuchzeria, Spiraea, Aquilegia.* (For all taxonomic uses, it can be so called.) Fusion by stigmas only, as in the Asclepiadaceae and Apocynaceae, is rare. Fusion among the units of the syncarpous gynoecium is congenital in most taxa. It may be, in part, ontogenetic; the distal parts of the carpels may arise from separate primordia and, all the primordia later uniting, form a ring-shaped meristem which develops the basal part.

Carpels adnate laterally to the receptacle have been described as forming a *false syncarpous gynoecium;* if the carpels are also connate, the gynoecium is truly syncarpous. Some so-called false syncarpous gynoecia have been misinterpreted, as in *Nigella,* where the central column of tissue between the carpels is interpreted as receptacular. The "receptacle" in this description is actually the fused margins of the carpels, as demonstrated by anatomy.

Septal Nectaries. Where lateral union between carpels is incomplete, the unfused areas commonly become secretory, and well-defined nectaries are formed. These nectaries are common in the Liliaceae and related families and occur occasionally in the palms and other genera. The secretions of these glands may exude along longitudinal slits between carpels or at the top of the ovary when the carpels are fused along the dorsal margin. These nectaries may have elaborate form and exude the secretion through a specialized canal leading to a small opening on the top or side of the ovary—*Cocos, Ananas*. The septal nectary seems to be restricted to monocotyledons and has probably arisen independently in members of this taxon.

The presence and position of the unfused areas of carpel walls in the syncarpous gynoecium is related to the history of carpel connation. The transformation of the semienclosed pockets between the carpels into nectaries and the elaboration of their form constitute an excellent example of adaptation of flower structure to insect pollination.

Coenocarpous and Paracarpous Gynoecia. Apocarpous and syncarpous are terms long in common use to describe gynoecia consisting of free and of fused carpels, respectively. But attempts have been made to supplant "syncarpous" by introducing a new term, *coenocarpous,* reviving an old term, *paracarpous,* and redefining *syncarpous*. In the new treatment, gynoecia made up of united carpels are coenocarpous; those with a single ovarian chamber, with placentation parietal, free central, or basal, are paracarpous; only those with two or more separate chambers and longitudinal fusion incomplete are syncarpous. Syncarpous gynoecia, under this definition, may show three structural zones: a fertile base, the ovary, which is syncarpous; a median part, the style, which is paracarpous; and a stigma, apocarpous. These terms, so used, are not good, because coenocarpous and syncarpous have much the same meaning, and the classification is valueless, because "paracarpous" and "syncarpous" forms are considered morphologically distinct types. No line can be drawn between paracarpy and syncarpy; in some families, all transitional stages occur, and, in some taxa, the ovary is "syncarpous" in the basal part and "paracarpous" in the upper part. Yet the new terms have been accepted and used in publications in continental Europe.

The term *false coenocarpy* has been proposed to describe gynoecia where adnation of carpels to a cup-shaped receptacle ties together carpels otherwise free—Hydrocharitaceae, Pomoideae, *Butomus*. The term seems to be applied also to gynoecia where the carpels are held together by ventral union with a central receptacle. Inclusion of *Nigella* and *Butomus* among examples of false coenocarpy is based on misinterpretation of the "core" of the gynoecium, which is called the receptacle. But anatomy shows that this center consists of the fused ventral margins of the carpels. Anatomy also shows that the carpels of the Pomoideae are not adnate to a hollow receptacle. The gynoecia of these taxa are "coenocarpic," not "falsely coenocarpic."

Comparative study of the gynoecia of syncarpous taxa shows that these new terms are invalid as descriptive of morphological types. It may be helpful to use the adjectives paracarpous and coenocarpous to describe the parts of a gynoecium that differ in structure from base to tip. The term false coenocarpy is superfluous.

Types of Carpel in Syncarpous Gynoecia. Carpels of all closing types, from typical conduplicate to extreme involute (Fig. 74) and from widely open to completely closed, have apparently entered into the formation of syncarpous gynoecia. Union of closed carpels with lateral fusion forms a multilocular, syncarpous ovary, with the placentae close together at the center. The grouping of the placentae about the floral axis is *axile placentation* (Fig. 88*A*). This placentation type has been generally considered the basic type, from which have been derived the other types in syncarpous ovaries. The major vascular supply of the typical carpel is a median (midrib) bundle, the *dorsal* bundle—which is occasionally double—and two submarginal bundles, the *ventral* bundles. In the closed carpel, the ventral bundles are inverted, in comparison with the dorsal. The inversion is brought about by the bringing of the carpel margins together ventrally.

It has been generally assumed that, in the evolution of syncarpy, connation developed only among closed carpels, that the open carpels of types with parietal placentation were secondarily open. But the anatomy and ontogeny of many taxa—especially the Parietales and others that have parietal placentation—indicate that connation took place not only between closed carpels but between those that were still open. The carpel at pollination time is still unsealed in members of many lines and is definitely open in a few; connation, well advanced in some of these lines, must have come about in open, as well as closed, carpels. The variety of degree of carpel closure existing in taxa not closely related suggests that connation of carpels, like other types of fusion, is a general angiosperm tendency, expressed independently in many lines, and that it occurred among carpels in all stages of closing, even in widely open types.

The common chamber enclosed by open carpels varies in form with the degree of closure of the carpels; it is rounded in cross section where the carpels are widely open (Fig. 84*E*), strongly lobed where the carpels are nearly closed (Fig. 84*A*, *C*, *D*). Where the locule is strongly lobed, the ovules commonly project into the lobe formed by the carpel that bears them, but all may project into the central part of the locule. Rarely, ovules are borne inside the open locule of a carpel to which they do not belong—*Viburnum*.

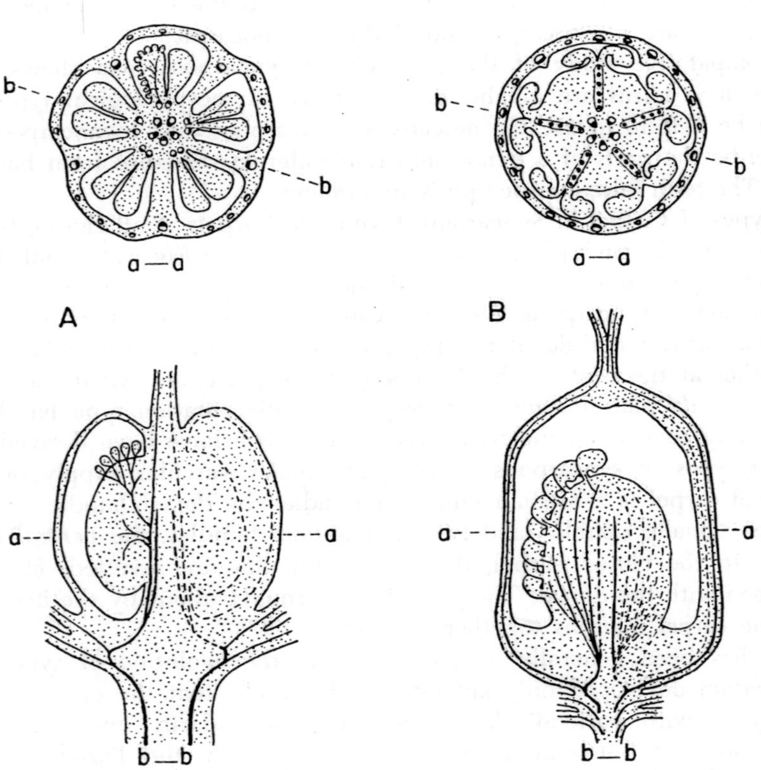

Fig. 88. Diagrammatic cross and longitudinal sections of ovary structure showing in *A*, *Rhododendron indicum*, axile placentation and in *B*, *Lychnis dioica*, free central placentation. Carpels, adnate to prolonged receptacle; vascular bundles of receptacle (in the center), normally oriented; the ventral bundles of the carpels, inverted. (*From Eames*, 1951.)

Placentation in Syncarpous Gynoecia. In syncarpous taxa in which the individual carpels have laminar placentation, the term used for free carpels, *laminar*, is satisfactory; but, where the placentation of the individual carpels is fundamentally submarginal, modifications in ovule number and position have made necessary other descriptive terms. Several of these terms have long been in use, but without uniformity in descriptive or morphological application.

Where closed or nearly closed carpels, with their fertile areas submarginal, are connate laterally or by their ventral margins, the placentae lie near together around a central point (Fig. 84*B*)—sometimes around the prolongation of the receptacle (Fig. 88*A*). This is *axile placentation,* frequently incorrectly called *axial.* The term axial implies that the placentae are borne on the axis or are a part of the axis, an old view still sometimes maintained, but, as shown by evidence from anatomy and ontogeny, and comparisons with related taxa, this is incorrect. No placentae are morphologically cauline.

Axile placentation has been commonly considered the primitive type —in some interpretations, more primitive than laminar—from which other types have been derived. It is surely *a* primitive type in syncarpous gynoecia, the type from which free central and some forms of basal and suspended placentation have been derived; the type commonly called parietal has probably not been derived from axile.

The term parietal has long been applied to placentation where the ovules are borne "on the wall" of the ovary, as the word implies. Though usually applied only to placentation in syncarpous ovaries, it has also been applied to placentation in free carpels, and its use in this way has been considered desirable for placentation in free, follicular carpels. But all placentae are borne morphologically "on the wall" and are foliar, not cauline, and terms indicating merely a laminar position are not sufficiently specific. In syncarpous, unilocular gynoecia, the ovules are usually borne in longitudinal rows—a typical parietal placentation. There are two interpretations of the origin of this placentation, which is obviously double in nature: (1) that the placentae represent axile placentae, withdrawn from their original central position by reopening of the carpels and retraction of their margins; (2) that the placentae have been formed by the bringing together of margins and submarginal placentae of *open* carpels in the development of syncarpy.

In syncarpous gynoecia that are formed by the union of open carpels, margin to margin, the submarginal placentae of adjacent carpels are brought close together and fuse to form a two-lobed placenta or, when fusion is intimate, a structurally simple (morphologically double) placenta. This fusion of parts of different carpels to form a single structure that may show no histological evidence of this compound nature is repeated in the vascular tissue. The two ventral bundles may lie near together, so that the placenta has a double vascular supply; they may fuse to form a two-lobed bundle or may form a simple bundle. All stages in the union of the bundles are found in cross sections of typical syncarpous ovaries with parietal placentation. All these bundles, like the dorsal bundles, are normally oriented with xylem toward the inside, the vascular structure to be expected if fusion of adjacent carpels is, morphologically, margin to margin. Under the theory that the unilocular, syncarpous ovary with parietal placentation has been derived from the multilocular with axile placentation, the lateral walls, with their placentae, are considered retracted; in this way parietal placentation has been claimed to have been formed from axile. In axile placentation, the ventral bundles are inverted—in contrast with the dorsal bundles (Fig. 88). If the margins and placentae are simply withdrawn by narrowing of carpel sides, the bundles would remain inverted. But in many—perhaps all—taxa with this type of ovary, the ventral bundles are normally oriented like the dorsals—Dilleniaceae, Violaceae, Caricaceae, Passifloraceae, Hypericaceae. The orientation of the bundles of the carpels is the same as that of open carpels.

The evidence from ventral-bundle orientation supports the fusion-of-open-carpels theory of origin of this parietal type of syncarpous gynoecium. An opening up of closed carpels to form this type of ovary would be a return to an ancestral condition, a step not taken in evolutionary modification as interpreted under the "law of irreversibility."

The existence of syncarpy among open carpels is evidence that connation in the gynoecium began, in the history of angiosperms, before closure of the carpel, or that the two kinds of fusion arose together. Apparently, evolutionary advance in connation and carpel closure is continuing in living plants.

Connation of open carpels suggests that the taxa with parietal placentation are less advanced than those with axile placentae where the carpels are closed. In most classifications of angiosperms, the Parietales are placed among the lower orders on the basis of other floral characters. The recognition of parietal placentation as a primitive, not a highly advanced, type, supports the generally accepted position of this order. (Close relationships within the Parietales are not necessarily suggested.)

Where, in a syncarpous unilocular ovary, the ovules are borne on a mound or column in the center at the base of the common locule, free from the lateral ovary walls, the placentation is well termed, descriptively, *free central* (Fig. 88*B*). The nature of this type of placentation has been much discussed, with strongly divergent opinions that involve the nature of the carpel itself and the question of basic ovule position—cauline or appendicular. It has been suggested that free central placentation should be called *free marginal.* If the term were modified to *free submarginal,* it would be morphologically correct, because it suggests the correct morphological condition—a placenta consisting of carpel borders freed from the lateral walls and united in the center of the gynoecium (sometimes clothing the projecting tip of the receptacle). But the long-used term free central is simpler, and also suggests the freedom of the borders from the lateral walls and the median position.

The major differences in interpretation of the morphology of free central placentation relate to the nature of the placenta. Are the ovules borne on a prolongation of the receptacle, that is, are they cauline in nature—borne on the stem, not on the carpels, which merely enclose them? Does the placenta represent the ventral strips of all the carpels, freed from the dorsal section and connate, with the ovules borne as in typical free carpels? Is the central placenta a compound structure, the

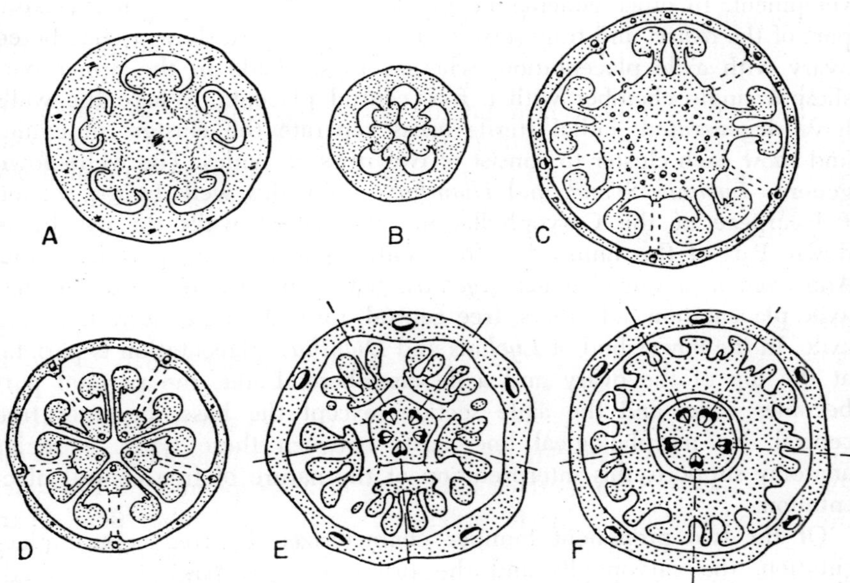

Fig. 89. Cross-sectional diagrams of ovary structure showing free-central placentation. *A* to *D*, *Lychnis*, showing ontogeny: *A*, *B*, early stages, lateral walls present; *C*, *D*, later stages, lateral walls degenerated. *E*, *Lysimachia*, and *F*, *Steironema*, showing lateral walls absent, placentae suspended from central column of carpel margins. Dotted lines indicate limit of carpels. (*A*, *B*, *after Lister; C*, *D*, *after Van Tieghem; E*, *F*, *after Douglas; F*, *modified.*)

tip of the receptacle coated by the fused ventral margins? Ontogeny and comparative and anatomical studies all demonstrate clearly that this kind of placentation is derived from axile placentation, that the central placenta consists typically of the fused placentae and ventral margins of the united carpels, freed from the dorsal section by loss of the lateral walls which connected them. This loss is ontogenetic in several families (Fig. 89*C*, *D*)—Caryophyllaceae, Primulaceae—and readily recognized, but phylogenetic, in others. In some genera, the receptacle extends upward into the placenta, as it does between carpel margins in some types of axile placentation (Fig. 88). If this receptacle tip is present, it is sheathed by the carpel borders on which the ovules are borne.

Abundant proof of the nature of free central placentation is found in both ontogeny and vascular anatomy. In some genera of the Caryophyllaceae and Portulacaceae, axile placentation is transformed during development of the flower into free central; the lateral walls of the carpels break down, leaving the united placental areas free, except at the base (Fig. 89). Flower buds of many genera show all stages of this development. In other genera, the breakdown is incomplete in the basal part of the ovary, and transverse sections made there show a chambered ovary with axile placentation, whereas those made in the upper part show a single chamber with a free central placenta. Where the walls have degenerated, radial sterile areas separate the clusters of ovules and show each cluster to consist of two rows of ovules. The well-known genera—*Lychnis, Silene,* and *Dianthus*—show this well. In the section Polycarpeae of the Caryophyllaceae, the carpel walls do not break down. Part of this family has free central placentation; part has axile. And even in a single genus, *Lychnis,* both types occur: *L. alpina* has axile placentation; all others, free central, formed, ontogenetically, from axile. In the flower bud of *Lychnis* and *Dianthus,* placentation is parietal at the top of the ovary and axile elsewhere. Later, the parietal part becomes axile, and the axile section, except the base, becomes free central. No morphological line exists between these types. Vascular anatomy supports this interpretation of the nature of free central placentation.

Of the two prominent families characterized by free central placentation, the caryophylls and the primroses, the latter is more advanced in placental type, because, in most genera, there is little or no evidence of ontogenetic loss of carpel walls; a few genera—*Dodecatheon, Lysimachia, Steironema, Samolus,* and probably others—show ontogenetic loss of carpel walls; in most genera, free central placentation is congenitally established.

Placentation with the ovules borne at the base of the locule in syncarpous ovaries is *basal.* Basal is a reduction type, derived from free central placentation, with the placenta reduced in size and the ovules in number, to few or one. No line can be drawn between free central and basal placentation; if the ovule or ovules are borne directly on the floor of the ovary without a definite projecting placenta, the placentation is considered basal. It may be derived from axile, as in the Fagaceae; it may also be derived from parietal, as shown by the Droseraceae. *Drosera* has parietal placentation; *Dionaea,* basal; and in *Drosophyllum,* where the ovules are borne at the base of the ovary walls, the placentation is intermediate between parietal and basal. Other transitional stages are occasionally seen. In *Salix*, the ovules are borne low down in a parietal position. Basal placentation derived from parietal seems to be rare, but, in the absence of transitional stages in related taxa or evidence from vascular structure, the derivation of basal placentation may not be evident.

Where the surviving ovule or ovules are borne on the base at one side of the center or low down on the side wall of the ovary (Fig. 81), they have been called *subbasal* or basal, as in free carpels. The ovule of the grasses and that of *Ranunculus,* for example, are subbasal, as shown by position and by anatomy.

The nature of the solitary ovule is syncarpous gynoecia has aroused continued discussion. Because of its central position, it has been considered borne on the apex of the receptacle and, therefore, cauline. Evidence from anatomy and ontogeny, with comparisons in related taxa, demonstrates that all these ovules are appendicular. (See "Phyllospory and Stachyospory.")

Suspended placentation in syncarpous gynoecia occurs where reduction in ovule number in submarginal placentae may leave one or a few ovules surviving at the top of the locule. These ovules may be attached on the ventral wall of the carpels or "on the roof" of the loculus. Often they hang downward on long funicles; this arrangement is similar to that of subbasal, but at the other end of the locule. Anatomy shows that many, perhaps all, of these are surviving members of submarginal placentation. Some suspended ovules are, perhaps, survivors from parietal and even laminar placentation. Little is known of the vascular supply of suspended ovules.

The "gentian type" of placentation (Fig. 80) has sometimes been described as laminar, but laminar placentation, characteristic of only primitive taxa, is not to be expected in so advanced a family as the Gentianaceae. Comparative study of placental form and of carpel anatomy throughout the family shows that ovule arrangement in the gentians is an elaborated and complex form of submarginal. From the evidence of anatomy, the placentae are best interpreted as much enlarged, with flanges sometimes intruding far into the locules and the fertile areas sometimes extended laterally over the inner ovary wall. The placentation of this family, though sometimes closely resembling the primitive laminar type, is a highly specialized type. Relationships of placental type are shown by Fig. 90.

Reduction in Carpel Number in the Gynoecium. In both apocarpous and syncarpous gynoecia, carpel number has been reduced from an ancestral many to few and to one. Reduced numbers and stages in the reduction are seen in many taxa: in the Ranunculaceae, within the genus *Delphinium,* three to five in perennial species, one in annuals; in

Cimicifuga, several to one; and one in the closely related *Actaea.* In the Rosaceae, reduction has been from many to one in *Sanguisorba,* to one in the Prunoideae, where one or two abortive carpels frequently develop. In the Proteaceae, the solitary carpel is the survivor of two; one is lost in ontogeny, as seen in *Manglietia.*

Syncarpous gynoecia, with carpels reduced to one, have been well termed *pseudomonomerous.* These gynoecia may show external evidence of their compound nature in the form of the ovary, the number of styles or stigmas, or the presence of abortive carpels adnate to the fertile one. Other evidence is in internal structure of the ovary and receptacle and in the gynoecia of related taxa. All available evidence

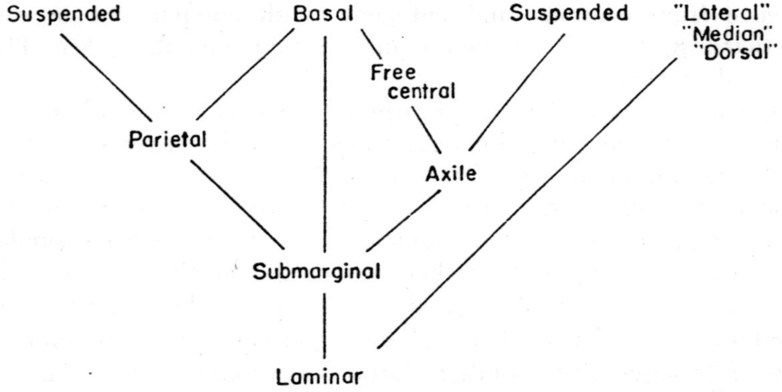

Fig. 90. Chart showing phylogenetic relationships of placentation types based on evidence from comparative form, ontogeny, and vascular anatomy.

must be used, because reduction may have gone very far, even to complete external loss of the component carpels. Ontogeny and the vascular skeleton will usually add to the evidence from superficial form and related taxa. Reduction in ovule number accompanies that in carpel number, and this type of gynoecium usually has a single ovule. Some pseudomonomerous carpels have been described as achenes, which they may resemble in form and ovule number. Pseudomonomerous carpels are found throughout the angiosperms—in many genera of the Urticales, in *Sparganium, Phryma, Rhus, Dipsacus, Viscum, Smilax, Ceratophyllum, Nandina, Ficus, Pontederia, Cassytha,* the Gramineae, *Valeriana, Viburnum, Lindera,* many palms. In taxonomic treatments, some of these taxa are described incorrectly, on the basis of external structure, as having one carpel; others, such as *Ulmus,* are correctly described as having two or, as in the grasses, three. The structural form of these reduced syncarpous gynoecia shows great variety. The sterile carpel or carpels may be evident in external form as supernumerary styles (Fig. 91) or external ridges; they may be apparent in cross sections of the ovary, sometimes as obvious sterile units, as in *Pontederia* (Fig. 92*B*); sometimes as solid carpels, as in *Valeriana* (Fig. 76*D*), *Cocos* (Fig. 93), and related taxa. Where, in pseudomonomerous gynoecia, the style of a sterile carpel persists, it is sometimes shorter than that of the fertile carpel—*Celtis, Morus*—but may be of the same

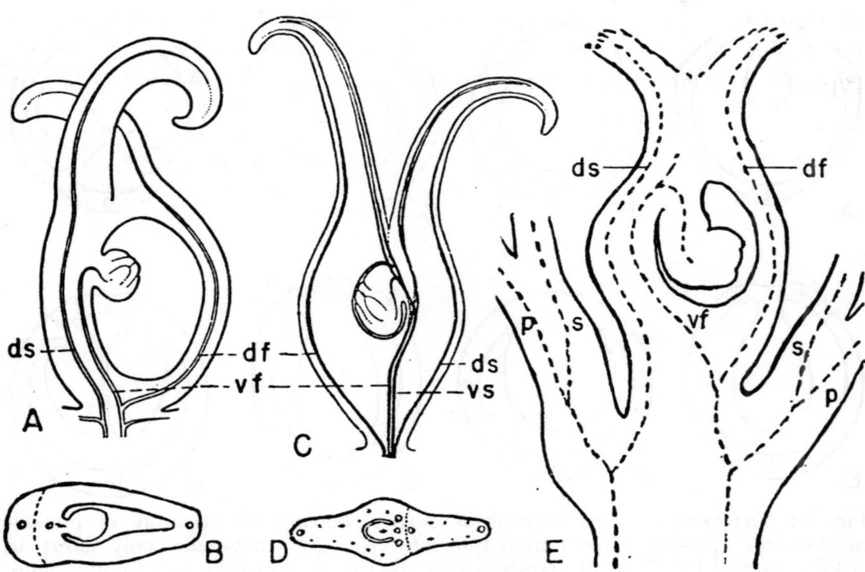

Fig. 91. Diagrams of lateral and cross sections of pseudomonomerous gynoecia, consisting of two carpels: one fertile, with ovules, and the other "solid," sterile, without locule. *A, B, Morus alba; C, D, Ulmus montana; E, Celtis occidentalis. df,* dorsal bundle of fertile carpel; *ds,* dorsal bundle of sterile carpel; *p,* petal bundle; *s,* stamen bundle; *vf,* ventral bundle of fertile carpel; *vs,* ventral bundle of sterile carpel. (*A to D, after Eckardt; E, after Bechtel.*)

length and may be even longer and take on the function of pollen reception. Sometimes the abortive carpel is represented only by a vascular bundle in the ovary wall. In the receptacle, evidence of carpels that have disappeared externally—seen as mere traces that end blindly—is occasional. But the interpretation of "blind" bundles in the receptacle tip is difficult, because they may be vestigial tips of stelar bundles extending beyond the uppermost carpellary traces, or they may be traces of lost carpels. Evidence of the nature of these carpels can be determined by number, origin, and position in relation to carpel traces below them. Misinterpretation of vascular bundles in the receptacle tip has been responsible for errors in understanding the morphology of carpels, gynoecia, and the inferior ovary. Under the theory of carpel polymorphism, bundles of much morphological significance in the receptacle tip were described as "discarded into the pith."

The presence of sterile carpels united with fertile carpels has been interpreted as representing a stage in the development of syncarpy from apocarpy. This interpretation is obviously a misreading of the process of evolutionary reduction.

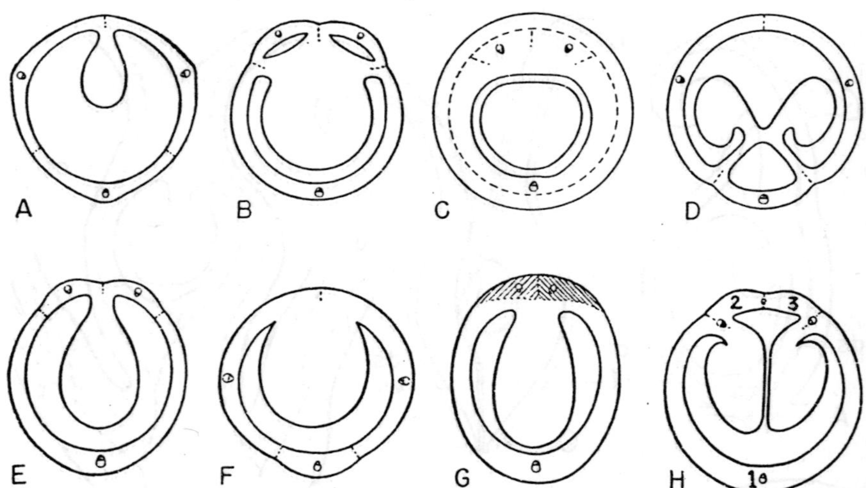

Fig. 92. Diagrams of cross sections of ovaries showing the structure of pseudomonomerous gynoecia derived from trimerous types. Dorsal bundles only shown in *A* to *G*; ventral bundles only shown in *H*. Carpel limits shown by radial dotted lines. *A*, Palmae, two placentae completely sterile, the third fertile on one side; *B*, Pontederiaceae; *C*, Marantaceae (*Thalia*); *D*, Caryophyllaceae (*Drypis*), the placenta between two carpels is completely sterile, the other two are fertile on one side only; *E*, Chenopodiaceae (*Beta*), origin of the ovule trace is directly under the median line of the fertile carpel; *F*, Gramineae, ovule borne on placenta between two large carpels, attached to one of them; *G*, Anacardiaceae (*Rhus*); *H*, Berberidaceae (*Nandina*). (*A to F, after Eckardt; G, from Payer; H, from Himmelbaur.*)

The "Solid" Carpel in Syncarpous Gynoecia. Where, in syncarpous gynoecia, reduction in carpel number is in progress, the vestigial carpels are sterile, commonly lack a locule, and are "solid" through part or all their length. Abortive carpels are seen in a number of families (Fig. 76). They are prominent in the ovaries of the Valerianaceae, Berberidaceae, and Pontederiaceae, some of the Palmae (*Cocos* and related genera), the Verbenaceae (*Blairia*), and the Caprifoliaceae (*Viburnum*). The sterile carpels may be evident as mere ridges on the sides of the ovary wall, as in *Pontederia* and *Valeriana* (Fig. 76*D*), but may be fused into the ovary wall and evident only internally, as in

Achlys (Fig. 76*E*), *Cocos*, and related genera. The vestigial carpels of the coconut merge into the common ovary wall—the "shell" of the coconut; and the position of the obliterated loculi is evident by compressed tissue (Fig. 93). This is evidence that the shell consists, morphologically, of the inner wall layers of a compound ovary, and is not the seed coat, as often described. Reduced carpels of the Cruciferae and some of the Papaveraceae (Fig. 76*C*) and Capparidaceae are structurally solid, though not commonly accepted as such. Evidence that these are "solid," in the morphological sense of loss of locule, not in the sense of the polymorphism theory, is in their vascular supply and in comparison of the gynoecia of these families. Accessory styles and stigma lobes may be external evidence of greatly reduced carpels. Evidence of carpels completely lost from the ovary may be present as vestigial vascular traces in the receptacle leading to the position of the lost organs—*Degeneria, Cimicifuga.*

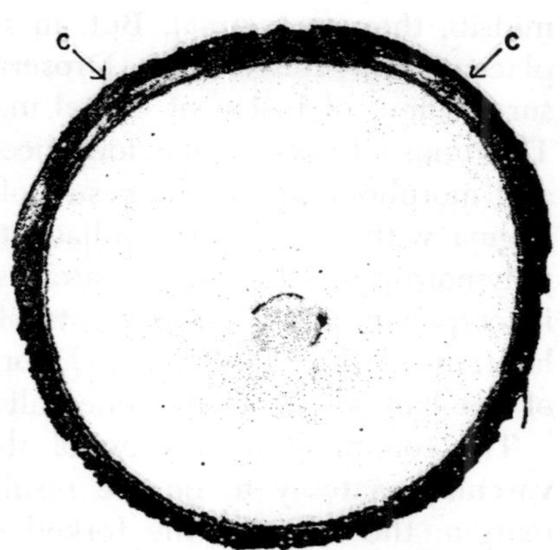

Fig. 93. Cross section of ripe "nut" of *Cocos nucifera*, showing at *C, C* remnants of the locules of the two sterile carpels embedded in the inner ovary wall ("shell").

Style and Stigma in Syncarpous Gynoecia. Connation among carpels may extend from base to tip or for only part of their length. Fusion by ovaries alone is common; that by proximal parts only of the ovaries is frequent. Fusion by stigmas alone is rare—Apocynaceae, Asclepiadaceae, Rutaceae, Simarubaceae. Free styles and stigmas often form a prominent part of syncarpous gynoecia. Fusion may be ontogenetic or congenital, or the distal fusion ontogenetic and the basal congenital.

Stigma form, where carpels are united, is, of course, related to number and intimacy of fusion in the uniting stigmas. Interpretation of carpel number in a syncarpous gynoecium is often based on number of lobes of the compound stigma, but this basis is unsound. The stigma of the free carpel may be simple or lobed, and, in reduced gynoecia, the number of stigmas is often less than that of the carpels present. Highly specialized stigmas—globose, cylindrical—may give no evidence, external or internal, of their compound nature; only structure of the ovary can determine this. Simple stigmas are not necessarily primitively simple.

Where, in syncarpous gynoecia, the number of carpels is two or three

—with an external appearance of one—the stigma of only one of the component carpels is usually present; the stigmas of the other carpels are abortive or lost. This is the common condition in pseudomonomerous gynoecia. Rarely, in these taxa, the stigma is borne on a sterile carpel over a placenta common to a sterile and the fertile carpels. In many taxa, number of stigmas or stigmatic lobes is not correlated with carpellary number; evidence from ontogeny and vascular anatomy and from the structure of the ovary and that of related taxa is essential in determining carpel number.

In syncarpous gynoecia, the stigmas commonly stand over the carpel midrib; they are *carinal*. But, in some taxa, notably those with parietal placentation—Resedaceae, Droseraceae—they stand over the "commissures," lines of fusion of carpel margins; these stigmas are *commissural*. This unusual position has long been correctly interpreted by taxonomists and morphologists as the result of the fusion of the lobes of a divided stigma with those of the adjacent carpels. Under the theory of carpel polymorphism, the commissural stigma is considered highly complex in carpellary nature; it consists of halves of two "solid" carpels, fused laterally with a "valve" carpel, forming a compound structure made up of one-half plus one plus one-half carpels.

The commissural position of the stigma is shown by ontogeny and vascular anatomy to be the result of the union of carpels, margin to margin; the lobes of the forked stigmatic tip of the carpel are fused laterally with the lobes of the adjacent carpels. The double stigmas so formed, each consisting of halves of the placentae of different carpels, stand alternately with the carpellary midribs. The vascular bundles that supply the stigma belong to different carpels—they are the ventral bundles of adjacent carpels. The midrib vascular bundles lead to the sinuses between the stigmas. The general form of the ovary top, in many taxa with parietal placentation, is misleading in its suggestion of carpellary position.

Syncarpy with Adnation to the Receptacle. The morphology of the syncarpous gynoecium is frequently complicated by adnation—internally, by fusion of carpels laterally or obliquely with the receptacle; externally, by fusion with the outer floral organs—perigyny and epigyny. In less specialized families, the receptacle is frequently prolonged beyond the bases of the uppermost carpels and may be adnate to the carpels. The extent of this adnation varies with the form of the receptacle and carpel base, and with the angle at which the carpels stand. If the adnate carpels are open, tissues of the receptacle may close the opening proximally. A line of demarcation between tissues of carpel and receptacle is difficult or impossible to determine.

In studies of flower structure, delimitation of the receptacle is often impossible, and interpretations of basic flower structure in syncarpous ovaries have differed widely. Some flowers, under one interpretation, have a receptacle prolonged beyond the distal carpels; under another interpretation, there is no receptacle beyond the carpels. But the approximate limits of receptacle and carpel are usually evident from the position of the ventral carpellary traces. Usually, the receptacle does not play a prominent part in the structure of the syncarpous ovary, though, under some interpretations, it has been considered to do so, as in the determination of the morphology of the inferior ovary.

The history of the receptacle in the evolution of the flower is that of shortening and elimination of internodes. But, in many taxa, especially in the more primitive families, a vestigial receptacle tip projects beyond the bases of the distal carpels. Here, the tip forms a definite but commonly obscurely limited part of the syncarpous ovary—superior or inferior. United ventral margins of the carpels have been interpreted as a prolonged receptacle tip and their vascular bundles as bundles of the receptacular stele. Critical comparative study of related taxa and an acquaintance with the vascular supply of floral organs and the derivation and course of their traces are necessary for the interpretation of vestigial receptacle tips.

The Inferior Ovary*

The morphology of the inferior ovary has doubtless been discussed more extensively than that of any other part of the plant body. Two theories of the nature of this ovary have received strong support since the early nineteenth century—the *appendicular* and the *axial* theories.

Under the appendicular theory—also called the *Candollean* and the *concrescence* theory—the inferior ovary is considered formed by the adnation of the bases of the outer floral whorls to the gynoecium; the axis forms no part of the ovary wall. Linnaeus, Goethe, de Candolle, Van Tieghem, Warming, and many other early botanists so interpreted the inferior ovary.

Under the axial theory—also called the *receptacular* theory—the inferior ovary is considered to consist largely of tissues of the axis, to represent an invaginated floral receptacle, which surrounds the gynoecium. Originally the theory of the German school of botanists, it was supported by Schleiden, Magnus, Payer, Eichler, Sachs, and many others.

Both theories, in essentials, go far back in the history of taxonomy and morphology, with great variety in interpretation. The axial theory was dominant in the later part of the nineteenth century and the first

* The anatomy of the inferior ovary is discussed in detail in Eames and MacDaniels: "Introduction to Plant Anatomy," 2d ed., New York, 1947.

quarter of the twentieth century, even though the appendicular theory had been strongly supported by Van Tieghem with anatomical evidence. In the later decades of the twentieth century, with much more information available from anatomy and ontogeny and a broader understanding of the effects of congenital fusion on anatomical structure and ontogenetic development, it has become clear that the inferior ovary has developed in two morphologically different ways—by adnation of floral appendages and by the hollowing of the axis tip.

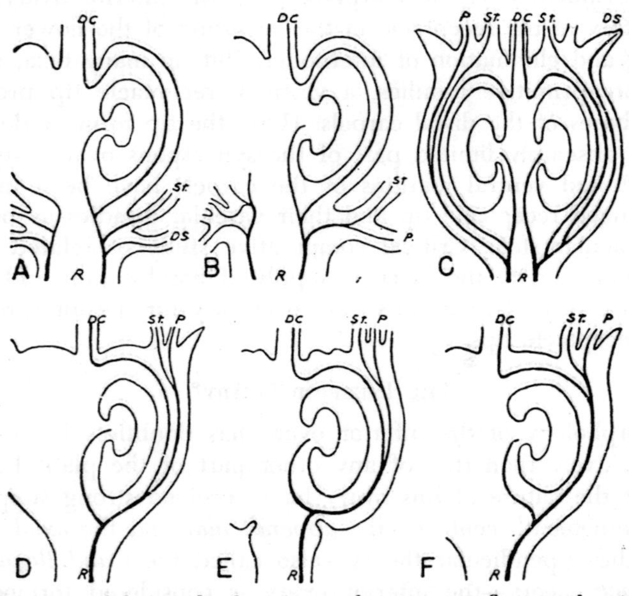

Fig. 94. Diagrams of longitudinal sections of ericaceous flowers showing evidence of appendicular nature of the inferior ovary by adnation of outer floral whorls to the ovary. Adnation increasing in extent, including fusion of the vascular supplies of all organs; the trunk systems so formed suggestive of stelar bundles. *A*, *Pyrola secunda; B, Andromeda glaucophylla; C, Gaylussacia frondosa; D, Vaccinium vacillans; E, V. pennsylvanicum; F, V. macrocarpon.* DC, dorsal carpellary bundle; DS, dorsal sepal trace; P, petal trace; R, receptacle; ST, stamen trace. (*After Eames,* 1931.)

Like many other structural modifications, the inferior ovary has developed independently in many taxa. Evidence of independent origin is found in many families. All stages in the adnation of the appendages are present in the Ericaceae (Fig. 94), Rosaceae, Amaryllidaceae; other families show both superior and inferior ovaries, with or without transitional stages. In the Iridaceae, *Isophysis* (*Hewardia*) has a superior ovary; other genera, an inferior ovary. In the Gesneriaceae, *Gloxinia* has a superior; other genera, an inferior ovary. In the Nymphaeaceae, *Nelumbo* and *Nuphar* have superior ovaries; *Nymphaea* and *Victoria,* inferior ovaries. As with other advances, the development of the inferior ovary has followed more than one line in structural modification. Fusion of adjacent organs is a common feature of specialization, and nearly all inferior ovaries seem to have been formed in this way; in only a few

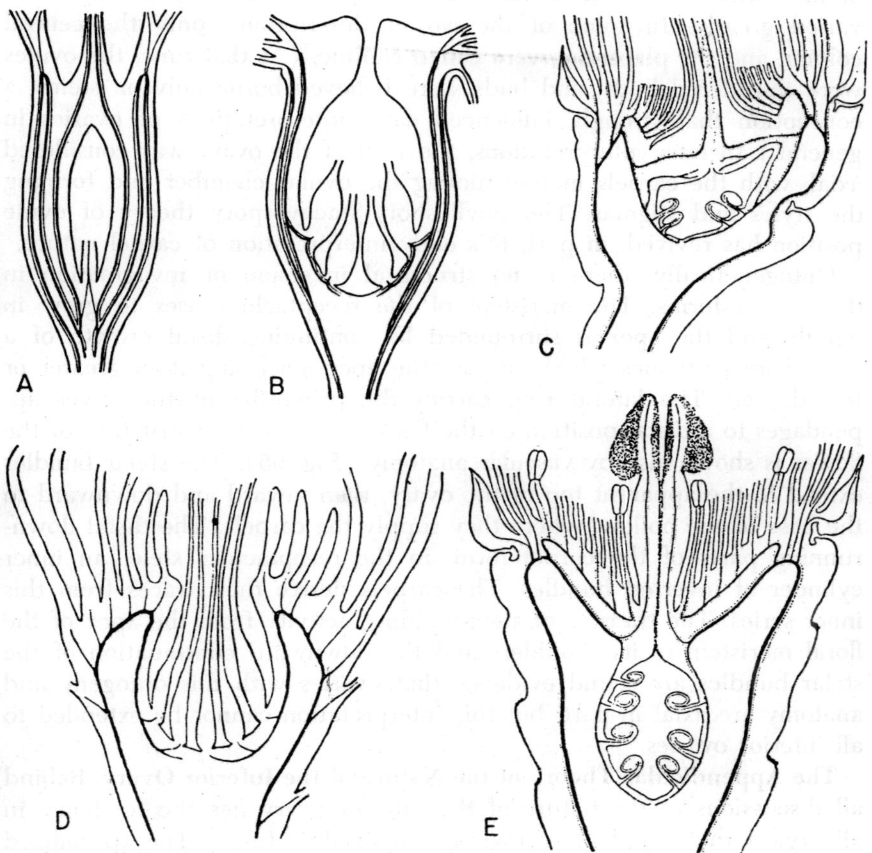

Fig. 95. Diagrams of longitudinal sections of inferior ovaries *A, C, E,* and pseudo-inferior ovaries, *B, D,* showing axial nature in vascular structure; stelar bundles continuing to top of ovary, turning inward and downward sharply and ending in center of floral cup; traces to ovules derived from down-running, inverted stelar bundles. In *B* and *D*, ovules incompletely closed by stelar sheath. *A, Darbya umbellulata; B, Rosa* sp.; *C, Pereskia bleo; D, Calycanthus occidentalis; E, Opuntia dillenii.* (*A, B, after Smith and Smith, B, adapted from Jackson; C, E, after Tiagi.*)

families has it been shown to be formed by a "sinking of the gynoecium in the receptacle"—Santalaceae, Cactaceae (Fig. 95).

Though the appendicular theory seems to be generally accepted in America as correctly interpreting the great majority of inferior ovules, the axial theory is still largely supported in other countries where evidence from anatomy and ontogeny is not considered or is interpreted as supporting the axial theory.

The Axial Theory of the Nature of the Inferior Ovary. Interpretations of the form and extent of the receptacle under the axial theory have varied greatly. In some of the earlier descriptions, only the central column and the placentae were called cauline. (At that time, the ovules were considered buds, and buds were believed borne only on stems, a conception that strongly influenced early interpretations of ovaries in general.) In later interpretations, the wall of the ovary was considered axial, with the carpels merely roofing the ovular chamber and forming the styles and stigmas. The Phyllospory-Stachyospory theory of ovule position has revived, in part, this early interpretation of carpel nature.

Ontogenetically, there is no structural inversion or invagination in these axial forms. The meristem of the receptacle ceases to grow in length, and the apex is surrounded by continuing distal growth of a ring of lower tissues which surpass the apex, enclosing it in greater or less degree. The lateral ring carries the primordia of the lower appendages to a distal position on the floral axis. The true structure of the flower is shown also by vascular anatomy (Fig. 95). The stelar bundles extend to the apparent top of the ovary, then inward and downward to the base of the hollow, where they supply the carpels. The distal downrunning parts of the carpel form, in the receptacular stele, an inner cylinder of inverted bundles. The carpels derive their traces from this inner series. The transfer of stem-building activity from the apex of the floral meristem to its shoulders and the downward continuation of the stelar bundles are sound evidence that ovaries with this ontogeny and anatomy are axial in part, but this interpretation cannot be extended to all inferior ovaries.

The Appendicular Theory of the Nature of the Inferior Ovary. Behind all discussions of the nature of the inferior ovary lies the tendency in all organs that stand close together to become fused. The specialized flower, with its appendages in close whorls, usually shows extensive fusion, both connation and adnation. Involved in the adnation are usually the calyx, corolla, and androecium, but, occasionally, the calyx remains free from the inner whorls and is free at the base of the inferior ovary—*Raspalia* (Brunoniaceae). The more primitive flowers, with some or all organs spirally arranged, have little fusion among their appendages. The inferior ovary is a characteristic of highly specialized taxa. Anatomy shows clear proof of fusion, especially of adnation. The vascular structure of epipetalous stamens shows all steps in the adnation of the stamen to the petal, from that where the two organs are merely appressed to that where all external evidence of fusion is lost and, internally, the vascular systems are united, with little or no evidence of the fusion. Similarly, organs in closely placed whorls are so fused that the vascular traces of organs in different whorls become united; the traces of sepals, petals, and stamens that lie in or near the same radius of the flower may be fused and form a single vascular bundle. This morphologically compound bundle may show structural evidence of compound nature, but is commonly simple. Comparisons of the vascular structure of the inferior ovaries of many families demonstrate stages in the fusion of the traces of the appendages and the appendicular nature of inferior ovaries in general.

In the genus *Rosa,* the basal part of the receptacle is an invaginated receptacle, with the carpels borne on the surface near the bottom of the hollow center; the distal part is appendicular, consisting of the bases of sepals, petals, and stamens (Fig. 95*B*). The location of the line between the two parts—where the stelar bundles bend downward—varies with the species.

The collarlike or cup-shaped structure on which perianth and androecium are borne and which surrounds or caps the gynoecium is commonly called in taxonomy the ***hypanthium.*** The term calyx tube—common in the older literature—is correctly applied only to the connate bases of sepals; as applied to the hypanthium, it is obviously morphologically incorrect, because it implies that, structurally, the hypanthium consists of appendages borne on other appendages. Under the axial theory of the nature of the inferior ovary, the hypanthium is considered basically axial, consisting of the rim of the receptacle; under the appendicular theory, it is considered to consist of the fused bases of sepals, petals, and stamens. Stages in the evolutionary development of the cup-shaped hypanthium are frequently seen. The Rosaceae provide a good example. In apocarpous genera, *Fragaria* and species of *Rubus,* the hypanthium is a flat disc around a convex receptacle; in other species of *Rubus,* in *Spiraea,* and *Physocarpus,* it is a shallow cup around a flat receptacle; in *Agrimonia,* it is a deep cup, with incurved top, around a flat receptacle. In syncarpous genera, the hypanthium is a shallow epigynous cup in *Aronia* and *Amelanchier,* a deeper epigynous cup in *Pyrus* and *Cydonia.* The Rosaceae also show stages in the adnation of the hypanthium to the carpellary walls. This fusion ranges from slight in *Spiraea,* to part way up in *Sorbus,* to the top of the ovary in most Pomoideae.

Morphologically, the term hypanthium is loose; it covers complex and varied structures. Usually it indicates a part of the flower that consists of the adnate bases of sepals, petals, and stamens. The fusion is not evident externally, and the hypanthium has been commonly interpreted as a part of the receptacle, but anatomy shows its true complex nature. It is receptacular only in those ovaries with invaginated receptacles. The fleshy sheath of the rose hip is receptacular in its basal part, appendicular in its distal part.

Among the Rosaceae, the Pomoideae have commonly been cited as showing inferior ovaries in which the receptacle takes a prominent part; the fleshy outer part of the apple and pear has been called cauline. But comparative studies of the floral anatomy of the Rosaceae show that this part of the fruit consists, morphologically, of the fused bases of the outer appendages, that it is morphologically homologous with the base of the hypanthium in *Fragaria, Rubus, Agrimonia.*

Some varieties of roses and pears frequently bear proliferating flowers in which a second flower grows from the center of the first. And, in some varieties of apples, the flower buds of scions grafted into vigorous seedlings develop into proliferating flowers and, rarely, fruits. Longitudinal sections of these proliferated flowers and fruits show, in their gross structure, evidence of the nature of the receptacle in these genera (Fig. 96). In *Rosa,* the vascular cylinder spreads out into the base of the fleshy cup, which bears the perianth and stamens on its rim, then dips down into the central cavity, contracting, and rises from the base into the pedicel of the secondary flower, above which its form is repeated. All appendages may be present in both flowers, with the upper stele arising between the carpels of the lower flower. In the pear, the vascular cylinder extends directly up through the center of the flower, without the spreading and invagination characteristic of the rose, and into the pedicel of the second flower, where it follows a similar course.

The flower buds of the apple scions develop branchlets that show stages transitional from typical flowers to leafy twigs. Some branchlets are flowerlike below, with sepals and petals, and leaves take the place of carpels; others are flowerlike throughout, but have the tip of the receptacle extending through the center of the carpel whorl, as a foliaceous twig. The anomalous structure of these flowers is the result of the grafting process and the stage of development of the flower buds when they became dormant. The grafting transferred the dormant buds, with flower primordia at different stages, from a slow-growing, fruiting branch to a vegetatively vigorous seedling. The primordia of floral appendages that were well formed in the bud continue development as floral organs; appendages still in early primordial stages develop as foliaceous appendages. The flower bud in which, as dormancy began, all the appendages had attained early primordial form, continued development as floral appendages when growth of the scion began. If the apex of the receptacle still retains a terminal meristem at this time, it may build a leafy stem in the center of the flower, between the carpels (Figs. 96 and 97).

Of special interest from the standpoint of the nature of the ovary in the apple are those flowers in which the flower is normal in every way, except for the proliferation of the receptacle as a normal leafy stem. Rarely, these flowers are pollinated, and apples mature with a leafy stem protruding from the apex (Fig. 97). Increase in diameter of the stem within the apple distorts and bursts open the fruit. Longitudinal

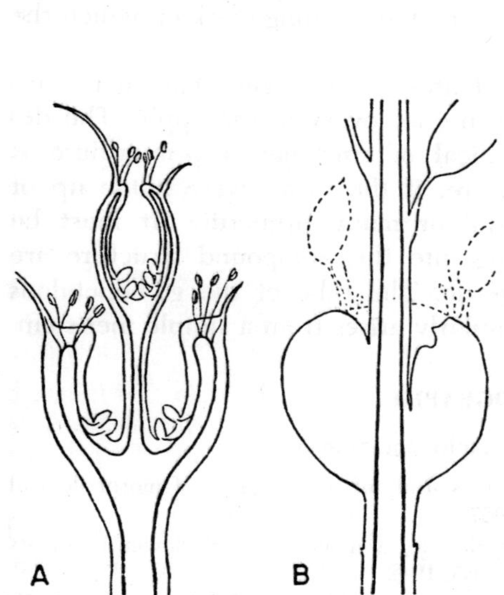

Fig. 96. Diagrams of longitudinal sections of proliferated flower of *Rosa cinnamomea* and fruit of *Pyrus Malus* showing by form and position of the vascular cylinder their morphological nature. In *A*, the receptacle of the flower is hollow; in *B*, the receptacular stele extends as a cylinder through the center of the fruit, showing that the flesh of the rose fruit is largely receptacular, that of the apple, appendicular. (Dotted lines in *B*, position of deciduous appendages.)

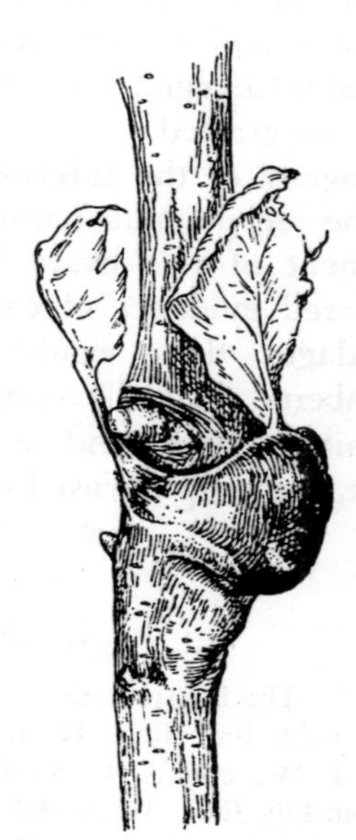

Fig. 97. Sketch of proliferated fruit of *Pyrus Malus,* 18 months after flowering, showing continuation of growth of receptacular stele through the center of flower and fruit. Fleshy fruit, representing bases of adnate sepals, petals, and stamens; the tops of two withered leaflike sepals still persisting. (*Specimen provided by Prof. K. D. Brase, N.Y. State Agricultural Exper. Sta., Geneva, N.Y. Drawing by Elfriede Abbe.*)

sections of these proliferated flowers and fruits (Fig. 96*B*) show the nature of the apple flower. The vascular stele continues through the flower and the fruit, without expansion or inversion, as in the rose. Only a slender central part of the "core," extending between the carpels, is cauline; the rest of the flower and fruit are appendicular. The interpretation of the apple ovary and the flesh of its fruit as axial in nature is inaccurate. If the fleshy part of the apple and pear were axial (cauline), the vascular stele would expand, as in the rose flower and fruit.

The objection may be raised that the proliferated flowers and fruits of the rose and pear are abnormalities and of little value in morphological interpretation. But each individual flower and fruit has the structure of the normal flower; the lower flower differs only in the proliferation of the receptacle. The proliferated apple flower and fruit is a floral axis, and its appendages are modified in the primordium stage by a stimulus to vegetative activity given by the seedling stock on which the scion was grafted.

Ontogeny of the Inferior Ovary. Ontogeny has been claimed to support the receptacular nature of the inferior ovary in the apple. The development of this ovary by a conical or ring-shaped primordium is considered evidence of cauline nature; if the ovary were made up of appendages, there would be several or many primordia. It must be remembered that the units entering into this compound structure are congenitally fused and arise en masse. The tube of the gamopetalous corolla, made up of fused petals, similarly arises from a simple meristem.

BIBLIOGRAPHY

See also Bibliography for Stamen

Arber, A.: The interpretation of the flower: a study of some aspects of morphological thought, *Biol. Rev.*, **12:** 157–184, 1937.

Bailey, I. W., and A. C. Smith: Degeneriaceae, a new family of flowering plants from Fiji, *Jour. Arnold Arb.*, **23:** 356–365, 1942.

——— and C. G. Nast: The comparative morphology of the Winteraceae. II. Carpels, *Jour. Arnold Arb.*, **24:** 472–481, 1943.

——— and B. G. L. Swamy: The conduplicate carpel of dicotyledons and its initial trends of specialization, *Am. Jour. Bot.*, **32:** 373–379, 1951.

Bancroft, H.: A review of researches concerning floral morphology, *Bot. Rev.*, **1:** 77–99, 1935.

Baum, H.: Über die postgenitale Verwachsung in Karpellen, *Oesterr. Bot. Zeitschr.*, **95:** 86–94, 1948.

———: Die Stellung der Samenanlagen am Karpel bei *Asclepias syriaca, Cynanchum vincetoxicum* und *Erythraea Centaurium, Oesterr. Bot. Zeitschr.*, **95:** 251–256, 1949.

———: Das Zustandekommen "offener" Angiospermengynözeen, *Oesterr. Bot. Zeitschr.*, **96:** 285–288, 1949.

———: Über die "primitivste" Karpellform, *Oesterr. Bot. Zeitschr.*, **99:** 632–634, 1952.

Blaser, H. W.: The morphology of the flowers and the inflorescences of the Cyperaceae, Thesis, Cornell University, 1940.

Bonnier, G.: Anatomie de la rose à proliferation centrale, *Bull. Soc. Bot. France*, **28:** 328–331, 1881.

Bosch, E.: Blütenmorphologische und cytologische Untersuchungen an Palmen, *Ber. Schweiz. Bot. Ges.*, **57:** 37–100, 1941.

Brouland, M.: Recherches sur l'anatomie florale des Renonculacées, *Le Botaniste,* **27:** 1–276, 1935.

Buchet, S.: La concrescence congénitale n'est pas une vue de l'esprit, *Bull. Soc. Bot. France,* **75:** 733–740, 1928.

Buxbaum, F.: Untersuchungen zur Morphologie der Kacteenblütte. I. Das Gynoeceum, *Bot. Archiv,* **45:** 190–247, 1944.

Capus, G.: Anatomie du tissu conducteur, *Ann. Sci. Nat. Bot.,* 6 sér., **7:** 209–291, 1878.

Chapman, M.: Carpel anatomy of the Berberidaceae, *Am. Jour. Bot.,* **23:** 340–348, 1936.

Chute, H. M.: The morphology and anatomy of the achene, *Am. Jour. Bot.,* **17:** 703–723, 1930.

Dickson, J.: Studies in floral anatomy. II. The floral anatomy of *Glaucium flavum* with reference to that of other members of the Papaveraceae, *Jour. Linn. Soc. London,* **50:** 175–224, 1935.

———: Studies in floral anatomy. III. An interpretation of the gynoecium in the Primulaceae, *Am. Jour. Bot.,* **23:** 385–393, 1936.

Douglas, G. E.: Studies in the vascular anatomy of the Primulaceae, *Am. Jour. Bot.,* **23:** 199–212, 1936.

———: The inferior ovary. II, *Bot. Rev.,* **23:** 1–41, 1957.

Dyal, S. C.: Studies in the family Valerianaceae, Thesis, Cornell University, 1941.

Eames, A. J.: The vascular anatomy of the flower, with refutation of the theory of carpel polymorphism, *Am. Jour. Bot.,* **18:** 147–188, 1931.

———: Again: "The New Morphology," *New Phyt.,* **50:** 17–35, 1951.

Eber, E.: Karpellbau und Plazentationsverhältnisse in der Reihe der Helobiae, *Flora,* N.F., **127:** 273–330, 1934.

Eckardt, T.: Untersuchungen über Morphologie, Entwicklungsgeschichte und systematische Bedeutung der pseudomonomeren Gynoeciums, *Nova Acta Leopold.,* N.F., **5:** 1–112, 1937.

———: Nachweis der Blattbürtigkeit ("Phyllosporie") grundständiger Samenanlagen bei Centrospermen, *Ber. Deutsch. Bot. Ges.,* **68:** 167–182, 1955.

El-Hamidi, A.: Vergleichend-morphologische Untersuchungen am Gynoecium der Unterfamilien Melanthioideae und Asphodeloideae der Liliaceae, *Arb. Inst. Allgem. Bot. Univ. Zürich,* ser. A, no. 4, 1952.

Fraser, M. S.: A study of the vascular supply to the carpels in the follicle-bearing Ranunculaceae, *Trans. Roy. Soc. Edinburgh,* **59:** 1–56, 1937.

Gauthier, R.: The nature of the inferior ovary in the genus *Begonia, Contr. Inst. Bot. Univ. Montreal,* **66:** 5–91, 1950.

Grant, V.: The protection of the ovules in flowering plants, *Evol.,* **4:** 179–201, 1950.

Grégoire, V.: Le morphogénèse et l'autonomie morphologique de l'appareil floral. I. Le carpel, *La Cellule,* **47:** 287–452, 1938.

Guéguen, F.: Anatomie comparée du tissu conducteur du style et du stigmate, *Jour. Bot.,* **16:** 300–313, 1902.

Hartl, D.: Morphologische Studien am Pistil der Scrophulariaceen, *Oesterr. Bot. Zeitschr.,* **103:** 185–242, 1956.

Hunt, K. W.: A study of the style and stigma, with reference to the nature of the carpel, *Am. Jour. Bot.,* **24:** 288–295, 1937.

Jackson, G.: The morphology of the flowers of *Rosa* and certain closely related genera, *Am. Jour. Bot.,* **21:** 453–466, 1934.

Joshi, A. C.: Carpel histogenesis and carpel morphology, *Jour. Indian Bot. Soc.,* **26:** 63–74, 1947.

Juhnke, G., and H. Winkler: Der Balg als Grundelement des Angiospermengynoeciums, *Beitr. Biol. Pfl.*, **25**: 290–324, 1938.

Junell, S.: Zur Gynaeceummorphologie und Systematik der Verbenaceen und Labiaten, *Symbolae Bot. Upsal.*, Vol. 4, 1934.

Lam, H. J.: A new system of the Cormophyta, *Blumea*, **6**: 282–289, 1948.

Leinfellner, W.: Das epidermale Randwachstum der Fruchtblätter, *Bot. Archiv*, **40**: 507–515, 1940.

———: Die U-förmige Plazenta als der Plazentationstypus der Angiospermen, *Oesterr. Bot. Zeitschr.*, **98**: 338–358, 1951.

———: Die Kelchblätter und underständingen Fruchtknoten und Achsenbechern, *Oesterr. Bot. Zeitschr.*, **101**: 315–327, 1954.

Leroy, J.-F.: Étude sur les Juglandaceae, *Mém. Mus. Nat. Hist. Bot. Ser. Paris*, **7**: 1–246, 1951.

Lignier, O.: Le fruit des Bennettitées et l'ascendance des angiospermes, *Mém. Bot. Soc. France*, **13**: 1–17, 1908.

Lindsey, A. A.: Floral anatomy in the Gentianaceae, *Am. Jour. Bot.*, **27**: 640–652, 1940.

Lister, G.: On the origin of the placenta in the tribe Alsineae of the order Caryophyllaceae, *Jour. Linn. Soc.*, **20**: 423–429, 1883.

MacDaniels, L. H.: The morphology of the apple and other pome fruits, *Cornell Univ. Agr. Exp. Sta. Mem.*, 230, 1940.

Morf, E.: Vergleichend-morphologische Untersuchungen am Gynoecium der Saxifragaceen, *Ber. Schweiz. Bot. Ges.*, **60**: 516–590, 1950.

Murty, Y. S.: Placentation in *Peperomia*, *Phytomorph.*, **2**: 132–134, 1952.

Newman, I. V.: Ontogeny of the angiospermic carpel, *Proc. Linn. Soc. N.S.W.*, **61**: 56–88, 1936.

Ozenda, P.: La nature morphologique du carpel, *Rev. Sci.*, **1946**: 393–404.

———: Recherches sur les dicotylédones apocarpiques, *Publ. Lab. École Sup.*, sér *Biol.* II, Paris, 1949.

Parkin, J.: The classical carpel and recent attacks, *Rpt. Bot. Exchange Club, Abroath*, **1933**: 598–606.

———: A plea for a simpler gynoecium, *Phytomorph.*, **5**: 46–57, 1955.

Puri, V.: The role of floral anatomy in the solution of morphological problems, *Bot. Rev.*, **17**: 471–553, 1951.

———: Floral anatomy and the inferior ovary, *Phytomorph.*, **2**: 122–129, 1952.

———: Placentation in angiosperms, *Bot. Rev.*, **18**: 603–651, 1952.

Raciborski, M.: Die Morphologie der Cabombeen und Nymphaeaceen, *Flora*, **78**: 244–279, 1894.

Rao, V. S.: The floral anatomy of some Verbenaceae with special reference to the gynoecium, *Jour. Indian Bot. Soc.*, **31**: 297–315, 1952.

Rauh, W., and H. Reznik: Histogenese becherförmiger Blüten- und Infloreszenzachsen, *Sitzber. Heidelb. Akad. Wiss. Math.-Nat. Kl.*, **1951**: 3–71.

Saunders, E. R.: Some morphological problems presented by the flower of the Nymphaeaceae, *Jour. Bot.*, **74**: 217–221, 1936.

Schaeppi, H., and F. Steindl: Vergleichend-morphologische Untersuchungen am Gynoecium der Rosoideen, *Ber. Schweiz. Bot. Ges.*, **60**: 15–50, 1950.

Smith, F. H., and E. C. Smith: Anatomy of the inferior ovary of *Darbya*, *Am. Jour. Bot.*, **29**: 464–471, 1942.

——— and ———: Floral Anatomy of the Santalaceae and Related Forms, *Oregon State Monographs, Studies in Bot.*, no. 5, 1942.

Sprotte, K.: Untersuchungen über Wachstum und Nervatur der Fruchtblätter, *Bot. Archiv*, **40**: 463–506, 1940.

Stauffer, J.: The floral anatomy of the Labiatae, Thesis, Cornell University, 1937.

Strasburger, E.: "Die Angiospermen und die Gymnospermen," Jena, 1879.

Swamy, B. G. L.: Further contributions to the morphology of the Degeneriaceae, *Jour. Arnold Arb.*, **30**: 9–38, 1949.

——— and I. W. Bailey: *Sarcandra*, a vesselless genus of the Chloranthaceae, *Jour. Arnold Arb.*, **31**: 117–129, 1950.

Thomas, H. H.: The nature and origin of the stigma: A contribution towards a new morphological interpretation of the angiosperm flower, *New Phyt.*, **33**: 173–197, 1934.

Thompson, J. M.: Studies in advancing sterility: The state of flowering known as angiospermy, with special reference to placentation and to the origin and nature of follicles and achenes, *Pub. Hartley Bot. Lab.*, no. 12, 1934.

———: On the gynoecial apex and the terminal legume of *Acacia longifolia* Willd. and *A. suaveolens* Willd., *Pub. Hartley Bot. Lab.*, no. 16, 1936.

Tiagi, Y. D.: Studies in floral morphology. II. Vascular anatomy of the flower of certain species of the Cactaceae, *Jour. Indian Bot. Soc.*, **34**: 408–428, 1955.

Troll, W.: Beiträge zur Morphologie des Gynaeceums, *Planta*, (I) **14**: 1–18, 1931; (II) **17**: 453–460, 1932; (III) **21**: 266–291, 1933; (IV) **21**: 447–485, 1933.

———: Die morphologische Natur der Karpelle, *Chron. Bot.*, **5**: 38–41, 1939.

Van Tieghem, P.: Recherches sur la structure du pistil, *Mém. Acad. Sci. Inst. Imp. France*, **21**: 1–262, 1875.

Uhl, N. W.: Studies in the floral morphology and anatomy of certain members of the Helobiales, Thesis, Cornell University, 1947.

Unruh, M.: Die morphologische Bedeutung des Karpells. Ein kritischer Sammelbericht, *Beitr. Biol. Pfl.*, **26**: 90–124, 1939.

———: Blattnervatur und Karpellnervatur, *Beitr. Biol. Pfl.*, **27**: 232–241, 1944.

Wilkinson, A. M.: Floral anatomy and morphology of some species of the tribes Linnaeeae and Sambuceae of the Caprifoliaceae, *Am. Jour. Bot.*, **35**: 365–371, 1948.

———: Floral anatomy and morphology of *Triosteum* and of the Caprifoliaceae in general, *Am. Jour. Bot.*, **36**: 481–489, 1949.

Winkler, H.: Verstehen wir das Gynözeum der Angiospermen schon? *Beitr. Biol. Pfl.*, **27**: 242–267, 1941.

Chapter 7

THE OVULE

The angiosperm *ovule* consists of a central body, enclosed more or less completely by one or two *integuments* and supported on a basal stalk, the *funicle*. The central body consists of a distal part, in which sporogenous tissue is borne, the *nucellus*, and a basal part, where funicle, integuments, and nucellus merge, the *chalaza*. The funicle may be long, slender, and curved, or short and stout; when absent, the ovule is *sessile*. The long funicle may be adnate to the ovule body, forming a ridge, the *raphe*. The integuments enclose the nucellus more or less completely and are commonly adnate to it, through part or all their extent. In some, chiefly more primitive, taxa they are free from the nucellus. The opening in the integumental sheath, where the tip of the nucellus is exposed, is the *micropyle*. Where the integuments do not reach the apex of the nucellus or spread distally, the micropyle is broad and shallow; where the integuments extend well beyond the nucellus or are thick, the micropyle is a long, slender canal, often constricted, sometimes irregular. The scar on the seed where it was attached to the funicle or the carpel wall is the *hilum;* the term is also applied to the swollen tip of the funicle.

Though there are, doubtless, only two true integuments, three and even four have been described for some taxa. Reduced ovules without integuments have been called *naked*. Where there are two integuments, the proximal, or *outer*, is usually the more massive and characteristically extends beyond the distal or *inner;* where there is only one integument, this is usually massive and is often thicker than the outer one is where two are present. Ovules with a prominent inner integument, which projects well beyond the outer (Fig. 98*C*), are occasional, as in the Annonaceae, Trapaceae, Proteaceae, some of the Cactaceae. Three integuments have been reported for several taxa, but the "third integument," in at least most of these taxa, is a modified part of one of the two normal integuments, usually a fleshy structure or aril. The term *aril* is applied rather loosely to fleshy parts of the ovule: to proliferations of the chalaza, the integuments, or parts of the integuments, that more or less completely envelop the ovule proper, extending upward from the base or downward from the integument tips (as shown by ontogeny); to outgrowths of the chalazal region; to fleshy funicles, as

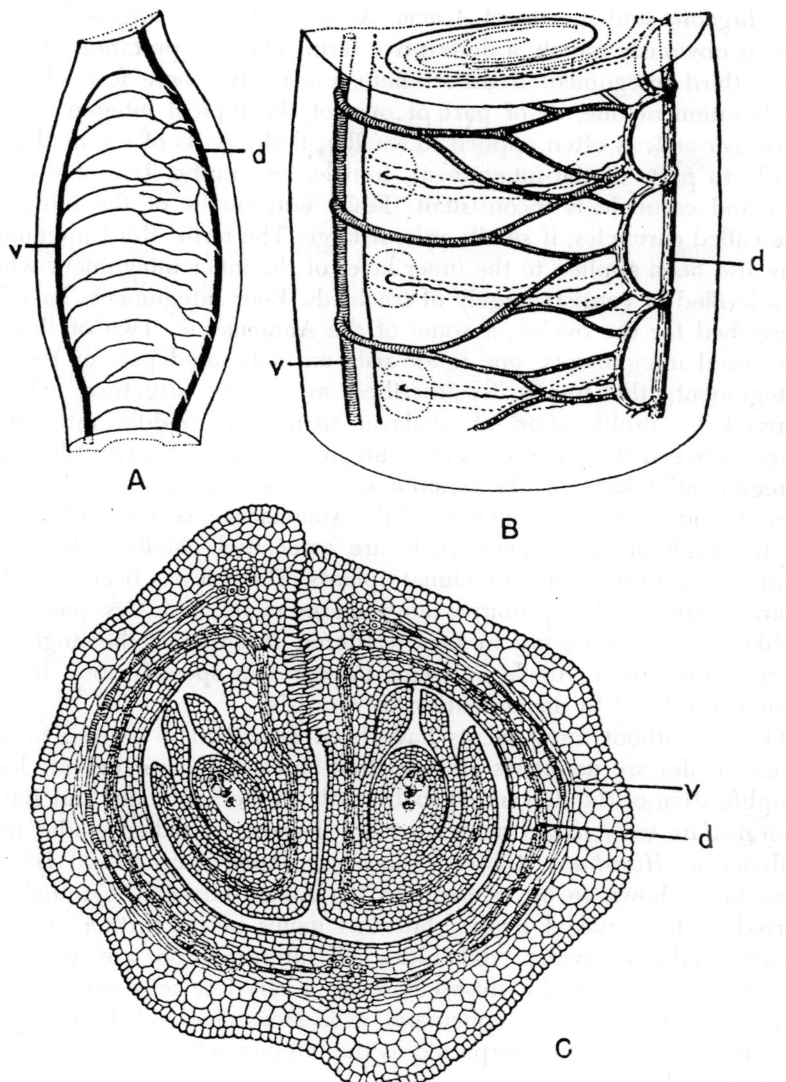

Fig. 98. Structure of the primitive conduplicate carpel, *Cananga*. *A*, *B*, sketches showing vascularization, origin of traces to the ovules from meshwork of branches connecting dorsal and ventral veins; *C*, cross section of mature carpel showing closure by interlocking of epidermal cells, inverted anatropous ovules, inner integument surpassing the outer, complete ovular vascular supply. *d*, dorsal vascularization; *v*, ventral vascularization; dotted circles in *B*, place of insertion of ovules. (*After Periasamy and Swamy.*)

in *Magnolia* and species of *Acacia*. As applied to the angiosperm ovule, aril is obviously merely a descriptive term. The interpretation of the aril as a third integument is unfortunate, especially where it is clearly an elaboration of one, or of part of one, of the typical integuments. The term *caruncle* is often applied to smaller, fleshy parts of the ovule, especially to parts of the integument, funicle, and raphe. Use of the terms aril and caruncle is inconsistent; fleshy outgrowths of the integument are called caruncles, if small; arils, if large. The term "third integument" has also been applied to the inner layer of the inner integument when it is sclerified or consists largely of tracheids. Four integuments have been described for the ovules of some of the Annonaceae. Two of these are the usual integuments; one is an aril—an elaborated part of the outer integument; the fourth is described as a new structure, which is formed by proliferation of chalazal tissues, a "middle integument," lying between the inner and outer integuments. In the seed, this "middle integument" takes over the function of the outer integument. This structure, found in only three genera of the Annonaceae, seems hardly worthy of the rank of integument. Arils are associated chiefly with tropical fruits, are rare in temperate-climate plants. The aril has been considered characteristic of the primitive angiosperm ovule, but this seems most unlikely, since it occurs in taxa scattered throughout the angiosperm system. Like the caruncle, it seems to represent specialization in ovule structure related to dissemination by animals.

Ovules without integuments have been described in several families. These ovules are doubtless all reduction types, with integuments lost in simplification of the flowers. Probably only in the Santalales and Balanophorales are truly naked ovules present. Reports of naked ovules in the Rubiaceae—*Houstonia, Coffea*—Olacaceae, Apocynaceae, and *Ipomoea* have been shown to be in error. In the Amaryllidaceae, *Crinum* is reported to have naked ovules, but this genus needs further study. In greatly reduced ovules, difficulties of interpretation are great; the nucellus may consist of a uniseriate layer of very few cells, even only one or two, and these may be ephemeral, as in *Houstonia* and the Olacaceae. The nucellus has been interpreted as the integument and the integument as the nucellus in the Apocynaceae.

Enlargements of the carpel wall adjacent to the ovule, of the chalazal region, and of the top of the outer integument that overtop and block, or appear to block, the micropyle to some extent are called *obturators* (Fig. 103). The term is best restricted to protuberances of the carpel. The term obturator, like caruncle and aril, has no morphological value. Good examples of obturators are seen in many of the Rosaceae—*Pyrus, Prunus, Spiraea.* In this family, the obturator consists of loosely arranged parenchyma and may serve as an aid in pollen-tube conduction to the micropyle, rather than obstruct its entrance there, as has been suggested. Basal appendages are present on ovules of some taxa, as in *Tetratheca,* where these projections are prominent, and lateral projections characterize some of the borages.

Fusion among the parts of the ovule—as among other organs—is common and is recognized as an advanced condition. Commonly, the integuments are fused to one another and to the nucellus for various distances, and the fusion is congenital. Lines of demarcation between these fused parts are usually absent. Freedom of the nucellus from the integuments, as in *Casuarina, Myrica, Juglans, Juliania, Cleome, Cananga* (Fig. 98*C*), is probably rare. Freedom of the integuments from one another—though they may lie appressed together—occurs in some of the more primitive dicotyledons, especially certain genera of the Rosaceae, Leguminosae, Capparidaceae, Ranunculaceae, some of the Amentiferae. All stages of fusion of the integuments are seen in other genera of these families—*Spiraea, Lupinus, Helleborus.* Rarely, integuments are separated by a space on the chalaza. The palms show all stages in fusion of the two integuments—from two entirely separate structures to one with no evidence of the fusion. Also, rarely, the outer integument fuses over the nucellus, closing the micropyle except in the inner integument, as in *Quercus, Alchemilla, Sibbaldia.* This closure has been interpreted as the result of loss of function under chalazogamy or parthenogenesis. Anatropous ovules with two integuments usually show only one integument on the side adjacent to the funicle; fusion of the outer integument and the funicle is congenital, and histological evidence of the union is lacking.

Two integuments clearly represent more primitive structure than one. The single integument in many—probably the great majority—of taxa represents, morphologically, a fusion of the two integuments. Most taxa with two integuments show connation at least in the lower parts. Many with a single integument show evidence of double nature at the tip, but the fusion is usually congenital and there is no histological evidence of the union. There are rare exceptions, as in *Helleborus,* where cell arrangement shows doubleness of the integument. It has been argued that the single integument represents the primitive condition, probably because it is the simpler condition and because, in the higher dicotyledons, it is derived, ontogenetically, "from the epidermis alone," whereas the outer integument is subepidermal in origin. But fusion, where congenital, produces false simplicity of form, and reduced structures do not retain the same histological origin as the unreduced form. Ontogeny alone does not give proof of morphological structure. The single integument may represent either the two integuments intimately fused or, probably infrequently, one surviving from two. Number of integuments is not a valuable character in the determination of natural relationships, as shown by differences in many undoubtedly related groups: most species of *Populus* have two integuments, but two species and the related genus *Salix* have only one; in the Piperaceae, *Peperomia* has one integument and the other genera, two.

Ovules with two integuments characterize the more primitive dicotyledons and the monocotyledons; the gamopetalous dicotyledons have only one. But there are marked exceptions to this distribution. In the Ranunculaceae and Rosaceae, there are genera with one integument; in the gamopetalous dicotyledons, the Primulaceae and Cucurbitaceae have two integuments. Closely related genera in the Ranunculaceae, Rosaceae, and Ericaceae have different numbers of integuments. Achene types usually have one integument. The single integument seems clearly to represent the advanced condition. The presence of one massive integument in the Sympetalae has been considered support for a monophyletic origin for this taxon.

The outer integument has normal stomata similar to those on the ventral carpel wall and, rarely, may have chlorophyllaceous tissue.

In some taxa, epidermal cells of the inner integument become specialized and form a nutritive layer, the *endothelium* or *integumentary tapetum,* around the archesporium and, later, around the embryo sac, taking the place of disorganizing nucellar tissue. During seed development, the inner integument is usually destroyed; the "seed coat" is formed by the outer integument alone. The endothelium is well developed in the Rosaceae and Hamamelidaceae and has been called a tapetum, the equivalent of the microsporangiate tapetum. Unfortunately, the term endothelium has been applied loosely to tissues of both the nucellus and the integument—to the uniseriate sheathing nucellus, which disorganizes as the mother cell divides and the megaspore germinates, and to the histologically modified and prominent inner epidermis of the integument which surrounds the embryo sac and developing embryo. The integumentary layer may become thick-walled and a part of the protective layers of the seed.

The ovule was early described as *crassinucellate,* where the nucellus is massive (Fig. 99) and the megaspore mother cells are deep in the distal tissue; and *tenuinucellate,* where small and delicate and the spore mother cells (usually one) are directly below the epidermis (Fig. 100). All gradations between these types are found. In tenuinucellate types, even in the same species, the spore mother cell may lie directly below the epidermis or be separated from it by one or two cells which belong to the subepidermal tissue or have been formed by periclinal divisions of the epidermis itself. The more massive nucelli, especially those of the Amentiferae, often show, at the base and center, an area of specialized parenchyma. This "central strand" has been considered possible evidence of an ancestral vascular supply to the nucellus (see Anatomy of the Ovule). At the base of the nucellus a few cells may become somewhat thick-walled and lignified, forming a basal area below the sporogenous tissue, the *hypostace*. It becomes more prominent as the embryo sac develops. Its functional significance is unknown. The use of terms applied to the various parts of the nucellus is loose and inconsistent. "Calotte" is applied not only to the cap of tissue set apart by radially arranged cells but to all tissues above the archesporial cells.

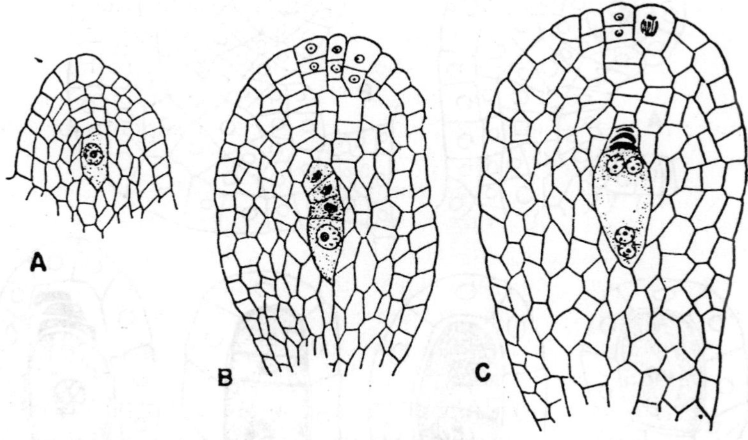

Fig. 99. Longitudinal sections of developing crassinucellate ovules, *Cercidiphyllum*, showing *A*, megaspore mother cell; *B*, linear tetrad of spores, the three outer degenerating; *C*, germinated innermost spore, forming a 4-nucleate embryo sac, the outer spores vestigial. (*After Swamy and Bailey.*)

The evolutionary history of the nucellus is chiefly that of reduction from the numerous cells of the massive type, in which the several archesporial cells are deeply sunken, to the extreme tenuinucellate type, in which a very few epidermal cells overlie one archesporial cell. Transitional forms are frequent.

Cells of the nucellus that lie below the epidermis and surround the mother cell laterally, and later the embryo sac, are called *parietal cells, cover,* and *wall cells.* Parietal cells is the best term, because they are probably homologous with the parietal cells of the anther sac. Unfortunately, they have also been called "tapetal" cells, but within the nucellus there is probably no definitely specialized nutritive layer like that of the microsporangial tissue; the endothelium is integumentary.

Modifications of the micropylar end of the ovule are many. The nucellus may project into the micropyle, as in some of the Caryophyllaceae and species of *Polygonum,* or even extend beyond it, as in *Rhodotypus, Malva, Hibiscus,* some species of *Euphorbia.* These proliferations of the nucellus may fuse with the integuments or with the obturator, as in *Manihot.* The distal epidermal cells of the nucellus may become greatly elongated and prominent, or develop thickened outer walls. Though the nucellus is always terminal, it may not appear so in

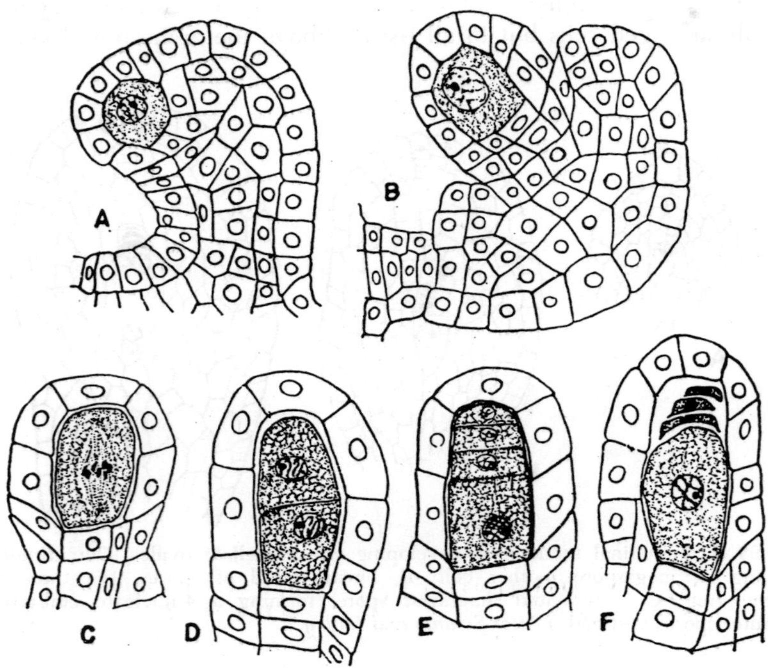

Fig. 100. Longitudinal sections of tenuinucellate ovules showing megasporogenesis, *Lobelia cardinalis. A,* ovule primordium showing hypodermal archesporial cell; *B,* archesporial cell enlarged, integument developing; *C,* first meiotic division in megaspore mother cell; *D,* daughter cells of megaspore mother cell; *E,* linear tetrad of megaspores, the proximal enlarging; *F,* the proximal spore mature, the others aborted. (*After Cooper.*)

tenuinucellate forms, where it consists of very few cells and is overtopped by a massive integument.

An important step in the reduction of the nucellus is seen in the Apocynaceae. In the more primitive taxa, the nucellus consists of a uniseriate sheath about the spore mother cell and the spore tetrad (Fig. 101*A*); in some genera, the lateral, sheathing (parietal) cells collapse and disappear, so that the nucellus is reduced to a single layer of a few cells on the top, directly below the micropyle (Fig. 101*B*). In

Apocynum, one of the highest genera, the nucellus persists only as a few covering cells; the sheathing layer is no longer distinguishable (Fig. 101*C, D*). This type, described as "vestigial," characterizes the Asclepiadaceae (Fig. 101*E*) and many of the Sympetalae. The Commelinaceae have been reported to have a "filamentous and constricted" nucellus, consisting of a small uniseriate row or cluster of cells surrounded by an epidermis. In the orchids, the nucellus (Fig. 102) has been called "insignificant," and interpretation of its structure is difficult. The presence of hypodermal cells here is in doubt, as is the reported epidermal origin

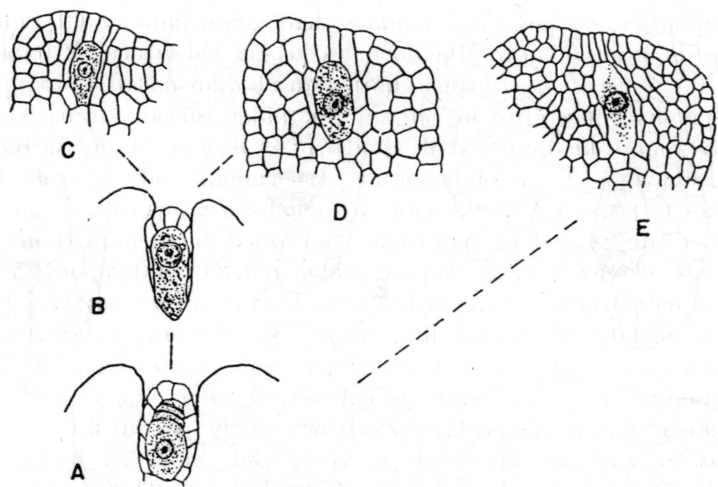

Fig. 101. Sketches of reduced nucelli in the Apocynaceae. *A, Cerbera lactaria,* nucellus, a uniseriate sheath; *B, Vinca rosea,* lateral cells of nucellus crushed by integument; *C, V. minor; D, Apocynum cannabinum; E, Asclepias curassavica. C, D, E,* lateral cells of nucellus absent, distal cells continuous with epidermal cells of integument. (*After Guignard.*)

of the spore mother cells. In taxa where no ovule is formed, as in the Balanophoraceae and Loranthaceae, cells of the placenta or even of the ovary wall function as a nucellus. In *Arceuthobium pusillum,* where there is no ovule, there is only a hump of undifferentiated placental tissue, in which two embryo sacs develop (Fig. 108*E, F*). Even in well-developed ovules, a nucellus may be absent at flowering time, as in *Paeonia* and *Nandina,* where the nucellus is wholly absorbed before pollination and an integumentary nutritive layer becomes palisadelike (a "mantle tissue").

In the massive nucellus of some amentiferous genera, a central core of tissue with a stalklike base is set apart histologically. Within this core, the sporogenous tissue develops. This core has been considered possible

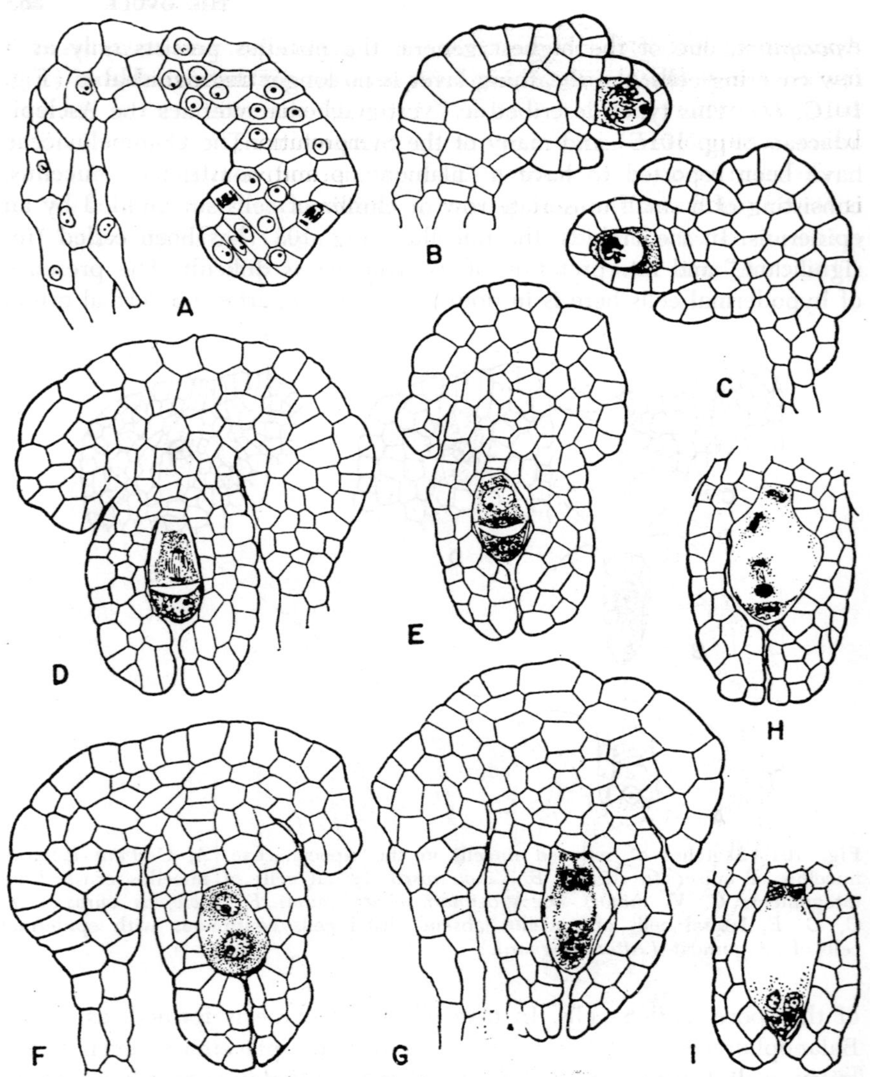

Fig. 102. Ontogeny of a simple (reduced) ovule, *Vanilla planifolia*, tenuinucellate, with vestigial integument. *A*, ovule primordium before development of archesporial cell; *B*, archesporial cell and primordium of integument; *C*, spore mother cell, nucellus and integument enlarging; *D*, dyad cells, integument complete; *E*, two megaspores and one micropylar dyad cell; *F*, 2-nucleate embryo sac; *G*, 4-nucleate embryo sac; *H*, nuclei of 4-nucleate sac dividing; *I*, mature (6-nucleate) embryo sac. (*After Swamy.*)

evidence of lost ancestral structure, of a very large sporogenous mass enclosed in a heavy "megasporangium wall," originally vascular at its base; the presence, in a few genera, of "tracheids" within this core was believed supporting evidence for this interpretation.

The massive nucellus is undoubtedly the primitive type (Fig. 103). It is found chiefly in the more primitive families, some of which, like the

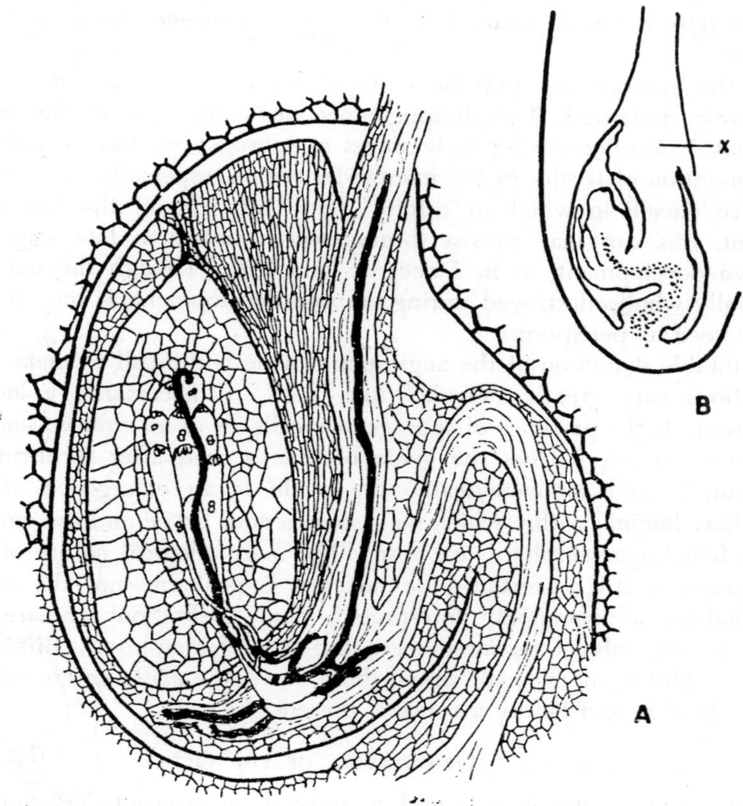

Fig. 103. Inverted, anatropous ovule of *Casuarina equisetifolia* in longitudinal section showing obturator *x* and course of pollen tube through obturator and chalaza to developing embryo sac. (*After Swamy.*)

Rosaceae, show stages in its reduction. The story of specialization of the nucellus closely parallels that of the integuments; the massive nucellus generally accompanies two integuments. The lower dicotyledons and the monocotyledons are, in general, crassinucellate; the gamopetalous dicotyledons, tenuinucellate; but there are many exceptions in both groups. Some of the exceptions are, perhaps, phylogenetically significant. The presence of a massive nucellus in the gamopetalous Cucurbitaceae supports other evidence that this family is more closely related to polypetalous than to other gamopetalous families. This view of the Cucurbitaceae is now frequently expressed.

Nucellar type is fairly constant in families—probably more frequently so than ovule type—but cannot be used generally as a family character; some families have more than one type of nucellus—Araliaceae, Cornaceae, Saxifragaceae, Ranunculaceae, Dilleniaceae, Caryophyllaceae, Potamogetonaceae, Araceae, Liliaceae, Amaryllidaceae, Iridaceae, Gramineae.

As the embryo sac and the embryo develop, the nucellus is progressively destroyed. The disappearance, in some taxa, of the small, few-celled nucellus during early stages of sporogenesis has caused misinterpretation of tissues in the micropylar region, especially in fruits. In massive nucelli in which a "calotte" (see Ontogeny of the Ovule) is present, this cap may persist through early and even late stages of embryo development, as in *Elaeagnus* and *Vitis*. Though the nucellus is usually wholly destroyed during embryo development, it may persist in the seed as perisperm.

A suitable definition of the angiosperm ovule is difficult to make, and definitions vary greatly. "Ovule," like "seed," is necessarily a loosely used term. Is the presence of sporogenous cells, or of an embryo sac, the essential character? Does the ovule become a seed after fertilization? Structurally and ontogenetically, the ovule is an emergence of the carpellary lamina, within which megasporogenous tissue and the embryo sac, a female gametophyte, are borne. The morphological nature of this emergence is still uncertain—the nucellus is probably not the megasporangium, as commonly interpreted. Some definitions require the presence not only of an integument but of a gametophyte. But, as a recognizable organ, an ovule primordium, the ovule exists before integuments develop and before a spore is present.

Form and Orientation of the Ovule

Variations in general form and in position of micropyle distinguish several types, fundamentally much alike and representative of evolutionary modifications of a basic type. Transitional forms are frequent. The ovule is *orthotropous* where it is straight and upright on the placental surface, the micropyle distal, and the funicle short or absent; *anatropous* where it is reflexed at the chalazal region and appressed to or adnate to the funicle and the micropyle usually faces the placenta; *campylotropous* where the ovule is more or less bean-shaped and the micropyle faces the placenta and is attached near the middle of the ovule; *amphitropous* where the funicle is adnate to the ovule proper for about half its length and the micropyle faces laterally. The term *atropous* is a synonym for orthotropous but has also been used for some intermediate types. The anatropous ovule is typically inverted; rarely, it is turned through much more than 180°, even 360° (Figs. 98*C* and 104*D*), as in *Sassafras, Plumbago, Opuntia, Scleranthus.* Inversion brings the ovule body and filament together, and adnation commonly occurs. The fusion is congenital; the funicle and the outer integument, where adjacent to the funicle, are so merged that the outer integument

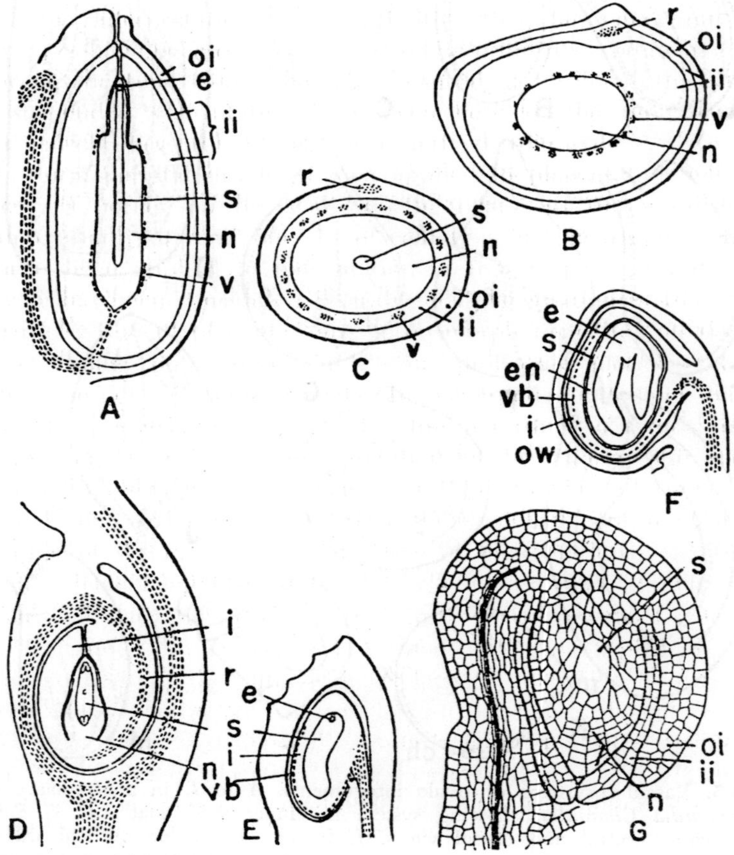

Fig. 104. Vascular supplies of ovules. *A, B, C,* showing vascular sheath in inner integument: *A, B, Ricinus communis* in longitudinal and cross sections, respectively; *C, Jatropha integerrima* in cross section. *D, Anemone narcissiflora,* longitudinal section of carpel base and ovule, showing inverted ovule united to carpel by obturator. *E, F, G,* simple vascular systems of anatropous ovules seen in longitudinal sections: *E, Echium plantagineum; F, Anchusa sempervirens; G, Delphinium ajacis. i,* integument; *ii,* inner integument; *oi,* outer integument; *e,* embryo; *en,* endosperm; *n,* nucellus; *ow,* ovary wall; *r,* raphe bundle; *s,* embryo sac; *v,* vascular sheath; *vb,* vascular bundle. (*A to F, after Kühn; G, after Lonay.*)

may appear absent on that side (Figs. 104*A* and 105*H*). The ridge formed by this fusion, the *raphe,* may give internal evidence of the fusion by the presence of the vascular strand of the funicle running along the side of the ovule body to the chalaza. In fruit development,

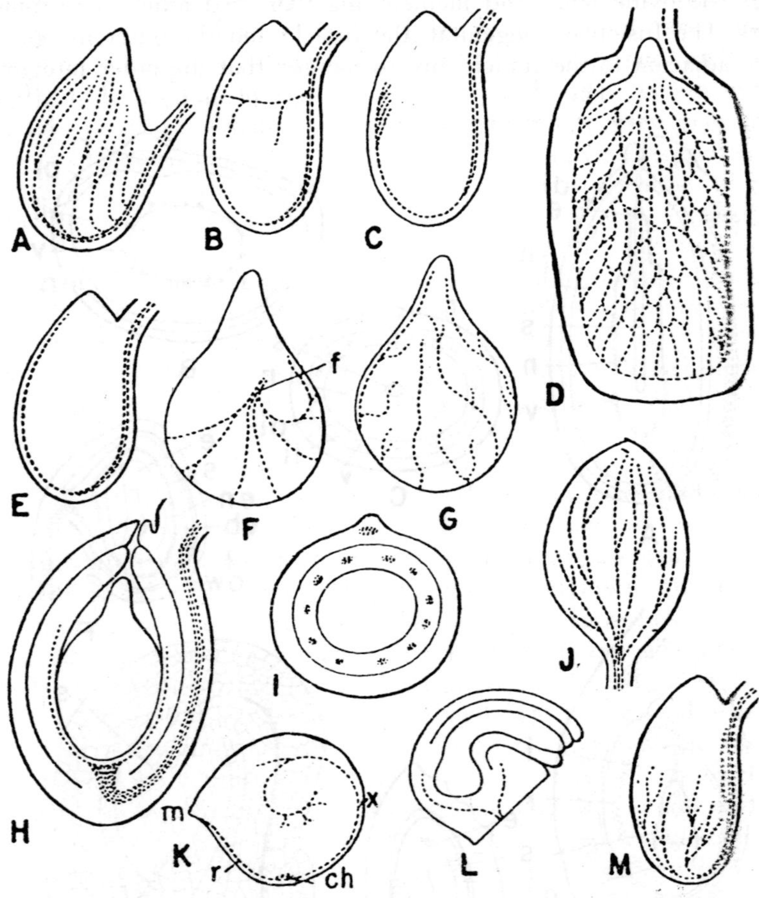

Fig. 105. Vascular systems of ovule integuments. *A* to *I*, in anatropous ovules: *D*, *Mormordia Charantia,* showing system in "inner seed coat"; *F*, *G*, *Echium plantagineum,* ventral and dorsal views, *f*, funicle; *H*, *I*, longitudinal and cross sections of same ovule showing vascular strands in inner integument. *J*, in orthotropous ovule. *K*, *L*, *M*, in campylotropous ovules: *K*, *Cicer arietinum; ch*, chalaza, *m*, micropyle, *r*, raphe bundle, *x*, point of branching of vascular bundle. (*After Kühn.*)

there may be further fusion; the ovule may become adnate to the ovary wall, as in some achenes and nut fruits.

The anatropous ovule is the most common type and is characteristic of many taxa, especially the more primitive taxa—Ranales, Helobiales,

Nymphaeaceae. The orthotropous ovule is found chiefly in the more highly specialized families where placentation is basal and the ovule solitary—Juglandaceae, Polygonaceae, Piperaceae, Restionaceae, Najadaceae. The campylotropous ovule characterizes some prominent families—Capparidaceae, Caryophyllaceae, Geraniaceae, Apocynaceae, Verbenaceae. Families may be uniform in ovule type or may have more than one type; the Palmae have both anatropous and orthotropous ovules. The direction of curvature of individual anatropous ovules may vary, depending upon space available. Occasionally, more than one type of ovule may occur even in the same carpel. Changes may occur in ovule type in later stages; an ovule anatropous at pollination may become campylotropous after fertilization, as it becomes adnate to the funicle.

Variety of ovule form doubtless represents adaptation of an ancestral type to conditions within the ovary—number of ovules, available space, placental type, location of pollen-tube transmitting tissue. All these conditions have doubtless been important in the modification of an ancestral form. The important problem is the determination of this primitive form. Modifications in form in minor ways are sometimes related to lack of space for normal development, but lack of space plays no important part in the determination of shape, because all types occur where the ovarian chamber is large, and beginnings of inversion occur in early stages. Relation to course of the pollen tube may influence ovule orientation in highly specialized ovaries where the ovule is solitary. Ancestral form probably plays the important part in ovule type. Both the orthotropous and the anatropous types have been considered primitive. The orthotropous ovule was perhaps selected as primitive because of its simplicity of form, but its correlation with high specialization in the carpel, gynoecium, and flower seems to show that its simplicity is derived. The anatropous type characterizes the most primitive families of both dicotyledons and monocotyledons. It is typically crassinucellate, and its archesporial tissue is frequently multicellular. The inversion of the anatropous ovule has been interpreted as evidence of specialization, as an adaptation to the course of the pollen tube, a placing of the micropyle closer to the carpellary surface along which the pollen tube grows. On the other hand, resemblance of this ovule type has been seen to fertile psilophytalian telomes. (A similar inversion of ovules is present in Paleozoic conifers.) But the ovule is a complex structure, and it seems most unlikely that the simple telomes of the early land plants should persist in the angiosperms. The anatropous ovule seems to be the basic type, one inherited from ancestral stock. The dominance of this type and its presence in the most primitive orders of both dicotyledons and monocotyledons is strong evidence for this theory. Some of the other types develop ontogenetically from anatropous primordia. The basis for the opinion that the orthotropous is the primitive type seems to be that it is apparently the simplest type and that its usual position—basal and "terminal" on the receptacle—is also simple. But it accompanies high specialization in flower structure. The Urticales show, in their various genera, the derivation of the orthotropous from the anatropous ovule. In the Ulmaceae and Moraceae, the ovule is anatropous and suspended; *Urtica* and *Boehmeria* have orthotropous, basal ovules; *Laportea* and other genera are transitional.

Ovule Number

The primitive ovule number was doubtless several or many, as in the Nymphaeaceae, Hydrocharitaceae, *Butomus, Drimys, Eupomatia.* Reduction in number occurred with the restriction of the broad fertile area of laminar placentation to two submarginal rows and, still further, within the rows to one or two ovules, as carpel size was reduced. Rarely, reduction to few or solitary ovules was directly from laminar, without the intermediate submarginal stage—Cabombaceae (Fig. 85), Helobiales, *Nelumbo,* some Annonaceae, Winteraceae. Elaboration of the placenta, with great increase of fertile surface—the development of large emergences within the locule—was accompanied by multiplication of ovules to high numbers (Fig. 80), many more than in the primitive carpel. Great numbers of ovules borne on large placentae are present in the Ericaceae, Scrophulariaceae, Orchidaceae, Cucurbitaceae, Gentianaceae, and other families. It has been estimated that the number of ovules in an ovary of *Cymbidium* may be two million.

Reduction in a single carpel to one or two ovules, with transitional numbers, is seen in both laminar and submarginal placentation in many families—Ranunculaceae, Rosaceae, Proteaceae, Nymphaeaceae, Leguminosae; similarly, within a locule in syncarpous ovaries—Liliaceae, *Claytonia,* Caryophyllaceae; within a group of related genera—*Cydonia, Pyrus, Sorbus.* Reduction of the total number of ovules in a syncarpous ovary to one ovule, usually basal, is frequent. These ovules represent surviving members in syncarpous ovaries of complex structure. They have usually been derived by reduction of free central placentae—which, in turn, represent fused and modified axile placentae. Rarely, basal ovules represent surviving members of parietal placentae, as in the Droseraceae.

The one or two ovules surviving in a follicle or achene are usually proximal ones, but they are sometimes distal—some Proteaceae; rarely, they are median members of the row. Evidence for the position of these ovules is the frequent presence of vestigial ovules and, when these are lacking, of ovule traces leading from ventral bundles to the position of lost ovules—*Caltha, Trollius* (Fig. 42). Where there is a pair of surviving ovules—*Hydrastis, Calycanthus,* many Proteaceous genera—one ovule persists from each submarginal row, as evidenced by the origin of its trace. In many achenes, the single ovule has become subbasal—has moved phylogenetically from a higher position during the shortening of the carpel. Evidence of this change of position is in the course of the ovule trace, which extends from the ancestral position of the ovule and bends back to the present position, as in *Piper, Ranunculus,* the Urticaceae.

Ontogeny of the Ovule

The ovule arises on the placental surface as a hemispherical projection, which is initiated by cell multiplication, chiefly in the first and second cell layers beneath the epidermis. The initiating cells may be many or few—one to three in very small ovules. In early stages, divisions in the epidermis are largely or wholly anticlinal and, in some taxa, remain so.

The ovule of some orchids consists of a file of cells formed by one initial covered by the epidermis. Ovules of the Balanophoraceae (Fig. 109), *Monotropa,* and perhaps other taxa may be even simpler. No ovules, as such, are formed in some parasitic families. Continued cell multiplication at the distal end of the primordium builds up a more or less elongate structure, which early shows, near its enlarging tip, evidence of archesporial cells. Differentiation within the ovule is basipetal —in sequence, the distal nucellar tip, the inner integument, the outer integument. Rarely, the outer integument precedes the inner in formation. The integuments are built up by annular meristems; the nucellus is enlarged chiefly by proliferation of the hypodermal initials, but, in some taxa, the epidermis plays an important part in the later stages of development by tangential as well as anticlinal divisions. In some families, such as the Vitaceae and Elaeagnaceae, a prominent *epidermal cap* or *calotte* is formed in this way, which covers the entire distal end of the nucellus, often appearing sharply distinct from the rest of the nucellus. This cap has been described as the megasporangium wall, with the implication that the rest of the nucellus is potentially sporogenous tissue. The term calotte has also, unfortunately, been applied to the parietal tissue above the archesporium, where the fertile tissue is deeply sunken; in these nucelli, both an epidermal cap and a calotte may be present. The archesporial tissue is formed from hypodermal cells directly below the epidermis or at some distance below this layer. The cells between the epidermis and the archesporial cells, the *covering* or *parietal cells,* may also multiply and aid in increasing the size of the nucellus. In tenuinucellate ovules, there are often no parietal cells; the spore mother cells lie directly below the epidermis. In small ovules, the few parietal cells may degenerate as the sporogenous tissue matures. Massive nucelli may be built up largely by the parietal cells, as in *Sassafras*, *Calycanthus*, and many of the Ranunculaceae, or by both hypodermal and epidermal initials, as in the Rosaceae, Urticaceae, Vitaceae. Sequence in development of ovules on the placenta is apparently correlated with the ontogeny of the carpel.

Time of Ovule Development. Development of the ovule is usually closely correlated with that of the flower, especially the stamen, but there are marked departures from this. Ovules may not have begun to develop at time of pollination, as in *Quercus* and *Fagus;* in some of the orchids, the ovules are not mature for weeks or months after pollination. In *Quercus* (black oaks), pollination occurs a year before the ovule is mature; similar conditions obtain in some of the conifers and cycads; in the Araucariaceae, there is no evidence of the ovule at the time of pollination, not even when the pollen tube reaches the position where the ovule will develop.

Anatomy of the Ovule

Ovule Traces and Venation. The distribution of vascular tissues in the ovules of the lower seed plants has received considerable attention, because of its possible phylogenetic significance. The absence of vascular bundles in the ovules of the conifers—with the exception of *Torreya* and *Cephalotaxus*—together with the supposed general absence of vascular tissue in the ovule integuments of most angiosperms, led to the theory that the vascular system of angiosperm ovules is greatly reduced, as a part of reduction from ancient gymnosperm seeds. But integumentary vascular systems are now recognized as fairly characteristic of angiosperms. These systems seem to be in process of reduction; the smaller and more delicate ovules of many families have none, and vestigial bundles have been described in others. Vestigial ovules usually have traces, and the placental position of lost ovules is often marked by the presence of their traces within the carpellary tissues. These traces are a well-marked feature in the anatomy of follicles in many families (Fig. 42*B*) and are part of the evidence that the follicle is the primitive type of carpel, rather than the achene, as is urged under the peltate theory of the nature of the carpel.

The ovule-trace supply in angiosperms is commonly a single bundle, derived at various places on the vein system of the carpel lamina. In a few taxa, the ovule has two traces—*Magnolia, Michelia, Schisandra, Austrobaileya;* more than two traces are rare. The origin of the traces varies with placentation type—from ventral veins, the laminar meshwork of small veins, the midrib vein (rarely); the place of origin is morphologically significant from the standpoint of the evolution of placentation. In laminar placentation, ovule traces are derived from all parts of the vein system; few or none are derived from the midvein in some taxa. In submarginal placentation, the traces are from the ventral veins. Most solitary, suspended, or basal ovules also have traces from ventral bundles, though the morphological origin may be obscure. Examples are seen in the "basal" ovules of some achenes, where the trace seems to be derived from the midvein—lateral and midveins are reduced and fused (Fig. 42*K, L, M*)—and in syncarpous ovaries, where the terminal ovule is a surviving member from a free central placenta and its trace is derived from the fused ventrals of all the carpels—Cyperaceae, *Polygonum* (where the ovule has six traces or one compounded of six). Rarely, in two-trace ovules, one of the traces has similarly been "taken over" functionally by a nearby ovule—*Ilex, Urtica gracilis. Ilex verticillata* has two basal ovules, one abortive, each with one trace; *I. crenata* has one basal ovule, with two traces derived from the ventral traces of different carpels. In the Cornaceae, the ovules of *Cornus* and *Corokia* have been shown to have more than one trace. In *Cornus,* the ovules of some species have two traces, one from each ventral bundle of one carpel, and other species have several traces—the extra traces apparently "taken over" from lost ovules. (Ovule traces are sometimes confused with carpellary traces, with resultant misinterpretations of carpel structure. Ovule traces are branches of the vascular system of the carpel and not derived from the receptacular stele. Under great reduction in the carpel, the ovule trace may be the distal continuation of the carpel trace.)

The presence of vascular bundles in the integuments of amentiferous genera—long considered some of the most primitive dicotyledons—and in the primitive gymnosperms was considered, at the beginning of the twentieth century, evidence that ovules with vascularized integuments are primitive. That vascular bundles occur in many other taxa, including such gamopetalous families as the Cucurbitaceae, Oleaceae, Caprifoliaceae, and Compositae, seems, surprisingly, to have been overlooked. Ovule integuments with vascular bundles have been described in more than one hundred species in over thirty families, scattered throughout the angiosperm system (Figs. 104 and 105). Their distribution is erratic; they are present in many genera of some large families such as the Compositae, Leguminosae, Euphorbiaceae; they may be absent in some species of a genus, though present in others; no vascular tissue has been reported in the integuments of the Primulales, Rhoeadales, Ericales, Rhamnales, Proteaceae, Polygonaceae. In families that generally lack them, isolated genera may have them—in the Ranunculaceae, only *Anemone* and *Glaucidium,* and in the Rosaceae, only *Prunus* are known to have them.

Ovular vascular systems seem to be most frequent in certain major taxa—the Ranales, Leguminosae, Amentiferae, Compositae, Contortae, Tubiflorae. They are rare in the monocotyledons—some palms and amaryllids. Further detailed study may change this picture, but a broad distribution throughout angiosperms surely exists. The history of our knowledge of ovular vascular systems parallels that of chalazogamy, which was, at first, believed restricted to so-called primitive taxa, the Amentiferae, but was later found widely distributed among angiosperms. The presence of a vascular supply in the integuments has been believed to be a primitive character, because of an assumption that this vascularization was correlated with freedom of the nucellus from the integuments, but this correlation is rare, and is found perhaps only in the Amentiferae.

Integumentary bundles are continuations or branches of the ovular trace which extends through the funicle—and the raphe, if the funicle is adnate to the ovule proper—to the chalaza and toward the micropyle. The simpler conditions are those where the funicular bundle is continued, unbranched, to the chalaza (Figs. 104*A*, *G* and 105*E*), or beyond the chalaza to the micropylar region. Where it is branched, the branching may be simple or complex—of successive, more or less pinnate, laterals on a main bundle, or the bundle may break up into a few or several branches in the chalaza; a "crown" or "nest" may be formed below the nucellus, with few or many major branches extending up into or through the integuments. A meshwork of fine bundles may sheathe the nucellus (Fig. 105*D*). Within genera, and often within families, the pattern of the integumentary vascular system is apparently constant; in other families, such as the Leguminosae, Oleaceae, Compositae, there is a series from no vascular bundles to an abundant, complex supply. But when angiosperms, as a whole, are considered, there is no evidence of consistency in number of bundles, type of branching, or distribution around the ovule.

The vascular bundles are usually restricted to the outer integument where there are two integuments; very rarely are they known to occur in the inner integument (Figs. 104*B*, *C*, and 105*H*, *I*), as they do in the Euphorbiaceae. Where vascular bundles are present in the inner integument, they have been described as running first through the outer integument and then entering the inner. If this description is accurate, it provides evidence that the vascular tissue of the integuments has no morphological significance. The bundles are simple, histologically; the xylem consists of a few annular and spiral elements, and there is little or no distinguishable phloem. The bundles end freely in the median or distal part of the integument or, less commonly, form a meshwork, as do veinlets in leaves and sporophylls.

There is little correlation between type of vascular system in the integument and type of ovule, except that, in most orthotropous ovules, there is little vascular tissue except a short bundle to the chalaza. Rarely, the vascular supply is strong, as in *Juglans*.

In the vascular system beyond the chalaza, no pattern is determinable. The basal strand may continue unbranched to the tip of the ovule, or may branch pinnately, palmately, or irregularly. The vascular tissue may lie on one side of the ovule only. This distribution of vascular tissue suggests that the strands in the integuments are minor bundle ends, related in number and position to function, and have little morphological significance.

Suggestions have been made that the vascular system of angiosperm ovules is a remnant of that of the ovules of the pteridosperms and cycads, but no structural similarity is apparent; there is no basic pattern in the angiosperm ovule as in the other taxa. A vascular supply is present as frequently in the higher taxa as in the lower; a double supply—considered in the gymnosperms as probably more primitive than a single one—occurs only very rarely. The vascular supply of angiosperm ovule integuments seems rather to represent, at least in most taxa, an extension of the veinlet system of the carpel lamina, related to the nutrition of the embryo (note its sheathing of the nucellus in some families) in ovules in which the nucellus is small and can itself provide little food for the embryo. (Compare the abundance of stored food and the large gymnosperm nucelli with the absence of stored food and the small angiosperm nucelli.) The opinion that the presence of vascular tissue in the integuments is a primitive condition probably rests on its early discovery in the Amentiferae. Absence of vascular tissue in the integuments is not necessarily an advanced character; its presence, not necessarily a primitive one.

Vascular Tissue in the Nucellus. Tracheids have been rarely reported in the nucellus—*Casuarina, Asclepias,* the Fagaceae, Capparidaceae. The tracheids are annular or spiral, very small and slender, usually isolated or in small clusters. They occur largely in the sporogenous region, sometimes so placed as to suggest remnants of longitudinal series extending upward from the base of the nucellus. In *Castanea,* they form a loose cluster about the base of the embryo sac; some tracheids extend downward toward the chalaza. In three genera of the Capparidaceae, a few tracheids have been found scattered through the nucellus. The report of tracheids in the nucellus of the Thymelaeaceae is probably in error. Tracheids described as "on the periphery of the nucellus" and "between the seed coat and the embryo" are perhaps in the inner layer of the inner integument, where integumentary tracheids may occur. In *Casuarina,* isolated tracheids occur occasionally among the sporogenous cells. The function of these tracheids and their morphological significance are unknown. Resemblance in form and structure to the elaters of liverworts has been suggested. Correlation of these tracheids with the central core of a massive nucellus, like that of *Casuarina,* might be used to support the hypothesis that the massive nucellus is a surviving remnant of an ancestral, stalked sporangium, with elaters among the spores, but this suggestion is farfetched. The histological nature of these "tracheids" needs critical study; they may be neither tracheids nor elaters—in the strangely sclerenchymatous flower of *Casuarina,* perhaps merely supporting cells.

The Hypostace. In the ovules of many taxa, there is a platelike or cup-shaped area of specialized cells below or around the base of the embryo sac, long ago termed the *hypostace* (Figs. 115A and 138A). Histologically, it consists usually of sclerenchymatous cells and is considered supportive or protective in nature, but layers of secretory and apparently nutritive cells in this position have also been called a hypostace. The hypostace has been confused with the area of branching vascular tissue at the base of the integuments. The term *epistace* has been applied to layers or groups of thick-walled cells at the top of the nucellus. A thickened cuplike tip of the integuments around the micropyle has been called an *operculum.* These three terms have no apparent morphological significance.

A projection of nucellar or chalazal tissues, not always structurally distinct from the hypostace, may extend into the base of the embryo sac. These projections are sometimes prominent features of embryo-sac structure, as in *Zostera, Ribes.* Various terms have been given to them: *podium, pedestal, postament.* The term podium is sometimes restricted to projections that are nucellar. Morphologically, these projections may be remnants of basal nucellar tissue, where disorganization accompanying embryo-sac growth has progressed on the sides more rapidly than below, or they may be an upgrowth of chalazal tissue. The antipodal nuclei or cells are sometimes sunken in the top of this projection. Embryo sacs with these basal intrusions are found in many families throughout the angiosperm system.

Reduction in the Ovule

Reduction of the integuments to one—by union of the two or loss of the inner one—and of the nucellus to a single layer of cells above the sporogenous tissue is common in the more specialized families. Greater reduction, accompanied by loss of the integuments and the nucellus, is characteristic of taxa with greatly reduced body habit—parasites, saprophytes, and some hydrophytes and epiphytes. In extreme reduction, the ovule itself is lost, and the archesporial tissue is borne in the placenta or in the carpel wall (Fig. 106*i*). In some parasites, the gynoecium—usually syncarpous, with inferior ovary—is so greatly simplified that its structure is determined only with difficulty. The importance of the recognition that some extremely simple angiosperm flowers may have "naked ovules" (without integuments) or no obvious

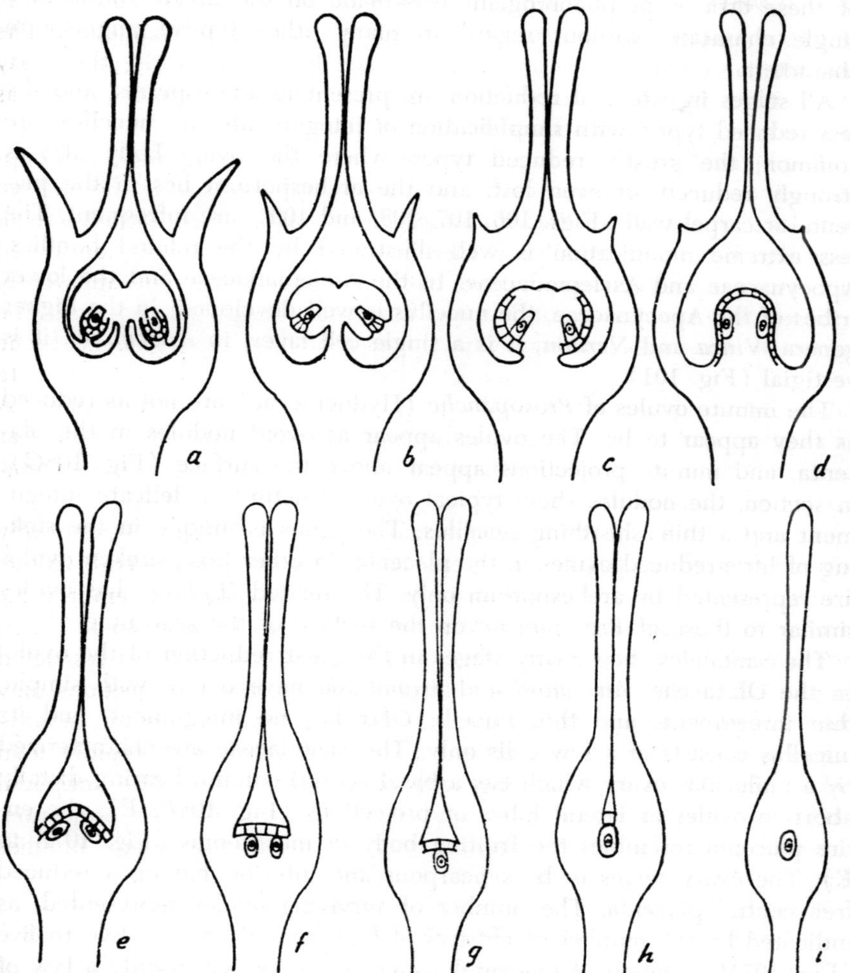

Fig. 106. Diagram illustrating hypothetical derivation of the *Balanophora* flower. *a*, normal, unreduced flower with free central placenta; *b*, ovules reduced in size, nucellus small, flat—*Thesium; c*, ovules not differentiated on placenta, their nucelli merged with placental epidermis—*Osyris, Santalum, Myzodendron; d*, placenta reduced to mound or nipple, nucellus lost—*Arceuthobium, Helosis; e*, nipple reduced —*Korthalsia; f*, embryo sacs on floor of ovarian chamber—*Viscum, Dendropthoë; g, h*, embryo sacs reduced to one at base of carpellary wall—*Scurrula; i*, solitary embryo sac embedded in reduced carpel—*Balanophora*. (*After Fagerlind.*)

ovules is emphasized by the claim—by others than morphologists—that some of the Sarcopodales are not angiosperms but belong in a lower group, "prephanerogams," intermediate between gymnosperms and angiosperms. (These plants have been shown to be true angiosperms, with reduced, not primitively naked ovules; the classification of these taxa as prephanerogams was made on the invalid basis of a single character, without regard to many other typical angiosperm characters.)

All stages in extent of reduction are present in angiosperms, and the less reduced types, with simplification of integuments and nucellus, are common; the greatly reduced types, where the ovule body also is strongly reduced, or even lost, and the archesporium lies in the placenta or carpel wall (Figs. 106, 107, 108, and 109), are infrequent. The less extreme modification is well illustrated by the related families, Apocynaceae and Asclepiadaceae. In the Asclepiadaceae and the lower tribes of the Apocynaceae, the nucellus is well developed; in the higher genera, *Vinca* and *Nerium,* it is a single cell layer; in *Apocynum,* it is vestigial (Fig. 101).

The minute ovules of *Prosopanche* (Hydnoraceae) are not as reduced as they appear to be. The ovules appear as ovoid nodules in the placenta, and minute projections appear above the surface (Fig. 108*G*); in section, the nodules show typical ovule structure, a delicate integument and a thin, sheathing nucellus. The genus is unique in the sinking of little-reduced ovules in the placenta; in other taxa, sunken ovules are represented by archesporium only. The related *Hydnora* has ovules similar to those of *Prosopanche* on the surface of the placentae.

The Santalales show many stages in the great reduction of the ovule. In the Olacaceae, *Anacalosa* and *Strombosia* have ovules with simple, thin integuments and thin nucelli; *Olax* has no integument, and its nucellus consists of a few cells only. The Santalaceae are characterized by a unilocular ovary which has a basal central column bearing distally abortive ovules as lateral lobes or projections (Fig. 108*E*, *F*); the entire placenta resembles the fruiting body of mushrooms (Fig. 107*A* to *E*). The ovary seems to be syncarpous and inferior, having a reduced free central placenta. The number of surviving ovules represented, as indicated by the number of clusters of archesporial cells, is two to five (Fig. 107*M*). Details of placental structure are known in only a few of the genera in the family. *Comandra* has ovules of anatropous form (Fig. 107*C*), with remnants of a nucellus and integument. The ovules of *Santalum* and *Thesium* are "outgrowths of the central column, pointing downward" (Fig. 107*A*, *B*); in *Osyris* and *Scleropyrum*, "the lobes point upward" (Fig. 107*E*, *F*, *G*). *Myoschilos* (Fig. 107*N*), *Arjona,* and taxonomic descriptions of *Quinchamalum* show that this peculiar plac-

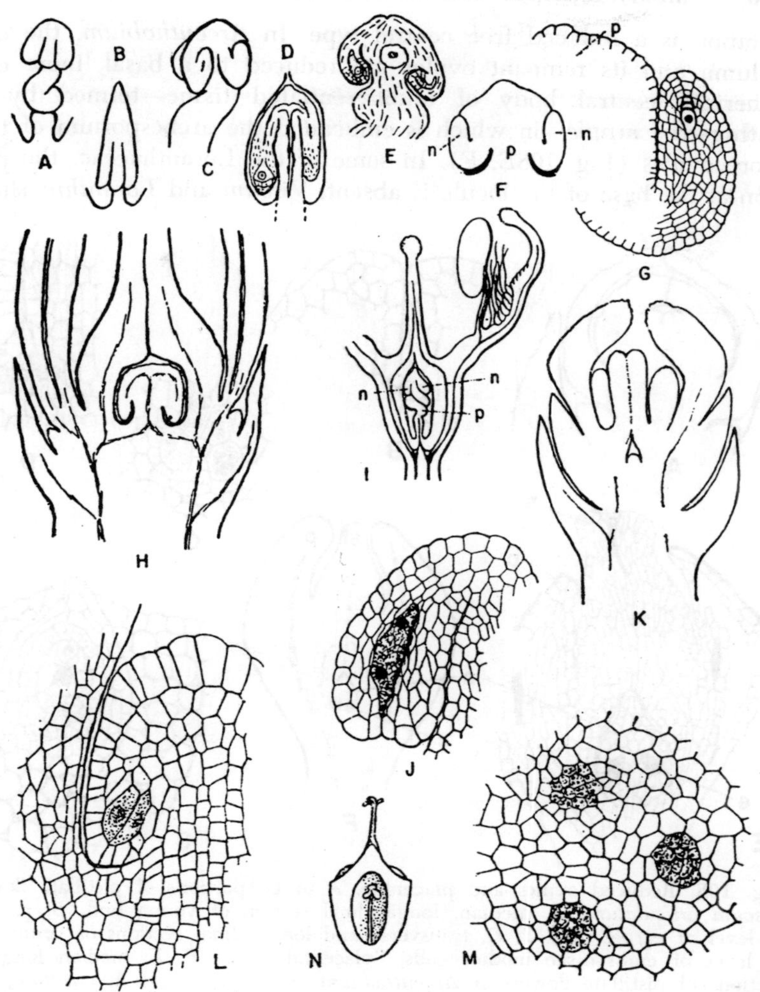

Fig. 107. Reduced ovules and placentae in parasitic plants; placentae modified, free central types; ovules reduced to suspended or erect lobes; archesporial tissue or embryo sacs near apex of lobes. *A* to *G*, showing variety in form of placentae and ovules; *A*, *Thesium divaricatum; B*, *Santalum album; C*, *Comandra umbellata; D*, *Thesium wrightianum; E*, *Scleropyrum* sp.; *F*, *G*, *Osyris alba*. *H*, *I*, *N*, longitudinal sections of flowers showing relation of placentae to ovary walls: *H*, *Olax imbricata; I*, *Thesium divaricatum; N*, *Myoschilos oblonga*. *J*, longitudinal section of an ovular lobe of *I* showing 2-nucleate embryo sac. *K*, *L*, *M*, *Loranthus sphaerocarpus*, flower bud with young placenta (nipple), longitudinal and cross sections of portion of the nipple showing archesporial cells. *n*, nucellus; *p*, placenta. (*A*, *B*, *after Schaeppi and Steindl; C*, *after Ram; D*, *E*, *after Rao; F*, *G*, *I*, *J*, *after Guignard; H*, *after Fagerlind; K*, *L*, *M*, *after Treub; N*, *after Miers.*)

entation is a reduced free central type. In *Arceuthobium*, the central column and its remnant ovules are reduced to a basal, more or less spherical, central body of undifferentiated tissue—termed by some authors the *nipple*—in which is embedded the archesporium of two or more ovules (Fig. 108*E, F*). In some of the Loranthaceae, the projection at the base of the locule is absent; *Viscum* and *Loranthus* show no

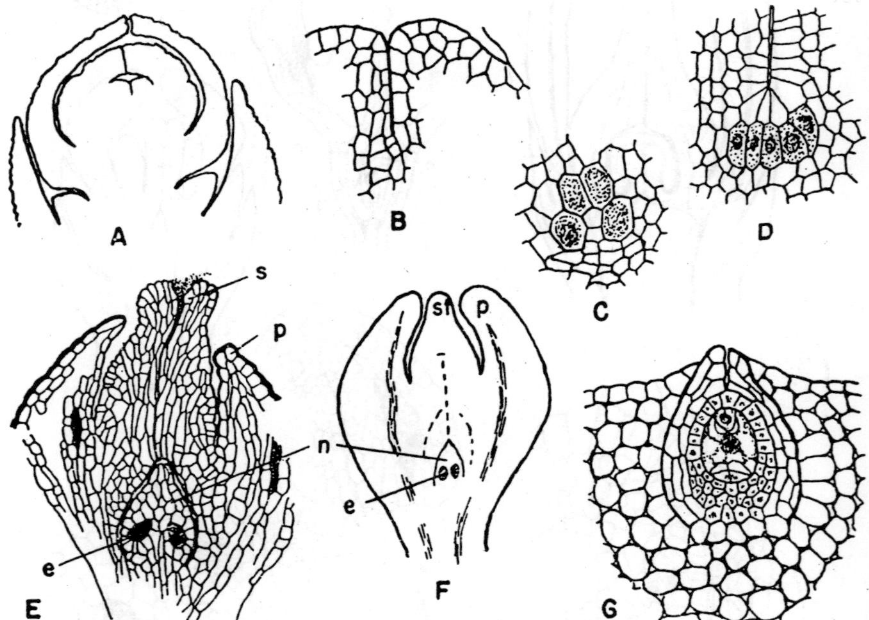

Fig. 108. Reduced ovules and placentae. *A* to *D*, portions of pistillate flowers of *Viscum articulatum: A*, median, longitudinal section of young bud; *B*, part of *A*, at level of carpel tips; *C, D*, transverse and longitudinal sections of young carpels at level of embryo-sac mother cells, "placental area." *E, F*, median longitudinal sections of pistillate flowers of *Arceuthobium americanum* and *A. pusillum*, respectively, showing reduced, cone-shaped placenta, nipple containing embryo sacs. *G*, transverse section of part of placental lobe showing longitudinal section of ovule with integument, *Prosopanche burmeisteri*. *e*, embryo sac; *n*, nipple; *p*, perianth; *s*, stigma; *st*, style. (*A to D, after Treub; E, after Dowding; F, after Thoday and Johnson; G, after Van Tieghem.*)

remnant of a central column (Fig. 108*A* to *D*). In the cleft (Fig. 108*B*) between the appressed bases of the reduced, connate carpels, a group of hypodermal cells, set apart by their richer cytoplasm, become the archesporium (Fig. 108*C, D*).

The Balanophoraceae present the extreme reduction of the ovule, as a part of the remarkable reduction of the entire gynoecium, a reduction to a small, slender structure with weak differentiation of ovary, style, and stigma, without stylar canal and ovular locule, and with almost no histological differentiation throughout (Fig. 109). The gynoecia of *Balanophora* and *Langsdorffia* are mere slender "emergences," with an enlarged median section, in which is borne a single archesporial cell (Fig. 109*A*).

In all families of the Santalaceae, reduction of the ovule to archesporial tissue is the ultimate stage. In the Balanophoraceae, this reduc-

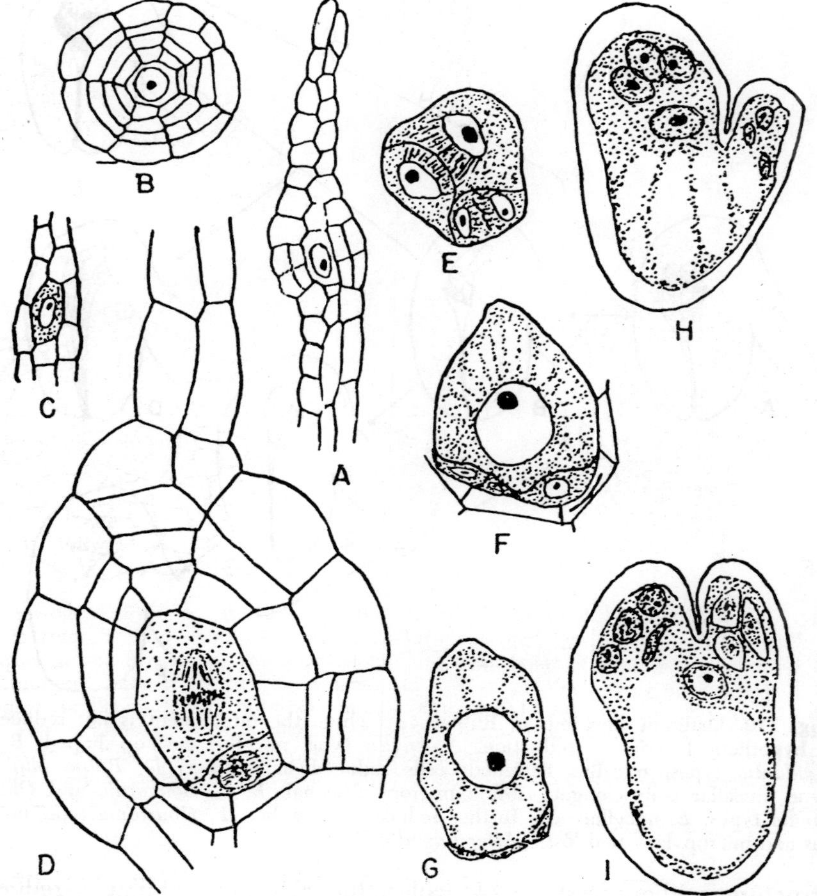

Fig. 109. Greatly reduced apogamous flower. *A* to *G*, *Balanophora elongata: A*, longitudinal section of young pistillate flower consisting of a single simplified carpel with a single embedded archesporial cell; *B*, cross section of an older flower; *C*, longitudinal section of ovary with young archesporial cell; *D*, megasporogenesis, second meiotic division; *E*, tetrad of megaspores; *F*, tetrad, with three spores degenerating; *G*, mature spore. *H*, *I*, *B. globosa*, 8-nucleate embryo sac and older embryo sac showing disorganization of antipodals, egg, and synergids. (*A to G, after Fagerlind; H, I, after Lotsy.*)

tion accompanies extreme reduction of the carpels to mere strands of undifferentiated tissue, showing little resemblance to gynoecia. In other families of the order, though flower structure is much reduced, the gynoecia show evidence of carpel number and limits, and vestiges of ovules are usually present. The Loranthaceae have retained carpellary

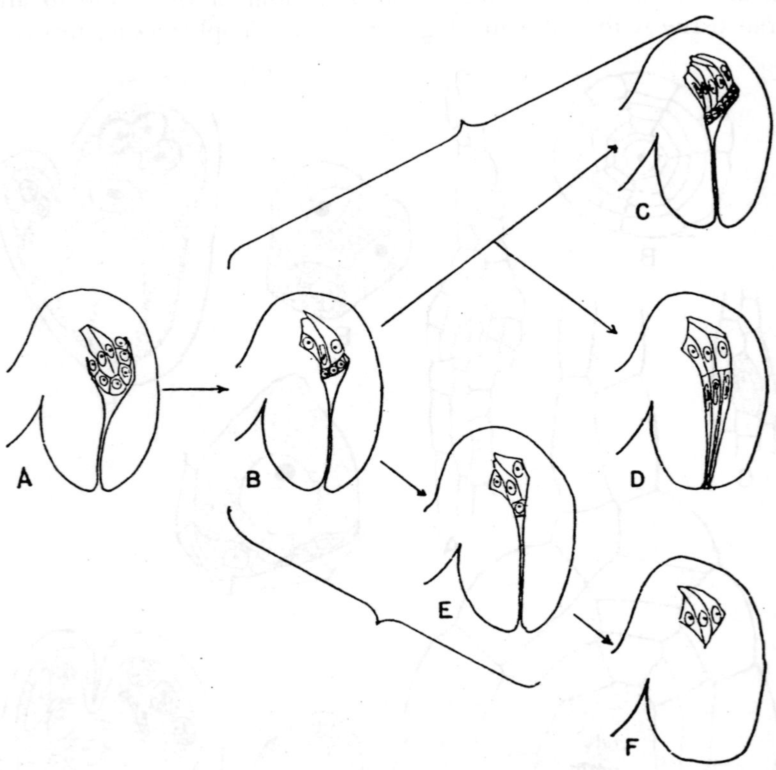

Fig. 110. Diagram showing, in Rubiaceae (where the integument is not reduced), a hypothetical series in reduction. *A*, *Phyllis* type, nucellus cushion-shaped; *B*, *C*, *Vaillantia* types, nucellus flattened, cells reduced in number; *D*, *Rubia ollivierii* type, nucellar cells elongate, filling micropylar canal; *B*, *E*, *Bouvardia* and *Oldenlandia* types; *E*, nucellar cells further reduced in number; *F*, *Houstonia* type, nucellus and micropylar canal lost. (*After Fagerlind.*)

structure, without vestiges of ovules; the placental column is reduced to a few cells at the base of the carpels.

The structure of the ovule of the Rubiaceae, especially that of *Coffea* and *Houstonia*, has been much discussed. In this family, though reduction of the nucellus is strong, the integuments remain well developed. Figure 110 shows theoretical steps in this reduction. The uniseriate nucellus *A* is reduced in size and cell number in *B*, and *E*; its cells become elongate, filling the micropylar canal in *D*; as a last step, the nucellar cells and the micropylar canal are lost.

Nature of the Angiosperm Ovule

The angiosperm ovule has been interpreted morphologically in many ways. It was long considered a bud, because, teratologically, a shoot sometimes develops from it; because it seemed to resemble an axis with a growing point and leaves; because it was borne on placentae which were considered axial; because, from it, as a seed, came the axis of a new plant. It has more commonly been considered of foliar nature—a sporophyll, a leaflet, a lobe of a leaf (simple or three-part), the tip of a leaf, an emergence, a trichome. It has been defined as a "monosporangiate megasporophyll"; under this interpretation, it has been suggested that "phylogenetic connections seem to be possible between some angiosperms and the Lycopodiales through the conifers and Gnetales." All difficulties in interpreting the ovule were supposedly avoided by calling it a "new structure, an organ *sui generis,*" "a complete and independent organ inserted on the stem." In most of these interpretations, it is recognized that, fundamentally, it is a structure that bears sporogenous tissue in which megaspores develop. In the twentieth century, it has nearly always been called a megasporangium or megasporangium with accessory tissues. But there have been occasional returns to the bud theory, and the interpretation of the integuments as leaves. The position of the ovule—on the surface of the carpel—and its form and structure have formed the basis for the theory that it is a sorus, with or without an indusium.

Similarly, the parts of the ovule have been variously interpreted. The nucellus has been considered the growing point of the bud, the "core" of a sorus, the axis of the bud, the megasporangium wall. The integuments have been called the leaves or scales of a bud, the indusium of a sorus, an "involucre," lobes of the sporangium wall, "protection about the sporangium." In an interpretation of the ovule as a leaf, the inner integument is identified as a terminal leaflet, the outer integument as two lateral leaflets.

Homology of the ovule with the stamen has been suggested; the nucellus and integuments were interpreted as equivalent to the anther, the funicle to the filament. Homology of the ovule with the fern sorus has been perhaps the most common interpretation. In this theory, the many sporangia of the sorus are considered reduced to one, and the sporogenous tissue of this surviving sporangium reduced to one spore mother cell. (Stages in reduction of the spore mother cells are common in living ferns.) The integuments represent the indusium of the sorus.

Many of these concepts of the nature of the ovule were shown to be morphologically impossible, because they violate principles of underlying form; they assumed that leaves are borne on other leaves or that stems are borne on leaves; somewhat similar interpretations have continued to the present time.

The various theories of the nature of the ovule are in agreement only in interpreting the ovule as a structure that bears the megaspores. The usual interpretation today is that it is primarily a megasporangium —that the nucellus is the sporangium, and the integuments, protective structures. The structural relation of ovule to carpel is important but is rarely mentioned. The ovule, with its integuments, funicle, and vascular supply, is much more than a sorus. Among the theories of the nature of the angiosperm ovule, the concept that it is an emergence of the carpel lamina seems best supported, when all evidence is considered. Ontogenetically, it arises and develops like an emergence, an emergence built up of tissues around and below an area in the primordium of the lamina where fertile tissue appears early. The position of the ovules on the lamina, primitively scattered, and the relation of the ovule traces to the vascular system—largely or wholly derived, primitively, from the minor veins of the vascular meshwork (derivation from strong, lateral veins is secondary)—support the emergence interpretation. The vascular supply of the ovule in origin, form, and structure is the equivalent of the supply of a minor part of the lamina. It is, essentially, a vein ending, with indefinite and variable branching, which is absent if the tissues are few, and extensive, forming a meshwork within the ovule, if they are many. The vascular system of an ovule resembles in amount and branching the supply to accessory tissues in enlarged, fleshy parts of fruits.

Unlike the typical emergence, the ovule has elaborate form: protective lobes, the integuments; a stalk, the funicle and raphe; minor adaptations undoubtedly related to enclosure in the carpel and course of the pollen tube. Though there has been no general acceptance of the theory that the ovule is an emergence of elaborate form, this interpretation was strongly supported in the last part of the nineteenth century. The better understanding of type and position of the angiosperm sporangium obtained in the early decades of the twentieth century seems to have given even stronger support to this theory. The recognition that the microsporangium is wall-less and sunken in the tissues of the sporophyll makes it probable that the megasporangium is similar in structure and position, and gives a basis for the interpretation of the ovule as an elaborate emergence.

Enclosure of the sporogenous tissue in the nucellus and the frequently rounded form of the nucellus have naturally suggested that the nucellus is the megasporangium wall. But comparison with the microsporophyll makes this interpretation unsatisfactory. Similarity between anther and ovule has long been remarked; "anther sac and nucellus are homologous"; the parietal tissue of the anther is like that of the ovule; "the ovule is more than a sporangium, as the anther is more than four sporangia." In the anther, the sporogenous tissue is enclosed by sterile tissues of the sporophyll; there is no evidence of a sporangium wall as such. If so important a structure as a sporangium wall is absent in the microsporophyll, the same condition would be expected in the megasporophyll; homology of basic structure should exist in the sporophylls. In both types of sporophylls, the sporogenous tissue arises hypodermally and is buried below varying amounts of sterile tissue—commonly called parietal cells in the anther and sometimes so called in the ovule. In the anther, this tissue seems to be a remnant of the mesophyll of the sporophyll. If it is of similar nature in the carpel—part of the tissues of an emergence—it is represented by the nucellus. Close similarity exists in details of development and structure of anther-sac wall and nucellus. Specialization in both anther and carpel is the progressive reduction of the surrounding sterile tissues until only one cell layer encloses the sporogenous tissues externally. In position and function, anther-sac wall and nucellus are alike, but histology shows no distinction of a sporangium wall. If this wall were present in ancestral angiosperms, some evidence of it might be expected in delimitation of sporogenous tissue in such primitive ovules as those of *Casuarina* and *Calycanthus,* but this delimitation is not present, although there are massive nucelli and many spore mother cells.

The freeing of the fertile-tissue-bearing parts and their change to a position more favorable for pollination—clearly seen in the stamen in the bringing of the microsporangia from a sunken position, midway on the sporophyll, to an exposed position, distal on the sporophyll—are paralleled in the carpel, by the raising of the fertile tissue, sunken among sterile cells of the lamina, to form an emergence consisting of the fertile tissue still enclosed by sterile laminar tissues. Protection of the sporogenous cells is given by surrounding cells (the nucellus) and by adjacent laminar tissues including parts of the vascular meshwork of the lamina (the integuments). In the more primitive families, the ovular emergence is complex, including massive nucellus and two integuments; in more advanced families, the nucellus is reduced to a very few cells, even a uniseriate layer, and the integuments to one. A parallel specialization is present in the highest types of anther, where the anther-sac wall is reduced to a uniseriate cell layer and the connective is greatly reduced.

The discarding of the fairly simple and convenient interpretation of the nucellus as the megasporangium may seem to make the teaching of morphology and relationships of the higher plants more difficult, but the obvious interpretation is often not the correct one. Recognition that angiosperm sporogenous tissue is basically deeply sunken and lacks an enclosing, protective wall opens new lines of search for ancestral forms.

A critical interpretation of such obviously highly specialized organs as the ovule and the anther sac has been retarded by expressions of the viewpoint that structure should be accepted for what it seems to be in the light of its location and apparent relationships. For example, statements like the following are sometimes made: the ovules in free central placentation are cauline, because they are borne on an apparent elongation of the receptacle; where congenital fusion has occurred and no histological or ontogenetic evidence remains, as in many inferior ovaries, the structure should be accepted as simple and not dissected into component parts, no longer recognizable; "although an ovule is more than a sporangium, as the ordinary anther is more than four sporangia, the distinction is theoretical, rather than practical." (If only interpretations where there is direct proof and those that are "practical" were undertaken, morphology would make little progress. When critical detailed comparisons of organ with organ in related taxa are not considered, false interpretations, such as the presence of both foliar and cauline ovules in the angiosperms, are made.)

Cauline and Foliar Ovules—Stachyospory and Phyllospory. The position of ovules—whether they are borne on stem or leaf—has been controversial since the earliest days of morphological description. In the eighteenth and often in the nineteenth centuries, all ovules were frequently called cauline; in the twentieth century, interpretations have been chiefly that all are foliar or some cauline and some foliar. The interpretation that all are foliar has been said to be "interesting because of its ingenuity" of explanation. It can as well be said, when ovules are generally accepted to be all foliar, that the explanations of cauline nature for some ovules are remarkable for lack of ingenuity in interpretation. When evidence from anatomy, ontogeny, and comparison of flower and carpel structure in clearly related taxa is considered, all ovules are foliar. The presence of stachyosporous and phyllosporous ovules in different genera of the same family, even in different species of a genus—*Lychnis*—can hardly be considered morphologically sound, and these claims have been made only by those not experienced in morphology. In all major vascular taxa, spore-bearing tissues are restricted to morphologically similar parts of the plant body, not borne indiscriminately on stem or leaf, as required by the stachyospory-phyllospory theory.

The angiosperm ovule need not be interpreted in terms of ovule structure in other taxa; it is probably not a modification of the ovules of any living gymnosperm and perhaps not of any known fossil taxon. If its nature as an outgrowth of the sporophyll lamina is accepted, it is morphologically unlike any other known ovule. As has often been suggested, it has resemblances, in position and structure, to a fern sorus and, in its eusporangiate development and embedded sporogenous tissue, to the eusporangiate ferns. The sporangia of the Ophioglossaceae especially suggest the angiosperm ovule in some characters: they are modified distal parts of the fertile leaf, and the sporogenous cells are embedded in sterile tissues that are not set apart as a "sporangium wall"; the vascular supply is a single strand, which branches below the sporogenous tissue; development of the sporangium is eusporangiate. (These resemblances suggest possible relationship between eusporangiate fern stock and an ancestral angiosperm stock.)

Since the seeds of the pteridosperms have become well known, resemblances have been seen between the ovules and seeds of the angiosperms and these ancient structures. Emphasis has been placed on the presence of two integuments in both pteridosperms and angiosperms. But these resemblances seem superficial, consisting merely of protecting structures, two in number, with vascular supplies. And the integuments of the pteridosperms, like those of the cycads and *Ginkgo*, have strong vascular supplies in definite positions, in contrast to those of the angiosperms, which are weak and indefinite in form and position. Under one interpretation of the integuments, the outer one consists of bracts; only the inner is a true integument. The morphological nature of the angiosperm ovule is still uncertain.

The nucellus of the pteridosperms is structurally complex; that of the angiosperm is simple. There seems to be little morphological resemblance between the ovules of the angiosperms and those of the pteridosperms, cycads, and *Ginkgo*. The ovule has probably developed independently in several lines, with sporogenous tissue borne in ovules of different nature; those of the pteridosperms, cycads, and *Ginkgo* represent, at least in part, lobes of the fertile leaf; those of the angiosperms, parts of lesser rank, emergences.

BIBLIOGRAPHY

Bechtel, A. R.: The floral anatomy of the Urticales, *Am. Jour. Bot.*, **8**: 386–410, 1921.

Benson, M.: Contributions to the embryology of the Amentiferae. I, *Trans. Linn. Soc. London*, ser. 2, **3**: 409–424, 1894.

Berlese, A. N.: Studi sulla forma, struttura e sviluppo del seme nelle Ampelidee, *Malpighia*, **6**: 482–531, 1892.

Blaser, H. W.: Studies in the morphology of the Cyperaceae. I. Morphology of flowers. A. Scirpoid genera, *Am. Jour. Bot.*, **28**: 542–551, 1941.

Cooper, G. O.: Microsporogenesis and development of seed in *Lobelia cardinalis*, *Bot. Gaz.*, **104**: 72–81, 1942.

Corner, E. J. H.: The annonaceous ovule and its four integuments, *New Phyt.*, **48**: 332–364, 1949.

Coy, G. V.: Morphology of *Sassafras* in relation to phylogeny of angiosperms, *Bot. Gaz.*, **86**: 149–171, 1928.

Dahlgren, K. V. D.: Die Morphologie des Nuzellus mit besondere Berücksichtigung der deckzellosen Typen, *Jahr. Wiss. Bot.*, **67**: 347–426, 1927.

Dowding, E. S.: Floral morphology of *Arceuthobium americanum*, *Bot. Gaz.*, **91**: 42–54, 1931.

Eames, A. J.: Again: "The New Morphology," *New Phyt.*, **50**: 17–35, 1951.

Eckardt, T.: Nachweis der Blattbürtigkeit ("Phyllosporie") grundständige Samenanlagen bei Centrospermen, *Ber. Deutsch. Bot. Ges.*, **68**: 167–182, 1955.

Fagerlind, F.: Embryologische, zytologische und bestäubungsexperimentelle Studien in der Familie Rubiaceae, *Acta Horti Berg.*, **11**: 195–470, 1937.

———: Bildung und Entwicklung des Embryosacks bei sexuellen und agamospermischen *Balanophora*-Arten, *Sv. Bot. Tidskr.* **39**: 65–82, 1945.

———: Bau der floralen organe bei der Gattung *Langdorffia*, *Sv. Bot. Tidskr.*, **39**: 197–210, 1945.

———: Blüte und Blütenstand der Gattung *Balanophora*, *Bot. Not.*, **1945**: 330–349.

———: Gynöceummorphologie und embryologische Studien in der Familie Olacaceae, *Bot. Not.*, **1947**: 207–227.

———: Beiträge zur Kenntnis der Gynäeciummorphologie und Phylogenie der Santalales-Familien, *Sv. Bot. Tidskr.*, **42**: 195–229, 1948.

Guérin, P.: Le tégument séminal et les trachées nucellaires des Thyméléacées, *Compt. Rend. Acad. Sci. Paris*, **156**: 398–400, 1913.

Guignard, L.: Observations sur les Santalacées, *Ann. Sci. Nat. Bot.*, 7 sér., **2**: 181–202, 1885.

———: L'ovule chez les Apocynacées et Asclepiadacées, *Mém. Acad. Sci. Inst. Fr.*, **55**: 1–34, 1917.

Haan, H. R. M. de: Contribution to the knowledge of the morphological value and the phylogeny of the ovule and its integuments, *Rec. Trav. Bot. Néerl.*, **17**: 219–324, 1920.

Hagerup, O.: On the origin of some angiosperms. III, *Kgl. Dansk. Vidensk. Selsk. Biol. Medd.*, **14**: 1–34, 1938.

Heinricher, E.: Über Bau und Biologie der Blüten von *Arceuthobium oxycedri* (DC.) MB., *Sitzber. Kais. Akad. Wiss. Wien Math.-Natur Kl.*, Abt. I, **124**: 481–514, 1915.

Kershaw, E. M.: The structure and development of the ovule of *Myrica gale*, *Ann. Bot.*, **23**: 353–362, 1909.

Klebelsberg, R.: Über die Samenanlagen von *Quercus robur* und intraseminale Gefässe, *Oesterr. Bot. Zeitschr.*, **60**: 329–335, 378–393, 1910.

Kühn, G.: Beiträge zur Kenntnis der intraseminalen Leitbündel bei den Angiospermen, *Bot. Jahrb.*, **61**: 325–385, 1928.

Kuijt, J.: Dwarf mistletoes, *Bot. Rev.*, **21**: 569–620, 1955.

Lam, H. J.: A new system of the Cormophyta, *Blumea*, **6**: 282–289, 1948.

Lonay, H.: Analyze coördonée des travaux relatifs à l'anatomie des téguments séminaux, *Arch. Inst. Bot. Liège*, **4**: 3–146, 1907.

———: Contribution a l'anatomie des Renonculacées, structure des péricarpes et des spermodermes, *Arch. Inst. Bot. Liège*, **3**: 3–152, 1901.

Lotsy, J. P.: *Balanphora globosa* Jungh., eine wenigstens örtlich verwittwete Pflanze, *Ann. Jard. Bot. Buitenzorg*, **16**: 174–186, 1899.

Miers, J.: On some genera of the Olacaceae, *Jour. Linn. Soc. Bot.*, **17**: 126–141, 1878.

Monnier, G. Le: Recherches sur la nervation de la graine, *Ann. Sci. Nat. Bot.*, 5 sér., **16**: 233–305, 1872.

Orr, M. Y.: The occurrence of tracheids in the nucellus of *Steriphoma cleomoides* Spreng., *Notes Royal Bot. Garden Edinburgh*, **12**: 241–242, 1919–21.

Periasamy, K., and B. G. L. Swamy: The conduplicate carpel of *Cananga odorata*, *Jour. Arnold Arb.*, **37**: 366–372, 1956.

Peter, J.: Zur Entwicklungsgeschichte einiger Calycanthaceen, *Beitr. Biol. Pfl.*, **14**, 59–86, 1920.

Ram, M.: Morphological and embryological studies in the Santalaceae I. *Comandra umbellata* (L.) Nutt., *Phytomorph.*, **7**: 24–35, 1957.

Rao, L. N.: Studies in the Santalaceae, *Ann. Bot.*, n.s., **6**: 151–175, 1942.

Rauch, K. von: Cytologisch-embryologische Untersuchungen an *Scurrula atropurpurea* Dans. und *Dendrophthoë pentandra* Miq., *Ber. Schweiz. Bot. Ges.*, **45**: 5–70, 1936.

Rutishauser, A.: Entwicklungsgeschichtliche und zytologische Untersuchungen an *Korthalsella dacrydii, Ber. Schweiz. Bot. Ges.*, **44**: 389–436, 1935.

Schaeppi, H., and F. Steindl: Blütenmorphologische und embryologische Untersuchungen an *Osyris alba, Ber. Schweiz. Bot. Ges.*, **47**: 369–392, 1937.

Steindl, F.: Pollen- und Embryosackentwicklung bei *Viscum album* und *V. articulatum, Ber. Schweiz. Bot. Ges.*, **44**: 343–385, 1935.

Swamy, B. G. L.: A contribution to the life history of *Casuarina, Proc. Am. Acad. Arts and Sci.*, **77**: 1–32, 1948.

———: The comparative anatomy of the Santalaceae: Node, secondary xylem, and pollen, *Am. Jour. Bot.*, **36**: 661–673, 1949.

———: On the life history of *Vanilla planifolia, Bot. Gaz.*, **108**: 449–456, 1947.

——— and I. W. Bailey: The morphology and relationships of *Cercidiphyllum*, *Jour. Arnold Arb.*, **30**: 187–210, 1949.

Thoday, D., and E. T. Johnson: On *Arceuthobium pusillum* Peck, II. Flowers and fruit, *Ann. Bot.*, **44**: 813–824, 1930.

Treub, M.: Observations sur les Loranthacées, *Ann. Jard. Bot. Buitenzorg*, **2**: 54–76, 1885; *Ann. Jard. Bot. Buitenzorg*, **3**: 1–12, 1883.

———: Sur les Casuarinacées et leur place dans le systèm naturel, *Ann. Jard. Bot. Buitenzorg*, **10**: 145–231, 1891.

———: L'organ femelle et l'apogamie du *Balanophora elongata, Ann. Jard. Bot. Buitenzorg*, **15**: 1–25, 1898.

Van Tieghem, P.: Sur les divers modes de nervation de l'ovule et de la graine, *Ann. Sci. Nat. Bot.*, 5 sér., **16**: 228–232, 1872.

———: Sur les phanérogames à ovule sans nucelle formant le group des innucellées ou Santalinées, *Bull. Soc. Bot. France*, 3 sér., **43**: 543–572, 1896.

———: Structure de l'ovule et de la graine chez les Hydnoracées, *Jour. Bot.*, **11**: 233–238, 1897.

———: Structure des quelques ovules et parti qu'on en peut tirer pour ameliorer la classification, *Jour. Bot.*, **12**: 197–220, 1898.

———: L'hypostace, sa structure et son rôle constante, sa position et sa forme variables, *Bull. Mus. Hist. Nat.*, **7**: 412–418, 1901.

Vermoesen, C.: Contribution a l'étude de l'ovule, de sac embryonnaire et de la fécondation dans les angiospermes, *La Cellule*, **27**: 115–157, 1911.

Walton, J.: L'évolution des téguments et de la protection du sporange, *Année Biol.*, sér. 3, **56**: C129–C132, 1952.

Warmung, E.: De l'ovule, *Ann. Sci. Nat. Bot.*, 6 sér., **5**: 175–266, 1878.

Worsdell, W. C.: The structure and morphology of the 'ovule': An historical sketch, *Ann. Bot.*, **18**: 57–86, 1904.

Chapter 8

ARCHESPORIUM

Early in the development of the ovule primordium, just before, or accompanying, the appearance of the integument primordia, one to several hypodermal cells at the apex of the primordium are set off from surrounding cells by increase in size, larger nuclei, and denser cytoplasm. These cells form the *archesporium*—in crassinucellate ovules, a small cluster or plate of cells; in tenuinucellate ovules, usually a single cell, terminating an axial row of cells (Fig. 102*B*). In crassinucellate ovules, the archesporial cells divide, chiefly periclinally, and form, externally, *primary parietal cells* and, internally, *sporogenous cells* (Fig. 111). In tenuinucellate ovules, the archesporial cells commonly become sporogenous cells without division and later mature as *megaspore mother cells* (Fig. 102*C*), the cells that form the tetrads of megaspores. The terms archesporium, sporogenous tissue, and primary sporogenous cells have been loosely used; they apply to early stages in the elaboration of the meristematic tissue from which fertile cells are differentiated, and lines of separation cannot be drawn. In the ovules of many taxa that have little or no nucellar tissue and one or two archesporial cells, the archesporial cells become the spore mother cells.

Parietal tissue, developed from the primary parietal cells, varies from few to many cells; if there is only one cell, there may be no proliferation, as in many specialized ovules. Divisions may be periclinal only, or both periclinal and anticlinal, and massive nucellar caps may be built up. In these caps, dominant periclinal cell divisions may differentiate a calotte. Periclinal divisions in the epidermis are infrequent and add little to the bulk of large nucelli. Where the ovule primordium is small, only one archesporial cell is usually formed, and this is terminal on a median row of cells sheathed by the epidermis; parietal tissue is absent or consists of very few cells.

The sporogenous cells are commonly few, but range in number from one or two to ten or more. *Casuarina* and *Calycanthus* have eight to ten in a central, more or less columnar, cluster, which is not clearly limited (Fig. 111). The large number, together with cells apparently transitional to parietal cells, suggests derivation from an ancestral multicellular sporogenous cluster. In many primitive and some advanced families, the number is two to several; in most advanced families and those with highly specialized ovules or gynoecia, only one is usually formed. Where there are several of these cells, they may form a plate, a cluster, or a longitudinal row. Larger numbers of sporogenous cells —some of which may not mature as spore mother cells—and of parietal cells doubtless represent primitive ovule structure; the solitary sporogenous cells and one or no parietal cells, the advanced structure. Within

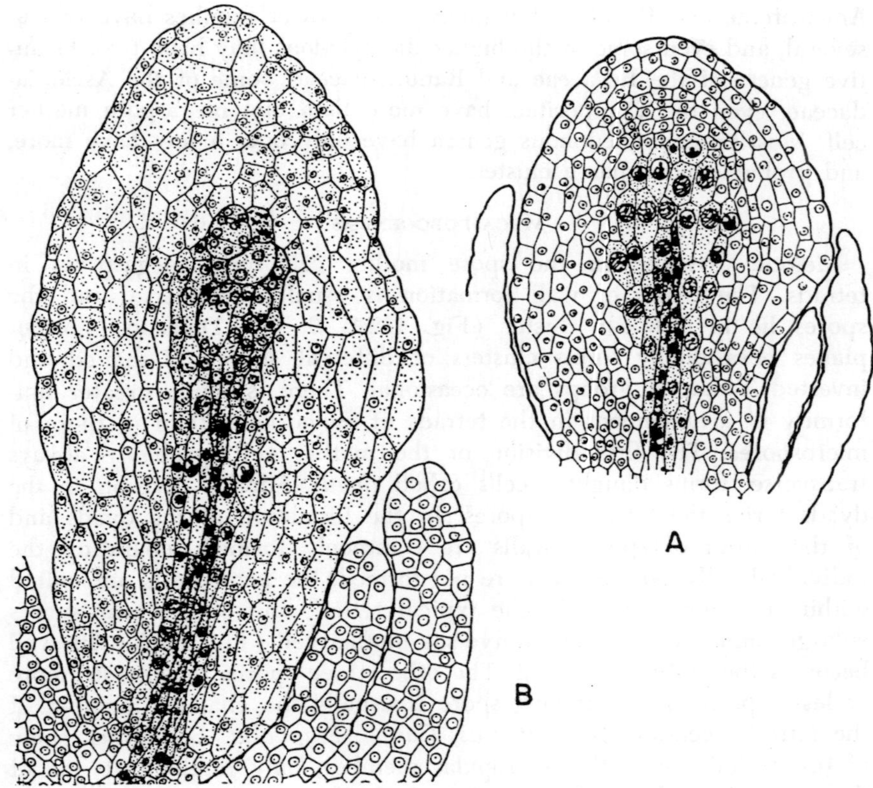

Fig. 111. Longitudinal sections of young crassinucellate ovule of *Casuarina montana* showing ontogeny of multicellular archesporium, parietal cells, and free integuments. *A*, early stage, sporogenous cells with large nuclei; lower cells still maturing; parietal cells formed by tangential divisions. *B*, later stage, megaspores mature, some germinating; procambial strand at base of ovule. (*After Swamy.*)

families, there are occasional marked differences in amount of parietal tissue; in the Ranunculaceae, parietal tissue may be abundant or scanty; in the Rosaceae, genera with achenes—*Fragaria*—may have massive nucelli. Few or no parietal cells characterize most of the gamopetalous taxa; massive nucelli are largely restricted to the lower polypetalous taxa.

Megaspore Mother Cells

The megaspore mother cells are much larger than the surrounding cells, with very large nuclei, dense cytoplasm, and thin walls (Fig. 102*C*). It has been generally implied in descriptions that the presence of more than one or two megaspore mother cells in an ovule is rare and restricted to a few primitive taxa—the Casuarinaceae, some of the Amentiferae and Ranales—but many of the lower families have two to several, and even some of the higher dicotyledons may have two. Primitive genera of the Rosaceae and Ranunculaceae, some of the Asclepiadaceae, even the Compositae, have more than one megaspore mother cell. Most monocotyledonous genera have only one; a few have more, and probably none has a cluster.

Megasporogenesis

Meiotic divisions in the spore mother cell form megaspores in tetrads. The plane of wall formation is usually transverse, and the spores lie in a linear cluster (Fig. 112*A*, *B*, *C*). Division in other planes occasionally forms clusters of different shapes; T-shaped and inverted T-shaped groups are occasional. There is much greater uniformity of arrangement in the tetrads of megaspores than in those of microspores. The first division of the mother cell, probably always transverse, forms daughter cells called *dyads*. Similar division in the dyads forms the tetrad of spores. In the formation of the dyads and of the tetrad of spores, walls are commonly formed separating the individual cells; where walls are not formed, the protoplasts are united within a common wall. In the typical tetrad, one of the four spores enlarges and becomes the embryo-sac mother cell (Fig. 112*C*). Its wall becomes the embryo-sac wall. The other three spores degenerate more or less rapidly. The surviving spore is usually the chalazal member of the tetrad; occasionally—*Butomus, Canna, Styphelia, Loranthus,* some of the orchids—it is the micropylar member; only rarely is it one of the median members—*Putoria, Aristolochia.*

THE EMBRYO SAC*

Development of the embryo sac begins by germination of the megaspore, accompanied by enlargement of the spore and increase in cytoplasmic content and in size of the nucleus. Increase in size and content is, in part, at the expense of the other spores, which are crushed and

* Description of the embryo sac in detail can be found in the full and excellent treatment by P. Maheshwari, to which the reader is referred: "An Introduction to the Embryology of Angiosperms," New York, 1950.

largely absorbed, and often, in part also, of adjacent nucellar tissue (Fig. 112*E*). Nutrition of the embryo sac comes also, in some taxa, from a uniseriate palisade layer formed by the epidermis of the enclosing integument (Fig. 112*E* to *H*). Structural development of the typical 8-nucleate, monosporic embryo sac is by successive divisions: into two; each of these two into two; and each of the four so formed into two.

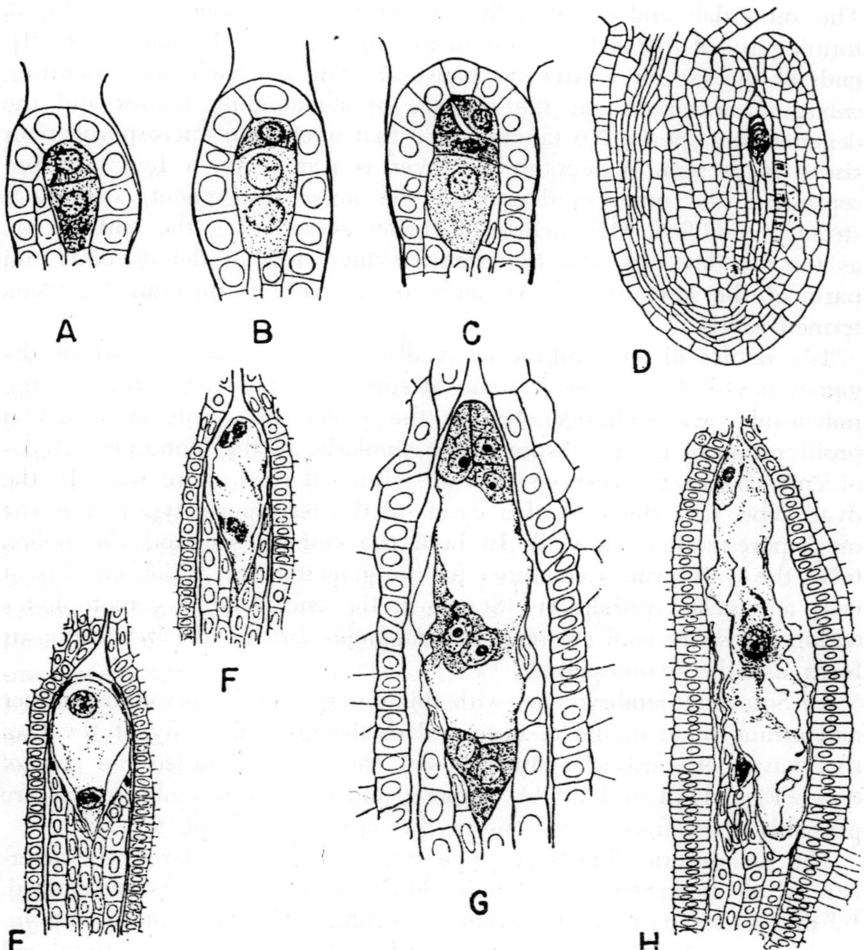

Fig. 112. Megasporogenesis in *Sarracenia purpurea*. *A*, megaspore mother cell, dyad forming; *B*, basal dyad divided, micropylar dyad degenerating; *C*, tetrad of megaspores, the chalazal spore enlarging, the micropylar spores degenerating; *D*, longitudinal section of ovule with tetrad of spores surrounded by uniseriate nucellus; *E*, 2-nucleate embryo sac with collapsed nucellar cells; *F*, 4-nucleate embryo sac; *G*, mature, 8-nucleate embryo sac, one antipodal cell out of plane of section; *H*, embryo sac showing polar nuclei fused. (*After Shreve.*)

Of the eight nuclei so formed, three lie in the micropylar end, two near the center, and three in the chalazal end. The three micropylar nuclei, with surrounding cytoplasm—the *egg* and the two *synergids*—become walled by delicate membranes. The three at the chalazal end, the *antipodals,* usually also become walled, but the two median nuclei, the *polar nuclei,* remain unwalled. The synergids and the antipodals are usually ephemeral, but the antipodals may persist and even multiply. The antipodals and occasionally the synergids may become haustorial, forming tubes that extend into surrounding tissues. In some taxa, the ends of the sac form extensive haustoria. The sac itself, as it matures, enlarges haustorially, at the expense of surrounding tissues and the degenerating spores. No tapetum like that about the microsporogenous tissue is present. A tapetumlike layer is present in a few taxa but represents the inner epidermis of the inner integument, sometimes strongly modified. Destruction of tissues surrounding the embryo sac as the sac develops may be extensive, involving all the nucellus and parts of the integuments. An embryo sac without nucellus has been termed *naked.*

The nature of the embryo-sac wall—whether or not a part of the gametophyte—has received some attention, as has the nature of the pollen-tube wall. Ontogenetically, the pollen-tube wall is clearly a proliferation of the microspore wall; similarly, in the monosporic types of embryo sac, the embryo-sac wall is an enlarged spore wall. In the dyad type, it is the wall of a dyad; in the tetrasporic types, it is the megaspore-mother-cell wall. In both the embryo sac and the pollen tube, the wall forms a container for the gametophyte, which consists of cells and naked protoplasts. Stages in the enclosure of gametophytes within the spore wall are seen in *Selaginella, Isoëtes,* the heterosporous ferns, and the gymnosperms.

An 8-nucleate embryo sac, with this ontogenetic history and number and arrangement of nuclei is typical of the great majority of the taxa that have been critically studied, and was, at first, called the *normal* type (Figs. 113*A* and 118*A*). Bisporic and tetrasporic embryo sacs are probably specialized variations of this apparently basic type.

The monosporic, 8-nucleate type was, at first, considered characteristic of all angiosperms, and only slowly were other types recognized. In the earlier part of the twentieth century, the description and interpretation of sacs with more or less than eight nuclei stimulated critical study of many new taxa and the reexamination of supposedly well-known forms. The existence of bisporic and tetrasporic types was established, and reexamination of the embryo sacs of some supposedly well-known taxa made changes in their interpretation necessary. The embryo sac of *Lilium,* often used in teaching as an example of the

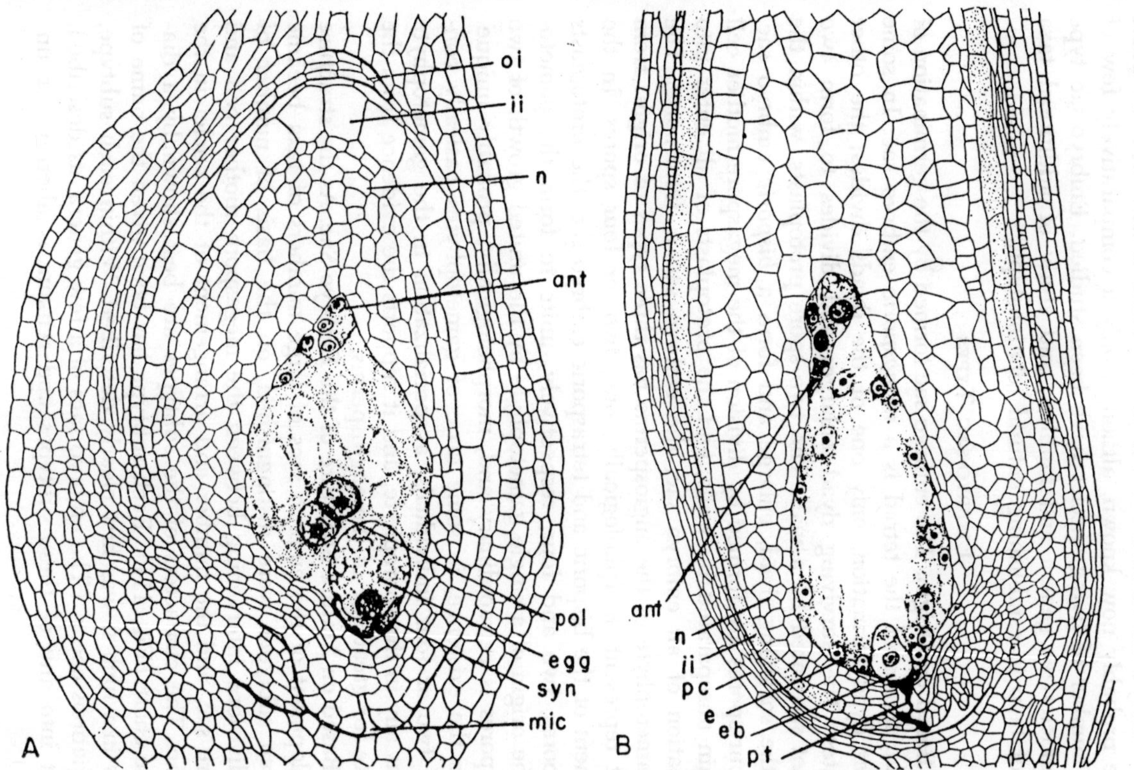

Fig. 113. Longitudinal sections of ovules of *Sorghum* sp. (sugar cane) showing *A*, mature embryo sac, and *B*, embryo sac with two-celled embryo and free-nucleate endosperm. *ant*, antipodals; *e*, endosperm; *eb*, embryo; *ii*, inner integument; *mic*, micropyle; *n*, nucellus; *oi*, outer integument; *pc*, pericarp; *pol*, polar nuclei; *pt*, pollen tube; *syn*, synergid. (*After Artschwager, Brandes, and Starrett.*)

typical sac, monosporic and 8-nucleate, was shown to be tetrasporic. Study of embryo-sac structure continues to some extent, but the basic types are probably now known, although only a comparatively few of the vast number of living species have been studied. Embryo-sac type often characterizes genera and families, and study of additional taxa may be of help in establishing natural relationships.

Embryo-sac Types

Types Based on Origin. The embryo sac formed by the germination of one of the spores of the tetrad is a *monosporic* embryo sac. In some taxa, after dyad formation, only one of the dyads divides; the other degenerates. The surviving dyad enlarges and divides to form two spores, not separated by a wall. The two spore protoplasts, within the wall of the surviving dyad, enlarge and form a *bisporic* embryo sac. Where four spores are formed by division of the megaspore mother cell but remain unseparated by walls, and the protoplasts of all enter into the formation of an embryo sac, the embryo sac is *tetrasporic.* The female gametophyte of the angiosperms has three types of origin: from cells that represent, morphologically, one, two, or four spores. In the development of the bisporic and tetrasporic embryo sac, the protoplasts of the spores, two and four, respectively, unite to form the gametophyte. The origin of a single gametophyte by the united growth of two or four spores is extraordinary and, morphologically, probably unique. It implies the existence, in some taxa, of gametophytes basically consisting of two or four individuals, a specialization in the gametophyte as great as any in the sporophyte, and it supports the evidence from the sporophyte for great age of the angiosperms.

Types Based on Mature Structure of the Embryo Sac. As the structure and development of the embryo sacs of large numbers of taxa became known, ten subtypes were distinguished, on the basis of number of nuclear divisions intervening between the megaspore mother cell and the mature sac, and the number and arrangement of the nuclei in the mature sac. The characters of the subtypes are best shown by a diagrammatic chart (Fig. 114). Each subtype has been given the name of the genus in which it was first clearly described. Within each subtype, minor variations occur. (The embryo sacs of a few genera, as described, do not fit into any of these subtypes and remain as aberrant or unclassified.)

The common type, long- and well-termed the *normal,* is now termed the *Polygonum type.* Here, the egg is formed at the fifth division, the third from the spore. The eight nuclei are distributed as follows: three at the micropylar end of the sac, those of the egg and two synergids; three at the chalazal end, those of the antipodals; and two in the central

Type	Megasporogenesis			Megagametogenesis			
	Megaspore mother cell	Division I	Division II	Division III	Division IV	Division V	Mature embryo sac
Monosporic 8-nucleate Polygonum type							
Monosporic 4-nucleate Oenothera type							
Bisporic 8-nucleate Allium type							
Tetrasporic 16-nucleate Peperomia type							
Tetrasporic 16-nucleate Penaea type							
Tetrasporic 16-nucleate Drusa type							
Tetrasporic 8-nucleate Fritillaria type							
Tetrasporic 4-nucleate Plumbagella type							
Tetrasporic 8-nucleate Plumbago type							
Tetrasporic 8-nucleate Adoxa type							

Fig. 114. Diagram showing important types of embryo sacs in angiosperms. (*From Maheshwari.*)

region, the polar nuclei. The *Oenothera type,* a variation of this monosporic type, differs from the Polygonum type in that the egg is formed after the fourth, rather than the fifth, division. The mature sac has only four nuclei. The antipodals and the chalazal polar nuclei are not formed. These two types are monosporic.

The one bisporic type is called the *Allium type.* Here, as in the Oenothera type, the egg is formed at the fourth division. After the first division, one of the dyads degenerates; the other dyad, by successive divisions, forms eight nuclei. The mature sac has eight nuclei, which are arranged as in the Polygonum type. Morphologically, the sac is formed by two spores.

Seven tetrasporic types are now known. As the name implies, in these seven types, the protoplasts of all four spores formed by the megaspore mother cell build together the embryo sac. The gametophyte is, morphologically, a quadruple structure.

In three of the tetrasporic types, 16 nuclei are formed at the second division from the spore (the fourth division from the mother cell). Differences in the arrangement of the 16 nuclei and in the function of some of them distinguish the three types. In the *Peperomia type,* the egg and one synergid lie at the micropylar end of the sac; eight nuclei form a central cluster; six nuclei lie at the periphery of the lower part of the sac. In the *Penaea type,* clusters of three nuclei lie at each end and medianly on opposite sides of the sac, and the remaining four form a central cluster. In the *Drusa type,* three nuclei, those of the egg and two synergids, lie at the micropylar end of the sac, two in the center, and the other eleven at the base.

The fourth of the tetrasporic types in which the egg is formed at the fourth division, the *Fritillaria type,* has an 8-nucleate sac, in contrast with the 16-nucleate of the other three types. The formation of the mature sac is complicated by the union of three of the four spore nuclei in the base of the young sac in a multiple-spindle division, which forms two triploid nuclei. The four nuclei now present in the sac divide, and the eight nuclei become arranged as in the Polygonum type, but the antipodal nuclei and one of the polar nuclei are triploid.

In the fifth of the tetrasporic types—the *Plumbagella type*—the embryo sac is 4-nucleate; the four nuclei, formed at the second division of the megaspore mother cell, represent the nuclei of the four spores. Three of the four spore nuclei unite in a triple-fusion division in the base of the sac, as in the Fritillaria type, forming two basal triploid nuclei. The mature sac, however, differs from that of the Fritillaria type, in that the egg is alone at the micropylar end—there are no synergids—and one of the two central polar nuclei and the solitary antipodal are triploid. The last two tetrasporic, 8-nucleate types differ in the arrangement of the nuclei in the sac. In the *Plumbago type,* the nuclei of the four spores become isolated around the periphery of the sac—one at each end, the other two near the center. Division of the spore nuclei gives an 8-nucleate sac—and nucleus at each end, two along the sides, and four in a central cluster. The *Adoxa type* has the nuclei arranged as in the

Polygonum type, but, because the nuclei of four spores are involved, the number is attained by the third, rather than the fifth, division. Mature embryo sacs of the Adoxa and Polygonum types are much alike in nuclear arrangement.

(For the sake of simplicity in the discussion above, no distinction has been made between nuclei and cells in the sac.)

Minor variations in number of nuclei in the sac are occasional, especially in the antipodals, where chalazal nuclei may fail to divide, but some reports of lesser numbers are perhaps in error, because of failure to observe all the nuclei. Nuclei from surrounding nucellar cells are reported to enter the sac, suggesting a higher number of nuclei.

Specialization of the embryo sac has followed that seen in the gametophyte—both male and female—of the major taxa of vascular plants; reduction of the vegetative tissue and earlier and earlier formation of gametes occur in the ontogeny of the gametophyte. In angiosperms, the female gametophyte is reduced from the many cells of the gymnosperm gametophytes to eight as a basic number—the Polygonum type. In the Oenothera type, the monosporic gametophyte shows reduction from three to two divisions in its development, and an accompanying reduction of prothallial nuclei from seven to three. In the specialization of the embryo sac, the Oenothera type is the least modified of the advanced types. Still monosporic, the number of cell divisions occurring between the spore mother cell and the mature gametophyte is four, a reduction of one from that in the basic Polygonum type; the egg is formed at the second division from the spore [the fourth division from the megaspore mother cell (Fig. 114)]. The reduction in number of divisions naturally reduces the remaining (prothallial) cells from seven to three.

In the Allium type—bisporic but 8-nucleate, like the basic type—the gametophyte is mature after the fourth division from the spore mother cell, as in the Oenothera type, and, similarly, the egg is formed at the second division from the spore. The Allium and Oenothera types are alike in the number of cell divisions intervening between the spore mother cell and the spore and the formation of the egg, but the mature sac of the Allium type has seven prothallial cells, in contrast to three in the Oenothera type, because two spores build the sac in the Allium type. The Allium type sac resembles the Polygonum type in form and number of cells, but has a different origin. As in the Oenothera type, the number of cell divisions occurring in the formation of the gametophyte has been shortened by one, though the sac resembles the basic type in structure and in number of prothallial cells.

In the 8-nucleate, tetrasporic types of embryo sac—Adoxa, Plumbago—the number of nuclear divisions before egg formation is shortened by two divisions. The egg is formed directly by the division of a spore nucleus (the third division from the spore mother cell). The gamete could not be formed earlier, unless the sterile tissue of the gametophyte were eliminated; the spore would then become the egg directly. The number of prothallial cells is seven in the Adoxa and Plumbago types; three, in Plumbagella, where two are triploid fusion nuclei. The embryo sac of the Adoxa type resembles that of the Polygonum type but is tetrasporic in origin, and the egg is formed two cell divisions earlier than in the Polygonum type.

The Fritillaria type is like the Plumbagella in the presence, at the chalazal pole, of the sac of triploid nuclei, at the third division from the mother cell—three haploid nuclei form two triploid nuclei. This type is unlike the Plumbagella type, in that one more division (the fourth division from the mother cell) occurs in the formation of the egg. It resembles the Polygonum type superficially in its eight nuclei, but four of these nuclei, the antipodals and one polar, are triploid.

The tetrasporic Penaea, Drusa, and Peperomia types introduce a 16-nucleate sac. The egg is formed at the fourth division from the spore mother cell, the second from the spore, as in the Allium, Oenothera, and Fritillaria types. The presence of sixteen nuclei in these sacs, in contrast to the eight of the Polygonum type—where one more division occurs and only eight nuclei are formed—is the result of the four-spore make-up of the sac. The four spores developing together form 16 nuclei in two divisions, whereas one spore of the Polygonum type forms eight in three divisions.

In summary, the various subtypes show reduction in the development of the mature gametophyte from five successive divisions to three; in the formation of the egg from the spore, from three divisions to one; in prothallial cells, from seven to three. The fifteen prothallial cells of the 16-nucleate types suggest an increase, rather than a reduction, but the additional cells belong, morphologically, to the additional spores that contribute them. The large number of nuclei in these types is not evidence that they are more primitive than the 8-nucleate types, as has been suggested.

The number, arrangement, and function of the nuclei and cells of the embryo sac are remarkably constant for so great a group as the angiosperms, which have many lines of specialization in other characters. It seems doubtful that other major types will be found, even though the embryo sac of only a small number of the many genera has been studied. (Variations in apomictic genera have recently been cited as possibly new types.) Evidence concerning the possible polyphyletic origin of the angiosperms can probably not be found in the embryo sac. The less common types of sac are not restricted to definite lines of relationship. The mature sac is closely the same in structure in the major types—8-nucleate, with the egg apparatus and antipodals at the two ends and other nuclei in the center—though derived from different numbers of spores.

The union of spore protoplasts to form the gametophyte introduces a unique feature in the elaboration of alternation of generations in vascular plants. Simplification is continued—spores or dyads abort—in the tetrasporic types; and, in the Adoxa, Plumbago, and Plumbagella types, the egg is formed at the germination (first division) of the spore. Interpretation of the reduction shown in the bisporic and tetrasporic types is not strictly comparable with that in the monosporic types, because of the morphologically compound nature of the bisporic and tetrasporic gametophyte.

Despite the remarkable morphological difference in origin, the three major types of embryo sac are much alike in general structure. The Adoxa, Allium, and Fritillaria types much resemble the Polygonum type, but the Polygonum is clearly the basic one from which the bisporic and tetrasporic types have been derived; the general structural similarity probably represents parallel development. In these specialized types, the number of successive divisions in the formation of the female gametophyte is reduced—in the Adoxa, Plumbago, and Plumbagella types, by two. Among seed plants, the angiosperms show the last stage in the reduction of the gametophytic generation. The female gametophyte has many fewer cells than that of any of the gymnosperms. The male gametophyte similarly reaches extreme reduction—to a generative and a tube cell—a simplicity attained in gymnosperms only in the highest conifers, the Cupressaceae.

Theories of the Morphological Nature of the Embryo Sac

Interpretation of the embryo sac in terms of the female gametophytes of lower taxa is difficult, because of its great reduction. Progressive reduction of the gametophytes is characteristic of the vascular-plant series, but the steps from the gymnosperms to the angiosperms is a big one, even if the multinucleate gametophytes of *Gnetum* and *Welwitschia* are included. Several theories of the underlying structure of the angiosperm embryo sac, based on comparisons with gymnosperm gametophytes, have been proposed and discussed. In all theories, the sac is looked upon as the result of extreme reduction in that it consists of "archegonia and their contents," or of "vestiges of archegonia," with few or no prothallial cells.

The most strongly supported theory is that the sac consists of two archegonia (axial cell rows only), without prothallial cells. The micropylar cluster (the egg, the ventral canal cell or micropylar polar nucleus, and the two synergids) and the chalazal cluster, a similar group, but with a nonfunctioning egg, are considered to represent two four-celled archegonia, located at the two ends of the sac. In support of this theory is the definite arrangement of the nuclei in polar clusters of four nuclei or cells—even in most of the specialized types—and the occasional "inverted polarity," where the functions of the clusters are reversed and the chalazal cluster contains the egg. It has been pointed out, in refuting this interpretation, that reduction of the archegonium in other taxa is by progressive shortening of the "axial row,"—the neck cells lost first, then the ventral canal cell, while the jacket and prothallial cells still persist. Under both these theories, it is apparent that the female gametophyte in the angiosperms is reduced to an egg and several associated cells, whatever their morphological nature may be.

A modification of the two-archegonia theory interprets the micropylar cluster of four cells as remnants of two archegonia and the chalazal cluster of four cells as wholly prothallial. The egg and one synergid is said to represent one archegonium; the polar nucleus and the other synergid, a second archegonium. In support of this theory is the interpretation of the fusion of the micropylar polar nucleus with a male nucleus as a true fertilization, which gives rise to endosperm. In refutation of this two-archegonia-at-one-end theory is the assumption (1) that the egg and the one synergid are sister cells and (2) that the formation of the primary endosperm nucleus is an act of fertilization, because a male nucleus enters into the triple fusion. The egg and polar nucleus have been shown to be one pair of sister nuclei and the synergids another pair. The triple fusion does not produce an embryo.

Another interpretation of the embryo sac is based on comparison of the sac with that of *Gnetum*. In *Gnetum*, the sac contains many free nuclei, all of which appear to be potentially eggs. The "gnetalean theory" assumes that the angiosperm sac is derived from a sac of this kind by reduction in number of nuclei. But the *Gnetum* sac also appears to be a highly modified gametophyte, specialized by the multiplication, rather than the reduction, of potential eggs. In angiosperms, the synergids have been called "new features, not present in the gymnosperms," just as double fertilization has been called a new feature. If the embryo sac of *Gnetum* provides a stage intermediate between that of most gymnosperms and that of angiosperms, it should show some evidence in transitional form, rather than an entirely different type of gametophyte. The angiosperm gametophyte is so greatly reduced that, apparently, no interpretation of its nature can be well supported, even if the sac is considered without the complication of "double fertilization," which enters the picture only after the sac is mature.

THE ENDOSPERM

The fusion of the second male nucleus with the two polar nuclei forms the primary endosperm nucleus, which typically divides to form tissue that serves to nourish the developing embryo and often, later, the germinating seed. This second fertilization was unknown until 1898, but, after its discovery, was soon found in many taxa and came to be accepted as a basic step in the development of the embryo. When first known, it was called "vegetative fertilization" and considered a stimulation in the development of nutritive tissue for the embryo. There have been reports that, in some taxa, endosperm may develop from one or both polar nuclei, without fusion with a male nucleus, but these taxa should be reinvestigated. In embryo sacs where supernumerary nuclei unite in the formation of the primary endosperm nucleus or where some of the nuclei are already triploid, as in the Fritillaria and Plumbagella types, the fusion nucleus and the tissue formed by it are polyploid.

Types of Endosperm Formation

Differences in the method of formation of early stages of the endosperm form the basis for a distinction of types. Where the first divisions are *free nuclear* (Figs. 113*B* and 115)—without immediate wall formation about the new cells—the endosperm is termed *nuclear* (Fig. 115). The number of free nuclear divisions ranges from one to several, and the free-nucleate tissue so formed may persist as such until destroyed by the embryo, or may become walled. In sacs where walls are formed after the first divisions and may be formed at some or all of the later divisions, the endosperm is *cellular* (Fig. 116). No clear line exists between these types. A combination of the two types—the first division dividing the embryo sac transversely into two parts, one of which remains free nuclear and the other with walled cells—is called the *helobial type,* because it occurs frequently in the Helobiales. Probably, no endosperm is formed in the greatly reduced and peculiar embryo sac of the Podostemonaceae; and, in the Orchidaceae, the endosperm is small in amount and ephemeral. There seems to be no evidence as to which of the two types—nuclear or cellular—is the more primitive.

Lobes of the sac containing free nuclei may develop into extensive haustoria, which provide nourishment for the rest of the sac and the embryo. And there are many other modifications of form and structure in the sac when the endosperm is developing.

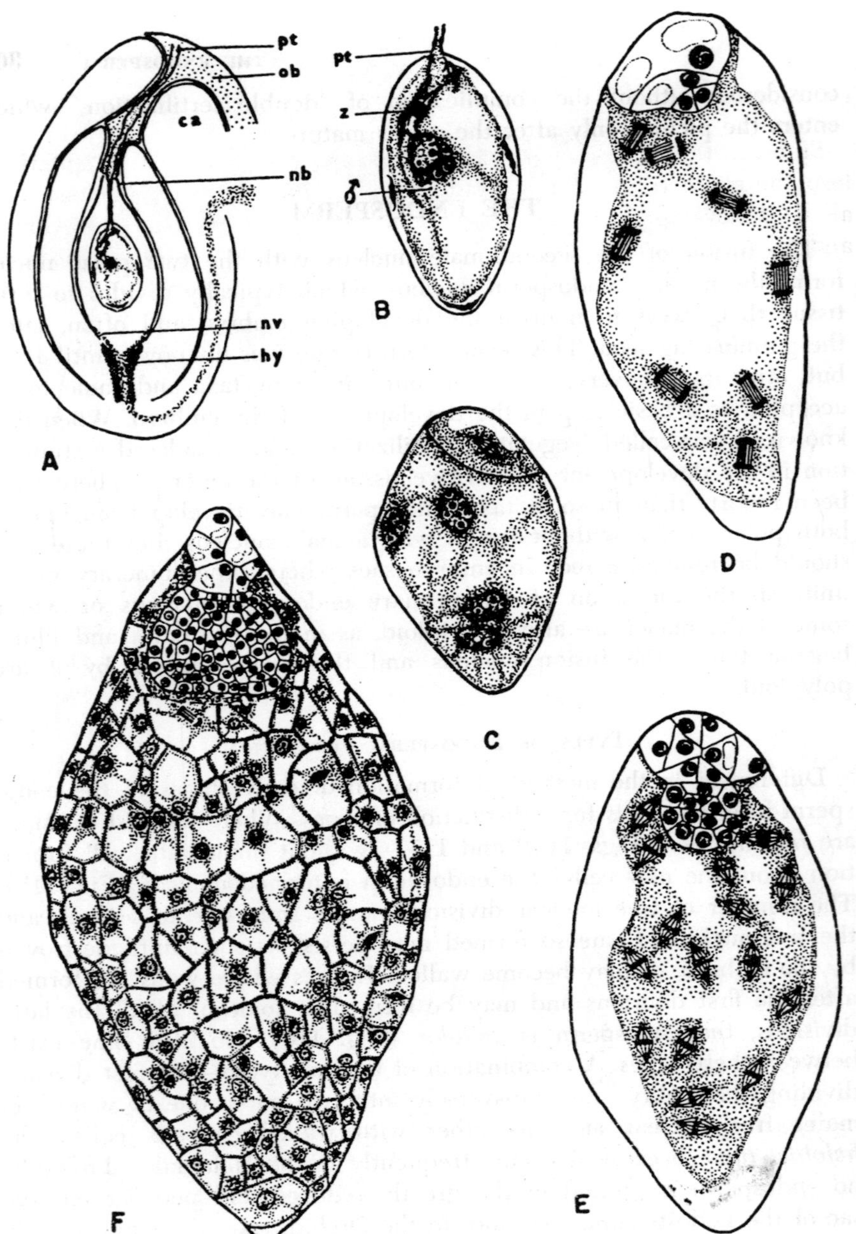

Fig. 115. Stages in the development of the embryo and cellular endosperm in *Acalypha indica*. *A*, longitudinal section of ovule at fertilization; *B*, embryo sac with pollen tube, zygote, fused polar nuclei beside male nucleus, vestigial megaspores; *C*, zygote dividing, four free endosperm nuclei; *D*, embryo four-celled, many free endosperm nuclei dividing; *E*, embryo multicellular with distinction of body and suspensor; *F*, embryo with two-lobed body and suspensor, endosperm cells walled. *ca*, funicular side of outer integument (in fruit, a caruncle); *hy*, hypostace; *nb*, nucellar beak; *nv*, nucellar vascular supply; *ob*, obturator; *pt*, pollen tube; *z*, zygote. (*After Johri and Kapil.*)

Cytological Make-up of the Endosperm

Since, in the basic Polygonum type of embryc sac, the nuclei entering into the triple fusion are haploid, the primary endosperm nucleus and all cells developed from it are triploid. The endosperm in the Allium and Adoxa types is also triploid. It is diploid in the Oenothera type and,

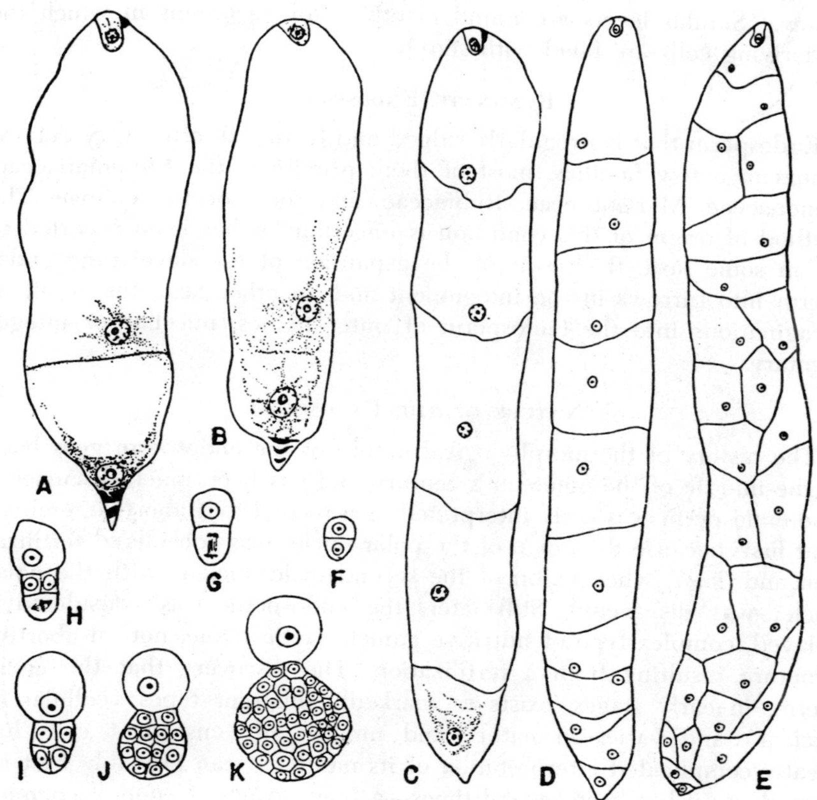

Fig. 116. Early stages in development of embryo and cellular endosperm in *Cercidiphyllum*. *A*, *B*, embryo sac, enlarged after fertilization, with zygote, vestigial antipodals, and two walled endosperm cells; *C*, *D*, *E*, successive stages in formation of the endosperm, the zygote still unicellular; *F* to *K*, stages in development of the embryo. (*After Swamy and Bailey.*)

in the other types, polyploid, varying in its chromosomal make-up. In the Fritillaria and Plumbagella types, one of the polar nuclei is triploid and, in the Peperomia, Penaea, and Plumbago types, more than two embryo-sac nuclei enter into the primary endosperm nucleus.

The endosperm is a food-storage tissue, usually of simple structure;

its protoplasts are filled with starch and other food substances, and its cell walls built up similarly of materials used in the development of the embryo. Modifications of its usually simple structure are frequent, especially in the outer layers, and the function of these outer layers is probably different from that of the inner. An *aleurone layer*, filled with aleurone grains, an outer layer of the endosperm in a few families, especially the Gramineae, consists of cells usually larger than those below. (Similar layers are found in other storage organs in which the underlying cells are filled with starch.)

Ruminate Endosperm

Endosperm that is irregularly ridged and furrowed, often very deeply, occurs in a few families, most of them primitive—the Eupomatiaceae, Annonaceae, Myristicaceae, Rubiaceae, and some of the Palmae. The method of origin of this condition is uncertain; it has been reported to be, in some taxa, the result of the expansion of the developing endosperm into furrows in the integument and, in other taxa, the result of invaginations into the endosperm of outer tissues, nucellar or integumentary.

Nature of the Endosperm

The history of the morphological nature of the endosperm goes back to the middle of the nineteenth century and has been much discussed.* The endosperm was early interpreted as a second, but abortive, embryo —at first, because the union of the polar nuclei was considered fertilization and, later, when union of the second male nucleus with the polar nuclei was discovered. Still later, the endosperm was considered a delayed, complex type of nutritive gametophytic tissue, not an abortive structure resulting from a fertilization. The discovery that the endosperm, in early stages, exists as markedly different types (cellular or nuclear) and varies in nature and number of constituent cells has greatly complicated interpretation of its nature. It can probably best be termed—as it has been several times—a "new structure," one of complex morphological nature, characteristic of the angiosperms only. Interpretation of the endosperm parallels that of the extraordinary bisporic and tetrasporic gametophytes, also "something new," found only in angiosperms.

BIBLIOGRAPHY

Artschwager, E.: Development of flowers and seed in the sugar beet, *Jour. Agr. Res.*, 34: 1–25, 1927.

* An excellent detailed discussion is presented in P. Maheshwari's "An Introduction to the Embryology of Angiosperms," 1950, pp. 424–425.

———, E. W. Brandes, and R. C. Starrett: Development of flower and seed of some varieties of sugar cane, *Jour. Agr. Res.*, **39**: 1–30, 1929.

Battaglia, E.: The male and female gametophytes of angiosperms—an interpretation, *Phytomorph.*, **1**: 87–116, 1951.

Brough, P.: Studies in the Epacridaceae. I. The life history of *Styphelia longifolia* R. Br., *Proc. Linn. Soc. N.S.W.*, **49**: 162–178, 1924.

Cooper, G. O.: Microsporogenesis and development of seed in *Lobelia cardinalis*, *Bot. Gaz.*, **104**: 72–81, 1942.

Coy, G. V.: Morphology of *Sassafras* in relation to phylogeny of angiosperms, *Bot. Gaz.*, **86**: 149–170, 1928.

Dahlgren, K. V. O.: Die Morphologie des Nuzellus mit besonderer Berücksichtigung der deckzellosan Typen, *Jahr. Wiss. Bot.*, **67**: 347–426, 1927.

Fagerlind, F.: Die tetrasporische Angiospermen-Embryosac und dessen Bedeutung für das Verständnis der Entwicklungsmechanick und Phylogenie des Embryosacks, *Arkiv Bot.*, **31A**: 1–71, 1944.

———: Embryologische, zytologische und bestäubungsexperimentelle Studien in der Familie Rubiaceae nebst Bemerkungen über einige Polyploiditätsprobleme, *Acta Horti Berg.*, **11**, 195–470, 1937.

Guignard, L.: Recherches sur le sac embryonnaire des phanérogames angiospermes, *Ann. Sci. Nat. Bot.*, 6 sér., **13**: 136–199, 1882.

Hjelmquist, H.: Studies on the floral morphology and phylogeny of the Amentiferae, *Bot. Not. Suppl.*, **2**: 5–171, 1948.

Johansen, D. A.: "Plant Embryology," Waltham, Mass., 1950.

Johri, B. M. and R. N. Kapil: Contribution to the morphology and life history of *Acalypha indica* L., *Phytomorph.*, **3**: 137–151, 1953.

Li, H. L.: The evolutionary significance of the endosperm and its bearing on the origin of the angiosperms, *Jour. Wash. Acad. Sci.*, **47**: 33–38, 1957.

Maheshwari, P.: "An Introduction to the Embryology of Angiosperms," New York, 1950.

Nilsson, H.: Die Homologie des angiospermen Embryosackes, *Bot. Not.*, **1941**: 50–58.

Peter, J.: Zur Entwicklungsgeschichte einiger Calycanthaceae, *Beitr. Biol. Pfl.*, **14**: 59–84, 1920.

Rao, L. N.: Studies in the Santalaceae, *Ann. Bot.*, n.s., **6**: 151–175, 1942.

Samuelsson, G.: Studien über die Entwicklungsgeschichte der Blüten einiger Bicornes-Typen, *Svensk. Bot. Tidskr.*, **7**: 97–188, 1913.

Sargant, E.: The reconstruction of a race of primitive angiosperms, *Ann. Bot.*, **22**: 121–186, 1908.

Schnarf, K.: Contemporary understanding of embryo-sac development among angiosperms, *Bot. Rev.*, **2**: 565–585, 1936.

———: Archegonium und Archegontheorie, *Biol. Gen.*, **16**: 198–224, 1942.

Schürhoff, P. N.: Über die Entwicklung des Eiapparates der Angiospermen, *Ber. Deutsch. Bot. Ges.*, **46** (8): 560–572, 1928.

Shreve, F.: The development and anatomy of *Sarracenia purpurea*, *Bot. Gaz.*, **42**: 107–126, 1906.

Smith, F. H.: Development of the gametophytes and fertilization in *Camassia*, *Am. Jour. Bot.*, **29**: 657–663, 1942.

———: Megagametophyte development in five species of *Erythronium*, *Am. Jour. Bot.*, **42**: 213–224, 1955.

Steffen, K.: Mehrzelligen Archespor bei Balsaminaceen, *Planta*, **36**: 203–213, 1948.

Swamy, B. G. L.: Inverted polarity of the embryo sac of angiosperms and its relation to the archegonium theory, *Ann. Bot.*, **10**: 171–193, 1946.

———: A contribution to the life history of *Casuarina*, *Proc. Am. Acad.*, **77**: 1–32, 1948.

———: The embryology of *Epidendrum prismatocarpum*, *Bull. Torrey Bot. Club*, **75**: 245–249, 1948.

———: Embryological studies in the orchids. II. Embryology, *Am. Midl. Nat.*, **41**: 202–332, 1929.

——— and I. W. Bailey: The morphology and relationships of *Cercidiphyllum*, *Jour. Arnold Arb.*, **30**: 187–210, 1949.

Chapter 9

FERTILIZATION*

The tip of the pollen tube, carrying the male gametes, breaks through the embryo-sac wall nearly always at or beside the egg apparatus. In chalazogamy, where the tube reaches the sac at or near the base of the sac, it passes along the outside of the sac to the micropylar end before entering (Fig. 103). In entering, it may pass through and destroy one, rarely both, of the synergids; it may pass between the egg and a synergid, or at one side of the group of cells. (In the Plumbago and Plumbagella types of embryo sac, there are no synergids, and, in some taxa, they may degenerate before fertilization.) The male cells are discharged into the sac, and the nucleus of one unites with the nucleus of the egg, and the nucleus of the other unites with the two polar nuclei in a triple fusion. These two nuclear fusions constitute "double fertilization." Some phenomena in the formation of the zygote suggest that the male cytoplasm enters into the formation of the zygote, but this is uncertain. (In araucarian conifers, the male cytoplasm takes a prominent part in the formation of the first cells of the embryo.)

The time elapsing between pollination and fertilization is controlled by many factors, especially structure of the style, temperature, and humidity. Length of the style is apparently of less importance than other factors. The pollen may germinate on the stigma "immediately," or only after some hours. The shortest time reported for the tube to reach the embryo sac is fifteen to forty-five minutes after pollination. The usual time is a few hours, but days, even months, may be required. In the Orchidaceae, where the ovules may not be developed at time of pollination, the period may be as long as six or seven months—*Cymbidium*. Where dormancy due to unfavorable climatic conditions intervenes, the period is several months or even a year—six months in *Hamamelis* and thirteen in the red and black oaks. (Similar variation in time from pollination to fertilization is present in the conifers—from a few days to about thirteen months in *Pinus* and *Agathis*.)

* For a detailed discussion of the cytological and genetical aspects of the fusion of the male and female gametophytes and of the second male gamete with the polar nuclei, the reader is referred to textbooks on cytology and genetics.

The long period between pollination and fertilization has been considered a primitive character, apparently, in part, because of the long period found in some of the Amentiferae. The shortest periods are characteristic of annuals—*Zea* and other grasses. The long periods of the red oaks, contrasting with much shorter periods in other oaks, are doubtless associated with the adaptation of the red oaks to the colder, temperate climates. Perhaps the very long, as well as the very short, periods represent specializations.

THE EMBRYO

The Proembryo. After fertilization, the zygote rests for a period that varies greatly in different taxa—from a few hours to several weeks, even over winter in some autumn-flowering plants. Development of the endosperm—division of the primary endosperm nucleus—usually begins immediately after fertilization, before division of the zygote, but may accompany or, rarely, precede syngamy.

Methods of development of the first few cells of the proembryo vary considerably. Six general types are distinguished, though most of these show variations that make clear distinctions among them not always possible. The types are based on the position of the first cell walls in the zygote and early proembryo, and on the extent and part of the embryo proper developed by the first-formed cells. In five of the types, division of the zygote is transverse; in the sixth type, which is rare, it is longitudinal, or obliquely so. (Of the two cells formed by a transverse division, that toward the center of the embryo sac is called *terminal*; that toward the micropylar end, *basal*.) In the group of types in which the first division is transverse, the position of the next division, which is in the terminal cell, determines two subgroups. This division, like that in the zygote, is either longitudinal or transverse and forms subgroups based on the part played by the basal cell in continuing the development of the embryo.

The terms applied to the types of early embryogeny—to proembryo stages—are based on the names of families in which members of each type are found. The term proembryo is variously used. It is commonly applied to the very early stages of the embryo—the first few cells—the stages on which the classification below is based. But it is also often used to cover all stages of the embryo while it remains radially symmetrical, before differentiation of cotyledons and axis begins. This is probably the better use; otherwise, there is no term for the intermediate stages. The following outline—in key form—is a convenient summary of the six types. It does not indicate relationships among the types.

Basic Types of Early Embryos

I. Division of the zygote transverse
  A. Division in terminal cell longitudinal
    1. Both basal and terminal cells take part in the development of the embryo .... *Asterad type*
    2. Basal cell plays little or no part in the development of the embryo .... *Onagrad type*
  B. Division in terminal cell transverse
    1. The basal cell plays an essential part in the development of the embryo proper .... *Chenopodiad type*
    2. The basal cell plays no essential part in the development of the embryo proper
      a. The basal cell usually forms a suspensor of two or more cells .... *Solanad type*
      b. The basal cell divides no further and becomes a large suspensor cell* .... *Caryophyllad type*
II. Division of zygote longitudinal .... *Piperad type*

The Asterad type is found in many families and is characteristic of most Compositae that have been studied. The Onagrad type is widely distributed in angiosperm families. It is frequently called the *Crucifer type,* but Onagrad is preferable, because it is characteristic of the Onagraceae and, in this family, is the simplest type known, with the exception of the reduction (Piperad) type. Many variations of the Onagrad type are recognized. The Chenopodiad type, based on *Chenopodium,* is known only in a few genera outside the Chenopodiaceae—*Polemonium, Myosotis.* The Solanad type, characteristic of the Solanaceae, is found in several other families. The Caryophyllad type is widely distributed in both dicotyledons and monocotyledons. Many variations are described in this type, as in the Onagrad type. The Piperad type stands apart from all the others, not only in the position of the wall in the division of the zygote but in the reduction in the number of cells in the proembryo and in the morphological structure of the mature embryo. Found in only rather few dicotyledons—Piperaceae, Santalales, Balanophorales, and, apparently, *Scabiosa* in the Dipsacaceae—it is clearly a reduction type, and, with the exception of *Scabiosa,* the genera possessing it are highly specialized and reduced. As a reduction type, the Piperad type has also been considered hardly a basic type, but rather as "unclassified."†

* Suspensors consisting of a few cells may be present in this type, but those additional to the basal cell are derivatives of the terminal cell.

† For further details of structure of the various types and of the varieties of each type, the reader is referred to D. A. Johansen, "Plant Embryology," Waltham, Mass., 1950; and P. Maheshwari, "An Introduction to the Embryology of Angiosperms," New York, 1950.

The absence of an early free nuclear stage sets the angiosperm embryo well apart from that of the gymnosperms. Primitive form is difficult to determine in the great variety of proembryos in the angiosperms. The presence of suspensors naturally suggests comparison with lower taxa, where suspensors play an important part in embryogeny, the transfer of the young embryo from a superficial to a deep-seated position in the nutritive gametophytic tissues. In the angiosperms, the suspensors play a similar, but less important, part, because the zygote is already close to and sometimes surrounded by the nutritive tissues. Progressive stages in the loss of the suspensor apparently exist. The great diversity in details of form of the early embryo is no greater than that in parts of the mature sporophyte—flower, leaf, ovule—and a correlation between what may be considered advanced or primitive embryo types with advanced or primitive taxa is hardly to be expected. Correlation can perhaps be seen in the presence of the reduced Piperad embryo in the parasitic Balanophorales and Santalales, with their reduced vegetative bodies.

Comparison of Proembryo Types. In comparison of types of early embryo development, emphasis has been placed on the degree of development of the suspensor. The suspensor is looked upon as a primitive feature of the embryo, probably because of its prominence in lower groups, especially the higher gymnosperms, and a series in reduction can perhaps be found within angiosperms, from massive, many-celled suspensors to those of one or two cells. The massive proembryo is commonly believed to be the primitive type. Differences in the embryos of dicotyledons and monocotyledons were long ago believed to appear in early-proembryo stages. It has been claimed that, if the longitudinal division of the terminal cell of the young suspensor forms two equal cells, two cotyledons are formed; if these two cells are unequal, only one cotyledon is formed. This conclusion has not been supported. (For comparative morphological study, the embryo of both mono- and dicotyledons must be morphologically mature—not merely at the stage found in the ripe seed before after-ripening; errors of interpretation have been made by the study of immature embryos—those in the seed at time of shedding.)

In general form and in the possession of a suspensor, the angiosperm embryo resembles that of gymnosperms. In the angiosperms, the suspensor seems to be a disappearing structure. In the gymnosperms, the suspensor carries the proembryo from the peripheral position of the archegonia into the central part of the endosperm; in the angiosperms, the suspensor similarly aids in the transfer of the zygote or young embryo from a position at one end of the sac to a more nearly median position, but the suspensor is a weaker structure than that of gymnosperms, and the endosperm develops around it. Within the angiosperms, the suspensor seems to show reduction from a massive, elongate structure to a few-celled filament or a single cell. Suspensors consisting of only one or two cells appear to be vestigial structures and function only in minor degree. They have often been overlooked, and the embryo described as without a suspensor.

In angiosperm embryos that have the greatest reduction of the suspensor, all, or nearly all, the tissues formed from the zygote go into the formation of the mature embryo. In the gymnosperms, a major part of the young embryo is tissue accessory to the embryo proper: massive, multicellular suspensors and, in some taxa—Araucariaceae, Welwitschiaceae, some of the Podocarpaceae and Taxaceae—protective caps about the embryo initials. In the angiosperms, the suspensors are mostly few-celled, and there is no cap.

Later Stages of Embryo Development. In early stages, the embryo is more or less clearly separable into suspensor and embryo proper. The first divisions commonly form a filament of cells which differentiates into a basal part, the suspensor, and a terminal, early-enlarging part, the embryo proper, from which the body of the mature embryo forms (Fig. 117). The suspensor varies greatly in size and form in different taxa—even in a single family, as in the orchids. It may consist of a single cell, a chain of cells, or a mass of cells not readily separable from the embryo proper. The embryo proper (Fig. 118*B*) develops, typically, by divisions of the apical, or apical and subapical, cells to form quadrants and octants (Figs. 119 and 120). By continuing division in many planes, there is formed a multicellular structure, which varies greatly in form in different taxa—pyriform, subspherical, ovoid, obovoid, columnar, flattened. On this undifferentiated mass, the suspensor often persists as a minor appendage (Figs. 120*N* and 121*P*). In some taxa, as in the Nymphaeaceae, the suspensor may develop late and has been called vestigial. The single-cell suspensor (Fig. 120) has also been considered a remnant, but functions by great enlargement. Some embryos have been described as having no suspensors, but, in interpretations of embryos in which only one or two basal cells are formed by the first transverse divisions in the proembryo (Fig. 119), these cells are not considered a suspensor, especially in many of the earlier descriptions of embryo development.

Differentiation of organs of the embryo—cotyledons and axis—may be immediate, brought about by continuous growth of the embryo, or a period of cessation of growth may intervene before the final stages of development. This interruption in embryo development may play a part—largely unrecognized—in the process of *after-ripening* of seeds.

As compared with the great number of descriptions of embryo (chiefly proembryo) development in taxa throughout angiosperms, there is a surprising lack of information concerning the stages intermediate between the proembryo and the mature embryo, especially the histology of the origin of cotyledons and stem and root apices, and the time and rate of development of these structures. Apparently, this stage has been considered unimportant, but some aspects of it are of much importance

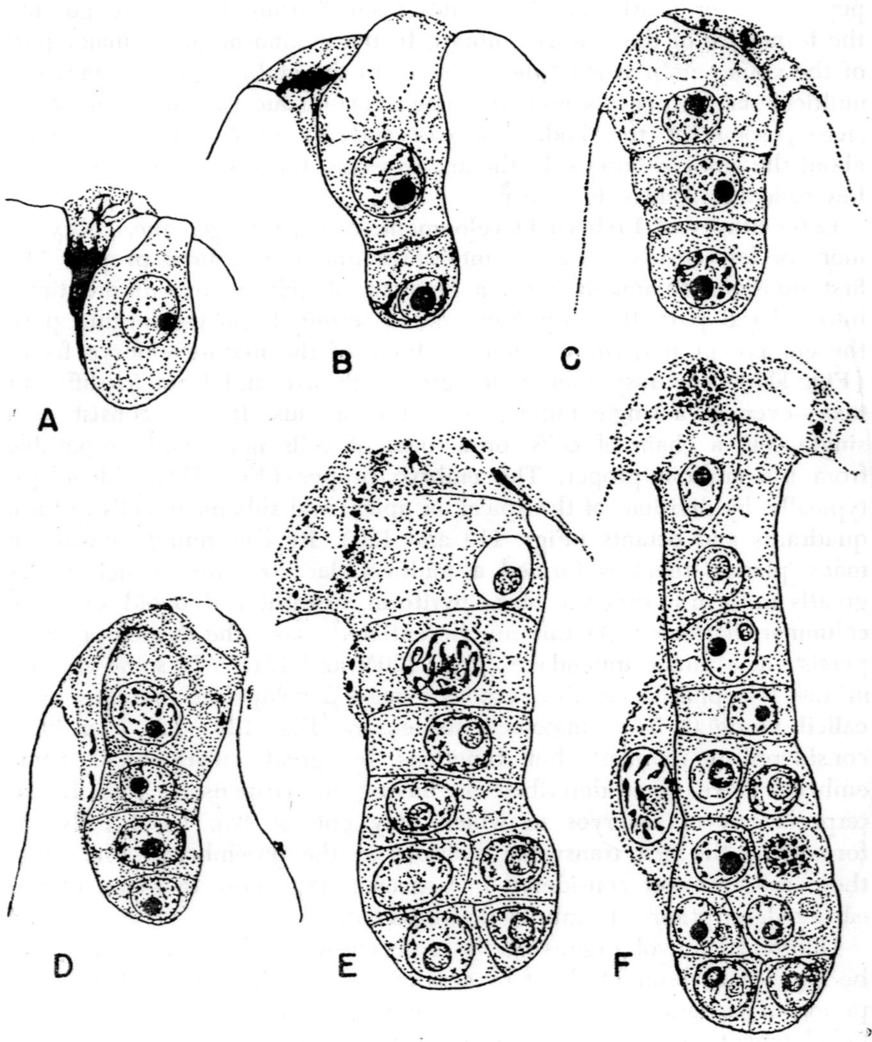

Fig. 117. Early stages in ontogeny of the embryo of *Beta vulgaris*. Filament of cells formed by transverse divisions; in *E*, and *F*, distal part enlarged by divisions in many planes. (*A*, *B*, *after Artschwager; C to F, after Artschwager and Starrett.*)

in studies of dormancy and after-ripening. For this stage, the earlier studies of embryo development provide most of the available information.

The beginning of differentiation of cotyledons and axis may be early in some taxa, even as early as the octant stage, but, usually, the origin of these organs is not apparent until the embryo is massive, and it may

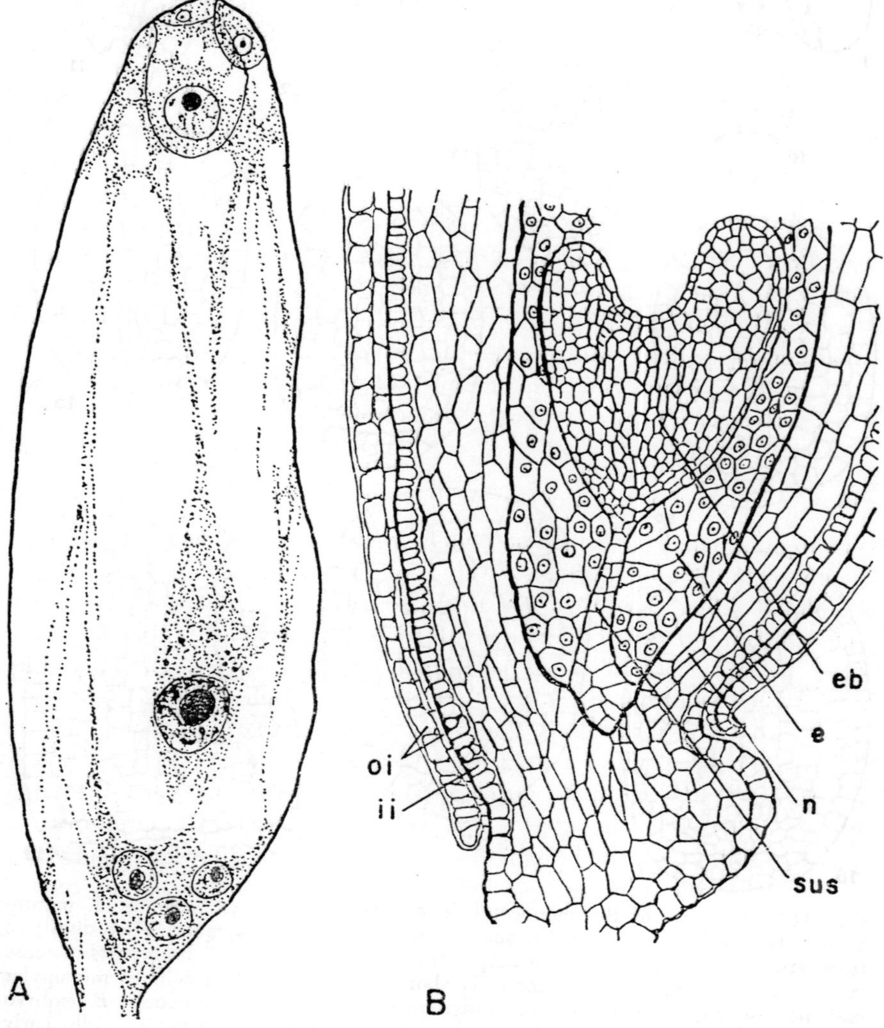

Fig. 118. *Beta vulgaris. A,* mature, 8-nucleate embryo sac, polar nuclei fused; *B,* mature embryo in micropylar end of the seed. *e,* endosperm; *eb,* body of embryo; *ii,* inner integument; *n,* nucellus; *oi,* outer integument; *sus,* suspensor. (*After Artschwager.*)

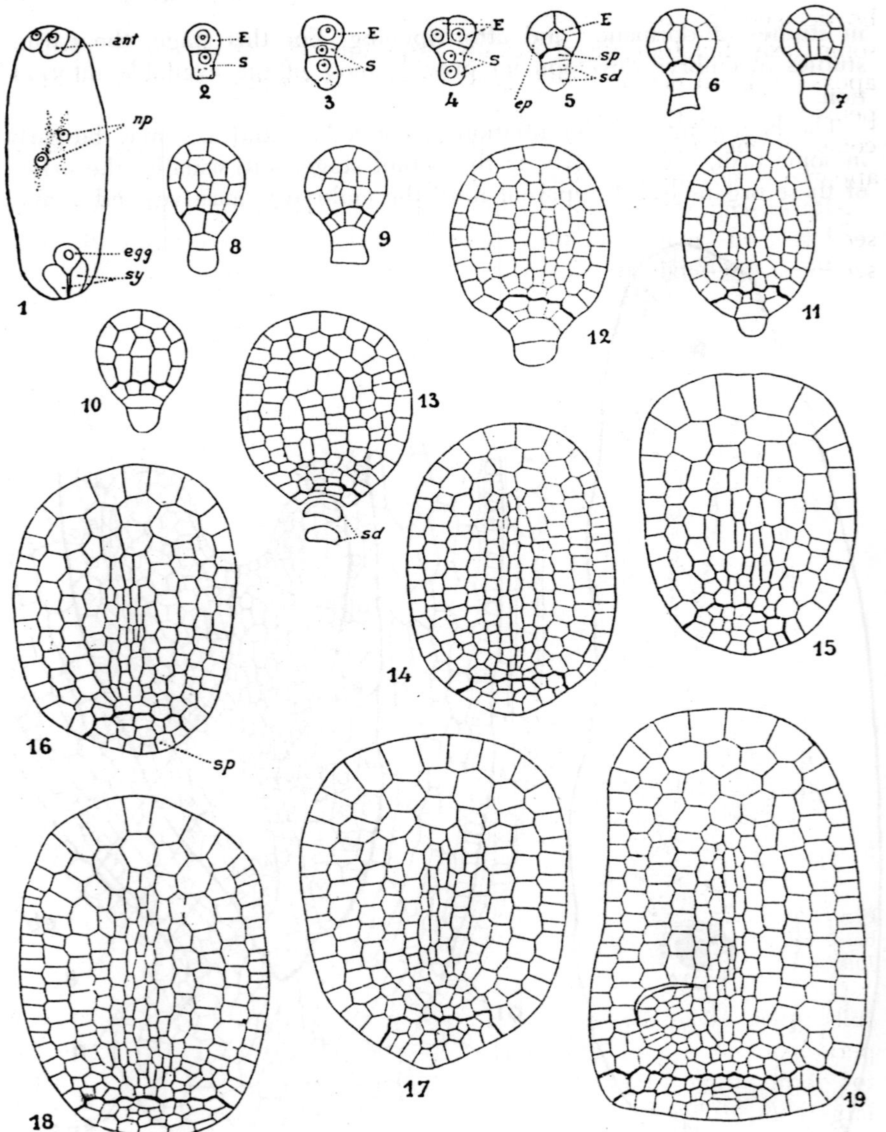

Fig. 119. Ontogeny of the monocotyledonous embryo, *Juncus bufonius*. 1, mature 8-nucleate embryo sac; 2 to 19, successive stages in development. First divisions, transverse, separating the embryo proper from suspensor; 4 to 12, both parts increase in size; 14 to 19, basal suspensor cells abort and are detached, cells of micropylar end increase greatly; root-meristem differentiation in 19. *ant*, antipodals; *E*, embryo proper; *ep*, epidermis; *np*, polar nuclei; *S*, suspensor; *sd*, suspensor cells early detached; *sp*, persistent suspensor cells; *sy*, synergids. (*After Laurent.*)

be delayed by an intervening period of dormancy. In the embryo of some taxa, the four distal octants form the cotyledons and the stem apex; the basal octants form the hypocotyl; the region of union of suspensor and embryo proper forms the root apex. The root apex may become distinct much later than the other parts of the embryo and is apparently not an appendage (Fig. 119, 19).

After-ripening. The embryo is generally considered mature when the seed is shed—mature in the sense of readiness to establish itself as a seedling. But seeds are not "ripe," morphologically, until the embryo is

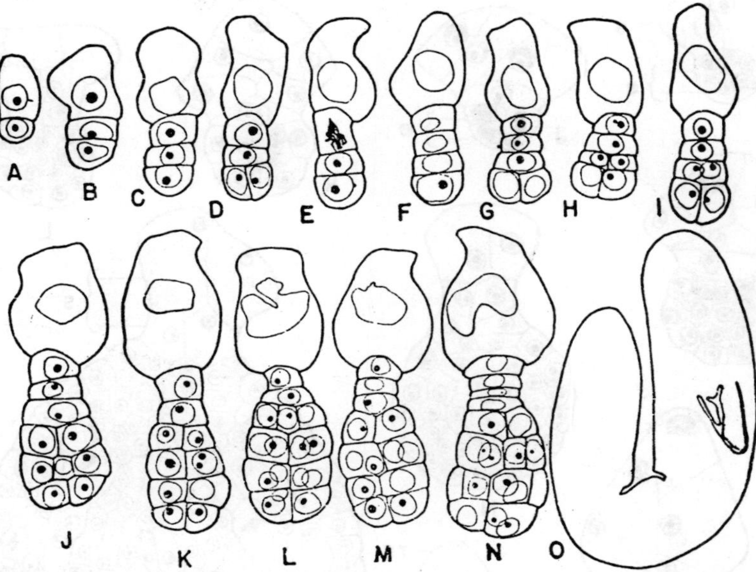

Fig. 120. Early stages in ontogeny of embryo of *Sagittaria guayanensis.* Enlarged basal cell of filament forms prominent part of suspensor. *O,* mature embryo. (*After Johri.*)

full-grown; at shedding, the seeds of many taxa are immature, and a period of intraseminal growth must ensue before the embryo is ready for germination. The existence of this concealed development has been largely overlooked in the twentieth century; in the many studies of dormancy, there has apparently been little or no correlation of stages of embryo development with external conditions. The term *after-ripening* has been used to cover changes necessary to germination—relations of temperature, humidity, and light—but, in many taxa, changes in internal growth are at least as important as these physiological relations.

Continuous development of the embryo is well known in some gymnosperms; it is characteristic of the cycads, *Gnetum,* and *Ginkgo;*

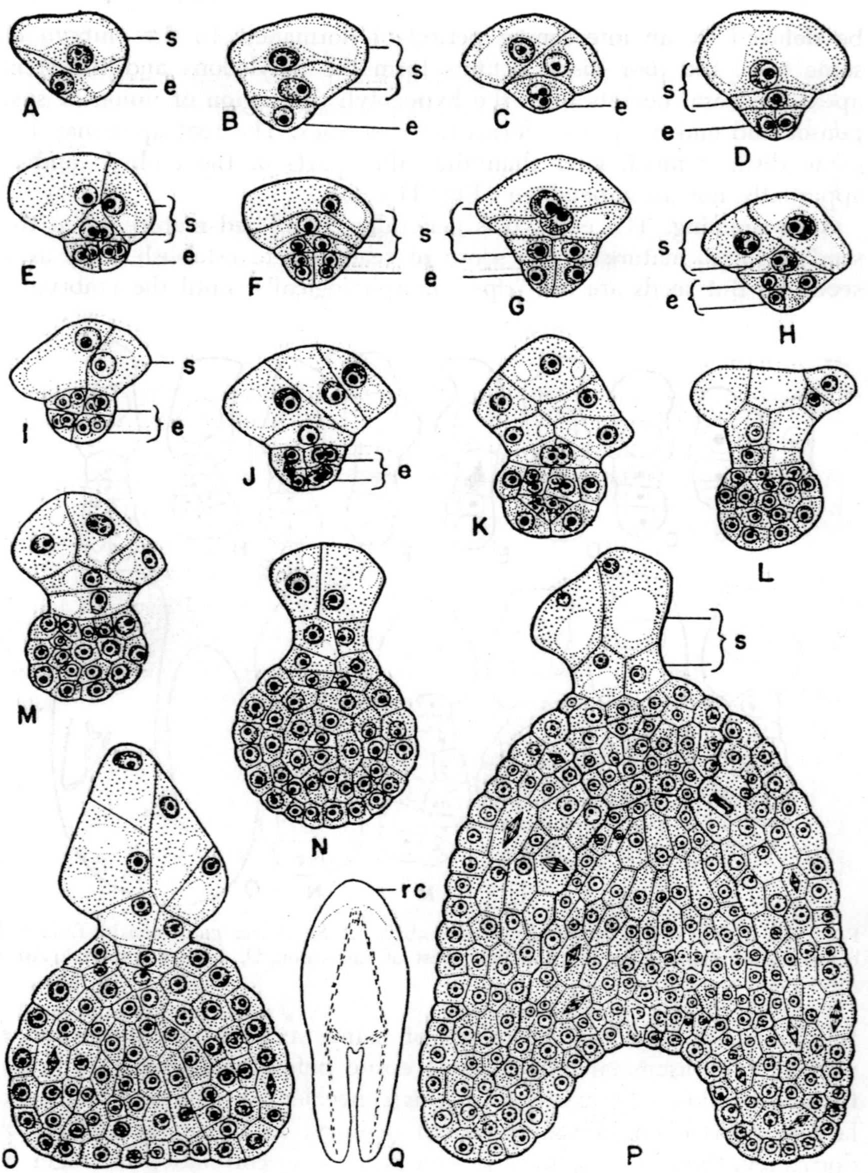

Fig. 121. Development of dicotyledonous embryo, *Acalypha indica*. *A* to *P*, stages showing histological origin of suspensor and embryo body; *Q*, diagram of longitudinal section of mature embryo. *e*, embryo; *rc*, root cap; *s*, suspensor. (*After Johri and Kapil.*)

even fertilization often occurs in the "ripe," fruitlike "seeds" of *Ginkgo,* as they lie on the ground. No dormant period is present in viviparous taxa and in many tropical genera; there is none in many aroids—*Peltandra, Orontium, Anthurium*—and in the Hydrocharitaceae—*Thalassia*—and in other aquatic taxa. The coconut palm, *Cocos nucifera,* has long been known as an example of embryo development in a seed on the ground. When the ripe coconut falls to the ground, the embryo is an undifferentiated, cylindrical body about 8 mm long. The embryo continues to grow, apparently without resting, and the plumule breaks through the husk after about four months. In *Cacao* and *Thalassia,* the embryo continues growth through the germination stage while within the fruit, and the germinating seeds are set free to continue growth when the fruit disintegrates. Dormant seeds are characteristic of most angiosperms, but, in many, the embryo is not dormant. Studies of this intraseminal growth are few, and its extent among angiosperms is little known. That the seeds of many genera of the Ranunculaceae have undifferentiated embryos at shedding is common knowledge, and probably many other herbaceous genera—both dicotyledonous and monocotyledonous—have seeds of this type. This type has been described in *Crocus,* several genera in the Fumariaceae, Papaveraceae, Umbelliferae, and, in woody plants, in *Fraxinus.* In *Anemone,* the embryos of different species are at various stages of development at seed shedding, and, consequently, seeds with well-developed embryos germinate early; those with undeveloped embryos, months later.

Continuation of embryo growth within the seed after shedding is apparently not related to soil type, but presence and absorption of water are usually necessary, and presence or absence of light and freezing may affect development. The seeds of many aquatic genera require continuous submersion in water. Seeds of some taxa continue this intraseminal growth unbroken at time of shedding; in others, there may be a long period when the immature embryo lies dormant before completing growth. Examples of time reported necessary to complete growth from seed-shedding stage to germinating stage are: *Fumaria,* eight days; *Caltha,* ten days; *Clematis,* seventeen days; *Actaea, Thalictrum, Hepatica,* two months; *Fraxinus, Cocos, Paris,* four months; *Crocus,* six months; *Corydalis,* ten months; *Trillium,* twelve months. The seed of *Ranunculus Ficaria* is described as not germinating until the second spring after it is shed. Measurements of length of embryo while still within the seed show rate of intraseminal growth: in *Eranthis,* May—100 μ, June—225 μ, July—430 μ, September—1000 μ, November—3000 μ; in *Corydalis,* May—150 μ, October—1000 μ; in *Ranunculus Ficaria,* May—100 μ, October—625 μ. Embryo development within seeds may be the continuation of growth in process at time of shedding or may be a renewal after dormancy for various periods, up to more than a year following seed shedding. Intraseminal growth, so far as known, seems to be restricted largely to the more primitive taxa and to herbaceous, geophilous genera.

After-ripening of the epicotyl in the seedling stage, after hypocotyl and radicle are developing, has received some attention. But this is not akin to after-ripening of the embryo in the seed; it is related to the dominance of growth activity elsewhere in the seedling—the elaboration of the cotyledons as foliage organs; the establishment of a root system; in some genera, the storage of food in the hypocotyl.

There is need for much more information concerning the later stages of embryo development. Studies of dormancy and after-ripening should include the full story of embryology. Hidden morphological changes may play important parts in the story of germination.

Germination

The bursting of the seed coats in germination results from the enlargement of the embryo, either throughout, or restricted to definite parts. The term *epigeal* is applied to germination where the seed coat is carried up into the air, regardless of place where germination begins (seeds on the surface of the soil or somewhat below may be carried up into the air); the term *hypogeal,* where seed coat and cotyledons remain in the soil. Differences between the epigeal and hypogeal methods of germination and early seedling development are generally great, though transitional conditions are frequent; the hypogeal method is clearly advanced. Most dicotyledons (like the gymnosperms) are epigeal; monocotyledons are largely hypogeal. The papilionaceous legumes show many transitional forms.

The root system varies greatly in time and vigor of development in both types of germination. In epigeal germination, the root tip and hypocotyl commonly grow rapidly; their elongation pulls the cotyledons from the seed coats. Exceptions, where the main root grows little or not at all, are occasional—*Sempervivum, Phyllodoce, Anemiopsis.* In these forms and in those monocotyledons in which the main root is weak or does not develop, the first root hairs form a crown at the base of the hypocotyl. In hypogeal development, the main root usually develops rapidly and becomes thick and strong. The taxon "Megarrhiza" of Darwin and Asa Gray was based on possession of development of this type. Root hairs are abundant on the roots from earliest stages. Here belong the Fagaceae, Nymphaeaceae, Juglandaceae, Hippocastanaceae. In the monocotyledons, several types of germination have been distinguished. In the majority of taxa—Palmae, the advanced Liliaceae, Amaryllidaceae—the root tip appears very early, and the root develops rapidly; the cotyledon tip remains within the seed coats, at least for some time. In the Helobiales, Pandanales, the primitive Liliales, and some other taxa, the main root develops only weakly or not at all, and the embryo is at once freed from the seed coats. Epigeal germination is doubtless the primitive type. Both types and many transitional forms are present in the Liliaceae. Some genera have epigeal; others, hypogeal; the Scilleae have both types and also transitional types.

The Angiosperm Embryo at Germination and in Early Seedling Stages

Interpretation of the structure of the embryo at time of germination is difficult, because the various organs are incompletely differentiated and the vascular system is immature. The morphology of the embryo, especially that of the monocotyledons, can be determined only by comparison with early seedling stages. Although the embryos and seedlings of angiosperms have received much study since about 1800, there is still disagreement in interpretation of the nature of some of the parts; terminology is much confused, and definitions often conflict seriously.

The term mature embryo has been loosely applied—to the embryo in the ripe seed; to the embryo at time of germination (often very different from that in the ripe seed); and to embryos with organs distinct (in many taxa, a doubtfully recognizable state). Obviously, the embryo that continues growth within the seed during an after-ripening period should not be called "mature" when the seed is shed. No line can be drawn between proembryo and "mature" embryo, or between mature embryo and seedling. The structure of the later stages of embryos within the seed and that of seedlings are here discussed together, because, only in this way, can the morphological nature of all three be determined.

Few terms descriptive of the entire embryo have been used, but four applied to form of the monocotyledonous embryo need mention, because they are used in two ways. These terms were probably first applied in 1811. Two of these—*straight* and *curved*—have continued in occasional use; two others—*rémotive* and *admotive*—which apply chiefly to relation of embryo form to type of germination, are probably obsolete. Straight and curved were originally applied to the form of the vascular system of the embryo (straight and curved from root tip to plumule), but have largely been used to describe the embryo itself (Fig. 122). Used for general description, they have little significance, because embryo form is often correlated with seed form, and distinction between a straight and a curved embryo may be difficult to make (Figs. 123 and 124). An embryo was considered rémotive when the vascular axis of cotyledon-hypocotyl-radicle was straight and parallel with that of the plumule; admotive, when the plumule-hypocotyl-radicle axis was curved and the cotyledon axis at an obtuse angle with it. Distinction between these two types may also be difficult to make.

The Axis in the Embryo. Much of the confusion in the application of terms to the embryo is related to the axis. Under the older terminology,

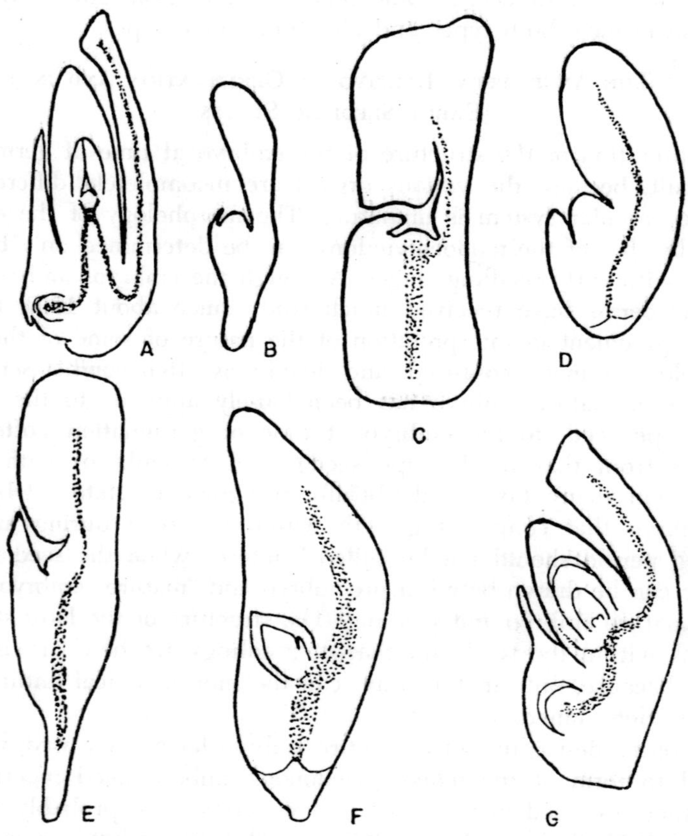

Fig. 122. Longitudinal sections of monocotyledon embryos showing pseudoterminal position of cotyledon attained by dominant growth of cotyledon and accompanying change of axis from straight to curved, and plumule from terminal to so-called lateral. *A, Leersia clandestina; B, Guzmannia tricolor; C, Costus; D, Aechmea miniata; E, Tillandsia vestita; F, Karatas amazonica; G, Oryza sativa.* (*A, G, after Schlickum; B, D, E, F, after Gatin; C, after Boyd.*)

the axis was considered to consist of the "caulicle," or "embryonic stem," and the "radicle," the "embryonic root." But these terms have been less used in recent years. In most embryos, no true stem has yet been formed; the part termed caulicle is usually the hypocotyl. Commonly, there is a root primordium, but usually no true root is present until germination. Some of the higher monocotyledons bring further difficulties into the interpretation of the axis of the embryo by the presence of a compound structure, the *mesocotyl*—the middle part of the axis in some grasses and other monocotyledons. It has been urged that this term be dropped as superfluous, since "this part is either

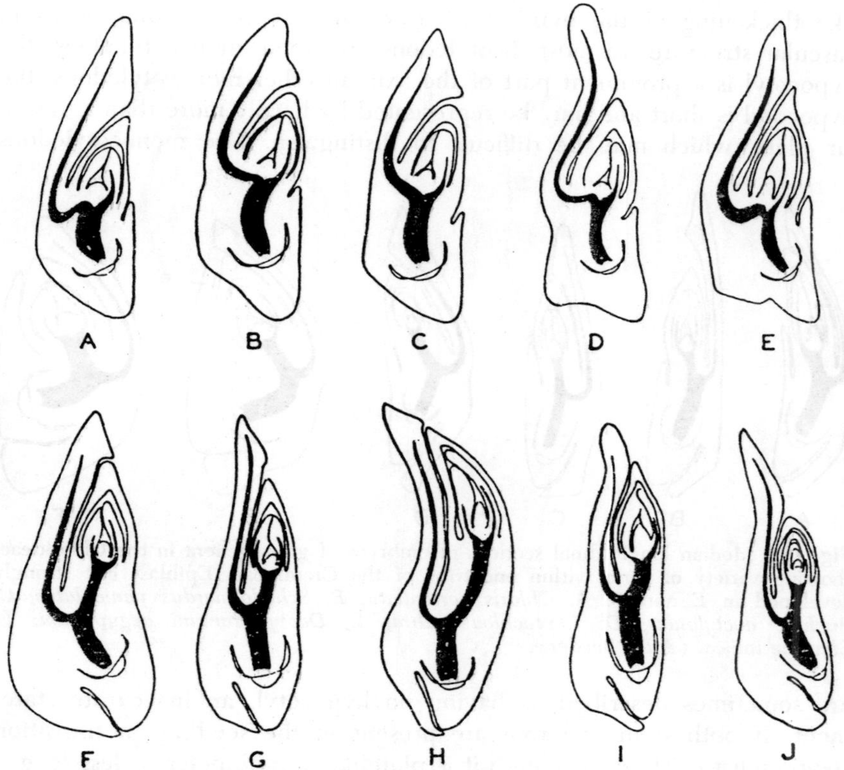

Fig. 123. Median longitudinal sections of embryos of grass genera showing variety of basic form. *A* to *F*, axis curved, cotyledon pseudoterminal, plumule so-called lateral; *G* to *J*, axis nearly straight and plumule terminal. *A*, *Festuca elatior;* *B*, *Dactylis glomerata;* *C*, *Deschampsia flexuosa;* *D*, *Phalaris arundinacea;* *E*, *Phleum alpinum;* *F*, *Panicum clandestinum;* *G*, *Brachiaria platyphylla;* *H*, *Echinochloa crusgalli;* *I*, *Setaria italica;* *J*, *Sorghum vulgare.* (*After Reeder.*)

hypocotyl or epicotyl." But study of anatomy and comparative form shows that the mesocotyl is a compound structure, the hypocotyl, with an adnate part of the cotyledon. (Discussion later in this chapter.)

The Hypocotyl. Morphologically, the hypocotyl is the part of the embryo and seedling in which there is anatomical transition from root structure—with radial arrangement of alternating strands of phloem and exarch xylem—to stem structure, with vascular strands consisting of both phloem and endarch xylem. The hypocotyl is not set off sharply from root and stem; it is not an organ *sui generis*—though sometimes so called. In some taxa, there is an external line of limitation—the "collet"—between hypocotyl and root; the root is smaller in diameter than the hypocotyl, and its epidermis is piliferous. (The term collet has also been applied to the cotyledonary sheath, as well as to this collar-like thickening of the hypocotyl.) Internally, there are differences in vascular structure. In most dicotyledons and some monocotyledons, the hypocotyl is a prominent part of the axis; in other monocotyledons, the hypocotyl is short and may be represented by hardly more than a vascular plate, which may be difficult to distinguish. The monocotyledons

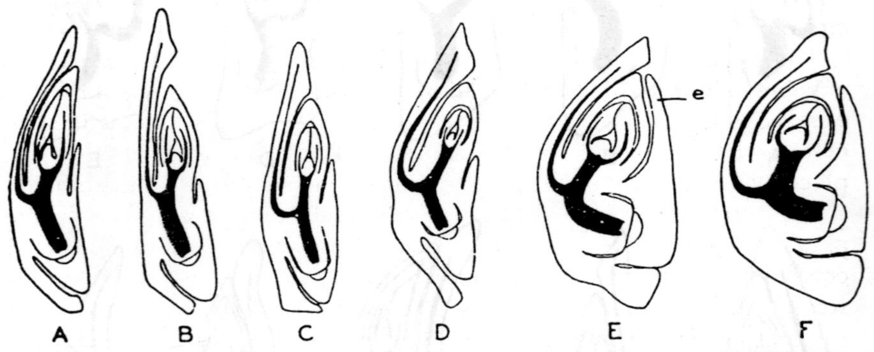

Fig. 124. Median longitudinal sections of embryos of grass genera in the Chlorideae, showing variety of form within one tribe of the Gramineae. Epiblast (*e*) strongly developed in *E* and *F*. *A*, *Chloris verticillata*; *B*, *Schedonnardus paniculatus*; *C*, *Buchloë dactyloides*; *D*, *Leptochloa scabra*; *E*, *Dactyloctenium aegypticum*; *F*, *Eleusine indica*. (*After Reeder.*)

are sometimes described as having no hypocotyl, an inaccurate statement. If both stem and root are present in the seedling, a transition region must exist, even though it is platelike, a millimeter or less long—a region perhaps detectable only with microtome sections. The hypocotyl in monocotyledons is perhaps typically only a few millimeters in length, but is long in some families, especially the Liliaceae and Amaryllidaceae.

The term *transition region* is applied to the part of the hypocotyl in which the changes in vascular tissue occur; the hypocotyl may have root structure below and stem structure above this region. The transition region may be *low,* that is, short; or *high,* that is, elongate, even sometimes several centimeters long. A high transition region usually accompanies elongate hypocotyls; a low transition region is usually present in massive hypocotyls.

In greatly reduced embryos, without clear distinction of parts, as in the orchids and some parasitic plants, the hypocotyl is not apparent. In many aquatic taxa and in some biennials, the hypocotyl, with adjacent parts of root and stem, serves as a storage organ and is greatly distorted. This distortion occurs, in some taxa, within the seed—Potamogetonaceae, Verbenaceae, Cactaceae, Myrsinaceae, Rhizophoraceae—and, in other taxa, after germination—some Cruciferae, *Beta, Raphanus, Brassica,* Umbelliferae, Amaranthaceae. Hypocotyls, similarly distorted, are common in taxa with cormous habit. In seeds in which the hypocotyl is greatly enlarged, the cotyledons are small. An enlarged hypocotyl, serving as a storage organ, characterizes the seeds of many aquatic genera, which have little or no endosperm and a reduced embryo.

The hypocotyl of most monocotyledons, together with the plumule, is commonly described as lateral and oblique. But there are some taxa, especially the more primitive, which have the hypocotyl median and erect and the plumule clearly terminal. (Too much emphasis has been placed on the grass embryo as typical of monocotyledonous embryos.)

The term hypocotyl has been more loosely used than any other applied to the embryo. It has been defined as "all the embryo below the cotyledons," "the primitive stem and root," "the nascent axis," "the first stem," "the axis of the embryo," "the caulicle," "the upper part of the caulicle." Confusion in the application of the term hypocotyl is, in part, related to the interpretation of the embryonic root, the radicle.

In vascular structure, tetrarchy seems to be the basic type; triarchy and diarchy, derivative types. Tetrarchy characterizes massive seedlings, and these are associated with large seeds and the tree habit. Since the tree habit is undoubtedly primitive, the tetrarch habit seems primitive in angiosperms. But vascular form in hypocotyls varies greatly; a type may be "stable," constant throughout a taxon, or "unstable," varying even in individuals of a species. Any type may be unstable; the derived types are especially so.

The Epicotyl. The part of the stem that develops above the cotyledonary node—the level of attachment of the one or two cotyledons—is the *epicotyl.* This part of the axis—with true stem structure—is developed by the axial growing point at the top of the hypocotyl; it is present in the plumule in meristematic stage. It is fairly well developed in the seeds of some dicotyledons, as in some Leguminosae, Fagaceae, Cruciferae, but is rarely present in monocotyledons until after germination. It is better developed in taxa with epigeal than in those with hypogeal germination, where the early stages of the seedling develop below ground.

The Plumule. The plumule is the bud of the embryo, the meristematic base of the epicotyl with leaf primordia. The sheathing coleoptile is often included as a part of the plumule, but this is inaccurate, because the coleoptile is a part of the cotyledon. Within the seed, the plumule is further developed in hypogeal than in epigeal taxa; the first foliage leaf also develops earlier in the hypogeal type. The plumule is prominent in most highly specialized embryos, as in grasses.

The Radicle. Compared with the other organs of the embryo, the radicle has received rather little attention. In the embryo within the seed, it usually is represented by a primordium only; a true root is formed only at germination. The histological origin of the root is difficult to determine; the primordium arises late in the development of the embryo and is clearly endogenous in many taxa. The relation of the root primordium to the differentiating axis of the embryo is obscure. The primary root, formed by the endogenous primordium, often appears somewhat lateral, but it is generally considered to form one end of the embryonic axis, because it appears to be attached directly to the end of the hypocotyl. Yet the view has sometimes been held that the root is, like the cotyledon, a lateral structure; that the primitive axis was rootless. The lateness of appearance of the root and its lateral position, as well as its failure to function in some genera—*Nelumbo*—have been believed to support its secondary nature. Root origin needs much critical study; the nature of the root may be important in the search for the ancestry of the angiosperms.

Characteristically, in monocotyledons, a radicle is not present before germination. At germination, the radicle may develop rapidly, slowly, or be greatly delayed. In some taxa, it is filamentous and ephemeral; in others, it persists as a taproot—palms, dracaenas, yuccas—and, rarely, it is not formed. In the higher monocotyledons, the primary root is late in development and seems to be a disappearing part of the embryo. Lateral, adventitious roots are formed where the primary root is weak or lacking, as in some orchids and some aquatic and parasitic families. It is difficult to distinguish the presence or absence of a root in the embryo before germination. It is frequently stated that, in germination, the root elongates and breaks through the seed coat, but, in some taxa, rupture of the seed coat results from elongation of the hypocotyl and of the cotyledonary sheath, rather than of the radicle. Some descriptions of embryo structure emphasize the radicle as an important part, inaccurately interpreting the axis as consisting of plumule at one end and radicle at the other, and omitting mention of the hypocotyl.

The Coleorhiza. In the highly specialized embryos of the monocotyledons, a sheath of tissue surrounds the base of the root (Fig. 130). This sheath has been interpreted in several ways—as a part of the scutellum, of the axis, and of the epiblast. It has, however, commonly been considered of no morphological value, because it is merely a part of the massive late stage of the proembryo, pushed aside by the endogenous root tip as it develops. The presence of similar, smaller sheaths about the bases of adventitious roots in young seedlings supports the view that it is not morphologically significant. A coleorhiza is absent in the simple, primitive monocotyledonous embryo.

The Cotyledons. The cotyledons have been much discussed, since the earliest years of descriptive botany—they were described by Malpighi in the seventeenth century. In the earlier years, interest centered in their morphological nature; in the twentieth century, the relationship of the two of the dicotyledons to the one of the monocotyledons has received most attention. Cotyledons were early called seedling leaves, storage leaves, organs *sui generis,* and haustoria "reminiscent of the suctorial 'foot' of pteridophytes." In the form of a scutellum, the cotyledon has been called merely a lobe or appendage of the radicle, or of the axis. The consistently paired arrangement of the cotyledons—in most taxa, with neither one apparently belonging in the phyllotactic spiral of the leaves—has been used to support the *sui generis* theory. But, in some monocotyledons, the cotyledon seems to fit into the spiral system of appendages.

Cotyledons are closely leaflike in relation to the axis—in vascular structure, in ontogeny, and in detailed histological structure when they become photosynthetic organs. Leaflike cotyledons that serve only temporarily as storage organs are doubtless the primitive type. Within the angiosperms, there has been added to their basic function of food storage the capacity to absorb the food from the endosperm and, with the endosperm as an intermediary, from the perisperm also. All stages in the development of this function and in the structural modification of the cotyledon accompanying this change are found. Transformation of the tip of the cotyledon into a temporary absorbing structure, held within the seed coats and later freed from the seed, is a first step toward the transformation of part or all of the cotyledon into a suctorial structure, permanently retained in the seed. Elaboration in form of the cotyledon is extreme in taxa where there is a highly specialized suctorial "organ"—the *scutellum.* This modification is an outstanding character of the higher monocotyledons. (The term scutellum is commonly restricted to the embryos of grasses and sedges, but is separable from the retained absorbing tips of cotyledons of other taxa only in degree of specialization.)

Apparent dichotomy of the cotyledons—bilobing in some taxa, and a "paired" vascular system in many—led to the theory that cotyledons are primitive organs retained in the embryo, not appendages of leaf rank. But bilobed cotyledons occur in taxa scattered throughout the angiosperms. Unequal bilobing of a remarkable type is present in the highly specialized embryos of grasses and a few other families.

The so-called terminal position of the cotyledon of the monocotyledons (Fig. 122) has been used to support the *sui generis* theory of cotyledon nature, on the basis that a terminal cotyledon cannot be interpreted as the morphological equivalent of a leaf, which is a lateral appendage, and, lacking stem structure, it cannot be a part of the stem. But both mono- and dicotyledonous taxa that have aberrant

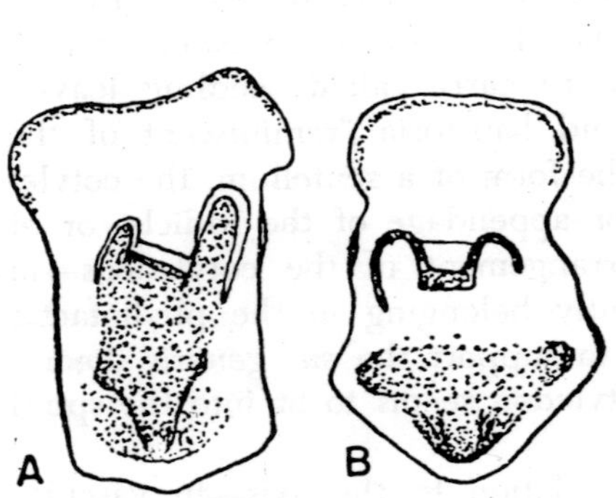

Fig. 125. Longitudinal sections of seeds showing monocotyledon embryos with two cotyledons and terminal stem apex. *A*, *Commelina karwinskyi; B*, *Tinantia erecta.* (*After Solms-Laubach.*)

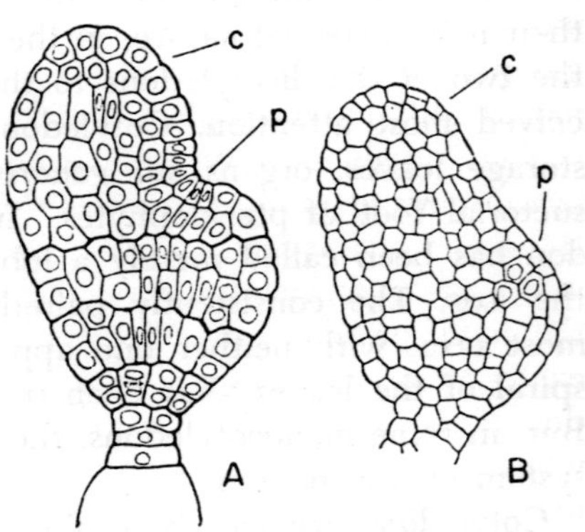

Fig. 126. Longitudinal sections of developing embryos. *A*, a monocotyledon, *Ottelia alismoides*, and *B*, a monocotyledonous dicotyledon, *Claytonia virginica*, showing similarity when terminal position of plumule growing point *p* becomes pseudolateral by dominant development of the pseudoterminal cotyledon *c*. (*After Haccius.*)

embryo structure demonstrate that all cotyledons are, morphologically, lateral organs. In the embryos of some monocotyledonous families—Commelinaceae, Dioscoreaceae—the stem apex is definitely terminal in position and in ontogeny (Fig. 125). It is apparent that the single cotyledon, where dominant in the embryo, has crowded the axis and its growing point to a lateral position (Fig. 126).

The interpretation of the angiosperm cotyledon as basically "peltate" is an extension of the similar interpretation of floral organs as peltate. In dicotyledons, the cotyledonary petiole, like the leaf petiole, is frequently grooved, sometimes even tubular, and simulates ascidiform structure. A pair of cotyledons, united by their petioles, may also suggest "peltate" form, but the only cotyledons that show evidence of true peltate form are those of taxa where the leaves are strongly peltate, as in *Peperomia;* the cotyledon, like the leaf, is not fundamentally peltate. The true peltate leaf is a specialized form, developed by the union of basal lobes of the lamina.

The cotyledon is a simple organ in most of the dicotyledons (Fig. 127) and in the more primitive monocotyledons (Fig. 135). In the more advanced monocotyledons where, in specialization, it has assumed the functions of absorption of stored food (the scutellum) and of specialized protection for the plumule (the coleoptile), it is a complex organ, no longer recognizable as a simple lateral appendage. In its most extreme form—transverse division into two segregated parts, serving different functions—it is one of the most highly modified plant organs (Fig. 130).

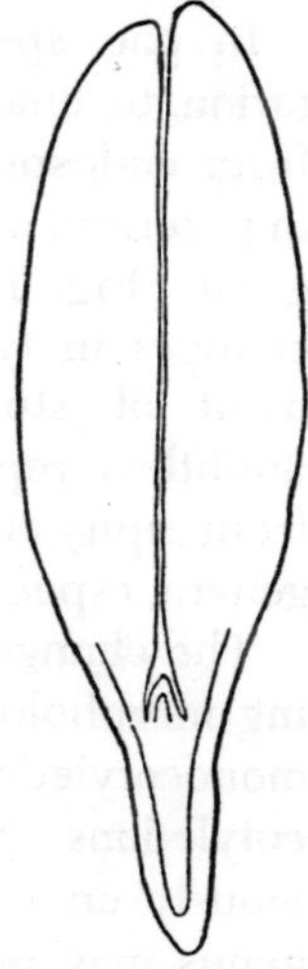

Fig. 127. Diagram of longitudinal section of mature dicotyledonous embryo, *Pyrus Malus,* showing cotyledons, plumule, hypocotyl, and root apex.

The cotyledon is typically simple in shape, laminar, conforming in most exalbuminous seeds more or less to the shape of the seed. It may strongly resemble the leaf and, after germination, function similarly, as in many of the more primitive monocotyledons. Where the seed is exalbuminous, or has very little endosperm, the cotyledon usually has much the form of the normal leaf and, when developed after germination, closely resembles the leaf—Alismataceae, Butomaceae, *Triglochin, Philydrum, Orontium, Allium, Trillium, Tofieldia, Narthecium.* The cotyledon may persist as a photosynthetic organ throughout the first growing season; in some monocotyledons, the cotyledon, or part of it, serves as the only assimilatory organ for the first year—*Erythronium* and other liliaceous genera. When not leaflike at germination, it may enlarge greatly and take on leaf structure and function, as in many dicotyledons and some of the more primitive monocotyledons—Liliaceae (especially the Scilleae), Araceae. Rarely, it is leaflike in form in the seed, as in some Dioscoreaceae, Euphorbiaceae.

Where serving primarily as storage organs, cotyledons are thick and often distorted in shape. They are bilobed, even divided, in some dicotyledonous families—Cruciferae, Tiliaceae, Juglandaceae, some of the Amentiferae. The cotyledon of the monocotyledons is typically elongate, with a sheathing base which is closed for some distance above the base. (The plumule is enclosed by this sheathing base, with, often, only a small, slitlike opening above its tip. The developing plumule may extend through this opening or break through the sheath, as in some of the palms and lilies.) In the dicotyledons, the bases of the cotyledons frequently fuse, forming a shallow cup or even a tube—*Podophyllum,*

Sanguisorba. This tube may become much elongated and petiolelike—*Thea, Quercus*—superficially resembling the "neck" of the single cotyledon in many monocotyledons.

The form of the cotyledon, especially the position of the scutellum, varies greatly, even within a family—Liliaceae, Palmae, Gramineae (Fig. 123).

In the specialization of the cotyledon, its modification in adaptation to change of function, from food storage to absorption of food from endosperm and perisperm, and to protection of the plumule during germination (coleoptile) has been varied and, structurally, very great (Fig. 130). The modifications are related largely to evolutionary changes in type of germination associated primarily with the development of storage endosperm and perisperm. Hypogeal germination doubtless represents an advance over epigeal germination. Transitions from epigeal to hypogeal germination are well shown by liliaceous genera, especially the Scilleae and *Lilium*, where both types occur.

The change from epigeal to hypogeal germination and its accompanying morphological modifications have occurred in both dicotyledons and monocotyledons but are more general and much greater in the monocotyledons. No line exists between the types, and the change is obviously an easy step, which has been taken many times. A family or genus may possess only one type, or both types, with transitional forms —*Lilium, Sophora, Erythrina, Phaseolus*. Probably in only a few dicotyledons has the absorbing function developed in the cotyledon. In *Peperomia*, the entire lamina of one cotyledon has become suctorial (Fig. 136); in the monocotyledons, a distal part or all of the lamina is transformed.

The structural changes—step by step—in the elaboration of the suctorial habit are well illustrated in the monocotyledons; the tip of the cotyledon is first modified, then the entire lamina is transformed into a specialized absorbing structure, ultimately retained permanently in the seed coats. The absorbing part of the cotyledon, in its most specialized form, is the *scutellum*. The change related to the protection of the embryonic bud is the transformation of a median part of the cotyledon into a caplike cover for the plumule, the *coleoptile* (Fig. 130). The distal and proximal parts of the cotyledon are separated by a long, slender median part, the "neck," which is prominent in many monocotyledons. This neck ("*Zwischenstück*" of some early authors) was generally recognized in the nineteenth century as a part of the cotyledon, connecting the two specialized ends. (The extended neck of the cotyledon is clearly correlated with hypogeal germination.)

In the highest grasses, the top of the neck becomes adnate to the hypocotyl and is no longer *externally* evident, except in those taxa in which it forms a ridge on the embryonic axis (Fig. 134*G*, *H*). The compound structure, so formed, is the *mesocotyl* (Fig. 134*A*, *B*). The externally complete separation of the two parts of the cotyledon, base and tip, is an extraordinary modification of an organ, the more so because the tip of the cotyledon, the scutellum, is lateral at or near the base of the axis, and the sheathing base of the cotyledon, the *coleoptile*, above it, covers the apex of the axis (Fig. 130). Only some of the highest grasses and sedges and some of the Bromeliaceae are known to have this extreme modification of the cotyledon. (With the great reduction of the flower and inflorescence and the anemophilous pollination of the grasses, this modification is substantiating evidence of the high specialization of the Gramineae.)

The term scutellum is commonly used to designate the highly specialized part of the cotyledon of many monocotyledons, the distal part, which is structurally and functionally modified as a structure that absorbs food materials stored in the seed outside the embryo (Figs. 129 *C–H* and 130). All stages in its evolutionary development are found, but the term is usually restricted to its most highly specialized form, an absorbing part, which, especially in the grasses, is more or less shield-shaped. There are many transitional forms between the earliest stages of development of an absorbing part and its highly elaborated form. Cotyledons, as well-defined absorbing organs, are present in many monocotyledonous families, but in only few dicotyledons. The term scutellum is sometimes defined as the absorbing cotyledon of the grasses. This is an inaccurate definition, because the scutellum is only a part of the cotyledon and is not morphologically distinct from the absorbing structures of other families from which it differs only in shape and size. If the term is to be restricted on the basis of form and prominence in the embryo, it should be applied also to similar structures in some of the sedges and, perhaps, of other families.

The Liliaceae show early stages in the transformation of the cotyledon tip into a suctorial structure. In *Paris,* the cotyledon apparently has no suctorial function; the entire cotyledon becomes photosynthetic (Fig. 128*E*). In the related *Trillium* (Fig. 128*G*), the tip of the cotyledon remains within the seed at germination, functioning as a suctorial structure; the rest of the cotyledon is withdrawn from the seed and becomes leaflike. As the cotyledonary petiole elongates and becomes erect, the suctorial tip is withdrawn but withers without becoming green. In *Erythronium* (Fig. 128*F*), the mesophyll cells of the cotyledon tip differ in form and arrangement from those lower down and lack chlorophyll, but the tip does not wither after withdrawal from the seed. Similar early stages in modification of the cotyledon tip are present in other primitive monocotyledonous families. In *Triglochin,* the tip is slightly enlarged, starchless, and retained in the seed but hardly differs histologically from the starch-bearing lower part of the cotyledon; in *Philydrum* and *Pontederia,* absence of starch distinguishes a weak, short-lived, suctorial tip. In the Juncaceae, only the tip is suctorial, but it is retained within the seed coats, not withdrawn, as in *Trillium.* The Bromeliaceae also show various stages in this modification in form and function; *Tillandsia* has no suctorial tip; in *Pitcairnia,* the foliaceous cotyledon has a short but thick absorbing tip. *Puya* and *Dyckia* have a thick cotyledon, which largely fills the seed—there is little endosperm—and remains within the seed coat for weeks, then emerges and becomes a photosynthetic organ.

Where the cotyledon tip is retained permanently within the seed coats, the epidermal cells that are in contact with the endosperm are palisadelike. The palisade layer may have a smooth or a papillose surface. The absorption of food from the perisperm is also by these specialized epidermal cells, apparently not directly, but through the partially collapsed endosperm cells as intermediaries.

The absorbing tip of the cotyledon varies greatly in shape. It is commonly cylindrical, club-shaped, or filamentous, but may be laminar or peltate; when very small, it is usually spherical. At germination, the scutellum may enlarge greatly, pushing deeply into the endosperm, as in the palms, sedges, Commelinaceae, Musaceae; its form is controlled by position of the endosperm and shape of the seed. In some primitive monocotyledons, especially the Helobiales, there is no storage endosperm and the cotyledon has no absorbing tip. In some highly specialized embryos also—those of orchids and some reduced aquatic taxa—there is no endosperm and no absorbing tip.

The elaboration of a simple photosynthetic cotyledon, such as that of *Paris,* to form a high type of suctorial structure, by transformation of the blade into an absorbing structure permanently retained within the seed coats, is only the first and simpler stage in the evolutionary history of the suctorial cotyledon. The division of the cotyledon into parts, with the distal part suctorial, a scutellum, and the proximal part elaborated as a protective structure for the plumule, a coleoptile, accompanied by adnation to other parts of the embryo, makes the story of change in form, function, and anatomy highly complex. Only comparison of stages—readily found in different major taxa, with attention to anatomy—makes the morphology of the highest types clear.

Critical studies of the mature embryo, of germination, and of seedling structure in the more primitive Liliales should be made to determine the phylogenetic development of the absorbing cotyledon. The Helobiales, which may also be important in these studies, are perhaps too greatly reduced by their aquatic and semiaquatic adaptations to give evidence along this line. There has clearly been parallel development of the cotyledon along several or many lines to form an absorbing structure.

The histological structure of the absorbing parts of the cotyledon among major angiosperm taxa varies greatly; specialization of this part of the cotyledon has followed more than one line. In dicotyledons that have endosperm, the cotyledons, with rare exceptions, remain only temporarily within the seed coats after germination. After the endosperm is absorbed, the cotyledons are freed from the seed coats and commonly become more or less leaflike, assimilating organs. Histologically, also, they resemble leaves, and their epidermal cells are, in form, like those of most dicotyledons.

In the monocotyledons, especially in those with hypogeal germination, histological modification of the absorbing tip is much greater than in dicotyledons. Modification consists primarily of change of form in the epidermal cells; the cells are elongated perpendicular to the surface of the cotyledon and, in some taxa, become papillose and form a distinct absorbing epithelium. Types of epidermal modification characterize major taxa. In the Liliales and Juncaginaceae, the epidermal cells are only slightly or not at all modified. In the grasses, elongation of the epidermal cells is great—up to several or many times their transverse diameter—and the tips of the cells may penetrate, brushlike, into the endosperm. Intermediate types are common in other families.

Other modifications related to absorption involve shape of the cotyledon tip itself (Fig. 128). The modified tip (scutellum) may have the general form of the seed itself, as in the grasses (greatly elongated in *Zizania*, for example), and may retain that form after germination, as in the Commelinaceae. The absorbing tip in many taxa is roughly spherical but, with germination, increases greatly in size, and changes in shape, filling the seed coats, as in the palms, where it may even become ruminate, following the convolutions of the endosperm. Club-shaped, cylindrical, and even filiform tips are frequent. Tips of this general form may become completely haustorial (Fig. 128*D, H*); all cells may be elongated parallel with the long axis, as in the Cyperaceae. In *Juncus*, the tip is pyriform; its large, little-elongated, epidermal cells resemble the lilialean type. (It is noteworthy that the cotyledon tips of the sedges differ greatly in type from those of the grasses, and that the tip in *Juncus* is not closely like either of these but resembles that of the lilies. Cotyledon structure fails to support the view that the Juncaceae are intermediate between lilialean stock and the grasses and sedges, or the view that grasses and sedges are closely related.)

In the dicotyledons, the cotyledon serves—in addition to storage—for absorption of food and often later for assimilation; in the lower taxa

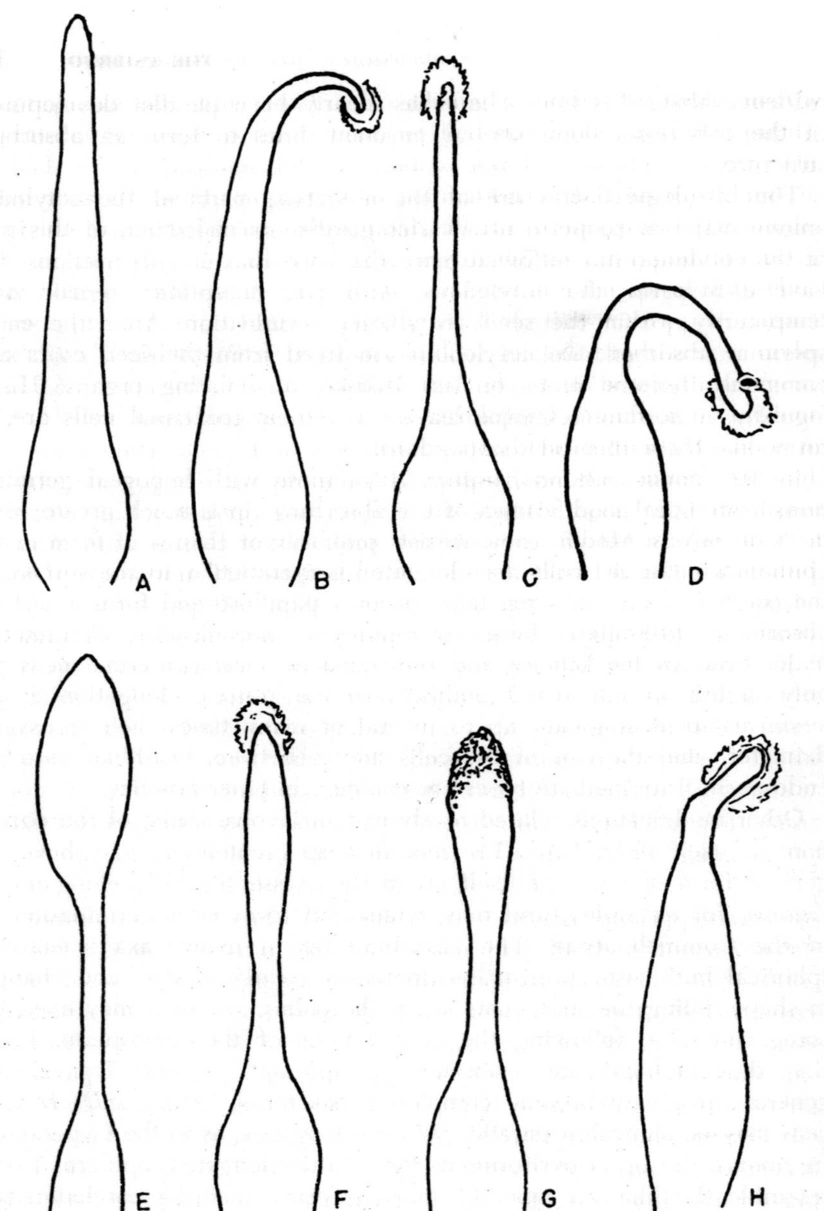

Fig. 128. Diagrams of primitive cotyledon types showing elongated "neck" and stages in retention of tip within the seed coats. *A* to *D*, without blade, showing stages in development of a suctorial tip: *A*, *Alisma* type; *B*, *Allium* type, tip not withered, temporary retention of seed coat; *C*, *Bowiea* and *Agave* type, tip withered, retention or not; *D*, *Narcissus* type, permanent retention. *E* to *H*, cotyledon with blade; *E*, *Paris* type, no retention; *F*, *Erythronium* type, temporary retention; *G*, *Trillium* and *Costa* type, temporary retention, tip strongly modified, collapsing after freed from seed coat; *H*, *Arum* type, permanent retention of seed coat, blade greatly reduced, permanent suctorial structure, a scutellum. (*Adapted from Boyd and Schlickum.*)

of the monocotyledons, it serves for absorption and assimilation; in the higher taxa, for absorption and—as the coleoptile—for the protection of the plumule. The elaboration of the cotyledon in histological structure, as well as in complexity of form—both adaptations primarily to hypogeal germination—supports the view that the monocotyledons are, all in all, more highly specialized than the dicotyledons.

The change in function of the tip of the cotyledon and the retention of this part within the seed (Fig. 128) is accompanied by increased development of the plumule within the seed; where there is no

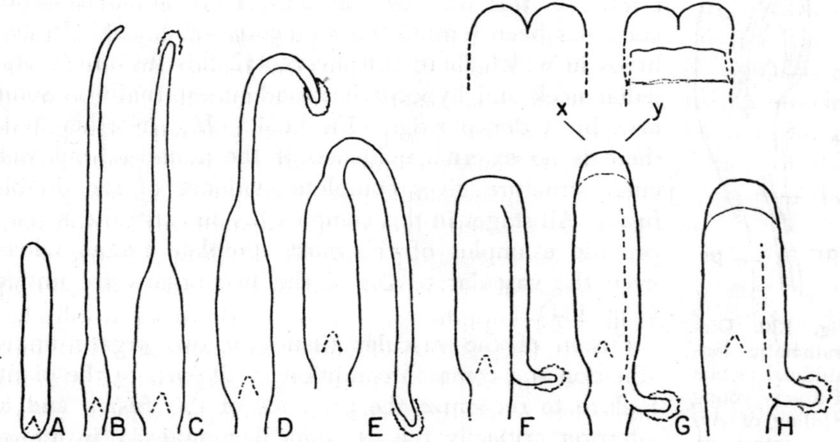

Fig. 129. Diagrams of cotyledons of monocotyledons showing evolutionary changes accompanying the change from epigeal to hypogeal germination. *A*, without a neck; *B*, *C*, elongated neck, seed coat retained in *C*; *D* to *H*, successive stages in the downbending and connation of the cotyledonary neck with the sheathing base: *G*, *H*, adnation of neck to hypocotyl forming a mesocotyl. *x*, *y*, two forms (in face view, split open) of prolonged shoulders of sheath margins at point of downbending, shown in *G*, and *H*, in side view; *x*, stipular, and *y*, stipular with addition of ligule.

photosynthetic activity in the cotyledon, early growth of the embryo is dependent upon the first foliage leaves. The cotyledon, not functioning in photosynthesis, tends to retain its tip in the ground; its base, sheathing the plumule, elongates as growth continues, the median, connecting part forms a slender uparched connecting "neck" (Figs. 128 and 129). Morphologically, the neck plays an important part in the structural changes involved in the formation of the most specialized embryos. The neck has the form of an inverted U or V, with the absorbing tip at the end of one arm and the sheathing base at the other (Fig. 129*E*). The distal arm is parallel with the sheathing base and, in evolutionary specialization, comes to be appressed to it and, ultimately, to fuse with it

(Fig. 129*F* to *H*). The cotyledon is doubled back upon itself, and its proximal and distal segments—except for the absorbing tip—may become laterally connate through part or most of their extent. Lateral fusion of the two arms of the neck may extend any distance, from the point of bending downward even below the base of the cotyledon to and along the hypocotyl (Fig. 129*G*, *H*). Complex structure is formed where fusion is extensive —connation between separate parts of an organ, the cotyledon, together with adnation of these connate parts to another organ, the hypocotyl. The compound structure formed in this way by adnation of cotyledon to hypocotyl has been termed the *mesocotyl* (Fig. 130). In embryos in which there is a mesocotyl, this fusion of cotyledon neck and hypocotyl is evident externally in some taxa by a dorsal ridge (Fig. 134*G*, *H*); in other taxa, there is no external evidence of the adnation, but vascular structure gives complete evidence of the double fusion. All stages in this complex fusion exist; the grasses provide examples of the most complete union, where even the vascular tissues of the two organs are united (Fig. 134).

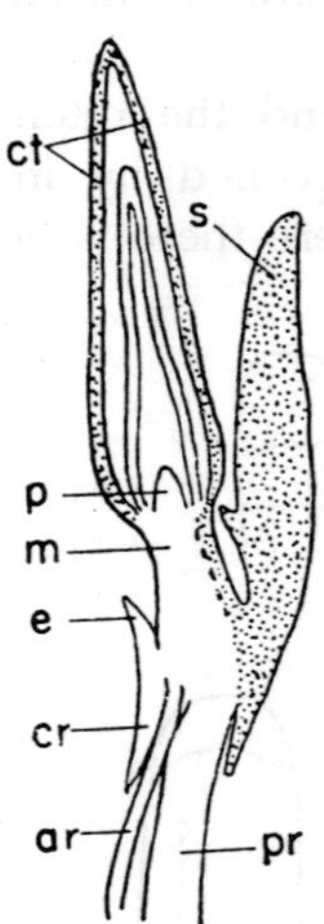

Fig. 130. Diagrammatic longitudinal section of a very young seedling of *Avena sativa* to show morphology of grass embryo, especially the several parts of the cotyledon, shaded. *ar*, adventitious root; *cr*, coleorhiza; *ct*, coleoptile; *e*, epiblast; *m*, mesocotyl; *p*, plumule; *pr*, primary root; *s*, scutellum. (*After Sargant and Arber.*)

Fusion of the vascular tissues of two organs under adnation is a common condition in all parts of the plant. Failure to recognize the presence of this fusion and to interpret critically the anatomical complexity in monocotyledonous embryos has been largely responsible for varied and often extraordinary interpretations of the embryo. If a series of embryos from more primitive monocotyledons—the Liliales, Helobiales—to the advanced Palmae and Gramineae is considered comparatively, the story of evolutionary change is evident, and the terminal form, that of *Zea* and other grasses, can be correctly interpreted. Much of the evidence for this interpretation of the highest forms of monocotyledonous embryo was available as early as the 1870s, and, at that time, the statement was made that "in the Zingiberaceae and palms, there can be no question but that the coleoptile, cotyledonary sheath, and scutellum represent the cotyledon or part of it." The whole story became clear in 1915, but various later interpretations, overlooking the evidence provided by the vascular skeleton and the existence of adnation, have obscured the true morphological make-up of the grass embryo. The sheathing base and scutellum (connected by the neck) have sometimes been called "lobes" and likened to the two lobes of the cotyledon in the dicotyledons, but division in the dicotyledons is into lateral, not basal and distal, parts. In the monocotyledons, the suctorial part is the distal part; the "neck" consists of the median part, often with lateral, stipulelike margins extending upward above the point of downbending (Fig. 129*G*, *x*). The upward-extended stipular margins have been considered to form a ligule like that of the grass leaf, and the sheathing base of the cotyledon has been commonly called a "ligular sheath." There has been much confusion in the terminology and descriptions of these prolongations. The terms stipular and ligular have both been used loosely for the distal parts of the sheath, which, as prolongations of the margins of the sheath, are surely stipular. The term ligular should be used only for the collarlike, transverse expansion on the ventral side of the cotyledon neck at the point of downbending (Fig. 129*G*, *y*); this structure is the morphological equivalent of the ligule of the grass leaf. That the coleoptile, commonly, is a stipular structure in part has not been generally recognized. The apparent structural separation of the coleoptile from the scutellum (Fig. 130)—often considered the cotyledon—has seemed to support the view that the coleoptile is the first leaf. The frequently bilobed tip of the sheath and its paired veins, in many taxa, are strong support for its two-stipule nature. (The bilobed form persists in coleoptiles with marked "shoulders"—*Bromus, Ammophila, Festuca,* some forms of *Zea.*)

There is great variety in form of the stipular sheath (Figs. 131 and 132). Its margins vary in extent of fusion; the sheath may be tubular to the top or ventrally open for short or long distances. Distal "appendages"—free tips of the stipules—may be prominent, lanceolate, auriculate, or absent. The sheath, as a whole, is commonly cylindrical but may be ovoid, or subspherical. Where there is lateral adnation to the neck, its form is related, in part, to the position of the scutellum and to the degree of completeness of the fusion (Fig. 131*B*, *F*, *H*).

Stipular sheath is a good term, morphologically suggestive of the nature of this structure. The stipules of the typical monocotyledon leaf are thin marginal wings, often with free distal parts; the cotyledon shows similar stipules in some taxa with lanceolate or auriculate tips. The point of dorsal downbending of the cotyledonary neck is the base of the free stipule tips and the tips remain erect (Fig. 129*G*, *H*). In this position, they shelter the plumule and, in further specialization of the cotyledon, fuse by their margins and form a protective cap over the plumule. This cap, together with adjacent parts of the sheath—the apex of the downbent cotyledon—forms the coleoptile (Fig. 130, *ct*). Morphologically, the coleoptile is a median part of the cotyledon, the "elbow" section of the bent "neck," together with the distal parts of the stipules.

A median part of the cotyledon, with its appendages, has become distal in the germinating embryo. In this high specialization, the sheath has become elongated—even an intercalary meristem has been reported for some taxa—and the stipular cap elaborated in form and tissue structure as a protective, "soil-penetrating" structure. (Protection of buds by stipules is well known in dicotyledons also, as in *Liriodendron.*)

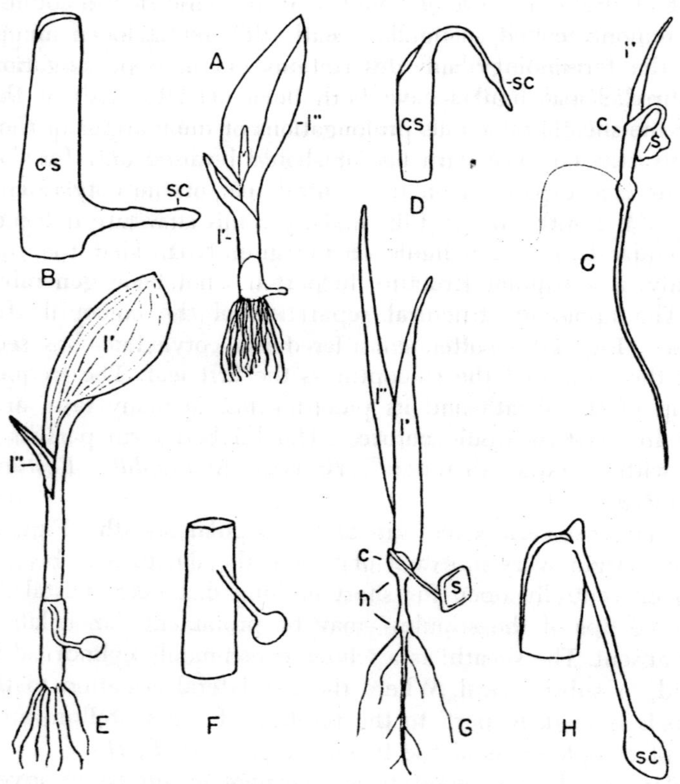

Fig. 131. Sketches of seedlings and cotyledons, showing neck of cotyledon free in *C, D,* and connate with cotyledonary sheath for various distances in the others. *A, B, Canna indica; C, D, Asphodelus luteus; E, F, Commelina coelestis; G, H, Iris pseudacorus. cs,* cotyledonary sheath; *h,* hypocotyl; *l', l'',* first two leaves; *s,* seed coat; *sc,* scutellum. (*After Schlickum.*)

The terms ligule and ligular sheath have been rather loosely used in descriptions of monocotyledonous embryos. Though the ligule of the embryo is undoubtedly homologous with the ligule of monocotyledon leaves, these terms obscure the underlying stipular nature of the structure. ("Ligules" are probably not always stipular in nature.) The term stipular—morphologically a better one—should supplant ligular, although ligular has been long established.

Division of the cotyledon into proximal and distal parts—parts that differ in function and are even separated in cotyledons by fusion of the middle section to the hypocotyl—is a structural result of the retention of the cotyledon tip within the seed coats. Retention of the seed coats in the ground involves early withdrawal of the plumule from the seed coats and the penetration of the soil by the plumule tip, as the plumule

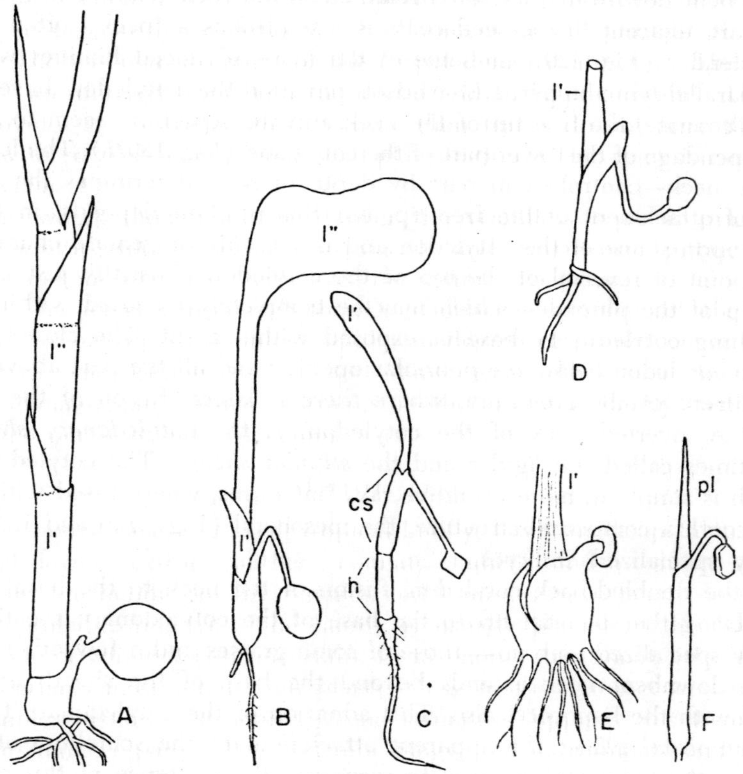

Fig. 132. Sketches of seedlings showing variety of cotyledon form, neck not connate to cotyledonary sheath, slight development of shoulders in some. *A, Dracaena draco; B, Arum italicum; C, Hyacinthus candicans; D, Iris pseudacorus. cs,* cotyledonary sheath; *h,* hypocotyl; *l′, l″,* first and second leaves; *pl,* plumule. (*A to D, after Lewin; E, Tschirch; F, Boyd.*)

elongates. The sheathing base of the cotyledon, which, in the more primitive epigeal taxa, surrounds and protects the plumule, becomes a more complete and effective protecting shield, the coleoptile.

The nature of the coleoptile (piléole of some older textbooks) has long been a highly controversial subject. It has been interpreted as the first true leaf, the ligule of the cotyledon, a "plumular scale," the sheathing base of the cotyledon, a ligular sheath, a stipular sheath made up of the united stipules of the cotyledon and stipules of the first leaf. Comparison of embryos throughout monocotyledons, together with their anatomy (Fig. 134), especially that of the various types among grasses, shows that the coleoptile is a part of the cotyledon. With the specialization of the tip of the cotyledon as an absorbing organ and the adoption of hypogeal germination, the cotyledon became bent downward. Its tip is retained in the seed below the ground; the part adjacent to the seed coats is extended as a "neck," often long and slender (Fig. 129). In some of the more advanced families where the parallel-lying upper and median parts of the cotyledon have become connate, the free tip of the neck and the scutellum seem to form an appendage of the lower part of the cotyledon (Fig. 129*H*). The length of the neck—controlled in part by depth of soil—determines the position of attachment of the free tip, but this position depends, in large part, upon shape of the cotyledon and is a family or generic character. The point of reversal of the top of the cotyledon is usually just above the tip of the plumule—which may be completely enclosed within the sheathing cotyledonary base or exposed within a slit. The closed part of the cotyledon below the plumular opening and all the part above the attachment of the scutellum, where there is lateral fusion of the base and the inverted neck of the cotyledon, is the *cotyledonary sheath,* sometimes called the *ligular* and the *stipular sheath.* The cotyledonary sheath is simple in more primitive taxa but forms, where fused with the hypocotyl, a compound structure, the mesocotyl (Figs. 130 and 134) in highly specialized embryos.

In the doubled-back cotyledon, fusion of the neck to the basal part may bring the suctorial tip to the base of the cotyledonary sheath; in highly specialized embryos—those of some grasses and a few other taxa—the downbent neck extends beyond the base of the sheath and is fused with the hypocotyl also. This adnation of the cotyledonary tip to the hypocotyl places the apparent attachment of the scutellum at the base of the hypocotyl. Failure to recognize the existence of this adnation and the secondary position of the scutellum has been the cause of much of the difficulty in the interpretation of grass embryos.

Stages in the connation of the neck and the sheathing base are common, and fusion of the neck with the hypocotyl is evident in the presence of a ridge along the side of the hypocotyl (Fig. 134*G, H*). In more advanced taxa, external evidence of the fusion has disappeared, but downward-extending, *inverted* vascular bundles of the connate neck are present in the outer tissues of the lower sheath and in the cortex of the hypocotyl. Anatomy (Fig. 134) gives internal proof of the adnation in at least three unrelated families; the fusion must have occurred independently more than once. In most taxa that have mesocotyls, there is little or no external evidence of this adnation; only critical comparison of vascular structure demonstrates the fusion—*Zea, Sorghum, Coix.* The body of the embryo, though apparently simple, is, morphologically, more than an axis; in the genera that have mesocotyls, it is partly cotyledon.

The coleoptile was probably first considered cotyledonary in 1849, when it was noted that it differed from the plumular leaves in having two vascular bundles, rather than the odd number characteristic of the leaves. This interpretation was general in the middle of the nineteenth century, when prominent botanists recognized that scutellum and coleoptile constituted the cotyledon. The scutellum was called the "terminal part" of the cotyledon, "retained in the seed as an organ of absorption," even though it was noted that, in some of the grasses, this organ is borne "*above* the first internode." The early recognition of the basic morphology of the cotyledon is an example of the keen insight in morphological interpretation of the botanists of the middle and later nineteenth century.

The Mesocotyl. The hypocotyl, with the downbent, adnate top of the cotyledon, forms the *mesocotyl* (Fig. 130). (See also description under coleoptile.) The axis of the embryo of some highly specialized grasses, sedges, the bromeliads is, except at its tips, a mesocotyl. It is especially long and well developed in *Leerzia, Oryza, Coix, Panicum, Carex, Zizania, Cyperus.* It has been suggested several times that the term mesocotyl be dropped as superfluous and meaningless. Under the interpretation of the coleoptile as the first leaf, and of the axis below it as the first internode of the stem—one interpretation of some grass embryos—the term would be superfluous, but the coleoptile is surely not the first leaf, and this structure is not a simple axis. A term is necessary for this compound structure, and, though "mesocotyl" may be morphologically inappropriate, because it is more than the middle part of the cotyledon, it has long been in use. The complex nature of the mesocotyl was recognized as early as 1872, when, on anatomical evidence, a part of the axis was shown to be neither root, stem, nor hypocotyl and was described as an "elongated node." In the 1890s, it was called by some authors the "*Zwischenstück,*" a part of the axis with an accessory vascular supply. Its true nature—the result of adnation of part of the cotyledon to the hypocotyl—was demonstrated by anatomical structure in 1915. More recent interpretations have not given critical attention to comparative anatomy and the results of fusion of adjacent vascular bundles; the critically important vascular structure of the mesocotyl has been overlooked or dismissed as meaningless.

A mesocotyl is characteristic of several genera of grasses, a few sedges—*Carex, Cyperus*—and some of the Bromeliaceae—*Aechmea, Billbergia.* In all these families, and perhaps in others, more examples will doubtless be found.

Difficulties of interpretation of the monocotyledonous embryo are indeed great, and confusion in description is not surprising. The extraordinary form of a cotyledon that is divided into two parts isolated in their attachment on the axis, with the distal part of the cotyledon attached below the proximal part, is, in itself, remarkable. And with this separation in position goes difference in form and function.

Failure to recognize the presence of adnation—regardless of the clear anatomical evidence (Fig. 134)—is largely responsible for the delay in recognition of the basic structure of these embryos. Yet textbooks of the nineteenth century, especially those published from 1860 to 1880, described the cotyledon of the monocotyledons as consisting of the coleoptile and the scutellum. This division of the cotyledon was demonstrated by comparative study of form and by anatomy in 1872, 1881, 1885, 1887, 1896, 1899, 1915, but has been otherwise interpreted since 1915. The earlier interpretation is the correct one.

Relation of the Cotyledon of the Monocotyledons to the Embryo Axis. Although the cotyledons are recognized as lateral appendages of the axis in the dicotyledons, the cotyledon of the monocotyledons is commonly called terminal. If the solitary cotyledon is, morphologically, the equivalent of one of the two in dicotyledons, its apparently terminal position must be secondary, and comparison of embryos throughout the monocotyledons shows this to be so. In some families—the Dioscoreaceae and Commelinaceae—and in scattered genera elsewhere, the stem apex is clearly terminal and the cotyledon lateral (Fig. 125). In most monocotyledons, the surviving cotyledon has become dominant in the embryo and crowds the axis to one side. In some taxa, the lateral position of the axis is assumed ontogenetically, but the pseudoterminal position of the cotyledon has become established phylogenetically in most taxa. The "terminal" position has been used as evidence that the cotyledon is not the morphological equivalent of the leaf, but other organs—carpels, stamens, and ovules—similarly appear to be terminal when not so. It is argued that, in differentiation from the proembryo, the cotyledon develops first and terminally, and the axis appears as a later, lateral structure. But in some genera—*Sparganium* and *Pistia*—the first plumular leaves also appear before the stem apex. This has led to morphologically remarkable interpretation that, in these genera, the first leaves are cauline, the later ones, lateral appendages—a morphological absurdity.

Relation of the Solitary Cotyledon to the Pair of Cotyledons. One theory of the relationship of the one- and the two-cotyledon embryos is that the single cotyledon represents the primitive condition and the two cotyledons have arisen by a splitting of the ancestral solitary one. The presence of two traces and of two, often opposed, vascular bundles in the lamina of the single cotyledon of *Anemorrhea* and other genera has been used to support this theory. But this anatomical structure is found also in many dicotyledons and is characteristic of cotyledons in other seed plants. On the existence of a paired vascular system in the cotyledon of many monocotyledons, the relationship of cotyledon types in the angiosperms has been read in two directions: the single cotyledon has been considered a double structure, the result of the fusion of two; the two cotyledons have been considered formed by the splitting of the single cotyledon. Here also, the prevalence of a double vascular system in cotyledons of both major groups of angiosperms and of some gymnosperms renders both theories untenable.

The origin of the solitary cotyledon by the suppression or loss of one of a pair is supported by much evidence. All stages of reduction in size and loss of function of one cotyledon are present in the dicotyledons. In many taxa, there is some difference in size in the cotyledons, and, in a few taxa, the differences reach an extreme, with one nearly or wholly lost—*Cyclamen, Ranunculus Ficaria,* species of *Corydalis, Claytonia, Peperomia* (Fig. 136). Stages in the specialization of the one-cotyledon embryo are also found in some monocotyledons, where differentiation in form and function occurs in a pair of cotyledons. The tip of one of the pair may develop as a suctorial organ—*Trillium, Arisaema, Arum.* A second cotyledon is present in vestigial form in primitive genera in the Dioscoreaceae and Commelinaceae—*Tamus, Dioscorea, Commelina, Tinantia* (Fig. 125).

Ontogeny likewise seems to show evidence that the solitary cotyledon has been derived from two by the loss of one. During early differentiation of organs in the embryo of many taxa, a ridge appears around one end of the flattened proembryo. On one side of this ridge, the cotyledon develops; on the other, growth ceases early. In the dicotyledons, the two cotyledons develop on opposite sides of similar annular ridges.

The solitary cotyledon undoubtedly represents one of an ancestral pair; the other has been gradually reduced and lost. It should be remembered that, in the period when the solitary cotyledon was considered the ancestral condition, the monocotyledons were generally accepted as the primitive angiosperms. This concept doubtless formed, at that time, the background for theories of the relationship of the cotyledons of the two major groups.

Variations in Cotyledon Number. Within the dicotyledons, number of cotyledons varies in some genera: *Centranthus* has three, two, or one; *Calendula, Dimorphotheca, Ambrosia, Impatiens, Raphanus,* two or one. In these genera, the solitary cotyledon occurs only sporadically, and a vestige of the second cotyledon may be present. A third cotyledon is frequently present in many common genera—*Acer, Juglans.* Polycotyledony has been claimed to exist in some dicotyledons, especially in parasitic genera—*Loranthus, Nuytsia, Persoonia.* The cotyledons of the dicotyledons have a tendency to be divided, often so deeply as to suggest four or eight. This lobing is characteristic of some Cruciferae, Burseraceae, Loranthaceae, Lauraceae, Tiliaceae, some genera, *Amsinckia, Xeropetalum, Dombeya.* Some species of *Persoonia* and *Loranthus* have the normal two cotyledons; others, any number from three to eight. That the larger numbers represent divisions of a basic two has been shown for some taxa by both anatomy and ontogeny. The "four cotyledons" of *Ceratophyllum* have been shown to include the first two leaves of the plumule. Polycotyledony has been reported in occasional seedlings of many taxa—in as many as 2 to 4 per cent in "British plants." It is characteristically prominent in some ranalian taxa. The "magnolian line" in the woody Ranales has been reported to have a high percentage of polycotyledonary embryos—often as high as 87 per cent with three, and 13 per cent with four, none with two cotyledons; most seeds of *Degeneria* studied have embryos with three cotyledons. The embryo of the Juglandaceae is remarkable in the form of its cotyledons —deeply two-lobed and the pairs united face to face ventrally. Each of the apparent cotyledons consists of halves of two cotyledons. The fusion is not complete and is evident in germinating seeds. All abnormal cotyledonary conditions in the dicotyledons—polycotyledony, syncotyledony, schizocotyledony, monocotyledony—represent only extremes of modification of simple dicotyledony. All these specializations have doubtless appeared more than once. The loss of a cotyledon has clearly occurred many times; it is seen even within a genus—*Corydalis, Ranunculus, Claytonia.* The presence of a single cotyledon is not alone sufficient to characterize the monocotyledons, nor does it necessarily indicate that all members of this taxon have been derived from the same ancestral taxon.

The absence, in some dicotyledons, of one cotyledon has been termed "pseudomonocotyly," but this term has also been applied to dicotyledons that have the two cotyledons united in a sheathing base—*Podophyllum, Quercus.* Genera of the Berberidaceae show stages in the evolution of the cotyledonary tube by the connation of the lower margins of the cotyledons; stages showing increasing length of the tube are present in *Caulophyllum, Jeffersonia, Podophyllum.* The position of the Nymphaeaceae as dicotyledons was, at one time, discussed by morphologists; the embryo was interpreted by some as monocotyledonous. In this family also, especially in *Nelumbo,* the cotyledons, arising on a cotyledonary collar, form a tubular structure, which is strongly lobed in some genera. Connation occurs among cotyledons, as it does among leaves and floral organs.

The Epiblast. Except for the mesocotyl, the epiblast has been the most puzzling part, morphologically, of the grass embryo (Fig. 124*E, F*). The problem of its nature has been summarily disposed of several times by the statements that it is "inconsequential," a mere scale, that "it has no vascular supply and no significance." But the fact that, in some genera, it is a prominent part of the embryo—even as long as the cotyledon—and has received much attention, indicates that it is probably significant. It has been called a vestige of the second cotyledon, representing its sheath or its auricles—it is bilobed in some genera (*Danthonia, Stipa, Eleusine*); it has been interpreted as a lobe of the scutellum, of the coleorhiza (Fig. 130), of the hypocotyl, of the undifferentiated body of the embryo, freed by the breaking out of the endogenous root primordium. Granted that it is no more than a vestige, its position on the axis—if "directly opposite the cotyledon"—would support the view that it is a remnant of the second cotyledon. This interpretation was prominent in the middle of the nineteenth century. A similar, but usually larger, structure, is present in dicotyledons that have only one cotyledon, and stages in the reduction of the second cotyledon are present in some monocotyledons (Fig. 125). The epiblast is commonly described as small and scalelike, but varies greatly in size and form. It may even be of much the same size as the scutellum and is so described for some of the bamboos; it is prominent in the embryos of *Zizania* and *Leerzia*.

Although through much of the nineteenth century, the epiblast was considered the second cotyledon, in the twentieth century, this interpretation has been discarded, and it has been called a lobe or an extension of one of the other organs of the embryo or of the undifferentiated proembryo. The interpretation of the epiblast as a vestigial second cotyledon receives some support by the growing view that the grasses are related to the Commelinaceae, which have a reduced but well-marked second cotyledon. The epiblast has been considered unusual among grasses, but it is present in two-thirds of the genera; absence of an epiblast is unusual in the family. Its presence in the primitive grasses and its absence in the highest genera suggests that it is a primitive structure disappearing within the Gramineae. The strongest evidence of its cotyledonary nature would be its position directly opposite the scutellum, but *the attachment of the scutellum is not the attachment of the cotyledon*, which is just below the plumule (Fig. 130). The morphology of the epiblast is uncertain.

POLYEMBRYONY

The term polyembryony is commonly used to describe the presence of more than one embryo in an ovule or seed. The use of the term in this way is morphologically loose, because of the varied nature and origin of the multiple embryos. Some are sporophytic; others, gametophytic. The sporophyte types are derived from tissues of the mother sporophyte—nucellus or integument—or from the zygote, proembryo, or suspensor; the gametophyte types arise from cells of the embryo sac other than the fertilized egg. The term polyembryony—true polyembryony—is probably best applied only to the presence of more than one embryo derived from one fertilized egg. Embryos that arise by budding from tissues of the mother sporophyte are called *adventive.* (All supernumerary embryos are often called adventive.) In origin, true adventive embryos resemble adventive leaves and other organs. Embryos may arise by splitting from the body of the embryo; one or two develop laterally on the suspensor when it is massive. The origin of more than one proembryo directly from the zygote is rare.

Gametophytic embryos arise from synergids or antipodals; those from synergids may be of doubtful nature, for synergids, as sister cells of the egg, may have been fertilized. The place of origin of embryos developing within the embryo sac is often difficult to determine, because embryos that arise outside the sac protrude into the sac as they develop.

Polyembryony is common in some of the gymnosperms and has been found in many angiosperms in families scattered throughout the taxon. It may be a common character in a genus, as in *Citrus,* where, in some species, it is present in nearly every ovule, or it may occur in only an occasional ovule. Growth conditions perhaps have some influence on its development. In many families, no polyembryony has been reported.

Adventive embryos of nucellar origin are more common than those of integumentary origin; they are found in many dicotyledons and in a few monocotyledon families—Liliaceae, Amaryllidaceae, Orchidaceae, Araceae, Gramineae. Nucellar polyembryony has been described as present in about 19 per cent of the ovules of *Trillium undulatum* but the additional embryos usually degenerate. Integumentary embryos—derived from the integument only—are found mostly in the dicotyledons; in the monocotyledons, they are known only in the Liliaceae and Amaryllidaceae. Polyembryony is perhaps best known in *Citrus, Mangifera, Opuntia, Nicotiana, Funkia, Lilium, Erythronium, Allium, Poa.*

Occasionally, more than one embryo may be formed in an ovule in which two or more embryo sacs exist, either from two megaspores from the same mother cell, or from multiple megaspores developed from multicellular archesporium. Rarely, two ovules may be fused, simulating one. The term pseudopolyembryony has been applied to the development of multiple normal embryos from fertilized eggs of more than one embryo sac in an ovule, reported as occasional in *Trifolium, Rosa, Saxifraga.* The "polyembryony" of *Prunus, Quercus, Castanea,* and other drupes and nuts is based on the error of interpreting seeds as embryos. It has been suggested that polyembryony may be a primitive character, that the fertilization of synergids may be a retention from multiple-egg ancestors where more than one egg was fertilized. But the distribution of polyembryony scattered throughout the angiosperms gives this view no support.

ANATOMY OF THE EMBRYO AND YOUNG SEEDLING

Little vascular tissue is mature in most embryos; procambium is commonly distinct in some isolated areas, and a few mature protoxylem and protophloem cells may be present. The anatomy of the embryo proper is largely that of the histology of meristems. Anatomical relationships among the organs can be determined only in the young seedling where a vascular skeleton is present. The vascular system of the dicotyledon seedling (Fig. 133) follows closely a well-known single pattern; that of the monocotyledon, though basically the same, varies considerably (Fig. 134). The variations are related to adaptations to hypogeal germination and are chiefly those of fusions among the vascular bundles under reduction.

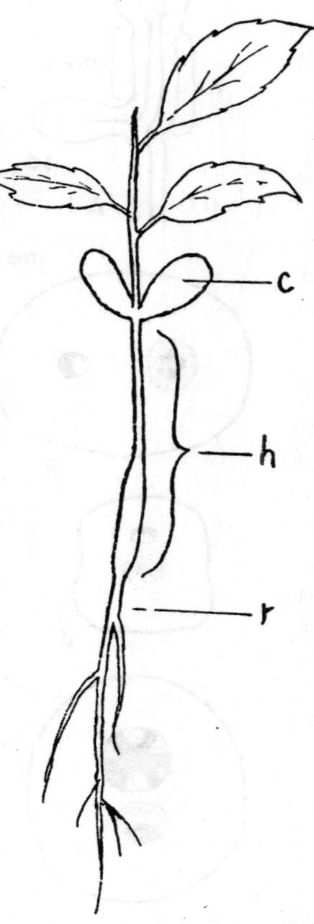

Fig. 133. Sketch of dicotyledonous seedling, *Pyrus Malus*. *c*, cotyledon; *h*, hypocotyl; *r*, root. (*Drawing by Dr. L. J. Edgerton, Cornell University.*)

Anatomy of the Cotyledon. The traces of the cotyledon vary in number; two is the common number in the lower taxa of both mono- and dicotyledons (Fig. 7); four is frequent, with transition to three by the fusion of the middle pair; three is common with the strong central bundle sometimes double (evidence of its derivation from the four-trace condition); larger numbers are probably uncommon—*Canna*, and some of the Araceae. Where there are four traces, the two median traces lie close together. The Liliaceae show all stages in the origin of an odd number of traces: two traces, four traces, a double trace, a median trace with one or more pairs of laterals. The Iridaceae have a median trace, simple or double, with one or two pairs of laterals; the Araceae, one to many traces. The genus *Scilla* shows stages in the origin of the median trace; the cotyledon of *S. siberica* has two traces, that of *S. peruviana*, one double

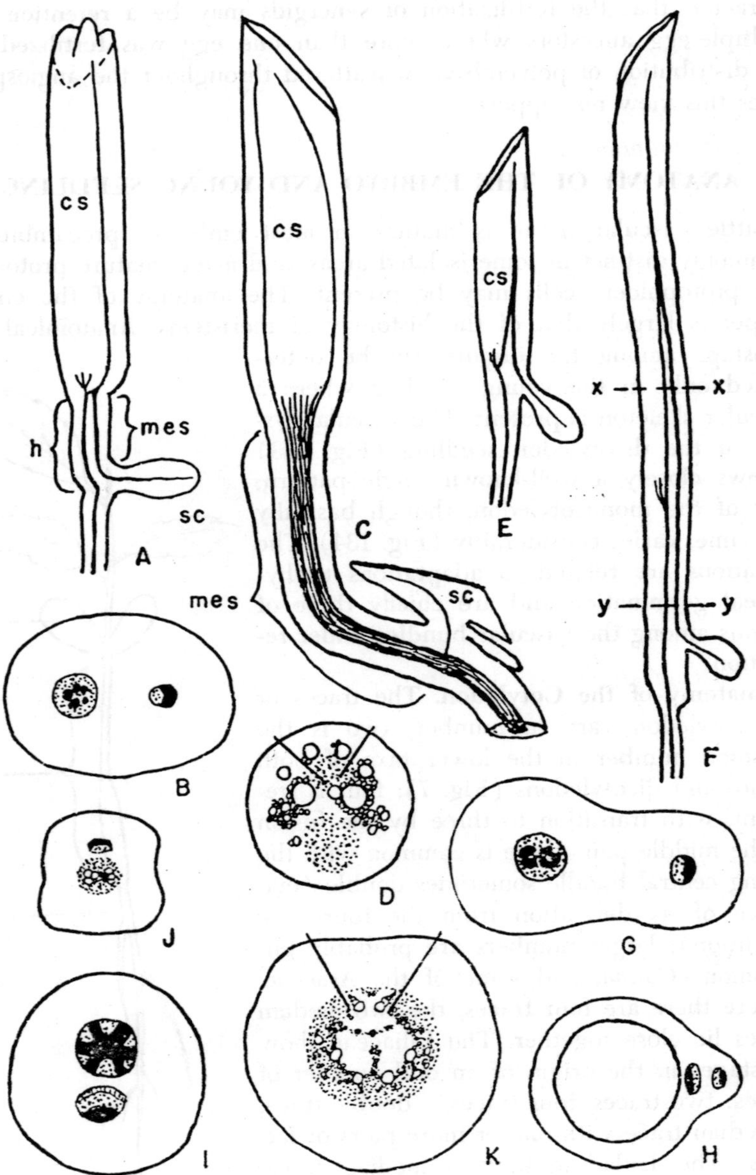

Fig. 134. Longitudinal and cross sections of monocotyledon seedlings showing vascularization. *A, B, Carex folliculata,* and *C, D, Panicum miliaceum,* showing neck of cotyledon connate to cotyledonary sheath and adnate to hypocotyl, forming the mesocotyl, and bringing the scutellum down to the level of the base or near the base of the hypocotyl. *B, D,* cross sections through the mesocotyl, showing in *B* the vascular stele and inverted cotyledonary bundle and in *D* the fusion of

trace. Odd numbers clearly result from the fusion of two or more traces. Number, position, and fusion of traces coincide with those of leaf traces. A single trace is characteristic of highly specialized cotyledons; all stages in lateral fusion are found—from fusion at the base only to fusion throughout the length of the bundles. Different taxa show various degrees of this fusion; commonly, the lateral bundles are free distally and, with the median bundle, form a three-veined cotyledon. The two free lateral bundles become major elements in the anatomy of many specialized cotyledons.

Characteristic of many monocotyledons is the presence of two strong lateral traces, which supply the sides of the sheathing leaf base. These are prominent in cotyledons in which the tip is bent downward. The median vein, which supplies the cotyledon tip, bends downward to the scutellum; the lateral veins may continue upward in the sheathing wings or, rarely, also continue downward to the scutellum. The tip of the median bundle, as it enters the scutellum, remains unbranched in small, cylindrical, and club-shaped forms, but becomes variously branched in large forms. In taxa with large amounts of endosperm, as in the palms, an extensive vascular system may be built up as the scutellum enlarges.

In most cotyledons that have two or three free traces, all the vascular bundles continue as veins well toward the tip (Fig. 7). In the down-bent cotyledons that have the tip transformed into a scutellum, all the bundles may continue into the scutellum, but usually, only the median extends to it; the lateral bundles are shorter, and their distal parts recurve more or less below the point of down-bending. In down-bent cotyledons with connate base and top, and where the sheath forms the prominent part of the cotyledon, the lateral bundles are also prominent; in those in which the median vein is reduced as a result of the connation, the lateral bundles may form the entire vascular supply of the sheath and, therefore, of the coleoptile, its distal "elbow."

---

cotyledonary bundle with the vascular stele. *E* to *H*, *Tigridia Pavonia*, *E*, a younger stage than *F*, both showing cotyledonary bundle going up into the sheath and back again (inverted) and on through the mesocotyl into the scutellum. Plumular traces showing separation of cotyledon from hypocotyl. *G*, cross section of *F* at *y–y* showing vascular stele and inverted bundle of adnate cotyledonary neck; *H*, cross section of *F* at *x–x* showing upward and downward (inverted) trace of cotyledonary bundle in the sheath. The prominent ridge in *H* showing evidence of connation of sheath and neck and in *G* of adnation of hypocotyl and neck. *I*, *J*, cross sections of mesocotyl, in *Oryza sativa* and *Phalaris canariensis*, respectively, showing vascular stele of hypocotyl and inverted cotyledonary bundle. *K*, *Zea mays*, cross section of mesocotyl showing histological union of stele and cotyledonary bundle. *cs*, cotyledonary sheath; *h*, hypocotyl; *mes*, mesocotyl; *sc*, scutellum. (*A* to *I*, *after Schlickum*; *J*, *K*, *after Van Tieghem*.)

An even number of traces is clearly basic for the angiosperm cotyledon; it also seems to be the primitive condition. The taxa now generally recognized as primitive—some of the woody Ranales, the lower Liliales, the Helobiales—have cotyledons with two strong traces, often with pairs of lateral traces. An even number of cotyledon traces is in strong contrast to the odd number characteristic of leaves; even-numbered leaf traces, as in *Austrobaileya* (Fig. 5) are rare. The significance of the double nature of the vascular supply of the cotyledon is obscure. In most gymnosperms, even-numbered traces are characteristic, and single traces seem to represent two fused traces. In angiosperms, the odd number of leaf traces is apparently, as with the cotyledons, a derivation from an even number; stamens and carpels occasionally have an even number of traces. The cotyledons retain strong evidence of an ancestral even number. The double vascular supply of cotyledons, together with frequent bilobing, has been looked upon as evidence in the cotyledon of dichotomy surviving as a primitive character in the embryo. But the traces are derived well apart on the hypocotylary stele, and there is no evidence of dichotomy in the embryonic axis. In the simple two-trace cotyledons of the Commelinaceae (*Tinantia*), the traces arise at various distances apart, even 120°, and are probably never directly opposite one another. This is also evidence against the view that the basic two-trace vascular supply and the frequent bilobing of cotyledons is a retention of ancestral dichotomy. In taxa with more than two traces, the traces are closer together, and two, or even more, may be derived from one gap.

The number of traces is fairly constant in families. Two traces characterize most of the lower monocotyledon families. The Liliaceae have two in the lower tribes; two, four, and six in others. The Zingiberaceae have two traces. The Amaryllidaceae have two, or one double trace; the Iridaceae, one, two, or three (*Crocus* has one in some species, two in others); the Araceae, one to several; the Cannaceae have several. Some of the woody Ranales have two cotyledonary traces. The vascular system of the cotyledon of monocotyledons has, characteristically, two major bundles or two with two additional lateral bundles. The dicotyledons have, commonly, three major bundles (the median equivalent to the two median bundles of the monocotyledons). Cotyledons with a pair of midveins are occasional in the dicotyledons—Austrobaileyaceae, some of the Chloranthaceae and Monimiaceae. In many grasses, the single trace breaks into two at the base of the cotyledon and forms a forked midvein—*Hordeum;* in other genera—*Triticum*—the median trace is double.

Anatomy of the Hypocotyl. The gross structure of the hypocotyl needs brief description, because of the occasional statement that the monocotyledons have no hypocotyl. The basis for this statement is, perhaps, the shortness of the hypocotyl of many of these plants, especially that of certain grasses that are commonly used to illustrate monocotyledonous embryos. But the grass embryos are far from typical of monocotyledon embryos; they represent highly specialized types. Interpretation of the hypocotyl as the caulicle has also formed a basis for the statement that a hypocotyl is lacking. Very short, almost platelike hypocotyls have received little attention anatomically, because of the complexity of structure where the transition occurs abruptly and the vascular strands lie almost horizontally. The hypocotyl of some monocotyledons has been described as "often platelike, hardly existent"; in a few, it is long and prominent in the embryo (Fig. 135*C*, *D*, *F*), especially where it is part of the mesocotyl (with adnate neck of the cotyledon). The shortening of the hypocotyl is a part of a general shortening at the base of the shoot in many monocotyledons in which the basal internodes are largely suppressed. The hypocotyls of dicotyledons are, in general, longer than those of the monocotyledons. The normal, cylindrical or platelike hypocotyl may be greatly distorted; it tends to be ovoid or spherical in seedlings of cormous taxa—*Crocus*—in fleshy annuals or biennials—*Raphanus*—and in poorly differentiated embryos—orchids, saprophytes, and some parasites.

The types of transition in vascular structure between root and stem—from radial and exarch to collateral and endarch—are not discussed here; they are well treated in textbooks of anatomy. But important in the interpretation of the morphology of the embryo is the relation of the vascular structure of hypocotyl to that of the adnate "neck" of the cotyledon in the formation of the mesocotyl. (Failure to recognize this adnation has been responsible for major errors in the interpretation of the nature of the coleoptile.)

Anatomy of the Seedling Root. The vascular cylinder of the primary root of seedlings is commonly diarch or tetrarch; other types seem to represent modifications of these basic types. Monarchy is rare in primary roots; polyarchy, frequent in the monocotyledons. It has long been argued that diarchy is basic, tetrarchy derived, and vice versa. Tetrarchy has been considered the basic type because it is associated with arborescent taxa (and diarchy is associated with herbaceous). (In the Leguminosae, tetrarchy is characteristic of woody taxa; diarchy, of herbaceous taxa.) The Ranunculaceae have diarch; woody families have tetrarch primary roots. The Compositae have both types and give no evidence that one is more primitive than the other. In the dicotyledons, there are few variations from diarchy or tetrarchy; in the monocotyledons, there are many variations; polyarchy—double or triple the dicotyledon numbers—is common in the monocotyledons. The simple transition region of the dicotyledons, like that of most conifers, has been viewed as derived from the complicated type of the cycads and the araucarian conifers. No conclusions can be drawn, however, as to the primitive type, because of the comparatively small amount of information about the transition region in angiosperms. A correlation of number of protoxylem poles in the hypocotyl with the number of cotyledon traces, in the light of fusion and reduction of traces, should be a help in determining the basic type or types in angiosperms.

Anatomy of the Mesocotyl. The mesocotyl is a compound structure made up of the hypocotyl and the adnate part of the cotyledon (Figs. 130 and 134*A*, *F*); it contains vascular tissues of both organs—the stele of the hypocotyl and one or more vascular bundles of the cotyledon (Fig. 134*B*, *G*). Stages showing the external and internal effect of adnation of the cotyledon neck to the hypocotyl are frequent. External evidence of the fusion is the presence, in some genera, of a longitudinal ridge on the embryonic axis (Fig. 134*G*); internal evidence is the presence of a vascular bundle running longitudinally *in the cortex* of the hypocotyl. The "superfluous" bundle belongs to the downbent adnate neck of the cotyledon (Fig. 134*A*, *E*, *F*); it is the vascular supply of the scutellum, the median bundle of the cotyledon continued downward from the point of downbending of the cotyledon neck. The vascular supply of the scutellum thus follows a "roundabout" course, like an inverted V (Fig. 134*E*, *F*). As a result of the bending downward of the cotyledon tip, the adnate vascular bundles, as seen in cross sections of the mesocotyl, are inverted, the xylem external to the phloem (Fig. 134*B*, *G*, *H*). (The presence of an inverted bundle in the cortex of the hypocotyl, apparently an anatomical anomaly, is a structural feature resulting from the adoption of hypogeal germination.)

Where adnation of the cotyledon and the hypocotyl is histologically more intimate, the cotyledonary bundle lies still deeper in the pericycle (Fig. 134*I*, *J*), or in contact with the vascular cylinder of the hypocotyl (Fig. 134*D*). In the most highly specialized embryos, it may even be fused with one of the bundles of the hypocotylary stele, forming a concentric bundle, or be united laterally with the vascular cylinder, losing its identity (Fig. 134*K*). (Histological union of vascular tissues of different organs lying side by side represents the highest degree of anatomical fusion; it is seen frequently in other parts of the plant—inflorescence to stem, flower to leaf, especially of floral organs to one another. But fusion of vascular bundles of different organs is nearly always between similarly oriented vascular bundles; in the mesocotyl, the vascular bundles are differently oriented.)

Where the "downrunning" cotyledonary bundle is histologically merged with the vascular tissue of the hypocotyl, its free tip, which supplies the scutellum, is attached at the base of the hypocotyl. This position is obviously secondary, as is shown by a series of taxa with adnate cotyledon neck (Fig. 129*F*, *G*, *H*). In a commonly held interpretation of the monocotyledon embryo, using *Zea* as an example, the scutellum represents the entire cotyledon; the mesocotyl is the first internode of the stem; the coleoptile, the first leaf; and the hypocotyl, a mere plate of tissue. But critical comparative study of monocotyledon embryos shows that the cotyledon is divided transversely by adnation into three parts; the free, distal part, the coleoptile; the free, basal part, the scutellum; and the median section, which forms the lateral part of the mesocotyl, is, superficially, lost by adnation. Failure to understand this complex structure has been responsible for the interpretation of the coleoptile as the first leaf and the hypocotyl as the first internode. The incorrect interpretation of homology is not surprising, because the morphological base of the cotyledon is distal on the embryo and the apex is borne proximally. So complete a "breaking apart" and apparent isolation of the parts of an organ probably occurs nowhere else in the angiosperms. It represents an extreme modification resulting from the adoption of hypogeal germination. Here, as in flower specialization, the monocotyledons seem to have surpassed the dicotyledons in specialization. The use, in teaching, of the embryo of *Zea* as a typical monocotyledon embryo is most unfortunate; this embryo is one of the most highly specialized in angiosperms.

The complex nature of the mesocotyl, internally obvious where the cotyledon bundle lies free in the "cortex" (Fig. 134*B*, *G*, *I*, *J*), is obscure or hidden where the bundle is united with the stelar vascular strands. The presence of the inverted bundle in the "cortex" of the hypocotyl was noted and illustrated several times in the later decades of the nineteenth and the first decades of the twentieth century, but, though well illustrated, it has been overlooked recently. If its presence had been evaluated, it could hardly have been urged that the term mesocotyl is superfluous.

Anatomy of the Cotyledonary (Stipular, Ligular) Sheath. The term cotyledonary sheath has been loosely applied both to the caplike or sheathlike structure, the coleoptile, which encloses the plumule, and to the basal part of the cotyledon. That the cotyledonary sheath is, as the term suggests, a part of the cotyledon, has been generally recognized, but it has rarely been made clear what part of the cotyledon it represents. Anatomy supports the interpretation that the sheath is the basal part of the cotyledon; the neck, the median part, with the angular "elbow" region differentiated as the coleoptile; and the apex, the scutellum. The cotyledons of taxa that have mesocotyls are, in part, "bottom-side-up" in relation to their apparent positions on the axes of the embryos.

In cotyledons in which there is little or no downbending of the distal part, all the vascular bundles, including those of the sides of the sheath, usually continue into the scutellum. These cotyledons are usually of simple, primitive type, with two or four vascular bundles. In cotyledons of advanced type, with one trace which divides to form a median and two strong lateral bundles, only the median bundle continues to the scutellum. The branches extend first upward and laterally and then commonly downward, on the opposite side of the sheath. The pair of arching, recurving bundles forms a prominent feature of the sheath in many taxa, and their continuation over the top of the coleoptile has entered into theories of the nature of the coleoptile. The course of the bundles has been considered remarkable and evidence that the coleoptile is an independent organ, not a part of the cotyledon. (The venation type of the specialized cotyledon—one trace, representing three fused laterally at the base, and forking to form a median and two lateral branches—is similar to that found in advanced types of carpels, petals, and leaves.) The position and course of the veins of the sheath support the interpretation that at least the upper part of the sheath is stipular (ligular). In the evolutionary development of the coleoptile, fusion of the "shoulders"—rounded stipule tips or auricles—continued distally the closed sheath of the primitive cotyledon, forming the cap-shaped protective "tip" of the coleoptile (Fig. 130). The arching veins show the make-up of the enclosing coleoptile. (Venation in auriculate stipules is frequently arching.)

Where there is adnation of the neck of the cotyledon laterally to its lower part, the midvein is bent back upon itself. The two parts are fused, and a bicollateral bundle (xylem both internal and external to the phloem) is formed. Stages in this fusion are frequent. Under continued specialization of this vascular fusion, where conduction "goes up and back" in a double bundle, the bundle is shortened distally and may become united with the vascular stele of the hypocotyl (Fig. 134*D*, *K*). The shortening and adnation may extend to the point of origin of the branches of the trace of the sheath, and the branches then appear to arise directly from the hypocotyl. The origin of the trace of the sheath, apparently directly from the hypocotyl, has been part of the evidence that the sheath and the coleoptile represent the first leaf. The understanding of the downbending and fusion of the distal part (neck) of the cotyledon to its basal part explains the "basal" (morphologically false) position of the scutellum and the inversion of the distal part of the cotyledon.

THE TWO MAJOR TYPES OF ANGIOSPERM EMBRYOS

The embryos of angiosperms have long been rather fully described in elementary textbooks.* The differences are chiefly those of ontogeny and of the mature embryos; the proembryos differ in no important way. Under phylogenetic reduction, the mature embryo may consist of only a few cells and show little or no differentiation into the usual organs. These reduced embryos characterize many parasitic and mycorrhizal taxa and also highly specialized taxa—Orchidaceae, Apostasiaceae, Burmanniaceae. The highly specialized embryo of the orchids shows little differentiation of organs. Similarly, in *Orobanche,* the embryo in the seed at shedding time is a mere sphere of cells. But this embryo, in contrast with that of the orchids, where there is no endosperm, is embedded in endosperm. In these very small embryos, where there is no after-ripening growth within the seed coats, differentiation of early stages of organs may accompany or follow germination.

The Dicotyledon Embryo. The mature embryo of the dicotyledons is simple in structure; its parts are distinct and usually readily recognizable—an *axis* with two lateral appendages, the *cotyledons.* At one end of the axis is the terminal bud, the *plumule;* at the other end, a root tip, the *radicle.* The axis between the cotyledons and radicle is the *hypocotyl* —the transition region between the markedly different structure of stem and root.† In dicotyledons, the hypocotyl is commonly prominent; the line between the hypocotyl and the root may be evident externally by differences in diameter and in nature of the epidermis, but examination of internal structure is usually necessary to delimit these parts. The radicle may be represented at germination by only a growing point or it may be well developed, a few millimeters long. The cotyledons are large and prominent, commonly alike and strictly opposite. During germination they become either epigeal or hypogeal. Well-developed plumules—the basal one or more immature internodes of the stem and their appendages—may be apparent within the seed; the stem within the bud (sometimes prominent) is morphologically the *epicotyl.* (The terms radicle, hypocotyl, and epicotyl have, unfortunately, often been loosely used, to the confusion of beginning students.) Endosperm is not so common in the dicotyledonous as in the monocotyledonous seed; the early-formed endosperm is commonly absorbed by the embryo as

* Unfortunately, the embryo usually chosen to represent the monocotyledonous type is that of the grasses, especially that of *Zea,* one of the most specialized embryos in the family; grass embryos, as a whole, are difficult of interpretation.

† The term *caulicle* was long applied to the axial part of the embryo, and considered the embryonic stem between the radicle and the plumule. But this part of the axis is hypocotyl, not stem, morphologically, and therefore the term caulicle is inaccurate and has become obsolete.

a whole, during development of the seed or during after-ripening. There are, of course, many modifications of this simple type; some forms are asymmetrical—one cotyledon smaller than the other or even lost. In the mature seed, the entire embryo may be very simple, as in reduced forms.

The Monocotyledon Embryo. Though monocotyledon embryos vary greatly in form, the axis is basically like that of typical dicotyledons. Some monocotyledons have the straight axis and terminal plumule, typical of the dicotyledons, but the great majority are described as having the stem tip (plumule) lateral. The so-called lateral position of the plumule is morphologically terminal. The shoot apex has been crowded out of its normal position by strong development of the huge cotyledon, survivor of an ancestral two. The displacement may be ontogenetic but has become congenitally established in the highly specialized embryos of the grasses and some other families. In these embryos, the cotyledon becomes pseudoterminal, even ontogenetically terminal, as are some other falsely terminal organs—leaves, carpels, stamens, and ovules. The axis is commonly strongly curved where the cotyledon is falsely terminal, and the plumule is described as lateral. The root may, similarly, be pushed into a lateral position. It is probably unavoidable that the morphologically obscure terms—scutellum, coleorhiza, epiblast, and coleoptile—be used in descriptions of the embryos of monocotyledons, but they are not helpful in the teaching of comparative form. In monocotyledons, the hypocotyl may be prominent but, in the higher taxa, is usually short. Though the transition region may be reduced to little more than a vascular plate in these embryos, as in some of the higher grasses, the statement that the hypocotyl is "nonexistent in the monocotyledons" is not correct. The *mesocotyl*, present in only highly specialized monocotyledon embryos, is a compound structure, consisting of the hypocotyl and an adnate part of the cotyledon.

Monocotyledon embryos commonly differ from those of the dicotyledons not only in number of cotyledons but also in the apparent position of the cotyledons. The two cotyledons of the dicotyledons are borne laterally, as an opposite pair at the top of the hypocotyl; the solitary cotyledon of the monocotyledons is commonly described as terminal, with the axis "lateral on the side of the cotyledon." The cotyledon appears to be terminal in most monocotyledons, because it extends beyond the plumule. Ontogenetically, it seems to be terminal in some taxa, but is obviously lateral in other taxa. Little is known of the histological details of cotyledonary origin in the monocotyledons; determination of point of origin on the rounded promeristem is difficult. In *Ottelia*, the apices of the cotyledon and the shoot are described as "arising side by side" in the proembryo. The "terminal" position of the cotyledon is attained by growth greater and more rapid than that of the shoot apex. The plumule is definitely terminal and the cotyledon lateral in some monocotyledons—several, probably all, genera of the Commelinaceae [*Commelina, Tinantia* (Fig. 125)], and the Dioscoreaceae (*Dioscorea, Tamus*), and in *Trillium* and *Guzmannia.* In the closely related genera, *Trillium* and *Paris,* fairly primitive monocotyledons, cotyledon and plumule position differ. In *Trillium,* the plumule is terminal and the cotyledon lateral, as in most dicotyledons; in *Paris,* the plumule seems to be lateral and the cotyledon terminal. In the seedling in both genera, the strongly developed, persistent, green cotyledon is erect; in *Trillium,* it stands parallel with the plumule. Ontogenetically, the plumule arises terminally in both genera, but, in *Paris,* it is pushed to one side by the thickening of the hypocotyl. These two genera show, in the relationships of hypocotyl, cotyledon, and plumule, a transition in cotyledon position from that of the dicotyledons to that of the monocotyledons. Though in *Paris,* the cotyledon arises, ontogenetically, before the stem apex, as in most monocotyledons, time of origin cannot be considered evidence that the plumule is lateral, because the cotyledon primordia of the dicotyledons also commonly arise first. Morphologically, the cotyledon is lateral, as is the solitary carpel of *Acacia, Prunus, Actaea, Bauhinia,* and the solitary stamen of *Euphorbia, Myristica, Najas.* (That these floral organs are truly lateral is shown in various ways—by comparison with the same organs in related taxa, by ontogeny, by vascular anatomy.)

Morphologically, the axis of a monocotyledonous embryo is not borne laterally on the cotyledon, though, in the young embryo, it appears to be initiated there and its development is later than that of the cotyledon. The pseudoterminal position of the cotyledon is the result of the dominance of a lateral cotyledon, which pushes the axis to one side by differential growth, as shown in *Paris;* in many taxa, the so-called terminal position is congenitally established. In those where a remnant of the second cotyledon is unquestionably present, the terminal position of the axis tip is clear.

The monocotyledonous embryo has long been a morphological puzzle and the subject of much controversy but, if the embryos of taxa now generally accepted as primitive—some of the Liliales and Helobiales, especially—are considered basic types, steps in the reduction and fusion leading to the extreme forms found in the grasses are clear.

Simplification is outstanding in the orchid embryo, where there is little evidence of separate organs; the bulbous body of the embryo seems to represent an enlarged axis, and the cotyledon only a minor projection. Very small, undifferentiated embryos in endospermless seeds characterize some parasites, such as *Orobanche*, and some saprophytes —*Monotropa, Pyrola, Bartonia;* these embryos are probably reduction types. In some aquatic taxa, such as *Hydrocharis* and *Ruppia*, both endosperm and scutellum appear to have been lost; in the ancestral stock of others, such as *Zostera* and *Halophila*, probably no endosperm and no absorbing cotyledon tip were present.

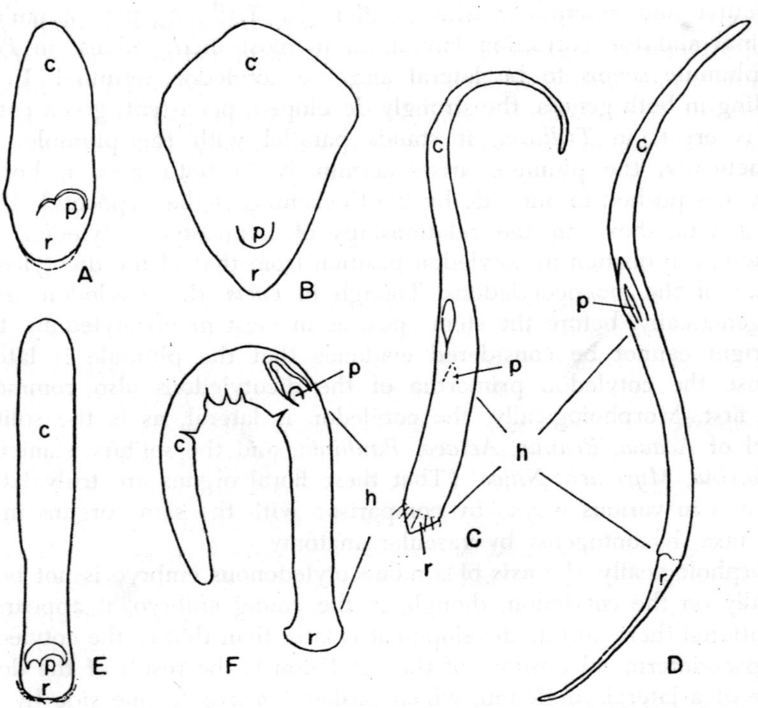

Fig. 135. Various forms of simple monocotyledon embryos; *A, B, E, F*, with endosperm; *C, D*, without endosperm. *c*, cotyledon; *h*, hypocotyl; *p*, plumule; *r*, root. *A*, *Arum orientale; B, Carex livida; C, Butomus umbellatus; D, Alisma Plantago; E, Sparganium ramosum; F, Sagittaria sagittaefolia.* (*After Tschirch.*)

The simple embryo of such monocotyledonous families as the Alismataceae, Butomaceae, and Scheuchzeriaceae has been considered reduced, but this is undoubtedly a primitive embryo. It is cylindrical and has little differentiation of organs. The sheathing cotyledon is cylindrical and seemingly terminal; the root is weakly developed; the plumule is also weakly developed and enclosed within the sheathing base of the cotyledon (Fig. 135*C, D*).

"Monocotyledony" in the Dicotyledons. Stages in the reduction of one of an ancestral pair of cotyledons, modified in function, are seen occasionally among the dicotyledons in taxa sometimes called "pseudomonocotyledons" and "aberrant dicotyledons"—*Trapa, Cyclamen, Abronia, Corydalis, Dicentra, Erigenia, Pinguicula, Carum, Peperomia, Claytonia, Ranunculus Ficaria* (the most thoroughly studied taxon), the Nymphaeaceae. In *Trapa, Eranthis,* and *Mammillaria,* one cotyledon is smaller than the other. *Cyclamen*—long known for its unusual germination—has one well-developed cotyledon and a second one represented by a small vestige; both cotyledons remain within the seed—the well-developed one serves as an absorbing organ; the embryo and germination are essentially those of a monocotyledon. *Corydalis cava* and *C. ochroleuca* differ from other species of the genus in their solitary cotyledon; a second cotyledon is represented by a "hump" opposite the well-developed one. In *Claytonia virginica,* one cotyledon is very small, and the embryo closely resembles that of the aberrant embryos of the Dioscoreaceae and Commelinaceae which have a second cotyledon. Species of *Pinguicula* show stages in the abortion of one cotyledon. *P. vulgaris* and *P. alpina* have only one cotyledon; *P. leptoceras* has one normal and one in which development ceases at various stages. The Umbelliferae perhaps provide, within one family, the most examples of suppression of the second cotyledon. Eight genera have species with one cotyledon. *Bunium* and *Scaligera* have some species with two, others with one cotyledon.

The embryo of *Ranunculus Ficaria* has been cited in support both of the view that, in the monocotyledons, one cotyledon has been lost, and also of the view that the one cotyledon of some embryos represents two fused cotyledons. In this species, most embryos have two cotyledon primordia, one of which aborts after forming a flattened appendage, and the embryo much resembles that of the monocotyledons. A few embryos in this species have fairly normal pairs of cotyledons. The structure of the two forms of embryo in this species was seen as evidence to support the view that the solitary cotyledon represents a union of two. The fusion of the cotyledonary petioles to form a unit structure, extending, in some individuals, to the cotyledon, which then is bilobed, was considered additional evidence. In refuting this theory, it was pointed out that each lobe is itself bilobed and that cotyledons of many angiosperms, even those of some monocotyledons, are bilobed. Sheathing cotyledonary bases are frequent in all the angiosperms, especially in geophilous taxa—*Eranthis, Podophyllum.* Fusion of the two cotyledons may extend so far into the cotyledons that the embryo has been described as having only one, bilobed cotyledon.

The existence of unusual cotyledonary structure in the water lilies has been much discussed. *Nelumbo* was described as having only one cotyledon. The genus was termed monocotyledonous, and the suggestion made that the family should be transferred to monocotyledons. But it was shown that *Nymphaea* has two cotyledons and that *Nuphar*, which has one of its two cotyledons reduced in size, shows stages in fusion of the cotyledons—basally in some species, by one margin in others. In *Nelumbo,* the single cotyledon consists of two, fused to form a cup-shaped, two-lobed structure surrounding the plumule. The lobing was interpreted as showing derivation of two cotyledons from one by division, but anatomy shows that the lobing is the result of basal fusion of the two cotyledons, a condition frequent in dicotyledons—*Podophyllum, Eranthis, Ranunculus.*

In the dicotyledonous genus, *Peperomia,* the geophilous species show stages in the origin of monocotyledony (Fig. 136). Most species of *Peperomia* have two cotyledons, strictly opposite, which, at germination, are withdrawn from the seed and become erect, photosynthetic organs. But the minute, cormous species show stages in the modification of one of the pair to form a suctorial organ. These stages are shown diagrammatically in Fig. 136. *P. pellucida* has two normal green cotyledons; *P. peruviana* has one photosynthetic cotyledon and the other, though well-developed, remains within the seed; in *P. parvifolia,* the cotyledon remaining within the seed has become club-shaped and forms a typical absorbent organ. Noteworthy in the early stage of retention of one cotyledon is the unreduced size of this cotyledon and the large cavity where the photosynthetic cotyledon lay (Fig. 136*D*). Where the retained cotyledon has become a specialized suctorial organ, the photosynthetic one is small, and the absorbing organ fills a small cavity in the endosperm. The elaboration of the suctorial organ is completed in *P. parvifolia.* The intermediate stage is similar to that in some araceous genera. Parallel stages in the development of the retained suctorial cotyledon exist in dicotyledons and monocotyledons. (The suctorial organ of most monocotyledons has an origin different from that of the dicotyledons.)

Aberrant Embryos in the Monocotyledons. Aberrant embryos seem to be less frequent in the monocotyledons than in the dicotyledons. From these monocotyledons comes some of the best evidence of derivation from dicotyledonous ancestors. In genera of the Dioscoreaceae—*Dioscorea, Trichopus, Rajania, Tamus*—and in the Commelinaceae [*Commelina, Tinantia* (Fig. 125)], the second cotyledon is present in reduced form. *Sagittaria* has been described as sometimes showing evidence of a lost, second cotyledon, a projection with a vascular trace on the hypocotyl opposite the normal cotyledon. Some taxa in the Araceae —*Arum, Arisaema*—and some Liliaceae—*Trillium, Paris*—have both cotyledons well developed, standing opposite each other at the top of the hypocotyl, one foliaceous and commonly interpreted as the first

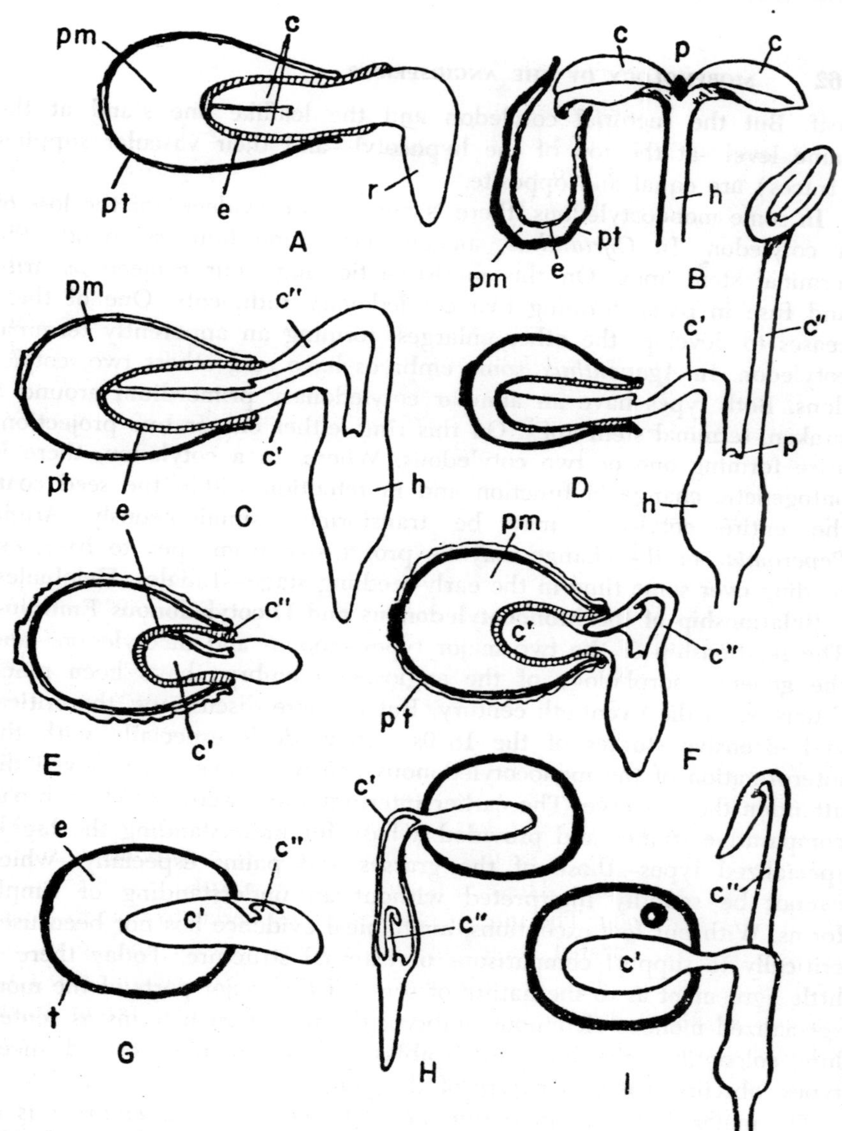

Fig. 136. Sketches to illustrate the theory of the origin of monocotyledony from dicotyledony by transformation of one cotyledon into a suctorial organ. *A* to *F*, germinating seeds and seedlings of *Peperomia* spp. and *G* to *I*, a monocotyledon showing stages in the evolution of a monocotyledonous embryo from a dicotyledonous embryo. *A*, *B*, *Peperomia pellucida*, two cotyledons withdrawn from the seed coats; *C*, *D*, *P. peruviana*, one cotyledon withdrawn, becoming an assimilating organ, the other remaining in the seed and becoming an absorbent organ; *E*, *F*, *P. parvifolia*, one cotyledon an assimilating organ, the other a club-shaped, absorbent organ in the seed; *G*, *H*, *I*, a monocotyledon: *G*, in germinating seed, one cotyledon an absorbing organ, the other a rudimentary first leaf; *H*, the embryo, except for the absorbent cotyledon, withdrawn from seed coats; *I*, the larger cotyledon remaining in the seed as a scutellum, the other forming a small leaf. *c*, cotyledon; *c'*, *c''*, absorbing and assimilating cotyledons, respectively; *e*, endosperm; *h*, hypocotyl; *p*, plumule; *pm*, perisperm; *pt*, pericarp and testa; *r*, root; *t*, testa. (*After Hill.*)

leaf. But the suctorial cotyledon and the leaflike one stand at the same level—at the top of the hypocotyl—and their vascular supplies (traces) are equal and opposite.

In some monocotyledons, there is ontogenetic evidence of the loss of a cotyledon. In *Cyrtanthus*, an annular primordium surrounds the terminal stem apex. On this meristematic ring, four projections arise and fuse in pairs, forming two cotyledonary rudiments. One of these ceases to develop; the other enlarges, forming an apparently terminal cotyledon. In *Agapanthus*, some embryos have one, others two, cotyledons. Both types have an annular cotyledonary primordium around a sunken, terminal stem apex. On this ring, either one or two projections arise forming one or two cotyledons. Where, in a cotyledon, there is ontogenetic change in function and in retention within the seed coat, the entire cotyledon may be transformed simultaneously—*Arum*, *Peperomia*—or the change may be progressive from apex to base, extending over some time in the early-seedling stage—Liliales, Helobiales.

Relationship of the Monocotyledonous and Dicotyledonous Embryos. The relationship of the two major types—mono- and dicotyledon—and the general morphology of the angiosperm embryo have been much discussed in the twentieth century. But in these discussions, the critical and extensive studies of the 1870s—which dealt especially with the interpretation of the monocotyledonous embryo—have not received the attention they deserve. The earlier interpretations were based on broad comparative studies and provided a base for understanding the highly specialized types—those of the grasses and palms especially—which cannot be soundly interpreted without an understanding of simple forms. With but few exceptions, anatomical evidence has not been used critically to support comparisons of external structure. Today there is little agreement as to the nature of some of the major parts of the more specialized monocotyledonous embryo; the use of such terms as scutellum, coleoptile, coleorhiza, and epiblast, applied to the most advanced types, obscures the real nature of the parts.

The major difference in the mono- and dicotyledonous embryos is in the number of cotyledons, a difference that forms an important part of the basis for separating the monocotyledons and dicotyledons. Yet, in this character, as in leaf venation and stelar structure, no clear line of separation can be drawn; among the dicotyledons, several taxa have one cotyledon, and some monocotyledons have vestiges of a second. Most of these aberrant forms have long been known, and described and cited in evidence of the derivation of one type of embryo from the other.

In the analysis of the evolutionary relationship of the two major types of embryo, two viewpoints have been supported: in one, the dicotyledonous type is considered primitive, the monocotyledonous derived from it by the fusion of two cotyledons or by the loss (repression) of one; in the other, the monocotyledonous type is considered primitive, the dicotyledonous type derived by the splitting of the one cotyledon. Anatomical evidence has been used in considerable part in support of these theories of evolutionary relationship, but usually without consideration of the fundamental anatomy of cotyledons as a whole—in all angiosperms as well as in gymnosperms. (A pseudodichotomous structure has long been recognized in many angiosperm cotyledons and has been used as evidence in the claim that cotyledons are organs *sui generis*, and not of leaf rank.)

Underlying the problem of whether the possession of one or of two cotyledons in the embryo is primitive, lies the almost universal presence of two cotyledons in the lower seed plants—even in *Selaginella*, there is a pair of opposite, cotyledonlike organs. (The polycotyledonary embryos of some conifers seem to have been derived from dicotyledonous types.) Comparison of angiosperm embryos with those of lower taxa should give evidence of the basic primitive number; none of the lower taxa has one cotyledon. The dominance of the two-cotyledonary state in embryos generally is evidence of the two-cotyledonary ancestry of the monocotyledons. The monocotyledons are not an isolated group of independent origin; their very close relationship with the dicotyledons—their derivation from ancient dicotyledons or from a common two-cotyledonous stock—is hardly to be questioned. The monocotyledons need not be called "syncotyledons," as has been suggested by those who believed there is evidence of fusion of two cotyledons to form the single cotyledon.

Evidence in support of the theory that the monocotyledons have been derived from dicotyledonous stock has been obtained from dicotyledonous taxa in which one cotyledon has obviously been reduced or lost. Furthermore, some monocotyledons have what appears to be a vestigial cotyledon opposite the normal one. In the mid-twentieth century, this structure has either been overlooked or considered "without significance." Evidence in support of the view that the monocotyledonous embryo is the primitive type has come chiefly from anatomical structure—from the double vascular supply of the cotyledon of some monocotyledons, especially those considered the more primitive. The single cotyledon is considered, under this theory, to be terminal and two lateral cotyledons to have been formed by the longitudinal splitting of the terminal one.

In the light of the large amount of information about cotyledons now available—much of it in the older literature—the dicotyledonous type is now generally accepted as primitive; the monocotyledonous type, as derived by the loss of one of the ancestral pair.

Nature of the Cotyledon of the Monocotyledons. At the beginning of the twentieth century, the cotyledons were described as "primarily haustorial organs originating phylogenetically as the nursing foot in the bryophytes and persisting throughout the higher plants." The presence of the scutellum in the embryo of the monocotyledon—a simple, foot-like, absorbing structure, basal in position—was considered evidence of primitiveness. The possession of the scutellum supported the view, held at that time, that the monocotyledons are the primitive angiosperms and that the grasses, with their large scutellum and "simple" flowers, are among the more primitive families. At the same time, some of the simple aquatic monocotyledons were considered highly primitive, resembling, in habit, some of the pteridophytes, especially *Isoëtes.* (That these aquatic monocotyledons have no scutellum was overlooked.) Recognition of the monocotyledons as highly primitive angiosperms gave support to the theory that the two cotyledons of the dicotyledons are the morphological equivalent of the one of the monocotyledons.

The cotyledon is surely not homologous with the "foot" of the pteridophytes. Resemblance between these structures is superficial only; both absorb food for embryos. The "foot" is one of the major parts of an embryo; the scutellum is a specialized part of a lateral appendage of the axis. [Furthermore, the scutellum is not the cotyledon; in some species of *Peperomia,* a dicotyledon, it represents an entire cotyledon (Fig. 136*D*), but, in monocotyledons, it is only the distal part.] In the dicotyledons, at least, the cotyledon is clearly an appendage of the axis of the embryo and, later, of the seedling. Much evidence supports the common view that the cotyledon is a leaf, modified in function as a storage and food-absorbing organ. In details of ontogeny and in vascular anatomy, it is a leaf; and in taxa in which it becomes a photosynthetic organ, it has typical leaf structure. The interpretation of cotyledons as organs *sui generis* is based, in part, on the fact that their traces are derived from the hypocotyl and not from the stem, as are those of leaves, and that usually, in position on the axis, cotyledons do not fall into the phyllotactic spirals. But they are lateral appendages of the axis, as are leaves; their traces are derived from the hypocotyl, as leaf traces are derived from the stem. The hypocotyl is transitional in structure from root to stem, and its upper part frequently has typical stem structure. The position of the cotyledons—at the base of the stem —is a feature of the crowding and condensation of the organs in the embryo. In position, form, structure, and ontogeny, the cotyledons are specialized leaves, not organs *sui generis.*

It has been argued that the cotyledon is not the equivalent of the leaf, because no leafbase or cushion is formed in its ontogeny. But the existence of leafbases is a controversial subject, and the recognition of them in the embryo would be most difficult, because all parts of the embryo are in the primordial stage and crowded together.

The interpretation of the cotyledon of the most advanced embryo types of the monocotyledons as consisting of scutellum plus the coleoptile goes back at least to the middle of the nineteenth century; it is so described in the textbooks of botany of that period. At the end of the nineteenth century, the monocotyledon embryo was remarkably well interpreted, but the early twentieth century saw the sound interpretation replaced by others now seen to be incorrect.

Two types of primitive cotyledon seem to be present in the monocotyledons: the cylindrical or ligulate type, without differentiation of lamina and base—Helobiales, Pandanales, some Liliales; and the leaflike type, with lamina and petiole—some Liliaceae (*Paris*, *Trillium*). In each of these types, absorbing structures have arisen independently by modification of the cotyledon tip. In the dicotyledons, the cotyledon is more leaflike in form, and only rarely is a suctorial structure formed. Absorption of food stored in the endosperm as a function of the cotyledon has doubtless arisen several or many times in angiosperms.

THE PRIMITIVE ANGIOSPERM EMBRYO

Both the minute, little-developed embryo and the polycotyledonary embryo have been considered the primitive type—the undeveloped embryo, because of its simplicity, and the polycotyledonary embryo, because of resemblance to supposedly primitive conifers with several cotyledons. Polycotyledony, syncotyledony, schizocotyledony, and monocotyledony all appear to represent modifications of dicotyledony. The minute, undifferentiated embryo, embedded in abundant endosperm, seems to be primitive; the well-developed embryo, with large cotyledons and no endosperm, specialized. The undeveloped embryo that continues development (after-ripening) in the ripe seed—under favorable conditions of water supply and temperature—resembles that of some gymnosperms—*Ginkgo,* the cycads—where the embryo grows continuously after seed shedding.

The seedlings of a few angiosperms, in contrast with the usual rapid development, remain entirely underground the first growing season after germination—species of *Veratrum* and *Arum*. In this character, they resemble the sporophytes of *Botrychium, Ophioglossum,* and *Lycopodium.* Slow development may be a primitive character or the result of after-ripening outside the seed.

BIBLIOGRAPHY

Artschwager, E.: Development of flowers and seed in the sugar beet, *Jour. Agr. Res.*, 34: 1–25, 1927.

———, E. W. Brandes, and R. C. Starrett: Development of flower and seed of some varieties of sugar cane, *Jour. Agr. Res.*, 39: 1–30, 1929.

——— and Ruth Starrett: The time factor in fertilization and embryo development in the sugar beet, *Jour. Agr. Res.*, **47**: 823–843, 1933.

Avery, G. S., Jr.: Comparative anatomy and morphology of the embryos and seedlings of maize, oats, and wheat, *Bot. Gaz.*, **89**: 1–39, 1930.

Bancroft, N.: A review of literature concerning the evolution of monocotyledons, *New Phyt.*, **13**: 285–308, 1914.

———: The arborescent habit in angiosperms, *New Phyt.*, **29**: 227–275, 1930.

Beccari, O.: Nota sull' embrione delle Dioscoriaceae, *Nuov. Giorn. Bot. Ital.*, **2**: 149–155, 1870.

Boyd, L.: Evolution in the monocotyledonous seedling: A new interpretation of the morphology of the grass embryo, *Trans. Proc. Bot. Soc. Edinburgh*, **30**: 286–303, 1931.

———: Monocotylous Seedlings, morphological studies in the post-seminal development of the embryo, *Trans. Bot. Soc. Edinburgh*, (I) **31**: 5–224, 1932.

——— and G. S. Avery, Jr.: Grass seedling anatomy: the first internode of *Avena* and *Triticum*, *Bot. Gaz.*, **97**: 765–779, 1936.

Brink, R. A., and D. C. Cooper: The endosperm in seed development, *Bot. Rev.*, **13**: 423–541, 1947.

Bruns, E.: Der Grasembryo, *Flora*, **76**: 1–33, 1892.

Bucherer, E.: Beiträge zur Morphologie und Anatomie der Dioscoreaceae, *Bibl. Bot.*, **16**: 1–34, 1889.

Buchholz, J. T.: Studies concerning the evolutionary status of polycotyledony, *Am. Jour. Bot.*, **6**: 106–119, 1919.

Cêlakovsky, L. J.: Ueber die Homologien des Grasembryos, *Bot. Zeit.*, **55**: 141–174, 1897.

Chouard, P.: Types de développement de appareil végétatif chez les Scillées, *Ann. Sci. Nat. Bot.*, 10 sér., **13**: 131–323, 1931.

Compton, R. H.: An anatomical study of syncotyly and schizocotyly, *Ann. Bot.*, **27**: 793–821, 1913.

Coulter, J. M.: The origin of monocotyledony in the grasses, *Ann. Mo. Bot. Gard.*, **2**: 175–183, 1915.

——— and W. J. G. Land: The origin of monocotyledony, *Bot. Gaz.*, **57**: 509–519, 1914.

Crété, P.: Lentibulariacées: Développement de l'embryon chez *Pinguicula leptoceras* Rchb., *Compt. Rend. Acad. Sci. Paris*, **242**: 1063–1065, 1956.

Davey, A. J.: Seedling anatomy of certain Amentiferae, *Ann. Bot.*, **30**: 575–599, 1916.

Duchartre, P.: Mémoire sur les embryons qui ont été décrits comme polycotylés, *Ann. Sci. Nat. Bot.*, 3 sér., **10**: 207–237, 1848.

Ebeling, M.: Die Saugorgane bei der Keimung endospermhaltiger Samen, *Flora*, **68**: 178–202, 1885.

Findeis, M.: Über das Wachstum des Embryos im ausgesälten Samen vor der Keimung, *Sitzber. Akad. Wiss. Math.-Natur Kl.*, **126**: 77–102, 1917.

Fletcher, J. J.: Illustrations of polycotyledony in the genus *Persoonia*, with some reference to *Nuytsia*, *Proc. Linn. Soc. N.S.W.*, **33**: 867–882, 1909.

Gardiner, W., and A. W. Hill: The histology of the endosperm during germination in *Tamus communis* and *Galium tricorne*, *Proc. Camb. Phil. Soc.*, **11**: 445–457, 1902.

Gatin, C. L.: Premières observations sur l'embryon et la germination des Broméliacées, *Rev. Gén. Bot.*, **23**: 49–66, 1911.

———: La morphologie de la germination avec la phylogénie, *Rev. Gén. Bot.*, **21**: 147–158, 1909.

Gressner, H.: Zur Keimungsgeschichte von *Cyclamen, Bot. Zeit.*, **32**: 801–814, 817–840, 1874.

Haccius, B.: Die Embryoentwicklung bei *Ottelia alismoides* und das Problem des terminalen Monokotylen-keimblatts, *Planta*, **40**: 443–460, 1952.

———: Verbreitung und Ausbildung der Einkeimblättrigkeit bei den Umbelliferen, *Oesterr. Bot. Zeitschr.*, **99**: 483–505, 1952.

———: Embryologische und histogenetische Studien an "monokotylen Dikotylen." I. *Claytonia virginica*, L., *Oesterr. Bot. Zeitschr.*, **101**: 285–303, 1954.

——— and E. Hartl-Baude: Embryologische und histologische Studien an "monokotylen Dikotylen." II. *Pinguicula vulgaris* L. und *P. alpina* L., *Oesterr. Bot. Zeitschr.*, **103**: 567–587, 1957.

Haskell, G.: Pleiocotyly and differentiation within angiosperms, *Phytomorph.*, **4**: 140–152, 1954.

Henslow, G.: A theoretical origin of endogens from exogens through self-adaptation to an aquatic habit, *Jour. Linn. Soc. Bot. London*, **29**: 485–528, 1891.

Hill, A. W.: Morphology and seedling structure of the geophilous species of *Peperomia*, together with some views on the origin of monocotyledons, *Ann. Bot.*, **20**: 395–427, 1906.

———: A revision of the geophilous species of *Peperomia*, with some additional notes on their morphology and seedling structure, *Ann. Bot.*, **21**: 139–160, 1907.

———: Studies in seed germination: Experiments with *Cyclamen*, *Ann. Bot.*, **34**: 417–429, 1920.

Holm, T.: Contributions to the knowledge of the germination of some North American plants, *Mem. Torrey Bot. Club*, **2**: 57–108, 1891.

Janczewski, E. de: Études morphologiques sur le genre *Anemone*, *Rev. Gén. Bot.*, **4**: 241–258, 289–299, 1892.

Johanson, D. A.: A critical survey of the present status of plant embryology, *Bot. Rev.*, **11**: 87–143, 1945.

Johri, B. M.: Studies in the family Alismaceae. III. *Sagittaria guayanensis* HBK and *S. latifolia* Willd., *Proc. Indian Acad. Sci.*, sec. B., **2**: 33–48, 1935.

——— and R. N. Kapil: Contribution to the morphology and life history of *Acalypha indica* L., *Phytomorph.*, **3**: 137–151, 1953.

Kennedy, P. B.: The structure of the caryopsis of grasses with reference to their morphology and classification, *Bull. U.S. Dept. Agr., Div. Agrostology*, no. 19, 1–44, 1899.

Klebs, G.: Beiträge zur Morphologie und Biologie der Keimung, *Untersuch. Bot., Inst. Tübingen*, **1**: 536–635, 1881–85.

Laurent, M.: Recherches sur le développement des Joncées, *Ann. Sci. Nat. Bot.*, 8 sér., **19**: 97–194, 1904.

Lewin, M.: Bidrag till hjertbladets anatomi hos monokotyledonerna, *König. Svensk. Vet.-Akad. Handl. Bihang.*, **12**: 1–28, 1886.

Li, Hui-Lin: The evolutionary significance of the endosperm and its bearing on the origin of the angiosperms, *Jour. Wash. Acad. Sci.*, **47**: 33–38, 1957.

Maheshwari, P.: The male gametophyte of angiosperms, *Bot. Rev.*, 15: 1–75, 1949.

———: The embryology of the angiosperms, a retrospect and prospect, *Curr. Sci.*, 25: 106–110, 1956.

McCall, M. A.: Developmental anatomy and homologies in wheat, *Jour. Agr. Res.*, **48**: 283–321, 1934.

Metcalf, C. R.: An interpretation of the morphology of the single cotyledon of *Ranunculus Ficaria* based on embryology and seedling anatomy, *Ann. Bot.*, **50**: 103–120, 1936.

Philipson, W. R.: The development and morphology of the ligule in grasses, *New Phyt.*, **34**: 310–325, 1935.

Reeder, J. R.: Affinities of the grass genus *Beckmannia* Host., *Bull. Torrey Bot. Club,* **80**: 187–196, 1953.

Sargant, E.: A theory of the origin of monocotyledons, founded on the structure of their seedlings, *Ann. Bot.*, **17**: 1–92, 1903.

———: The reconstruction of a race of primitive angiosperms, *Ann. Bot.*, **22**: 121–186, 1908.

——— and A. Arber: The comparative morphology of the embryo and seedling in the Gramineae, *Ann. Bot.*, **29**: 161–222, 1915.

Schlickum, A.: Morphologischer und anatomischer Vergleich der Kotyledonen und ersten Laubblätter der Keimpflanzen der Monokotyledonen, *Biblioth. Bot.*, **35**: 1–88, 1896.

Schmid, B.: Beiträge zur Embryoentwicklung einiger Dikotylen, *Bot. Zeit.*, **60**: 207–230, 1902.

Schnarf, K.: "Vergleichende Embryologie der Angiospermen." In Linsbauer, K. "Handbuch der Pflanzenanatomie," Abt. X, Teil 2, Berlin, 1929.

———: Die Bedeutung der embryologischen Forschung für natürliche System der Pflanzen, *Biol. Gen.*, **9**: 271–288, 1933.

Schneider, M.: Untersuchungen über die Embryobildung und -entwicklung der Cyperaceen, *Beih. Bot. Centralbl.*, **49** (I): 649–674, 1932.

Solms-Laubach, H. Graf zu: Ueber monocotyle Embryonen mit Scheitelbürtigen Vegetationspunkt, *Bot. Zeit.*, **36**: 65–74, 81–93, 1878.

Stebbins, G. L., Jr.: Cytogenetics and evolution of the grass embryo, *Am. Jour. Bot.*, **43**: 890–905, 1956.

Sterckx, R.: Recherches anatomiques sur l'embryon et les plantules dans la famille des Renonculacées, *Archiv Inst. Bot. Liège*, **2**: 3–92, 1900.

Swamy, B. G. L.: A contribution to the life history of *Casuarina*, *Proc. Am. Acad. Arts and Sci.*, **77**: 1–32, 1948.

——— and I. W. Bailey: The morphology and relationships of *Cercidiphyllum*, *Jour. Arnold Arb.*, **30**: 187–210, 1949.

Thomas, E. N. M.: A theory of the double leaf trace founded on seedling structure, *New Phyt.*, **6**: 77–91, 1907.

Tronchet, A.: Recherches sur les types d'organization les plus répandus de la plantule des dicotylédones: leurs principales modifications, leurs rapports, *Archiv Bot.*, **4**: 1–249, 1930.

Tschirch, A.: Physiologische Studien über die Samen, ins besondere die Saugorgane derselben, *Ann. Jard. Bot. Buitenzorg*, **9**: 143–183, 1891.

Van Tieghem, P.: Observations anatomiques sur le cotylédon des Graminées, *Ann. Sci. Nat. Bot.*, 5 sér., **15**: 236–276, 1872.

———: Morphologie de l'embryon et de la plantule chez les Graminées et les Cypéracées, *Ann. Sci. Nat. Bot.*, 8 sér., **3**: 259–309, 1897.

Wardlaw, C. W.: "Embryogenesis in Plants," New York, 1955.

Webber, H. J.: Polyembryony, *Bot. Rev.*, **6**: 575–598, 1940.

Winkler, H. Die Monocotylen sind monocotyl, *Beitr. Biol. Pfl.*, **19**: 29–34, 1931.

Worsdell, W. C.: The morphology of the monocotyledonous embryo and that of the grass in particular, *Ann. Bot.*, **30**: 509–549, 1916.

Yarborough, J. A.: *Arachis hypogaea:* The seedling, its cotyledon, hypocotyl, and roots, *Am. Jour. Bot.*, **36**: 758–772, 1949.

Chapter 10

THE SEED

The seed, like many other plant structures, cannot be rigidly defined. The term seed is loosely used, even in technical discussions; there is little agreement even on such important characters as the presence of an embryo and of dormancy in the embryo. The definition, "a seed is a fertilized ovule," is inadequate, because it is applied to fertilized ovules, both immature and mature. The presence of an embryo in some stage of development is usually implied, but many fossil seedlike structures that show no evidence of an embryo are termed seeds. And, in some living gymnosperms—*Ginkgo* and the cycads—there may be no embryo present when the seed is ripe and shed. (The ripe, plumlike "fruits" of *Ginkgo,* which fall to the ground with the leaves, may be either ovules or seeds, if the presence of an embryo determines a seed. Fertilization frequently occurs on the ground.) But, in angiosperms, probably all so-called seeds possess an embryo in some stage of development. Some definitions of the seed include dormancy of the embryo as an essential character. But, in many taxa, the embryo continues to grow within the seed following its shedding, without a dormant period; a seed may be "dormant" as far as external structure goes, but growing internally. Some angiosperm seeds apparently have no period of dormancy, suggesting a survival of a primitive stage in seed evolution—some tropical genera (*Myristica, Durio*); some aquatic genera (*Thalassia*). Dormancy of the embryo seems to represent an advanced stage in the evolution of the seed.

Seeds are commonly called "ripe" when they are fully grown and shed from the mother plant, regardless of the stage of development of their embryos; seeds have been called technically "ripe" only when ready to germinate. After-ripening—morphological and physiological—may delay readiness to germinate long after shedding. The term seed must remain in morphological use and will commonly be applied to matured ovules that contain an embryo at some stage. A loose use of the term for seedlike fossils that may be either ovules or seeds is necessary, as it is for the ripe "fruits" of *Ginkgo.*

Size of Seeds

The size of seeds has been considered as perhaps of phylogenetic importance. Both large and small seeds have been called primitive. Large size seems often correlated with the tree habit, especially with the trees of tropical forests. But large seeds often have the most highly developed embryos and are frequent in temperate-climate forests. Small seeds have been interpreted as most primitive because their embryos are usually little developed. But, in many seeds, after-ripening brings about greater development before germination. Seeds of medium size with well-developed embryos have also been called primitive. Size of seed seems less important, from the standpoint of primitiveness, than time of inception of dormancy and stage of specialization of the embryo. Since there is no period of dormancy in the seeds of some gymnosperms —cycads and *Ginkgo*—and of a few angiosperms, and since there is a long-continuing, after-ripening development in some primitive families, absence of dormancy seems probably primitive. (After-ripening may merge with germination.) Small seeds and after-ripening are associated chiefly with herbaceous plants and with a short period of seed maturation.

Structure of the Seed

The basic structure of the seed is that of the ovule, but, as the ovule matures, some parts may be lost and the nature of others obscured. In seeds formed from anatropous and campylotropous ovules, some evidence of ovule type is commonly present, but appendages and histological modifications may bury evidence of ovule type. In endospermless seeds, form may be determined by the presence of a large embryo, as in many legumes. Position of the micropyle may be evident as a minute opening or concavity, and the place of abscission from the funicle as a scar, the *hilum*. (In early descriptions of the ovule, the term hilum was sometimes applied to the enlarged tip of the funicle.)

The Seed Coats. The integuments of the ovule form the *seed coats* or *testa* of the seed. As the ovule enlarges, tissues of the integuments undergo great changes. The inner integument is commonly reduced and often lost; the outer may become massive and highly differentiated as a protective structure. Histologically, the coats may consist of few or many layers, which vary in different taxa in their relation to the ovular integuments (Fig. 137). Of the ovular integuments, some layers degenerate; others increase greatly in number and kinds of tissues. Misinterpretations of the integumentary make-up of seed coats have been made through failure to note loss of one integument in ontogeny. An inner integument that appears to be lost may be reduced to a single layer of delicate cells. Both integuments persist in some families—Rosaceae, Rutaceae, Euphorbiaceae—and remain more or less distinct in the mature seed. In the bamboos, absorption of the integument begins immediately after fertilization and is reduced to a single layer of cells. In *Zea*, only a few remnants of the integument persist in the fruit. Seed coats consisting of a single layer of cells are frequent in highly specialized seeds (Fig. 138*B*)—those of the orchids, some epiphytes and parasites (*Orobanche, Monotropa*).

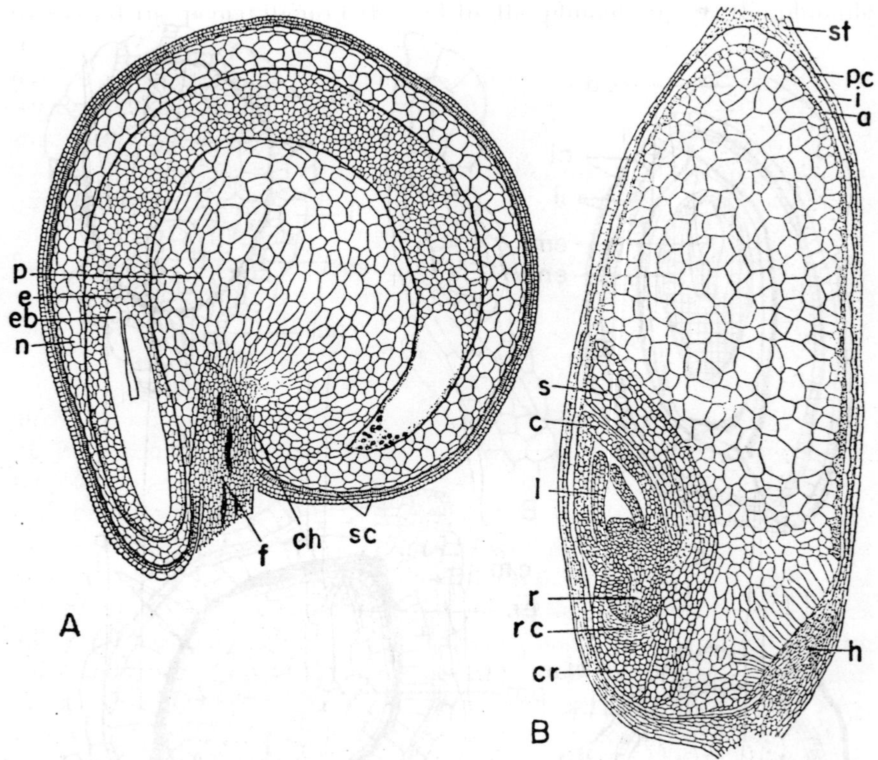

Fig. 137. Longitudinal sections of *A*, a dicotyledonous seed, *Beta vulgaris;* and *B*, a monocotyledonous fruit, *Sorghum* sp. (sugar cane). *a,* aleurone layer of endosperm; *c,* coleoptile; *ch,* chalaza; *cr,* coleorhiza; *e,* endosperm; *eb,* embryo; *f,* funicle; *h,* hilum; *i,* inner integument; *l,* plumular leaf; *n,* nucellar perisperm; *p,* perisperm; *pc,* pericarp; *r,* root primordium; *rc,* root cap; *s,* scutellum; *sc,* seed coats; *st,* base of style. (*A, after Artschwager; B, after Artschwager, Brandes, and Starrett.*)

Major and minor appendages may develop on the integuments—wings, spines, hairs, glands; these may be epidermal in origin or involve deeper tissues also. Type of appendages may be of some importance in indicating relationships. In the Meliaceae, the fleshy seeds of *Amoora* have winglike appendages, suggesting derivation from winged genera in the family. Appendages of the chalazal region are unusual, and their presence may suggest relationships.

The vascular supply of the seed coats is built up, as the seed enlarges, from the supply of the integument or, where the integuments have no vascular bundles, by extension from the bundles of the funicle. In delicate seed coats, the new tissues may form a prominent feature of seed structure. In the Capparidaceae and Thymeleaceae, a layer of

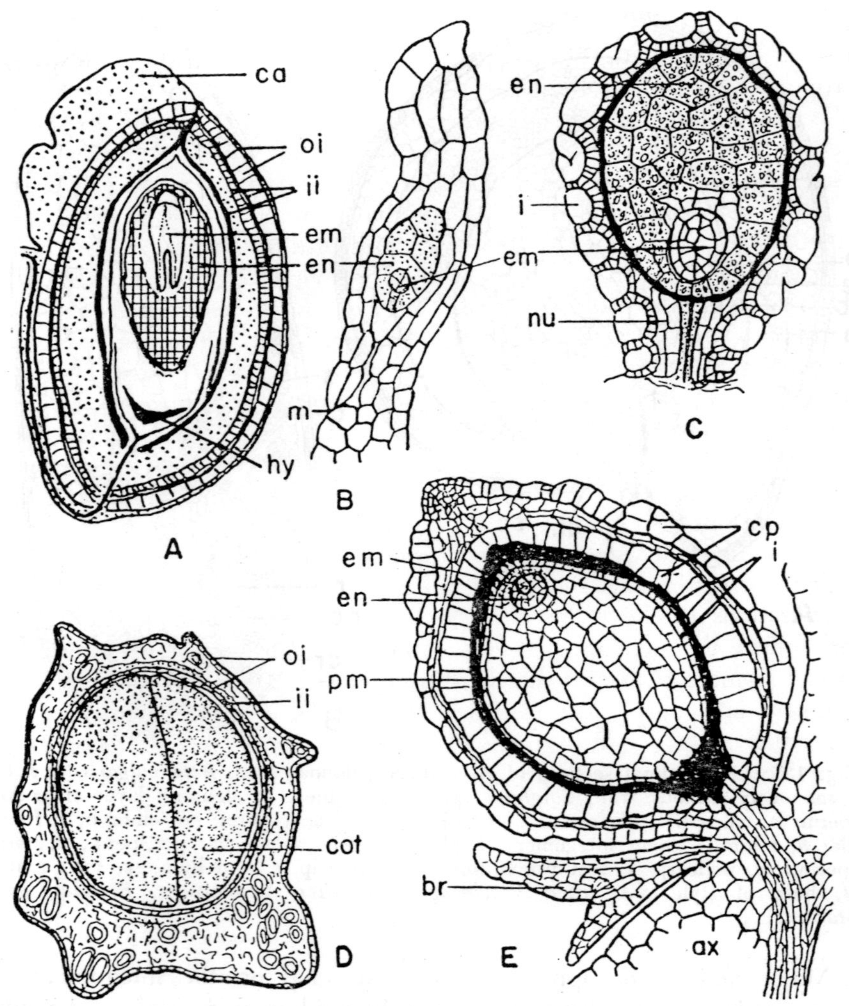

Fig. 138. Structure of seeds and a fruit *E; A, B, C, E,* longitudinal sections; *D,* cross section. Embryo well-developed in *A, D;* embryo in early stages in *B, C, E. A, Acalypha indica; B, Monotropa Hypopitys; C, Orobanche; D, Oenothera biennis; E, Peperomia pellucida. ca,* caruncle; *ax,* axis; *br,* bract; *cot,* cotyledon; *cp,* carpel (pericarp); *en,* endosperm; *em,* embryo; *hy,* hypostace; *i,* integument; *ii,* inner integument; *m,* micropyle; *nu,* nucellus; *oi,* outer integument; *pm,* perisperm. (*A, after Johri and Kapil; B, C, after Koch; D, after Kayser; E, after Johnson.*)

tracheidlike cells has been described as coating the embryo like a third integument. This tissue perhaps represents a sheath of vascular strands of the outer integument.

The Aril. The fleshy coats and appendages of some seeds have various origins. They may represent fleshy transformations of the outer cell layers of the outer integument; proliferations of parts of this integument, of the chalaza, or of the distal part of the funicle. Those that arise at the base of the ovule from annular primordia and develop acropetally, sheathing the integuments more or less completely, are called *arils* (see also under the Ovule). But the term aril is commonly used loosely to cover any fleshy, external part of the seed. Definitions usually restrict this term to sheathing coats developed from the base of the ovule. Those developing from the micropylar rim of the outer integument have been called *false arils* or *arilloids*. Some fleshy coats apparently develop from both ends of the ovule. Development of an aril may begin before fertilization but commonly begins later, while the ovule is enlarging. The term aril has sometimes been used to cover all modifications of the outer integument in the seed, even including some that are not fleshy, such as wings and spines. Where fleshy tissues are restricted to crests, as along the raphe, they are termed *strophioles;* where restricted to the base or to the apex of the seed, they are termed *caruncles* (Fig. 138A). These terms are confused in usage. Perhaps the only important distinction among external fleshy parts of the seed is between those that represent transformed layers of the integuments, chalaza, or funicle and those that are new parts of the ovule, secondary proliferations of the original ovule body, such as the so-called third and fourth integuments. (There is no convincing evidence that any taxon has more than two integuments.) Fleshy layers of the outer integument are conspicuous in *Lycopersicum, Passiflora, Punica, Carica.* Seeds that mature free from the ovary have integuments differentiated as in fruits; in *Caulophyllum,* the seed resembles a fruit in its fleshy outer and stony inner coats. Fleshy seed coats were long ago considered proliferations of the ovary wall or the placenta; those of the tomato have been called placental. Arils have been interpreted as third and fourth integuments. In the large genus of the Proteaceae, *Hibbertia,* arils, which are often called integuments, are merely elaborations of the outer normal integument; the various species show all stages in the development of an enclosing aril, from a fringed tip of the outer integument to a complete fleshy sheath.

The possession of arillate seeds has been considered a primitive angiosperm character, but this claim seems unlikely, because arils occur in taxa scattered throughout angiosperms. Arils appear to be ecological modifications related to dissemination.

The Endosperm in the Seed. Part or all of the endosperm formed in early stages of embryo development serves in nutrition for the embryo as it matures—both before the seed is ripe and, in many taxa, during after-ripening also. Seeds are commonly described as *albuminous*, with endosperm, and *exalbuminous*, without endosperm, but no line can be drawn between these types, because small amounts of endosperm are easily overlooked. Seeds described as exalbuminous, as in *Pyrus*, often have small amounts of endosperm. Although endosperm is usually formed at time of fertilization, the seeds of many families are described as without endosperm—Compositae, Geraniaceae, Cruciferae, Cucurbitaceae, Myrtaceae, Aceraceae, Helobiales; other families have genera with endosperm and others without—Araceae, Leguminosae, Rosaceae, Proteaceae, Betulaceae. The Proteaceae show loss of endosperm within the family; the more primitive genera have a little; other genera, none. The Eleagnaceae and Cactaceae have little or no endosperm. Parasites and saprophytes commonly have a uniseriate layer of endosperm cells. In angiosperms as a whole, albuminous seeds are more common in the monocotyledons than in the dicotyledons. Where there is no endosperm and the cotyledons are poorly developed, food may be stored in other parts of the embryo, especially the hypocotyl. Abundant endosperm, accompanying minute, undifferentiated embryos, characterizes many ranalian families. (Calycanthaceae are a prominent exception.) This ranalian seed structure is perhaps primitive for the angiosperms.

Histologically, endosperm varies greatly in different taxa, from soft and loose ("mealy") to hard and compact, with heavy cell walls. There is also great variety in the nature of the stored food—starch, oil, sugar.

The morphological nature of the endosperm has been much discussed. Its complex cytological make-up—with $3n$ nuclei—the result of a gametophytic union of two female nuclei and one male, has suggested three interpretations. One interpretation is that it is an abnormal sporophytic structure, formed by the addition of a second female nucleus to the union of egg and sperm nuclei. (The addition of an antipodal, vegetative nucleus to this triple union is considered unimportant.) The second interpretation is that the endosperm represents an abnormal gametophyte, a modified part of the vegetative tissue of the embryo sac, with the addition of a male nucleus. The third theory interprets the endosperm as a third phase—additional to sporophyte and gametophyte—in the life cycle, a phase represented in angiosperms by remnants of an ancestral structure present in thallophytes, but not found in other vascular plants.

The Perisperm. The term perisperm is applied to nutritive tissue belonging to the integuments or nucellus, or to both these tissues. It may consist of a single layer of cells, restricted in distribution, or may form a mass of cells, which enclose the embryo and endosperm. Perisperm is characteristic of many taxa of the higher dicotyledons and is prominent in the seeds of the Caryophyllaceae, Polygonaceae, Chenopodiaceae (Fig. 137*A*), and Phytolaccaceae; in the monocotyledons, integumentary perisperm is apparently unknown, but nucellar perisperm is characteristic of some taxa.

Perisperm is infrequent in angiosperms as a whole. Both perisperm and endosperm occur in a few families—Musaceae, Piperaceae, Zingiberaceae, some of the Nymphaeaceae, Chenopodiaceae (Fig. 137*A*), and related families. But endosperm without perisperm is the common condition.

Nucellar perisperm is uncommon, because, as the embryo develops, the nucellus degenerates and is commonly completely absorbed, but nucellar perisperm is characteristic of the Piperaceae and Scitamineae.

Integumentary perisperm is more common than nucellar perisperm. The integuments frequently serve as food storage regions; the tissues that contain the food form the integumentary perisperm.

The perisperm serves as an accessory nutritive tissue, supplementing the endosperm, which, when present, acts as an intermediary tissue transferring food from adjacent perisperm cells to the embryo. In some families, large amounts of perisperm food seem to be made available to the seedling through the activities of the endosperm—Cannaceae, Caryophyllaceae, Chenopodiaceae (Fig. 137*A*), Piperaceae (Fig. 138*E*). The transfer may take place during maturation of the seed, during after-ripening, or during germination.

The embryo and seedling seem to be nourished at all stages through tissues of the female gametophyte as an intermediary, not directly by the mother sporophyte.

THE FRUIT

In angiosperms, a *fruit* is the matured gynoecium, together, in many taxa, with other floral organs or parts of organs. A fruit has been called a "mature flower," in some respects a good definition, because, at pollination, the flower is an immature structure in axis and gynoecium. The pedicel and peduncle may form accessory parts of the fruit. Like many other terms, the term fruit is used loosely; it is applied to the separate carpels of the apocarpous gynoecium (sometimes called *fruitlets*)—*Fragaria*—and to fruiting inflorescences—*Ananas*. Descriptive terms for fruit types are many, but only in part have a morphological base. Classifications of fruits are largely or wholly "artificial"; some classifications have been called "natural," but these make use of such characters as "fleshy" and "dry." Most classifications of fruits have been prepared for taxonomic purposes and, as such, are important. Classification of fruits and descriptions of types are discussed here only in so far as they have morphological and phylogenetic value.

General structure of the gynoecium—*apocarpous,* carpels free from one another, and *syncarpous,* carpels united with one another—is an important basis for natural classification, because freedom of floral organs is clearly a primitive condition.

In an old, but still much-used, classification, fruits are classified as *simple, multiple,* or *aggregate.* Simple fruits consist of one, or of several united, carpels; multiple fruits consist of the gynoecia of more than one flower, cohering or connate on a common axis; aggregate fruits consist of the carpels of one gynoecium, free in the flower but coherent in the fruit (Fig. 139). (The terms multiple and aggregate are reversed in some definitions.)

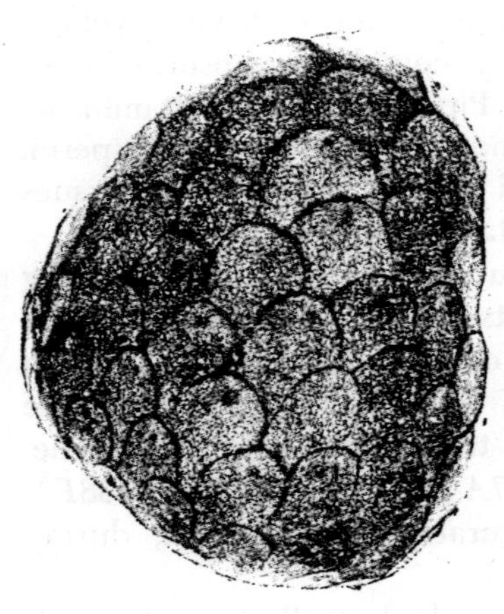

Fig. 139. Fruit of *Annona* sp.

Fruits in which organs other than the carpels are present—receptacle, pedicel, inflorescence axis, sepals, bracts—are *accessory* (Fig. 140). For example, the fruit of the pineapple, *Ananas,* is multiple and accessory, consisting of an inflorescence axis, with many connate ovaries fused with the bases of their subtending bracts and with the axis on which they are borne. The fruits of *Bromelia,* a related genus, are free and simple and without accessory parts. Opinions differ as to what constitutes the accessory parts of a fruit. In some descriptions of the pineapple fruit, the bracts are omitted, but the bracts are intimately fused with the flowers—even their vascular systems are partly united—and their bases form part of the flesh. On the other hand, the burrs of the beech and chestnut and the acorn cup—compacted, sterile branches of the inflorescence—are not considered part of the fruit, although, in *Quercus,* the acorn cup is adnate to the ovary in varying degrees. Morphologically, a limitation of "accessory" parts cannot be made. Terms based on histological structure—dry, fleshy, stony, etc.—and on method of dissemination have little or no morphological significance, and lines between types cannot be drawn. The range in fruit within a family may be great; in the Proteaceae, follicle, drupe, nut, and achene, and, in the Rosaceae, many types. The fruit of the Ericaceae ranges from dry to fleshy; that of *Epigaea* is a dry capsule, which opens at maturity, and an enlarged fleshy placenta protrudes. (The seeds have been described as having arils or caruncles.)

Dehiscence

Fruits are *dehiscent* when the tissues of the mature ovary wall—*pericarp*—break open, freeing the seeds; and *indehiscent* when the seeds

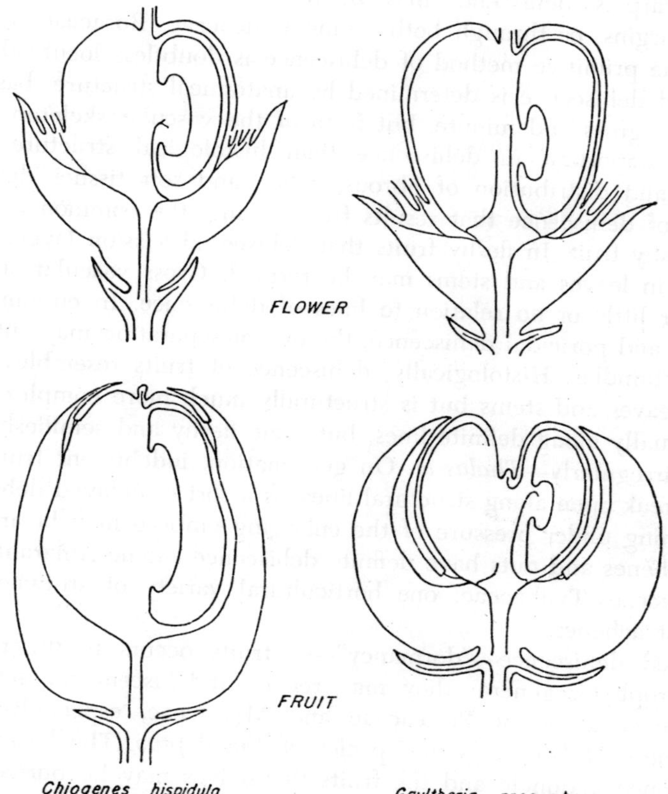

Fig. 140. Diagrams of longitudinal sections of flowers and fruits showing part played by floral organs other than carpels in the formation of an accessory fruit. Outer flesh of fruits consists of united bases of sepals, petals, and stamens: in *Chiogenes,* adnate to enlarged ovary; in *Gaultheria,* free from ovary. (*From Eames,* 1953.)

remain in the fruit after shedding. Dehiscence through the dorsal region of the carpel—a break opening directly into the locule—is *loculicidal,* and each carpel is split into halves; dehiscence ventrally through the septa—the common walls of adjacent carpels—separating the individual carpels and opening along the carpel margins, is *septicidal.* These two terms are commonly applied to syncarpous fruits. Free carpels, in fruit, may dehisce along ventral, dorsal, or both lines. Septicidal dehiscence consists, in part, of freeing of the individual carpels of a syncarpous ovary by splitting along the lines of carpel union. Opening of the carpels, so separated, is either along the ventral suture, as in *Tofieldia* and other primitive liliacean genera, by breakdown of the lateral walls in various degrees, or by a combination of these methods. In free carpels, dehiscence may be through either dorsal or ventral carpel margins, or through both—some Urticaceae, Rosaceae, Leguminosae. The primitive method of dehiscence is doubtless loculicidal.

Type of dehiscence is determined by anatomical structure, basic and secondary, gross and minute, but form of the vascular skeleton is usually less important in dehiscence than histological structure. Type, amount, and distribution of fibrous, stony, and soft tissues determine position of dehiscence that results from drying, the common condition in nonfleshy fruit. In fleshy fruits that dehisce, abscission layers similar to those in leaves and stems may be formed. Gross vascular structure may bear little or no relation to lines of dehiscence. In circumscissile, valvular, and poricidal dehiscence, the line of separation may cut across vascular bundles. Histologically, dehiscence of fruits resembles abscission of leaves and stems but is structurally much more complex. Opening is usually along definite lines, but some fleshy and semifleshy fruits dehisce irregularly—*Thalassia.* On germination, indehiscent fruits commonly break open along structural lines, as a sort of delayed dehiscence, but opening under pressure of the enlarging embryo may be irregular. Some achenes and nuts have definite dehiscence—some Amaranthaceae, Polygonaceae, Typhaceae; one horticultural variety of strawberry has dehiscent achenes.

Delayed dehiscence—"dormancy"—of fruits occurs in the fruits of some xerophytic genera; they may remain indehiscent for several or many years—those of Proteaceae and Myrtaceae, especially *Hakea, Callistemon, Melaleuca,* some species of *Eucalyptus.* The bases of the inflorescence peduncle and the fruits themselves may be buried in the bark and wood of the stems that bear them before the fruits open. (The cones of some species of *Pinus* become similarly somewhat sunken in branches before shedding.) Fruits of some species of *Eucalyptus* require over a year to mature and may not dehisce for many years, apparently not until their water supply ceases and the twig bearing them dies. In *Callistemon,* the pericarp, with some chloroplasts, continues slow growth for ten to eighteen years, dehiscing after three to twenty years. A cork cambium builds up a firm, corky surface. The cause of the abundant development, after forest fires, of seedlings of these xerophytes with long-delayed fruit dehiscence seems to be the killing of the twigs that bear them and the consequent shrinking and opening of the fruits.

Most indehiscent fruits are monocarpellary, but there are many exceptions, such as the fruits of the Fagaceae, Juglandaceae, Gramineae, Betulaceae, many Palmae.

Pericarp

Where the matured ovary wall, the *pericarp,* is differentiated into histologically distinct layers, these are distinguished as *exocarp, mesocarp,* and *endocarp*—outer, middle, and inner layers, respectively—or into exocarp and endocarp only. Distinction of these layers is generally sharper in fleshy than in dry fruits.

In the maturing of many fleshy fruits, as in *Rubus* and *Fragaria,* the rapid enlargement in ripening is the result of cell enlargement, not of new-cell formation. Apparently, in many kinds of fruits, cell multiplication in the pericarp ceases early, even at, or just after, flowering, as in *Rubus* and *Ananas.*

Stomata are frequent and lenticels occasional on fruits, especially on the fleshy types, appearing at different times in different taxa as the fruit enlarges, sometimes even in late stages of enlargement. They may form on only the outer surface of the pericarp; this is the common condition. But they may form on both outer and inner surfaces, as in capsules—*Papaver, Datura, Chelidonium, Hyoscyamus;* they are rare on the inner surface only—*Capsicum.*

The Integuments in Fruits

In indehiscent fruits and many dehiscent fruits, the inner integument is commonly much reduced or lost, the outer remaining thin, or much thickened and stony. Rarely, the inner alone survives; in the bamboos, the outer integument is absorbed immediately following fertilization, and the inner is reduced to a single layer of cells. Occasionally, both persist, as in the palms, where the outer contributes to the formation of the stony shell. In some achenes, especially in the caryopses of grasses, the integuments are represented merely by vestiges; in the millet, *Panicum milliaceum,* by the epidermis of the nucellus; in *Zea,* by a noncellular suberized layer, a remnant of the nucellar epidermis. In seeds maturing outside the carpel, as in *Caulophyllum,* the integuments become massive and pericarplike, with fleshy outside and stony inside layers.

Evolutionary Relationships of Fruit Types

The follicle is clearly the primitive fruit type, derived directly from the primitive carpel. Modification of the free carpel produces many types. Reduction of ovule number to one and loss of dehiscence produces the *achene* and other types. But the descriptive term achene is sometimes applied to fruits with two or more carpels—*Compositae*, *Typhaceae*—or with two seeds. No morphological line separates an achene from a *nut*, though the nut is typically syncarpous. The syncarpous structure of most nuts is obscure in the mature fruit, and errors of interpretation of carpel structure of nuts have been made. The fruit of *Corylus* has been described as monocarpellate, but is clearly bicarpellate, with one ovule in each locule. Morphologically, achenes consist of a seed with an enclosing ovary wall. The seed coats may be free from, or adnate to, the ovary wall. They may be reduced to a thin, papery layer or be represented by mere vestiges of tissue. Where there is intimate adnation, the two structures may be distinguishable only by ontogenetic studies.

Fruits may consist of seeds without an ovary wall; the seeds may ripen "naked"—*Caulophyllum, Lourya,* some parasites. (The naked seeds of certain parasites—Sarcopodales—have been considered to be primitively naked, and the taxon, therefore, not to be angiospermous. This interpretation provided part of the basis for the establishment of a taxon, the Prephanerogamae, intermediate between the gymnosperms and angiosperms.) In *Caulophyllum,* the developing fertilized ovules soon break through the ovary wall, which ceases to grow, and the ovule matures like a fruit.

Correlation has been seen between advanced characters—fruits of the achene or nut type—and anemophily. Some correlation has been seen between habit and fruit type. Trees, shrubs, and climbing herbs tend to have large and fleshy fruits; terrestrial herbs nearly always have small, dry fruits.

As interpreted under the durian theory, the primitive fruit is a fleshy follicle or cupule, with numerous arillate seeds. But arils are present in widely scattered taxa and seem to represent ecological adaptations in dissemination. The evidence from broad comparative studies throughout the angiosperms supports the view that the follicle with many seeds is the primitive fruit.

BIBLIOGRAPHY

The Seed

Artschwager, E.: Development of flowers and seed in the sugar beet, *Jour. Agr. Res.,* 34: 1–25, 1927.

———, E. W. Brandes, and R. C. Starrett: Development of the flower and seed of some varieties of sugar cane, *Jour. Agr. Res.,* 39: 1–30, 1929.

Billings, F. H.: Beiträge zur Kenntnis der Samenentwicklung, *Flora,* 88: 253–318, 1901.

Brandza, M.: Développement des téguments de la graine, *Rev. Gén. Bot.,* 3: 1–32, 71–96, 105–126, 150–165, 229–240, 1891.

Brink, R. A., and D. C. Cooper: The endosperm in seed development, *Bot. Rev.,* 13: 423–541, 1947.

Cooper, G. O.: Microsporogenesis and development of seed in *Lobelia cardinalis*, *Bot. Gaz.*, **104**: 72–81, 1942.

Corner, E. J. H.: The durian theory of the origin of the modern tree, *Ann. Bot.*, n.s., **13**: 367–414, 1949.

———: The anonaceous seed and its four integuments, *New Phyt.*, **48**: 332–364, 1951.

———: The leguminous seed, *Phytomorph.*, **1**: 117–150, 1951.

Eames, A. J.: The seed and *Ginkgo*, *Jour. Arnold Arb.*, **36**: 165–170, 1955.

Gatin, C. L.: La morphologie de la germination et ses rapports avec l'anatomie, *Rev. Gén. Bot.*, **20**: 273–284, 1908.

Godfrin, J.: Recherches sur l'anatomie comparée des cotylédons et de l'albumen, *Ann. Sci. Nat. Bot.*, 6 ser., **19**: 5–158, 1884.

Guilford, V. B., and E. L. Fisk: Megasporogenesis and seed development in *Mimulus tigrinus* and *Torrenia fournieri*, *Bull. Torrey Bot. Club*, **79**: 6–24, 1951.

Hill, A. W.: The method of germination of seeds enclosed in a stony endocarp. II. *Ann. Bot.*, n.s., **1**: 239–256, 1937.

Humphrey, J. E.: The development of the seed in the Scitamineae, *Ann. Bot.*, **10**: 1–40, 1896.

Johnson, D. S.: On the development of *Saururus cernuus*, *Bull. Torrey Bot. Club*, **27**: 365–372, 1900.

———: On the development of certain Piperaceae, *Bot. Gaz.*, **34**: 321–340, 1902.

Johri, B. M., and R. N. Kapil: Contribution to the morphology and life history of *Acalypha indica* L., *Phytomorph.*, **3**: 137–151, 1953.

Junell, S.: Die Samenentwicklung bei einigen Labiaten, *Svensk. Bot. Tidskr.*, **31**: 67–110, 1937.

Kayser, G.: Beiträge zur Kenntnis der Entwicklungsgeschichte der Samen mit besonderer Berücksichtigung des histogenetischen Aufbaues der Samenschalen, *Jahr. Wiss. Bot.*, **25**: 79–148, 1893.

Kennedy, P. B.: The structure of the caryopsis of grasses, *Bull. U.S. Dept. Agr., Div. Agrostology*, **19**: 1–44, 1899 (revised 1900).

Koch, L.: Über die Entwicklung des Samens der Orobanchen, *Jahr. Wiss. Bot.*, **11**: 218–261, 1878.

———: Die Entwicklung des Samens von *Monotropa Hypopitys* L., *Jahr. Wiss. Bot.*, **13**: 202–252, 1882.

Kuhn, G.: Beiträge zur Kenntnis der intraseminalen Leitbündel bei den Angiospermen, *Bot. Jahrb.*, **61**: 325–379, 1927.

Lonay, H.: Contribution à l'anatomie des Renonculacées. Structure des péricarpes et des spermodermes, *Archiv Inst. Bot. Liège*, **2**: 1–164, 1901.

Martens, P.: Le graine et le tube pollinique: réflexions sur les caractères des phanérogames, *Bull. Acad. Roy. Belg. Sci.*, sér. 5, **33**: 919–943, 1947.

———: Les préphanerogames et le problème de la graine, *La Cellule*, **54**: 105–132, 1951.

Martin, A. C.: The comparative internal morphology of seeds. *Am. Midl. Nat.*, **36**: 513–660, 1946.

Monnier, G. Le: Recherches sur la nervation de la grain, *Ann. Sci. Nat. Bot.*, 5 sér., **16**: 233–305, 1872.

Netolitzky, F.: "Anatomie der Angiospermen-Samen." In Linsbauer, K., "Handbuch der Pflanzenanatomie," Abt. II, Teil 2, Bd. 10, Berlin, 1926.

Orr, M. Y.: The occurrence of a tracheal tissue enveloping the embryo of certain Capparidaceae, *Notes Royal Bot. Garden Edinburgh*, **12**: 249–257, 1919–1921.

Péchoutre, F.: Contribution a l'étude du développement de l'ovule et de la graine des Rosacées, *Ann. Sci. Nat. Bot.*, 8 sér., **16**: 1–158, 1902.

Pfeiffer, A.: Die Arillargebilde der Pflanzensamen, *Bot. Jahrb.*, **13**: 492–541, 1891.

Pijl, L. van der: Ecological variations on the theme pod, *Indonesian Jour. Nat. Sci.*, **1/2**: 6–12, 1952.

Planchen, J. E.: Développement et caractères des vrais et des faux arilles, *Ann. Sci. Nat. Bot.*, 3 sér., **3**: 275–312, 1845.

Pritzel, E.: Die systematische Wert der Samenanatomie, insbesondere des Endosperms bei den Parietales, *Bot. Jahrb.*, **24**: 345–394, 1898.

Reeves, R. G.: Comparative anatomy of the seeds of cottons and other malvaceous plants, *Am. Jour. Bot.*, **23**: 291–296, 394–405, 1936.

Solms-Laubach, H. Graf zu: Über den Bau der Samen in den Famillien der Rafflesiaceae und Hydnoraceae, *Bot. Zeit.*, **32**: 337–342, 353–358, 369–374, 385–389, 1874.

Thomson, R. B.: Evolution of the seed habit in plants, *Trans. Roy. Soc. Canada*, 3 ser., **21**: 229–272, 1927.

———: Heterothally and the seed habit versus heterospory, *New Phyt.*, **33**: 41–44, 1934.

Ulbrich, E.: "Biologie der Früchte und Samen (Karpobiologie)," Berlin, 1928.

Walton, J.: L'évolution des téguments et de la protectrice du sporange, *Année Biol.*, **28**: 129–152, 1952.

Worsdell, W. C.: The morphology of sporangial integuments, *Ann. Bot.*, **16**: 596–599, 1902.

The Fruit

Artschwager, E., E. W. Brandes, and R. C. Starrett: See above under Seed.

Cave, C.: Structure et développement du fruit, *Ann. Sci. Nat.* Bot., 5 sér., **10**: 123–190, 1869.

Corner, E. J. H.: The durian theory extended. II. The arillate fruit and the compound leaf, *Phytomorph.*, **4**: 152–165, 1954.

Eames, A. J.: Floral anatomy as an aid in generic limitation, *Chron. Bot.*, **14**: 126–132, 1953.

Ewart, A. J.: The delayed dehiscence of *Callistemon*, *Ann. Bot.*, **21**: 135–137, 1907.

Fahn, A., and M. Zohary: On the pericarpial structure of the legumen, its evolution and relation to dehiscence, *Phytomorph.*, **5**: 99–111, 1955.

Farmer, J. B.: Contributions to the morphology and physiology of fleshy fruits, *Ann. Bot.*, **3**: 393–414, 1889.

Guérin, P.: Développement et structure anatomique du fruit et de la graine des Bambusées, *Jour. Bot.*, **17**: 327–331, 1903.

Hagerup, O.: The morphology and biology of the *Corylus* fruit, *Kgl. Danske Vidensk. Selsk. Biol. Medd.*, **17**: 3–32, 1942.

Hardy, A. D.: Delayed dehiscence in Myrtaceae, Proteaceae and Coniferae, *Proc. Roy. Soc. Victoria*, n.s., **38**: 157–158, 1926.

Janchen, E.: Versuch einer zwanglosen Kennzeichnung und Einteilung der Früchte, *Oesterr. Bot. Zeitschr.*, **96**: 480–485, 1949.

Johnson, D. S.: On the endosperm and embryo of *Peperomia pellucida*, *Bot. Gaz.*, **30**: 1–11, 1900.

Johri, B. M., and R. N. Kapil: Contribution to the morphology and life history of *Acalypha indica* L., *Phytomorph.*, **3**: 137–151, 1953.

Joxe, A.: Sur l'ouverture des fruits indéhiscents à la germination, *Ann. Sci. Nat. Bot.*, 9 sér, **15**: 257–376, 1912.

Kennedy, P. B.: See above under Seed.

Kraus, G.: Über den Bau trockener Perikarpien, *Jahr. Wiss. Bot.*, **5**: 83–126, 1866–67.

Leclerc du Sablon: Sur la déhiscence du fruits à péricarpe sec, *Ann. Sci. Nat. Bot.*, 6 sér., **18**: 5–104, 1884.

Meissner, F.: Die Korkbildung der Fruchte von *Aesculus*—und *Cucumis*—Arten, *Oesterr. Bot. Zeitschr.*, **99**: 606–624, 1952.

Pijl, L. van der: See above under Seed.

Randolph, L. F.: Developmental morphology of the caryopsis in maize, *Jour. Agr. Res.*, **53**: 881–916, 1936.

Sinnott, E. W., and I. W. Bailey: Changes in the fruit type of angiosperms coincident with the development of the herbaceous habit, *Rpt. Bot. Soc. Am.*, in *Sci.*, **41**: 179, 1915.

True, R. H.: On the development of the caryopsis, *Bot. Gaz.*, **18**: 212–226, 1893.

Ulbrich, E.: See above under Seed.

Wahl, C. von: Vergleichende Untersuchungen über den anatomischen Bau der geflügelten Früchte und Samen, *Bibl. Bot.*, **40**: 1–25, 1897.

Weberbauer, A.: Beiträge zur Anatomie der Kapselfrüchte, *Bot. Zentralbl.*, **73**: 97–105, 135–142, 161–168, 193–202, 250–257, 296–302, 1898.

Winkler, H.: Septizide Kapsel und Spaltfrucht, *Biol. Pfl.*, 24: 191–200, 1936.

———: Versuch eines "naturlichen" Systems der Früchte, *Beitr. Biol. Pfl.*, **26**: 201–220, 1939.

———: Zur Einigung und Weiterführung in der Frage des Fruchtsystems, *Beitr. Biol. Pfl.*, **27**: 92–130, 1940.

Chapter 11

NOTES ON THE MORPHOLOGY OF SELECTED FAMILIES

The following discussion is restricted to the most important morphological aspects of a few individual angiosperm families, chosen for their importance in showing evolutionary modification in form and probable phylogenetic relationships.

The more primitive families, especially those of the Ranales and Helobiales, have been studied intensively from 1920 to 1960, with attention to critical comparisons of all structure, both reproductive and vegetative. These studies, based on the viewpoint that inferences as to degree of primitiveness and closeness of relationship must be based on the summation of all characters, have resulted in the suggestion of many changes in the classification of the lower orders and families.

In making the following selection of families, no attention has been given to size of taxa or to possession of some unusual character. The descriptions are not morphologically complete, nor are they uniform in the various families, for only structure that illustrates change in form discussed in the earlier chapters is emphasized. Added to the commonly recognized primitive angiosperm taxa are a few families that present especially puzzling structure, such as the Casuarinaceae.

The sequence in treatment of the selected families is not intended to suggest phylogenetic relationships, though families that seem to be more or less closely related, as those of the magnolian plexus, are kept together. Linear sequences are not evident in these taxa. The living primitive families are obviously surviving remnants of an ancient stock or stocks, and a phylogenetic "tree" cannot be looked for. It is more and more clearly evident that there are several living stocks which have much the same rank in number of primitive characters.

RANALES

Winteraceae

The Winteraceae are outstanding among angiosperm families, because they show early stages in the history of many of the characteristic features of angiosperm morphology. Together with the other families of the Ranales, the Winteraceae provide a basis for many of the concepts and theories of phylogenetic relationships among angiosperms. Among the important features of the Winteraceae—in part additional to those commonly used in taxonomic characterization—are the following.

The flowers of *Drimys* show advance in the development of unisexual from perfect flowers: the New World species (section Wintera) are bisexual; the Old World species (section Tasmannia) are polygamodioecious or unisexual. A primitive perianth shows stages in specialization. The calyx is calyptrate; the sepals are bractlike but connate. The petals range from scalelike to petaloid, with reduction from several to two, and the position of the two produces zygomorphism. The stamens show stages in specialization from a broad, laminar type to wedge-shaped and subterete. The pairs of microsporangia, commonly submarginal and distal, seem to represent stages in the differentiation of an anther (Fig. 49). The pairs of sporangia are lateral or nearly so, an advance over the median position in laminar stamens. Dehiscence is latrorse to extrorse, with most genera extrorse. The pollen is borne in permanent tetrads—an advanced character—present elsewhere in the Ranales only in *Lactoris*. The stamens of most of the Winteraceae, in contrast to those of most other ranalian families, lack the protruding connective, characteristic of primitive stamens (only *Belliolum* has an extended connective). The absence of this primitive character, prominent in most ranalian families, is perhaps related to an unusual type of modification, in this family, of the ancestral laminar sporophyll—the transfer of the sporangia from a median to a subterminal or terminal position at the end of the connective. The pollen grains are of the primitive monocolpate type, with the germinal aperture reduced to a pore. Development within the androecium is centrifugal, a character only recently recognized in the Winteraceae. Nectaries are absent, and no description of the method of pollination has been found.

The carpels are conduplicate, not differentiated into ovary, style, and stigma, and range from long-stipitate to sessile. Evidence from comparative form and from anatomy in the Ranales and Rosales supports the view that the carpel is primitively stipitate. (The stipe is, anatomically, a part of the carpel, not of the receptacle, as sometimes described.) The carpels are "open"—not sealed—at pollination time in *Drimys*, section Tasmannia, and in species of *Bubbia* and *Exospermum;* in other taxa, they are open through only part of their length. The pollen is received on a longitudinal *stigmatic crest,* a ventral ridge formed by the approximated edges of the folded carpel lamina (Fig. 83). The somewhat flaring, unfused edges of the lamina form a two-lobed or "double" ridge. This ridge is covered on both dorsal and ventral surfaces by papillose hairs, among which the pollen germinates. In the more primitive species of *Drimys* and *Bubbia,* the stigmatic ridge runs the entire length of the carpel—is taxonomically described as "stigma decurrent"; in other taxa, it is much shorter, and a series exists in which the ridge is shortened and restricted to the distal part of the carpel. The shortening is by progressive, acropetal fusion of the carpel margins, and the remaining distal part of the stigmatic ridge becomes pseudoterminal by a "hunchback," abaxial distortion of the carpel. This shortened ridge seems to be the earliest "stigma," as a restricted, distal, pollen-receiving area—earliest because accompanying carpels that are hardly closed.

The leaves are simple and pinnately veined. The wood stands apart from that of most other angiosperms in the absence of vessels (Fig. 25). In other characters also, it is of the type considered most primitive for angiosperms. In ray structure, it is heterogeneous, with both uniseriate and multiseriate rays, and the parenchyma distribution is diffuse. The tracheids are long, with long-overlapping ends, and the pitting is frequently scalariform.

The centrifugal sequence in development in the androecium seems to be an evolutionary advance over the more common centripetal order and is surprising in the Winteraceae. The centrifugal sequence in the Dilleniales and the centripetal in the Ranales have been used as important characters in distinguishing these orders. But, in other general characters, the Winteraceae resemble the other ranalian families far more closely than the Dilleniales. The centrifugal sequence has probably arisen independently in these two orders.

In the gynoecium, the Winteraceae show stages in the development of syncarpy, a series from the free carpels of species of *Drimys* and *Bubbia,* through the coherent carpels of *Exospermum,* to the syncarpy of *Zygogynum.* Syncarpy in angiosperm flower specialization is commonly considered a late step, associated with specialized carpels, but connation is present even in this very primitive family. (It is present also in *Eupomatia* and the Magnoliaceae—*Pachylarnax.*) In placentation, the Winteraceae show early stages in the transition from laminar to submarginal.

Morphologically, the most important specializations in the Winteraceae are probably the transformation of the lateral stigmatic ridge into the terminal stigma and the narrowing of the laminar stamen, with transfer of sporangia to a near-distal position. In this evolutionary advance of the carpel, the stigmatic area is more favorably placed for pollen reception. And, in the parallel advance, taking place at the same time in the stamen, the microsporangia, sunken originally in the median or lower part of the laminar stamen, are transferred to distal or terminal positions and become protuberant. In this terminal position, dispersal of pollen grains is more free, whatever the agent.

The history of the long pollen tube of angiosperms begins, at least in part, in the change in place of pollen reception from proximity to the ovules—on the margins of open carpels, as in species of *Drimys* and *Degeneria,* where the tubes are very short, entering the carpel chamber all along the stigmatic ridge—to the distal stigma on a style. (In the gymnosperms, the pollen grain germinates in or near the ovule; there is either no tube or a short one, which is not a male-cell carrier. The Araucariaceae and *Tsuga* are notable exceptions.)

The Winteraceae show reduction of the inflorescence from a cymose cluster (*Drimys*) to a solitary terminal flower (*Zygogynum*). This series perhaps throws light on the nature of the solitary flower, which, in this family, clearly represents the surviving member of a determinate inflorescence. In the Ranales, there are many examples of solitary flowers —*Magnolia, Eupomatia, Himantandra, Zygogynum.* In these genera, there is little structural evidence of reduction of the solitary flower from an inflorescence, but the flowers of *Zygogynum* have surely been derived by reduction from inflorescences, and those of *Eupomatia* and *Himantandra,* in their position on short shoots that may bear more than one flower, suggest similar reduction. (The large size of many solitary flowers is not, in itself, evidence that this type represents a primitive condition; the flowers may be large because they are solitary. The theory that the solitary flower is primitive is, in part, a holdover from the old theory that the complex vascular sporophyte arose by the continued dissection and proliferation of an ancient simple sporophyte, with multiplication of fertile and vegetative parts.)

The phylogenetic relation of primitive solitary flowers to methods of pollination and to bisexual structure should be further explored. Most, if not all, such flowers are bisexual and lack nectaries. In some taxa, the method of pollination is unknown; in others, it is by beetles. Unisexual flowers seem to be associated with all types of pollination except the primitive type, that of beetles; this association seems to be evidence in support of the primitiveness of hermaphroditism.

Drimys is commonly considered the most primitive genus in the Winteraceae, partly because of its primitive carpels, but, when all characters are considered, *Belliolum* seems the most primitive. Its stamens, with the pairs of sporangia laminar rather than marginal, are much the most primitive in the family. (They suggest those of *Himantandra.*)

Lactoridaceae

The monotypic genus *Lactoris,* a shrub of Juan Fernandez Island, has been difficult to place, taxonomically; it has been placed in the Magnoliaceae, Saururaceae, Winteraceae, Piperaceae, Dilleniaceae, and "somewhere between the Piperaceae and the Annonaceae," but has more recently been generally accepted as constituting the monotypic ranalian family Lactoridaceae. Like other families of the Ranales, it possesses both primitive and advanced characters. Its solitary or paired flowers are polygamomonoecious; its perianth, a whorl of three sepals; its stamens, in two whorls of three stamens each; and its three carpels in one whorl. The stamens (Fig. 49*G*) are primitive—laminar, with four elongate, "somewhat remote" sporangia; the sporangia, abaxial and shown in illustrations as apparently protruding; the connective projects beyond the anther sac. The pollen is in tetrads, monocolpate, but elaborately sculptured. The carpels are follicular, with several suspended, anatropous ovules and a short, perhaps decurrent, stigmatic crest on the ventral margin. In illustrations, the carpels appear to be open in flower and fruit. The embryo is small, in abundant endosperm. Anatomically, the genus is insufficiently known, but the nodes are unilacunar, and the vessels are reported to be simply perforate, the wood rays all heterogeneous, and the fibers with bordered pits.

The family is definitely ranalian in its free floral organs, primitive stamens and carpels, and monocolpate pollen borne in tetrads. The stamen, without differentiation of filament and anther, and the open carpel are prominent primitive organs. The extent of protuberance of the microsporangia cannot be determined from the published figures. The abaxial position of the microsporangia on a laminar sporophyll adds one more family to the small list of those that have this character. If further study shows the carpel margins to be free and the stigmatic crest two-lobed, the carpel would resemble that of *Degeneria* and *Drimys* in primitiveness.

The Lactoridaceae are apparently not closely related to any other ranalian family. Though they resemble the Winteraceae in pollen characters, especially the tetrad clustering, they differ in their much greater specialization in flower structure—few organs in whorls—and in wood with vessels. They resemble the Piperaceae in their simplified flowers but, in many characters, are far less reduced. They seem to be one of the isolated families in the Ranales, with closer resemblances to the Winteraceae than to other families.

Magnoliaceae

The Magnoliaceae, best known of woody ranalian families because of the large size of its flowers and its prominence in cultivation, have frequently been considered the most primitive family in the Ranales, and, in some phylogenetic classifications, the most primitive among angiosperms. The family has long included several genera of doubtful affinities. But on the basis of recent broad morphological studies, several of these genera have been excluded—*Illicium, Schisandra, Tetracentron, Euptelea, Cercidiphyllum;* so circumscribed, the family seems to form a natural unit of ten genera, of which *Magnolia* and *Liriodendron* are best known. Though the Magnoliaceae are considered apocarpous, there is considerable fusion in the gynoecium. The carpels, at their bases, are adnate to the receptacle along their margins (they are closed in the proximal part in this way rather than by fusion of the margins) and, in some taxa, connate, with the union so intimate that the lateral traces of adjacent organs are fused. There is syncarpy in *Pachylarnax;* in *Talauma,* the fleshy carpels unite to form an *Annona*-like fruit. An important specialization in this primitive family is unisexuality in *Kmeria.* (Unisexuality is present in *Drimys* also.)

The family has been considered highly primitive because of its simple flower structure—polymerous throughout, with all organs spirally arranged on an elongate receptacle, with little connation and adnation. But, like the other woody ranalian families, this family is specialized in certain characters. In the perianth, there is reduction in number of tepals, with differentiation of a calyx and corolla, with sepals and petals in whorls of three. The stamens show transition from narrow-laminar to anther-filament types (Figs. 49, 50 and 51); the carpels show reduction from many to two or three, and the ovules from several to one or two. The leaves are simple and pinnately veined, as is characteristic of the woody Ranales. They have prominent stipules, variously attached, below or adnate to the petiole. The stipules are early deciduous, even from the petiole, where scars remain.

In the presence of occasional accessory flowers in several genera and of clusters of flowers ("axillary") in *Michelia,* there is evidence that the usually solitary flowers represent reduction from an inflorescence. (In the Winteraceae also, there is reduction of the inflorescence, from a complex type to a solitary flower—*Zygogynum.*)

The perianth provides an excellent example of the differentiation of a corolla from tepals—in contrast with the more common origin from staminodes—and of the origin of the whorled arrangement in floral organs. The androecium shows a series of stages in the phylogenetic modification of a laminar stamen. The long, adaxial microsporangia, characteristic of primitive families, are nearly median on the sporophyll in some taxa. With the narrowing of the sporophyll, they come to lie in pairs, closer and closer to the margins and, ultimately, to take a "marginal" position, with dehiscence extrorse (*Liriodendron,* species of *Michelia*). The various taxa show the development of a slender filament from the basal part of the broad primitive sporophyll, and of a terminal appendage from the distal part. The appendage is lost in *Liriodendron.* (Within the family, the stamen of *Liriodendron* is of high type in the possession of a long, subterete filament, the absence of a terminal appendage, and extrorse dehiscence. The laminar nature of the filament is still apparent; the filament is somewhat flattened and shows weak marginal growth.) Also shown in the family are stages in the change in position of microsporangia from sunken to protuberant (Figs. 51 and 52); these are anatomical evidence that the anther sac represents remains of the sporophyll lamina and that the sporangium has no sporangial wall. Both *Michelia* and *Talauma* show stages in the evolutionary development of the anther (Fig. 49).

The gynoecium of *Magnolia,* with its greatly elongated receptacle and closely placed, spiral carpels, is often cited as an example of the primitive gynoecium, but that of *Michelia,* with carpels laxly arranged and distinct, is more primitive. In ovule number also, *Magnolia* is more advanced than other genera in the family. *Magnolia* has the reduced number of two, *Michelia* has two to five, and *Manglietia,* three to several. Similar extreme reduction in ovule number directly from laminar, rather than from submarginal, placentation—the common origin of solitary and paired ovules—is present in *Nelumbo, Cabomba,* and *Brasenia* (Fig. 85). Study should be made of the detailed vascular supply of the carpel and ovules in *Manglietia* to establish more definitely the presence of laminar placentation in this line of woody Ranales.

The stamen traces in the family are commonly three, with five to seven in species of several genera. Reduction in number to one is clearly evident in both *Magnolia* and *Michelia,* where the lateral traces show all stages of loss. (Most species of *Michelia* have only one trace.)

The ovule traces have mixed origins—from placental vascular meshwork that is supplied by branches from *both* dorsal and ventral bundles (Fig. 141). This origin of ovule supply represents modified laminar placentation—reduction from a meshwork of connecting branches of the dorsal and ventral bundles. It closely resembles that of *Degeneria* and *Drimys* and is best seen in the closely related *Michelia.*

The vascular structure of the floral axis is complex because of the presence of a cylinder of cortical bundles, which join the stelar bundles in supplying the floral appendages (Fig. 41). (There is no cortical system in the vegetative axis, as in *Calycanthus.*) About twenty cortical bundles arise from the axial stele in the middle of the peduncle, enlarge, and become concentric. At the base of the receptacle, they branch and anastomose complexly, joining with the other higher branches from the stele, then form a meshwork throughout the inner cortex of the receptacle. The vascular supply to the appendages is insufficiently known for the family as a whole, and it is perhaps not uniform. The tepals seem to receive their supply wholly from the cortical bundles, the stamens from a double—inner and outer—series of the upward-continued cortical bundles. The median trace of the carpel is derived from the stele, the two lateral traces from the cortical system, where each is united in origin with a lateral trace of an adjacent carpel. The flowers of *Calycanthus,* the Himantandraceae, and the Annonaceae have a somewhat similar but less strongly developed, cortical system.

The nodes of the vegetative axis have 6 to 17 traces. In other Ranales, only *Eupomatia* approaches this high number of leaf traces.

The wood of the Magnoliaceae is more advanced than that of most other ranalian families. It has vessels of a fairly high type—scalariform, with a few widely spaced bars, and, in some genera, simply perforate—and terminal parenchyma.

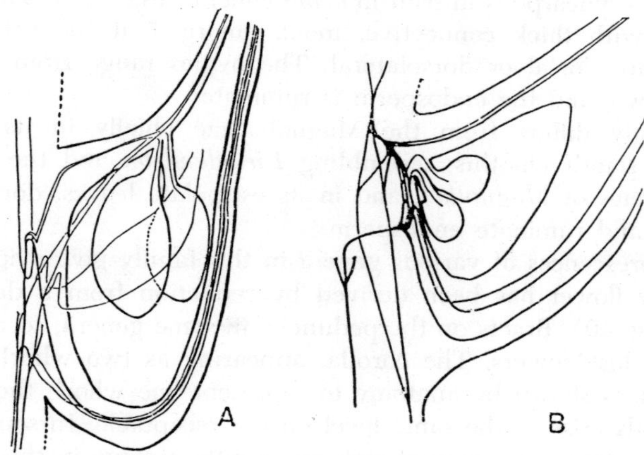

Fig. 141. Diagrams of longitudinal sections of carpels and ovules of the Magnoliaceae showing complex vascular supply derived from both stelar and cortical bundles of receptacle. *A, Magnolia acuminata; B, Liriodendron tulipfera.*

A prominent feature of flower structure in this family is the absence of staminodes, in strong contrast with the Himantandraceae and Eupomatiaceae, where they form prominent parts of the flower.

The Magnoliaceae provide an excellent example of lack of uniformity in advance from primitive characters among closely related taxa. *Magnolia* is commonly considered the most primitive genus and *Liriodendron,* one of the most advanced. But species of *Michelia* are more primitive in their deeply sunken microsporangia and in the prominence of the protrusion of the connective. If the solitary flower represents a reduced inflorescence, *Michelia* is more primitive in its clustered flowers. *Liriodendron* has an advanced type of stamen, with long filament and absence of protruding connective. *Michelia* has carpels of primitive form, with weak differentiation of style and stigma. *Kmeria* shows high specialization for ranalian families in its unisexuality, and *Pachylarnax,* in its syncarpy.

Annonaceae

The Annonaceae, a large, chiefly tropical family of about eighty genera of trees, shrubs, and vines resemble the Magnoliaceae in both vegetative and reproductive characters, especially in general flower structure. The flowers are large, bisexual (rarely unisexual); the perianth is whorled, of three series, the inner forming a "biseriate" corolla; the sporophylls are numerous and spirally arranged, but the carpels in some genera are reduced to few and definite in number. As in the Magnoliaceae, there are no staminodes. The gynoecium is apocarpous but becomes syncarpous in fruit in some genera (Fig. 139). The stamens are short, with thick connective, much enlarged at the tip, and the sporangia are dorsal or dorsolateral. The ovules range from numerous to one or two, and the endosperm is ruminate.

The family differs from the Magnoliaceae chiefly in its reduced, whorled perianth (in this, resembling *Liriodendron* and the more advanced species of *Magnolia*), and in its estipulate leaves, dorsal microsporangia, and ruminate endosperm.

The inflorescences of various genera in this family give evidence that the solitary flower has been derived by reduction from a determinate cluster (Fig. 40). Bracts on the peduncle, in some genera, seem to show position of lost flowers. The corolla, appearing as two whorls of three petals each, is shown by anatomy to represent one whorl; the traces of all the petals arise at the same level on the receptacle. In some genera, three of the six petals are vestigial, apparently a stage in the formation of a trimerous—one whorl—corolla. The thick, broad connective of the stamens may partly hide the sporangia, which, in some genera, may be semisunken, and the individual sporangia of each pair remain distinct at dehiscence, not fused as usual in the Magnoliaceae. (The stamens of this family, especially those of Australian and East Indian genera, should be studied for sporangial structure and position in the tissue of the lamina. In *Melodora, Polyalthea,* and *Ancana,* the connective is described as "concealing the cells," suggesting that the sporangia are completely sunken.) The Annonaceae may show, even better than the Magnoliaceae, stages in the evolutionary transfer of the microsporangia from a sunken to an apparently superficial ("protuberant") position. Pollination is apparently largely by beetles, though beetles do not feed extensively on the floral organs, as in *Eupomatia* and *Calycanthus.* The carpels are greatly reduced in number within the family, as are the ovules—from numerous to one—but reduction in ovule number does not accompany reduction in carpel number.

The ovules are anatropous, as in most ranalian taxa, and reduction in number from several to few parallels that in the Magnoliaceae. The ruminate endosperm suggests relationship to *Eupomatia* and the Myristicaceae, rather than to the Magnoliaceae, but is, perhaps, an ancient character retained in several lines. The receptacle ranges from convex to somewhat concave; in *Calycanthus* and *Rosa*, it is deeply concave.

The wood is remarkably uniform for so large a family; generic differences in the wood are difficult to determine. (*Asimina* stands apart from other genera in its ring-porous wood.) The vessels are simply perforate; the wood parenchyma is apotracheal, in fine, regularly spaced, concentric lines (a character that differentiates the wood of this family from that of all other families); the rays are chiefly uniseriate, a few, multiseriate. This wood is more specialized than that of the Magnoliaceae in vessel type and differs greatly in its parenchyma distribution. The phloem is stratified in some genera, an advanced character for a primitive family.

Anatomically, the flower resembles that of the Magnoliaceae in the presence of a cortical vascular system, but this system is less strongly developed than that of the magnolias. (The vascular system is less well known in detail than that of the magnolias; apparently, it has been described only in *Asimina* and *Annona.*) In *Asimina,* the cortical bundles are derived from the stele of the peduncle below the calyx and become inversely oriented or concentric. No uniform plan of arrangement is followed. The sepals receive all their several traces from the cortical system; the petals receive the median trace from the stele of the axis, their several lateral traces from the ascending cortical bundles. The stamens each receive a single trace. (This single-trace supply, the most advanced condition in the Magnoliaceae, is found only in a few species of *Michelia.*) The lower stamens are supplied from the cortical, the upper from the axial bundles. Some of the "stamen traces" branch and supply more than one stamen. At this stamen-trace level, other branches tie together the cortical and axial systems. The distal parts of the cortical system are used up in the formation of carpel traces, three to each carpel. (The vascular system of two species of *Annona* differs from that of *Asimina* in no essential way.) The floral cortical system of the Annonaceae, as compared with that of the Magnoliaceae, has been called both "only partially developed" and "reduced."

Only four examples of a cortical vascular system in the floral axis of angiosperms appear to have been described—those of the Magnoliaceae, Annonaceae, Calycanthaceae, and Himantandraceae; others may exist where cortical bundles are present in the stem, as in *Justicia* (*Dianthera*).

The patterns of the cortical systems of the Magnoliaceae and the Annonaceae are similar and differ from that of *Calycanthus*. The two types seem unrelated and can hardly be considered as indicating relationship between the Magnoliaceae and Annonaceae on the one hand, and the Calycanthaceae on the other. In *Calycanthus,* the cortical system, continued from the peduncle and stem below, is used up in supplying the lowest perianth parts; it does not continue in the receptacle, supplying higher organs, and does not unite upward with the vascular stele of the receptacle, as does the cortical system in the Magnoliaceae and Annonaceae.

The patterns of the cortical systems of the Magnoliaceae and Annonaceae are, however, much alike. The system in the magnolias is anatomically more complete and seems better developed. Two opinions have been expressed as to the relationship of the cortical systems in these two related families. One interpretation of the annonaceous pattern is that it represents a simplified reduction form of the more complex and better-developed magnoliaceous type; the other interpretation of the pattern in the Annonaceae is that it is an early stage in the elaboration of a cortical system, still incomplete and "unsettled" in pattern. The Annonaceae are, in several characters, the more advanced family, and it seems more likely that the cortical system is, in part, vestigial, modified in relation to the simplification of the flower, especially of the perianth and gynoecium. Studies of the cortical systems in the advanced and simpler genera of the Magnoliaceae should give evidence in support of one of these theories.

Among the important aspects of the morphology of the Annonaceae and Magnoliaceae, is the variety in microsporangial position. The Magnoliaceae seem to show a series in progressive change in the position of these sporangia from adaxial to lateral to abaxial (*Liriodendron*); the Annonaceae have abaxial sporangia. These related families perhaps provide evidence of the primitive position of the microsporangia, a major, but neglected problem in the morphology of the angiosperms. Apparently basic abaxial position is present only in primitive families—Degeneriaceae, Himantandraceae, Winteraceae, Lactoridaceae, Annonaceae, Ceratophyllaceae, Chloranthaceae, Lardizabalaceae—families that have many primitive characters and stamens of laminar or sublaminar type, with prominent, sterile distal part. Dehiscence in stamens with abaxial sporangia is, of course, extrorse, but it is difficult to determine whether extrorse dehiscence in higher families represents survival of a primitive type or is secondarily derived from introrse, as it appears to be in *Liriodendron.*

The Magnoliaceae and Annonaceae resemble one another in so many ways that the two families must be closely allied—doubtless derived from the same ancestral stock, with the Annonaceae more advanced in some characters than the Magnoliaceae. The Annonaceae perhaps represent the attainment of the advanced dorsal position; possibly they retain the dorsal position as ancestral, as they seem to retain the apparently primitive ruminate endosperm, which the Magnoliaceae do not have. The Magnoliaceae show stages in the attainment of the whorled perianth from spirally arranged tepals—*Liriodendron* and some species of *Magnolia* have a whorled corolla; the Annonaceae have a whorled perianth throughout.

Eupomatiaceae

Eupomatia, the only genus in the Eupomatiaceae, consists of shrubs and small trees of Australia and New Guinea with simple leaves and large, fleshy, solitary flowers of unusual appearance (see frontispiece). There are two species, *E. laurina,* with flowers terminal on short shoots, and *E. bennettii,* with flowers terminal on long, leafy, woody stems from tuberous roots. Taxonomically, the genus has been difficult to place; it has been placed in the Magnoliaceae, Winteraceae, Himantandraceae, Annonaceae. Family rank is well supported by its morphological structure, which was long insufficiently known.

Like other families in the Ranales, the Eupomatiaceae possess many primitive characters, combined with distinctive specializations that set it well apart in the order. Especially important are: an absence of a perianth; a syncarpous gynoecium of spirally arranged carpels; vessels of the most primitive type known in the angiosperms; a pseudoperianth *above* the stamens; inverted vascular bundles in the stamens.

Because published descriptions of this family are incomplete and, in part, inaccurate, the following somewhat detailed description* is added.

The flower is described as perigynous, with numerous stamens and staminodes inserted on the rim of the enlarged, turbinate receptacle, and with numerous free carpels "immersed in the receptacle, as in some of the Nymphaeaceae." This description is inaccurate in regard to the gynoecium, which is not apocarpous but syncarpous, with the carpels enclosed by the rim of the hollow receptacle, not embedded in it (Figs. 72*A* and 142*C, D*). There is no true perianth. The flower bud is covered by a "calyptra," morphologically a bract, a member of the phyllotactic spiral below the flower (Figs. 143 and 144). The calyptra is attached around the rim of the enlarged receptacle and encloses the bud completely. It is shed at anthesis (Fig. 143).

The numerous stamens, more or less connate at the bases of the staminodes above, are borne on the rim of the receptacle (Figs. 142*B, C*). Above the stamens are many petaloid staminodes which form a pseudoperianth (frontispiece)—a prominent part of the flower, because

* The description is based, in part, on observations in the field and on laboratory studies of living material made by Dr. A. T. Hotchkiss and the author. Dr. Hotchkiss is continuing detailed studies of the genus.

the stamens are strongly reflexed. The stamens have a broad, short filament with protuberant, adaxial sporangia (Fig. 142A). As the flower opens, they become strongly reflexed, exposing the anther sacs freely. The connective is protruding. The many fleshy petaloid staminodes, *above* the stamens, form a prominent part of the flowers in this genus

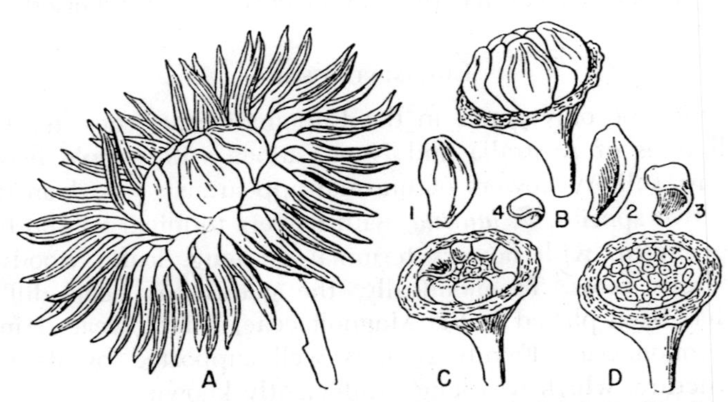

Fig. 142. *Eupomatia laurina. A,* flower showing petaloid staminodia above reflexed stamens; *B,* flower with androecium shed: 1, 2, 3, 4, series in form of staminodia, outermost to innermost; *C,* flower showing innermost staminodia enclosing gynoecial chamber; *D,* flower, stamens, and staminodia shed, showing floor of gynoecial chamber with tops of connate carpels. (*After Brown.*)

Fig. 143. *Eupomatia laurina.* Branch bearing four flowering short shoots. Above, buds showing stages of opening by shedding of calyptra. (*After Brown.*)

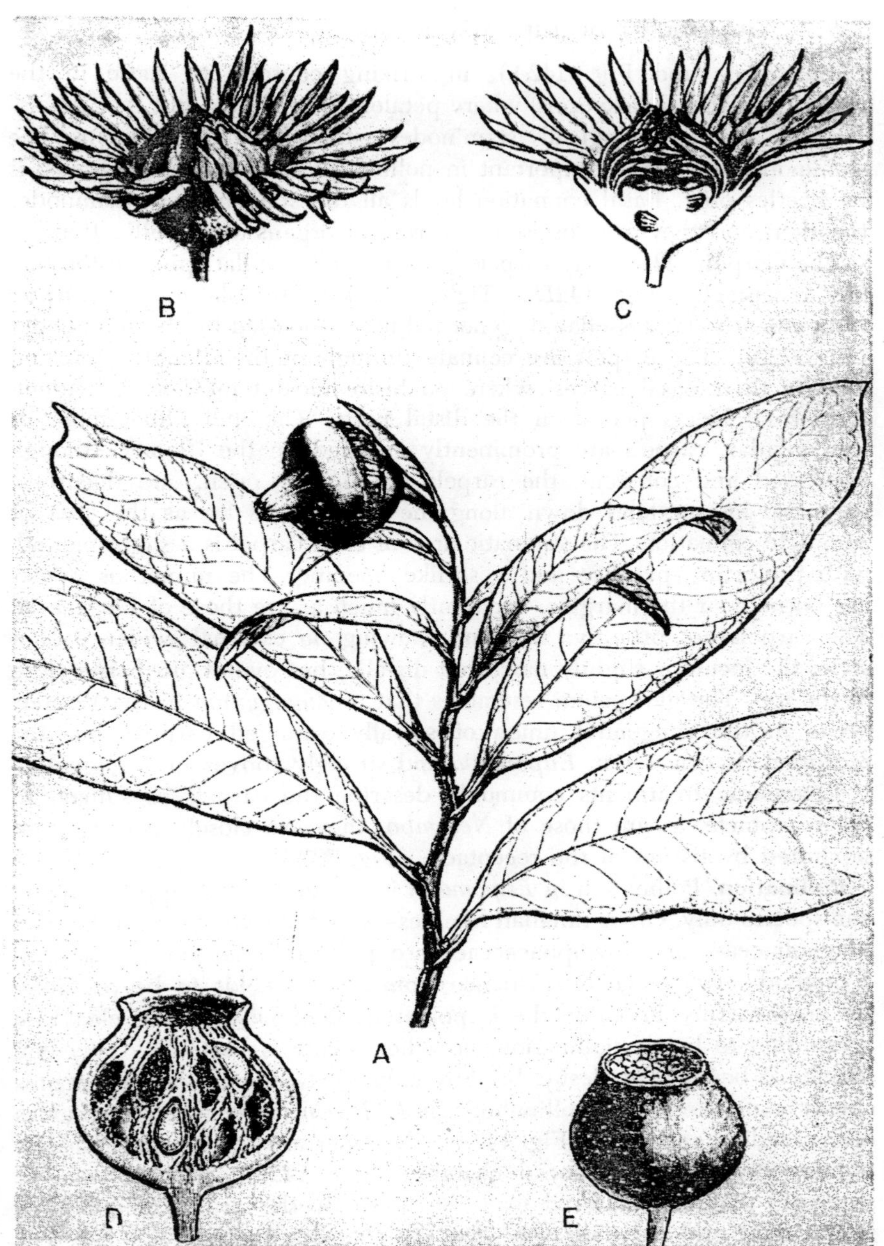

Fig. 144. *Eupomatia*. *A*, *E. bennettii*; *B* to *E*, *E. laurina*. *A*, flower bud terminal on long leafy shoot; *B*, flower showing stamens and overlapping staminodia; *C*, longitudinal view of flower showing loculi of carpels with ovules and gynoecial chamber enclosed by staminodia; *D*, *E*, fleshy fruits: *D*, in section showing seeds; *E*, showing floor of gynoecial chamber with spirally arranged carpel tips. (*From Baillon.*)

(frontispiece and Fig. 142*A*), in striking contrast to those in the majority of the angiosperms, where petaloid staminodes and the "petals" —which are morphologically staminodes—stand *below* the stamens. The staminodes are highly important in pollination, which, in this genus, is by beetles only. Basal connation holds all the stamens and staminodes together, and they are abscised as a ring of organs after pollination.

The Carpels. The many carpels are arranged in a flat spiral in the hollow receptacle (Fig. 142*D*). Tightly packed, the flat-topped carpels form a nearly flat, tessellated, gynoecial floor, enclosed by the staminodes (Fig. 72*A*). The carpels are connate throughout their length, congenitally in their basal halves, where no histological limitation is evident. Epidermal layers persist in the distal parts. The individual, more or less angular, carpels are prominently outlined on the "floor." Athough appressed and coherent, the carpels are slightly open, and decurrent stigmatic lines extend down along the margins, as far as the area of complete connation. The stigmatic area of each carpel is a slight mound, with projecting papillae and a slitlike opening. The numerous ovules are borne near the margins, apparently much as are those of *Degeneria*. The carpels are primitive in form, showing no external evidence of a style; the locule is slightly narrowed distally, but ovules are borne close to the top. The gynoecium resembles that of *Zygogynum* in the Winteraceae in the congenital union of spirally arranged carpels, but the receptacle is concave in *Eupomatia* and strongly convex in *Zygogynum*.

Taxonomic treatments commonly describe the carpels as sunken in the receptacle, as are those of *Nelumbo;* they are closely connate, not separated by tissues of the receptacle (Fig. 72*A*).

Pollination. Pollination in *Eupomatia* is by beetles and, apparently, by one species only. Other ranalian families—especially the Calycanthaceae, Magnoliaceae, and Nymphaeaceae—are pollinated largely by beetles, a method now shown by evidence from several fields to be primitive for angiosperms. Probably the Eupomatiaceae show the most elaborate adaptation to beetle pollination; no other family is visited by only one species of beetle. The fleshy, brightly colored staminodes are the organs chiefly connected with pollination. In *E. laurina,* the outer staminodes are spreading and erect (Fig. 142*A*); the inner are shorter and inflexed, tightly enclosing a gynoecial chamber (Fig. 142*C*), with clusters of stigmatic papillae scattered over the floor. Beetles are attracted by a strong odor and eat their way into the chamber through the bases of the staminodes. Within the chamber, they feed on the papillose margins of the staminodes and on padlike food bodies on the ventral surfaces of the staminodal tips which "roof" the chamber (Fig. 69*F*). The beetles eat these food bodies extensively and remain long within the chamber. In *E. bennettii,* the staminodes do not enclose the gynoecium chamber so completely as in *E. laurina,* and the food bodies are stalked, capitate bodies scattered over the margins and surfaces (Fig. 69G).

Germination of the pollen, development of the tube, and structure of the embryo sac are unknown. The embryo is small, and the endosperm is ruminate—apparently an ancient character.

The leaves are pinnately veined and develop conduplicately folded, remaining so until well grown, a character found also in *Himantandra.*

Anatomy. The foliar nodes are multilacunar, with 7 to 11 traces. The stamen has five to seven traces in *E. laurina,* three in *E. bennettii;* the carpel has several traces. In some of the stamens of *E. laurina,* the vascular bundles of the broad filament show, in addition to a median series, a dorsal inverted bundle, which is sometimes double. The presence of the occasional inverted bundles on the dorsal side is apparently the result of differential displacement and distortion of tissues in the phylogenetic narrowing of the sporophyll—an anatomical modification shown more clearly in the Nymphaeaceae. An adaxial position of the microsporangia is important in pollination in this genus, because, in the open flower, the stamens are strongly reflexed (frontispiece).

The wood is probably the most primitive among vessel-bearing angiosperms; the vessel elements are very long and long-tapering, with scalariform perforations over long, oblique ends in great numbers—from thirty to more than one hundred. And these perforations differ from the scalariform pits of the rest of the cell only in the absence of a closing membrane. The similarity of these vessel elements to those of *Pteridium* is close. The fibers have round, bordered pits; there are no libriform fibers. A histological character known elsewhere in the Ranales only in the Myristicaceae is the presence in the pith of a tanniniferous secretory system of branching tubes. The secondary phloem is "soft"; it is not stratified and fibers are absent. The sieve elements are long and slender, with slowly tapering ends. In contrast with these primitive characters in vascular tissues is the specialized, multilacunar node.

Summary. When all known characters are considered, *Eupomatia* is seen to be one of the most primitive living angiosperms. The flower has no perianth; protection before anthesis is by a bud scale. Sterile stamens serve as a corolla, but this "corolla" is *above* the stamens. The bracts of the peduncle, the calyptra, and the sporophylls and staminodia form a continuous spiral. The connation of spirally arranged organs is rare and remarkable where the number of organs is large. (Examples with fewer carpels are *Himantandra, Zygogynum,* and *Berberis.*) The stamens, with poorly differentiated anthers, have a flat filament and a protruding connective. The carpels resemble those of *Degeneria* and *Drimys* (section Tasmannia) in their decurrent stigma and unfused margins. (The connation of the carpels appears to restrict the entrance of pollen tubes to the top of the marginal opening, but the method of entrance of the tubes is unknown.) The ruminate endosperm is characteristic of several ranalian families. The xylem and phloem are primitive; only the vessel-less genera have more primitive wood, and only the companion-cell-less *Austrobaileya* has more primitive phloem.

There is advanced structure, however, in the perigyny of the flower, with connation of carpels and the bases of stamens and staminodes, and in the multilacunar node. The perigynous condition, related to the concave form of the receptacle, brings all the appendages closely together.

A resemblance of *Eupomatia* to *Calycanthus*, another primitive ranalian genus, is suggested by the concave receptacle, but similar receptacles are present in other orders, especially the Rosales. The presence of a tanniniferous tube system and ruminate endosperm suggests possible relationship to the Myristicaceae, and the presence of staminodes above the stamens and a solitary protective bud scale, to the Himantandraceae. The Eupomatiaceae are an isolated family, probably very old, with no surviving close relatives.

HIMANTANDRACEAE

The Himantandraceae, a small Australian family, consists of the genus *Himantandra*, with two species. The flowers are large, solitary, and terminal on short shoots, which may also bear one or two lateral flowers. The perianth has been interpreted in two ways: (1) as consisting of a calyx of two calyptrate sepals and several stamenlike petals; (2) as made up of a single, scalelike sepal, enclosing a whorl of four connate petals, with petaloid staminodia above. Anatomy supports the second interpretation. The stamens are laminar, narrow, and without distinction of anther and filament. The microsporangia are low down and sunken in the abaxial side of the sporophyll. The numerous staminodia are borne *both above* and *below* the stamens. The several carpels, spirally arranged, are differentiated into ovary and a "stigmatic style," a somewhat plumose distal part, with a decurrent, papillose surface. Slight syncarpy in the flower becomes strong in the fruit. The anatropous ovules are reduced to one or two at the base of the carpel. Peltate scales cover much of the plant.

In anatomy, the flower somewhat resembles in complexity that of *Magnolia*, with cortical bundles, which anastomose frequently and in various ways with the traces of the appendages. The calyx, of one sepal, seems to have several traces—like the calyptra in *Eupomatia*. The stamens, carpels, and, apparently, the petals have three traces. The wood is specialized. Its vessels are transitional from scalariform to simply perforate, and the lateral tracheary pitting is alternate.

The Himantandraceae resemble the Eupomatiaceae most closely. The calyptrate calyx, of a single appendage, the presence of staminodia both above and below the stamens, the conduplicate leaves, late in unfolding, suggest the Eupomatiaceae. But general flower type and anatomy—cortical bundles in the flower—suggest the Magnoliaceae. However, in the position of the microsporangia—adaxial or abaxial—the Himantandraceae differ from both the other families. The Magnoliaceae are stipulate; the other two, estipulate. The carpels of the three families differ greatly in their differentiation of a stigma and in their placentation. No close affinity between these families can be seen, but the Himantandraceae are clearly a member of the magnolialian plexus of families.

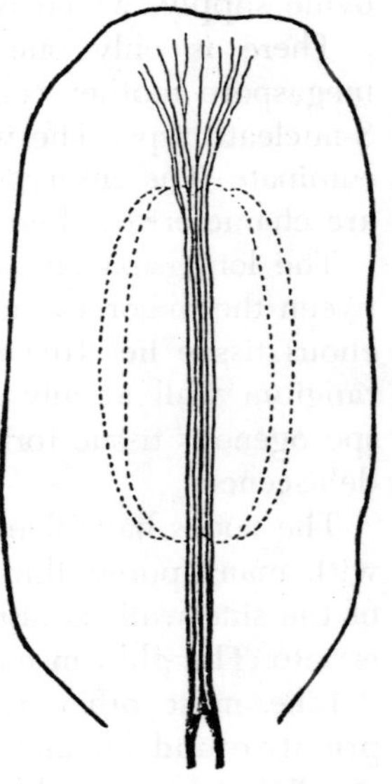

Fig. 145. Sketch of stamen of *Degeneria vitiensis* showing laminar form, outlines (dotted lines) of sporangia, and two-trace vascular supply. (*After Swamy.*)

Degeneriaceae

The Degeneriaceae, a monotypic family, is represented by the tropical Fijian tree *Degeneria vitiensis,* first collected in 1934, and described in 1942. The large simple leaves are pinnately veined. The solitary flowers are supra-axillary, borne on long pedicels. The perianth is differentiated into a calyx of three sepals and a corolla of about twelve petals in three or four series. The stamens are broad, laminar organs, with no suggestion of differentiation into anther and filament (Fig. 145). The pollen is monocolpate. Above the stamens are several staminodes, some with abortive sporangia. The gynoecium consists of a single carpel borne in a slight depression in the receptacle tip. The carpel is follicular, with many anatropous ovules in two rows; the fruit is indehiscent, remarkable for a free, multiovulate carpel.

The carpel is conduplicately folded, with its margins joined only by the interlocking of papillose, glandular cells. These hairs cover both ventral and dorsal surfaces of the cohering marginal areas, which flare slightly and form a two-lobed *stigmatic crest,* an unspecialized decurrent stigma, extending the length of the carpel (Fig. 83). The method of pollination is unknown. No nectaries have been described. The pollen germinates among the hairs of the crest, and the pollen tubes penetrate directly into the loculus and into nearby ovules.

Although the ovules are borne in two rows, the placentation is laminar, because the traces of the ovules are supplied by both the dorsal and ventral bundles (Fig. 83*A*, *B*). The majority of the proximal ovules are supplied by bundles from the ventrals; the traces of some of the distal ovules are branches from the dorsal bundles, and other ovules receive traces from a meshwork formed by branches from both dorsal and ventral veins. There is no constancy in the vascular plan of the ovule supply, which is basically that of laminar placentation.

There is only one archesporial cell, which becomes directly the megaspore mother cell. The embryo sac is of the normal monosporic, 8-nucleate type. The young endosperm is cellular and, when mature, is ruminate. The embryos are remarkable in their cotyledon number; three are characteristic, four frequent.

The long, abaxial microsporangia are deeply embedded in pairs between the major vascular bundles of the lamina (Fig. 67). The sporogenous tissue lies free in the mesophyll of the lamina; there is no sporangium wall at any stage, but the epidermal cells directly over the sporogenous tissue form a specialized thick-walled layer associated with dehiscence.

The nodes have five traces from five gaps. The vessels are scalariform with many pores, thin-walled and angular in cross section; the pitting of the side walls is also scalariform. The wood rays are narrowly multiseriate. The phloem consists of alternate layers of hard and soft cells.

Like most other ranalian families, the Degeneriaceae show both primitive and advanced characters. The leaves are like those of most ranalian taxa—simple, pinnately veined. The flowers are solitary, but the presence of two or three bracts along the flower stalk is perhaps evidence that lateral flowers have been lost. The supra-axillary position is probably not present in any other ranalian family.

The perianth is highly specialized for a family primitive in many characters; calyx and corolla are well differentiated, and their members whorled. The multiseriate corolla is an example of the development of the whorled arrangement before number of organs was reduced. (Corollas with two whorls are present in *Magnolia* and *Liriodendron;* the Annonaceae show reduction from two whorls to one.) The stamens and carpels are probably the most primitive in the order. The stamen in its laminar form seems to be the most primitive known among angiosperms. The sunken, wall-less microsporangia are doubtless the primitive angiosperm type. (The Magnoliaceae show many stages in the narrowing of the laminar sporophyll and the "freeing" of the sunken sporangia.) The monocolpate pollen is one of the simplest in angiosperms.

In this family, reduction of the many carpels of typical ranalian families to one is a prominent advanced character. Absence of fusion of the carpel margins, the consequent decurrent stigmatic crest—probably the earliest stigma—and the broad laminar stamen are the outstanding primitive features of the ranalian flower. The nondehiscent fruit with many seeds that are freed only by decay of the fruit walls is perhaps a first step in the development of the achene type of fruit.

In restriction of the area of ovule-bearing from the "allover" pattern of typical laminar distribution to broad lateral bands, *Degeneria* presents a stage in the evolutionary development of submarginal placentation. The vascular structure of the carpel demonstrates this. The ovule traces are derived from both the dorsal and the median veins. Though the ovules are restricted to two rows, their traces do not arise directly below the ovules, as in submarginal placentation. (This stage is seen also in *Drimys*.)

The wood, though advanced over the vesselless type, is primitive, with multiperforate, scalariform vessels and scalariform pitting. The nodes are multilacunar, with five traces from five gaps.

All characters considered, the Degeneriaceae are one of the most primitive angiosperm families, similar in degree of primitiveness to the Magnoliaceae, Eupomatiaceae, and Winteraceae. They seem to belong to the magnolian alliance, forming another isolated family, with specialized perianth and gynoecium.

Myristicaceae

The Myristicaceae, a small tropical family, have small, apetalous, unisexual flowers, extrorse anthers, connate stamens, and a solitary carpel. At least ten families of the Ranales and Laurales have been considered their probable relatives. The family is usually placed in the Ranales, with closer affinities still uncertain. In taxonomic treatments, this family has perhaps most often been placed close to the Lauraceae, which it resembles in flower structure. Wood structure supports this treatment. But its monocolpate pollen, its embryology, and its ruminate endosperm have suggested affinity with the Annonaceae. Its members have been called "apetalous annonads." In some opinions, the Annonaceae, Eupomatiaceae, Magnoliaceae, and Myristicaceae form a complex of related ranalian families, with Myristicaceae approaching the Lauraceae in some characters. But the wood of the Annonaceae and Eupomatiaceae differs greatly from that of the Myristicaceae.

In this family, the association of monocolpate pollen with unisexuality, a solitary carpel, connate stamens, and wood with both scalariform and simply perforate vessels is surprising. All genera have tanniniferous ducts in the wood rays, a character unknown elsewhere. (Similar tubes are present in the pith of *Eupomatia*.) Wood anatomy lends no support to the frequently made suggestion of relationship to the Annonaceae. The family appears to be close to the Lauraceae. A study of evolutionary advance in angiosperms, based on statistical correlations, lists the Myristicaceae as the third most primitive family, but, in this treatment, unisexuality is considered primitive.

SCHISANDRACEAE

Two genera of woody vines, *Schisandra* and *Kadsura*, each with over twenty species, constitute the Schisandraceae. On the basis of certain advanced characters, the genera have been removed from the Magnoliaceae (*sensu lato*) and considered to form an independent family.

The family has a combination of primitive and relatively advanced characters. Among primitive characters are the polymerous perianth (without distinction of calyx and corolla); many stamens and carpels, spirally arranged on an elongate receptacle; carpels not completely closed, with divided, decurrent stigmas; phloem without fibers and with long, slender, overlapping sieve elements, which have many sieve areas; vessels scalariform, with numerous narrow perforations and scalariform side-wall pitting. Advanced characters are the unisexual flowers; the tricolpate pollen; the unilacunar node with three traces; xylem ranging from relatively primitive to highly specialized; ovules reduced to from two to five. (Highly specialized xylem is associated with the vine habit.)

The carpels resemble those of the Magnoliaceae in reduction of ovules and in the complex ovular supply, but the origin of the ovular traces is clearer in the Schisandraceae than in the Magnoliaceae, because of the absence of a cortical vascular system. The placentation of this family, like that of the Magnoliaceae and Illiciaceae, is reduced laminar. The ovule supply is derived from either the dorsal or ventral bundles and may be double, with a trace from both dorsal and ventral bundles. This ovule supply indicates, as an ancestral ovular supply, a vascular meshwork, which supplies individual ovules from all parts of the vascular skeleton of the lamina. (Double ovule traces represent anomalous vascular morphology that results from placental reduction, where, as an ovule is lost, its vascular supply may be "captured" by an adjacent surviving ovule. This condition is occasional in basal ovules in syncarpous ovaries, where it is derived by reduction from submarginal placentation.) Reduction of ovule number to two or one, directly from laminar placentation is probably rare; this reduction is demonstrated by vascular structure of the carpel in these three families and in the Nymphaeaceae and Cabombaceae (Fig. 85).

The Schisandraceae, together with the Illiciaceae, seem to be related to the Magnoliaceae (*sensu stricto*) and to have been derived from the same ancestral stock, but are more specialized families than the Magnoliaceae.

Illiciaceae

Illicium, a genus of shrubs and trees, with bisexual flowers, many tepals, stamens, and carpels, has been variously placed taxonomically within the Ranales. It is considered to form the family Illiciaceae, with closest resemblance to the Schisandraceae.

The floral structure is primitive; the numerous organs are free and spirally placed, except the carpels, which are whorled. The nodes are unilacunar, with a single strong trace. The wood and phloem show a combination of primitive and advanced characters. The vessel members are long and slender, with long-overlapping ends, which have many scalariform perforations. The tracheary pitting ranges from scalariform to opposite-multiseriate. The sieve elements are like the vessel members in shape and have numerous sieve areas. There are no fibers in the secondary phloem.

The Illiciaceae resemble the Schisandraceae more closely than other ranalian families.

Calycanthaceae

The family Calycanthaceae consists of two small genera of shrubs, *Calycanthus* and *Chimonanthus.*

The flowers have a concave receptacle with numerous petaloid tepals borne spirally on its sides and few to many stamens, sometimes with staminodes above the stamens. The many carpels are borne, also in spirals, on the sides and at the bottom of the concave receptacle. There is no distinction of calyx and corolla. The stamens have a short filament and large anther, with broad and prolonged connective and elongate, laterodorsal anther sacs. The pollen is dicolpate. The carpels are achenelike, with a long filamentous style and a decurrent stigma, resembling that of *Cercidiphyllum.* The ovules are one or two. The archesporium consists of eight to ten cells. The embryo is large, and there is no endosperm.

Pollination is largely or wholly by beetles. The inner tepals, like those in *Eupomatia,* bend inward, forming an imbricate roof over a chamber in which pollination takes place. The tips of some of these tepals and the prolonged connectives of the stamens bear food bodies (Fig. 69). In some species, the stamen tips are conspicuous—white, on the dark red stamen, and lack the hairy coat of the anther; they are, perhaps, associated with pollination. The food bodies consist of delicate cells, rich in protoplasm and oil. There are no nectaries; nectar is secreted by the inner staminodia, and fragrant oils probably diffuse from the perianth, as in *Magnolia.*

The wood has simply porous vessels and paratracheal parenchyma; the phloem has sieve cells, with single, transverse end plates. These are advanced characters for a ranalian family, but the leaf node, with two traces from one gap, is primitive.

The family is characterized by a cortical vascular system of inverted strands, which supplies the ventral traces of the leaves. This system extends into the flower, but its relations with the traces of the floral appendages are not uniform. The lower tepals receive all their supplies from the cortical system; all the upper organs receive their supplies from the stelar system. The cortical bundles of the stem are inverted; those of the flower, normally oriented.

The morphology of the cortical vascular system in angiosperms is obscure. *Magnolia* has a strong, uniform cortical system in the flower; *Asimina* has a partial system, not uniform throughout the flower. Both these genera lack cortical bundles in the stem. *Calycanthus,* which has a weak cortical system in the flower, has a strong system in the stem. The cortical system of *Calycanthus* is unlike that of *Magnolia* and *Asimina,* and the two types are, or seem to be, unrelated. In *Calycanthus,* the peduncle, like the stem, contains a cortical system of vascular bundles, ending at a low level in the flower; the cortical bundles divide and all enter the lowest tepals—three or four traces to each tepal. The axial system, continuing upward, forms more bundles by radial division, some of which supply the upper tepals; others continue to the stamens—one trace each—on the inner rim of the hollowed receptacle; still others, becoming inverted, turn downward into the hollow of the receptacle wall. These inner inverted bundles may lie close to, and even fuse with, the "lower," outer, normally oriented bundles. Traces to the carpels—three each—are given off, in the usual manner, toward the inside by the descending strands, and the continuing strands form a weak meshwork beneath the floor of the hollow receptacle. The course of the vascular bundles—upward and downward and across the apex of the receptacle—explains the presence of inverted and concentric bundles and the "terminal" network of vascular tissue in the receptacle, and shows clearly the nature of the receptacle tip—an inverted apex. The passage of traces *inward* to some of the stamens and to the carpels is obviously normal.

The presence of cortical bundles in these primitive taxa has led to the opinion that cortical vascular systems are probably primitive for angiosperms. But this conclusion seems unjustified, because of differences in structure in the taxa, and because a somewhat similar system is present in the Melastomaceae, Myrtaceae, and Acanthaceae (*Justicia*). Cortical systems may represent vestiges of the ancestral structure of complex fern steles, but seem rather to represent specializations.

The primitive characters of this family are the numerous, free, spiral tepals, stamens, and carpels; lack of distinction of calyx and corolla;

anthers with broad and protruding connectives, which are transformed, distally, into food bodies; staminodes above the stamens; pollination by beetles; no nectaries; a decurrent stigma; a unilacunar, two-trace node; a many-celled archesporium.

In contrast with these many primitive characters, are the reduction of the carpel to achenelike structure, with one or two ovules; seeds with a large embryo and no endosperm; vessels with simple perforations; sieve tubes of high type. The concave receptacle, specialized over the primitive, is doubtless also an advanced character, a step in the structural protection of the ovules; it may represent an adaptation to beetle pollination. (The primitive *Eupomatia* has a similar concave receptacle.) The presence of a cortical vascular system is probably also an advanced character. Primitive characters in this family surpass in importance and number the advanced characters; the Calycanthaceae are surely ranalian.

The placing of the Calycanthaceae in or near the Rosales or Myrtales is not supported by pollen characters or by wood structure. When all characters are considered, the Calycanthaceae most closely resemble the Eupomatiaceae, Magnoliaceae, and Annonaceae.

Trochodendraceae and Tetracentraceae

The Trochodendraceae and Tetracentraceae, primitive, unigeneric families, are closely alike in several characters and unlike all other families in the Ranales. They are ranalian in their simple bisexual flowers and vesselless wood. The gynoecium, of several to many carpels in *Trochodendron* and four in *Tetracentron,* is syncarpous by the lateral union of the short carpels of unusual follicular type. In *Trochodendron,* there is further fusion in the adnation of the stamens to the backs of the carpels, an unusual type of perigyny. The form of the anatropous ovule has been called unique in angiosperms. (As the ovules mature, the chalazal region elongates greatly and forms a prominent basal appendage, with a "doubled-back" vascular bundle.) Another unique character in these two families is the structure of the parenchyma of the secondary wood. In gymnosperms and angiosperms generally, uniseriate wood-parenchyma cells are formed by transverse divisions of the fusiform xylem initials. In *Trochodendron* and *Tetracentron,* the transverse divisions are oblique, rather than at right angles to the long axis of the fusiform cell.

Taxonomically, the Trochodendraceae and Tetracentraceae have been variously treated. The placing of the Tetracentraceae in the Hamamelidales disregards its close similarity to the Trochodendraceae in ovule form and in vesselless wood, which has wood parenchyma with unique ontogeny. When all characters are considered, the two families seem to form two related, but distinct, families isolated in the loose, ranalian complex.

AUSTROBAILEYACEAE

Austrobaileya, a genus described in 1929, is a woody vine, native in northern Queensland. Because sufficient material was not available, its taxonomic position was at first uncertain; it was suggested by some taxonomists that it be placed in the Magnoliaceae; by others, in the Dilleniaceae; and by still others, in the Monimiaceae. Additional collections and critical study have shown that it deserves family rank among the more primitive Ranales.

The simple leaves are pinnately veined. Their traces are two, derived from the sides of a single gap (Fig. 5), the nodal structure now generally believed to be primitive for angiosperms. The perianth, consisting of several free tepals, is not differentiated into calyx and corolla. The stamens are laminar, with four median, protuberant, adaxial sporangia (Fig. 50). Above the stamens are several staminodia, some transitional to stamens. The several carpels are differentiated into ovary, style, stigma, and a stout stipe. All the floral appendages are borne in flattened spirals. The pollen is monocolpate, the ovules anatropous. The bracts, tepals, and stamens have two vascular traces; the stamens and staminodia show forms transitional to double and single traces. In the carpels, the median bundle is double at the base, where two independent traces unite.

The cambium is primitive, with long-overlapping fusiform initials. Its fairly primitive wood is unusual for a woody vine; its vessels are scalariform, with numerous perforations, and its tracheary pits are transversely elongate. The phloem is remarkable for absence of companion cells (Fig. 26). (The presence of companion cells in the phloem is an outstanding angiosperm character, contrasting with the absence of these cells in the phloem of gymnosperms.) The phloem is primitive also in its histological simplicity; it lacks sclerenchyma and typical sieve tubes. The sieve elements are long-overlapping cells, without the specialized terminal sieve plates characteristic of true sieve tubes. This simple phloem parallels in primitiveness the vesselless xylem of other families. Retention of this primitive vascular structure in this genus, a woody vine, is remarkable, because vines commonly have highly specialized xylem and phloem. It is further remarkable that, as a vine, it retains the most primitive nodal structure, not only at leafy nodes but in the floral nodes, those of bract, tepal, stamen, and carpel.

CERCIDIPHYLLACEAE

Among primitive dicotyledons, *Cercidiphyllum*, like *Euptelea*, has long been of doubtful position; it has been considered a member of the Magnoliaceae, the Trochodendraceae, and the Hamamelidaceae. But it has finally been shown to be worthy of family rank, with characters that set it far apart from other ranalian families and raise the question whether it does not deserve ordinal rank.

The genus is a dioecious tree of China and Japan. Its leaves show apparent transition from pinnate to palmate venation, and its simple flowers are in greatly reduced and compacted inflorescences. Fossils

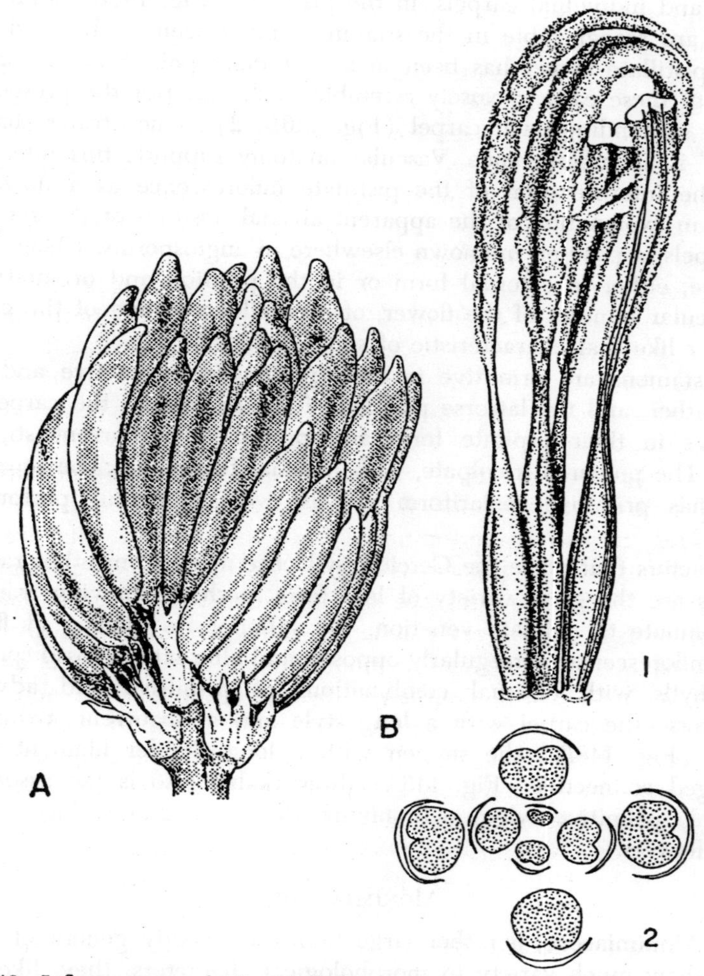

Fig. 146. Inflorescences of *Cercidiphyllum japonicum*. *A*, staminate, stamens in clusters, each cluster subtended by a bract; *B*, pistillate: 1, showing flowers consisting of single carpels, scar of bract shown at base of each carpel; 2, diagram showing arrangement of flowers and bracts in the inflorescence. (*After Swamy and Bailey.*)

of the genus—leaves, inflorescences, and fruits—show that it was represented in the Upper Cretaceous and the Tertiary by several species.

The flowers, without perianth, subtended by bracts, are crowded together on short axes, forming inflorescences which have frequently been interpreted as flowers. The fossil inflorescences show that the flowerlike inflorescences of the living species represent reduced racemes. Evidence that the "flowers" are inflorescences is the presence of bracts which subtend small clusters of stamens, in the staminate inflorescences (Fig. 146*A*), and individual carpels, in the pistillate (Fig. 146*B*). Individual flowers are recognizable in the staminate inflorescences, but reduction of the pistillate flower has been to a solitary carpel, and the compact cluster of these carpels closely resembles a flower. But the presence of a bract subtending each carpel (Fig. 146*B*, 2) demonstrates that the "flower" is an inflorescence. Vascular anatomy supports this interpretation. The interpretation of the pistillate inflorescence as a flower required an explanation of the apparent abaxial position of the suture of the carpels, a position unknown elsewhere in angiosperms. (There is no evidence, either in external form or in the position and orientation of the vascular bundles of the flower, of a possible twisting of the carpels or flower like that characteristic of some orchids.)

The stamens are primitive in their prolonged connective and basifixed anther, and the latrorse position of the sporangia; the carpels are primitive in their stipitate form and prominent, decurrent stigmatic ridges. The pollen is tricolpate, with unusual furrowing. The secondary wood has primitive scalariform vessels, with numerous perforations (Fig. 22).

Characters that place the Cercidiphyllaceae apart from other ranalian families are the great variety of leaf form, perhaps showing transition from pinnate to palmate venation; the subopposite placing of flowers in the inflorescences (irregularly opposite and alternate in fossil species); sporophylls with unusual combinations of primitive and advanced characters—the carpel with a long style and a decurrent, two-ridged stigma (Fig. 146*B*), the stamen with a long slender filament and a prolonged connective (Fig. 146*A*). Remarkable also is the association of highly primitive xylem with highly advanced inflorescence and reduced flowers.

MONIMIACEAE

The Monimiaceae, a rather large family of woody genera of varied habit, show much variety in morphological characters; they, like other ranalian families, combine primitive and advanced characters—some of the primitive characters of the Magnoliales and some of the advanced characters of the Laurales. The flowers range from bisexual to unisexual, mostly with a prominent hypanthium of varied, even extreme, form. The wood varies considerably in details of structure. The family has been considered close to the Calycanthaceae on the basis that, in both families, the hypanthium is greatly enlarged. But in other characters, the families are not alike. Similarity in wood structure and in stamens, which have lateral appendages and valvular dehiscence, suggests relationship to the Lauraceae. The Monimiaceae perhaps form a link between the Ranales and the Laurales.

Amborellaceae

The Monimiaceae were shown, in 1948, to contain a vesselless genus, *Amborella*, a shrub endemic to New Caledonia. The morphological characters are closely similar to those of the Monimiaceae but differ sufficiently to set *Amborella* apart from the other genera, as forming an independent family. Though the flowers are unisexual, the perianth is very primitive, with transition in spirally placed organs from bracteoles to tepals. The stamens have a broad filament. The wood is highly primitive: vessels absent; rays of the primitive heterogeneous type; tracheary elements very long, resembling those of other vesselless taxa (Fig. 24). The tracheary pitting is circular, occasionally scalariform. The nodes are unilacunar. Pollen morphology supports the segregation of *Amborella* as a separate family, the Amborellaceae.

Eupteleaceae

The relationships of *Euptelea*, which has been placed in the Trochodendraceae and the Hamamelidaceae, have been difficult to determine. The genus shows some primitive characters; the bisexual flowers have numerous free stamens, anthers with prolonged connective, and long-stipitate carpels (Fig. 73S). But the wood, as compared with that of *Trochodendron*, which is vesselless, has scalariform vessels; the pollen grains are tricolpate or polycolpate. In nodal anatomy, a modified, unilacunar type, this family differs from all other woody ranalian families. *Euptelea* is another isolated family with strongly contrasting, primitive and advanced characters; its primitive flowers and wood are ranalian, but the specialized pollen and nodal anatomy are not. It is surely not close to the Hamamelidaceae. *Euptelea*, with its peculiar combination of characters, seems worthy of distinction as an independent family.

Ceratophyllaceae

The morphology of *Ceratophyllum* has received much study, but there has been little agreement as to its interpretation or to the probable relationships of the family. Like other aquatic taxa, it shows both primitive and highly advanced characters, but some characters have been obscured by reduction in adaptation to an aquatic habit.

The stamens are primitive in form—no differentiation of anthers and filaments, and four elongate sporangia, median on the abaxial side (Fig. 49). The stamens, however, cannot be considered primitive on the basis of form alone, for simplicity of form may represent modification in adaptation to pollination under water. But the close similarity in form to stamens of the woody Ranales that are undoubtedly primitive is remarkable. In the abaxial position of the sporangia, the stamens further resemble some primitive ranalian types, but the significance of this character is unknown.

The gynoecium, in contrast with the androecium, shows high specialization—a solitary carpel with one ovule, which is borne distally on the dorsal side (Fig. 85*C*). The position of the ovule is evidence that reduction in ovule number has been directly from laminar placentation; this gynoecial character strengthens other evidence of primitiveness for the family.

The embryo has been described as "multicotyledonary, a remarkable type suggestive of the polycotyledonary embryos of some conifers." (Conifer embryos with more than two cotyledons have been shown to be specialized types.) But the embryo of *Ceratophyllum* has been shown to be a normal dicotyledonous embryo, with the plumule rather well developed in the seed, and the "extra" cotyledons to be the primordia of the first leaves.

The Ceratophyllaceae are an isolated family, with, apparently, little evidence of relationship to other dicotyledonous families. It has been frequently suggested that they are related to the Nymphaeaceae and Cabombaceae, but this relationship seems unlikely, because of the abaxial microsporangia and whorled leaves in the Ceratophyllaceae. Resemblance in placentation—dorsal attachment of ovules in *Cabomba* and *Ceratophyllum*—is unimportant; all have laminar placentation basically. The single integument of the ovule has been considered possible evidence of relationship to several families in the dicotyledons and to the Najadaceae among the monocotyledons, but ovules with one integument represent a frequent reduction type, especially in aquatic and parasitic taxa. A possibly significant character is the abaxial position of the microsporangia in *Cabomba* and *Ceratophyllum*, but *Brasenia* and the Nymphaeaceae—probably relatives of *Cabomba*—have adaxial microsporangia.

RANUNCULACEAE

The Ranunculaceae are remarkable for their many combinations of primitive and advanced characters. The flowers are primitive in their bisexuality, numerous appendages, spiral arrangement, actinomorphy, absence of fusion, follicular carpels, and anatropous ovules. They are advanced in the presence (in some genera) of unisexuality, few and whorled organs, zygomorphy, connation, and achenes. The ovule also shows strong specialization; the nucellus is absorbed before the flower opens, and the embryo sac is enclosed by the inner integument. This ovule structure is frequent in the Sympetalae but seems to be unknown elsewhere in the Ranales, except in *Nandina* (Berberidaceae). The wood, except in *Hydrastis,* is advanced, having simply perforate vessel elements. In the family, basically primitive, the advanced characters are doubtless related to the dominance of high specialization in habit—herbs and vines.

The family shows stages in the differentiation of a two-seriate perianth, of reduction in number of carpels, of connation in the gynoecium, of reduction of the follicle to the achene, with accompanying simplification of the carpellary vascular system (Fig. 42), of reduction in number of ovules from many to one, in the development of anemophily. The limitation of subfamilies has varied greatly, and *Paeonia* has been removed from the family to form a separate family of the Dilleniales. (A position beside the Dilleniaceae was long ago given it.)

The series in specialization in the gynoecium is perhaps most marked, with reduction in size and number of carpels, in number of ovules, and with the beginning of connation. *Cimicifuga* and the closely related *Actaea* show progressive reduction in carpel number—several to one within the genus *Cimicifuga* and one in *Actaea.* Reduction in number of ovules accompanies reduction in size of carpel; *Anemone, Clematis, Potentilla* (Fig. 42*N*), and *Adonis* show vestigial ovules at the top of the ovary, and *Trollius* (Fig. 42*B*) shows remnants of the vascular traces to lost distal ovules. The carpel of *Hydrastis,* with two ovules only and one of these often abortive, is intermediate between the typical follicle and typical achene (Fig. 42*D*).

Helleborus shows, in one species, connation of the carpels for half their length, and *Nigella* shows, in its various species, progressive stages in their union. The union of the carpels in *Nigella* has been interpreted as apparent only, the result of the upgrowth of receptacular tissue about the carpel bases, as occurs in *Nelumbo,* but no sound anatomical or ontogenetic evidence supports this interpretation.

The genus *Hydrastis,* the object of considerable attention by both taxonomists and morphologists, has been variously placed within the Ranunculaceae and in other families. It has been described as close to *Podophyllum* in the Berberidaceae. Strong resemblance has been seen to *Paeonia,* and it has been placed, with this genus, within the Ranunculaceae. Suggestions have been made that it, like *Paeonia,* be removed from the buttercup family and given family rank as the Hydrastidiaceae (a family apparently not technically described). Evidence from vascular tissue seems to show that the genus deserves family rank. The vessel elements, like those of *Paeonia,* are mostly scalariform; in all other genera of the Ranunculaceae, they are simple. The vascular bundles of the stem approach the amphicribral type, the phloem capping the xylem and like the xylem in amount, whereas the vascular bundles of other genera in the family approach the amphivasal type, where the phloem is almost surrounded by the arms of the xylem. Another difference in the xylem is the scalariform pitting of the lateral walls of the vessel elements, in contrast with the rounded pits of other genera. These vascular differences are as great as those that set *Paeonia* apart.

If this family is established, the question of the position of the Hydrastidiaceae must remain open until more information is available. Whether the androecium is centripetal or centrifugal is uncertain; the numerous stamens seem to mature simultaneously. Serological studies also leave the position of *Hydrastis* in doubt. Serum reactions of the genus show that it is apparently as close to *Podophyllum* in the Berberidaceae as it is to other genera in the Ranunculaceae.

The Ranunculaceae stand out in the Ranales as a divergent herbaceous family in an otherwise dominantly arborescent order. Most of the clearly primitive woody families in this order are characterized by the presence of "ethereal oil cells"; none of the Ranunculaceae has these cells. This histological character appears important as an indicator of phylogenetic relationship in the group of highly primitive families of this order. The absence of these cells from the Ranunculaceae widens the gap between this family and the woody families of the order.

The Ranunculaceae are sometimes cited as a good example of a natural family, although, at one time, they were described as "one of the most unnatural of natural families." The removal of *Paeonia* has been a first step in a possible further breakup of this family; *Hydrastis* will probably be removed, on the basis of the presence of scalariform vessels in the wood and differences in pollen-grain type.

The dominance of the herbaceous habit in the family strengthens the position of the family as probably the most advanced in the order.

Lardizabalaceae

The chief morphological interest in the small family Lardizabalaceae is in its primitive sporophylls: laminar stamens with deeply sunken sporangia (in at least some genera), massive connective with elongate apices; carpels with sessile stigma and laminar placentation. In contrast with these primitive floral characters, is the high specialization of the wood: vessel elements, simply perforate or scalariform with few bars; fibers, with simple pits, sometimes septate; parenchyma, absent or sparse; rays broad, primary only. This wood resembles that of the Berberidaceae.

In this family, which has primitive floral structure—absence of fusion and primitive sporophylls—the advanced floral character—unisexuality—and specialized xylem are doubtless related to the habit of the genera, which are chiefly lianas.

Sargentodoxaceae

The Sargentodoxaceae is a small unigeneric family, segregated (1926) from the Ranunculaceae. *Sargentodoxa* is a woody Chinese vine, clearly ranalian, but with characters resembling those of several families other than those of the Ranunculaceae. The staminate flowers resemble those of the Lardizabalaceae; the stamens, those of *Cercidiphyllum* and *Euptelea;* the pistillate flowers, in their gynoecia, are like those of *Schisandra.* The stamens have well-differentiated anthers and elongate connective. The many carpels are borne on an elongate receptacle. The seeds are ranalian, with minute embryos. Anatomically, little is known about this family. In some characters—simply perforate vessel elements and fibers with bordered pits—the xylem is advanced for a ranalian taxon, but the genus is a vine. The Schisandraceae are perhaps the closest relatives of the Sargentodoxaceae.

Berberidaceae

The Berberidaceae, a small family, commonly placed close to the Ranunculaceae, are heterogeneous in many characters. Their gynoecial structure is especially complex and obscure, and has had several interpretations.

The family shows several lines of specialization: herbs of different habits, and woody shrubs; some gynoecia with three, and others with two carpels; placentation of various not readily identifiable forms.

The gynoecium of *Berberis* has commonly been interpreted as unicarpellary, but anatomy shows that it is syncarpous and, in most flowers, consists of three intimately fused carpels. This interpretation is supported by study of occasional flowers that show their nature by external structure—where there is only one carpel, there is a ventral suture; where two, there is incomplete fusion distally.

Podophyllum peltatum, much used in teaching because of its apparently simple structure, is an unfortunate choice, because the gynoecium consists of two carpels, with one reduced, and the two placentae united to form one large, "parietal" placenta. External evidence of the syncarpy is the absence of the ventral suture which is present in unicarpellary gynoecia of the Ranunculaceae and Rosaceae and the unicarpellary flowers of *Berberis vulgaris*.

The presence of three carpels in the Berberidaceae is to be expected, because of the trimerous pattern of the flowers and because the related Lardizabalaceae have three carpels. The three carpels in *Berberis* are shown, by anatomical study, to stand at different levels, with the upper two commonly reduced, the uppermost strongly so. The two-carpellary gynoecia of most genera in the family represent a reduction from three. The different levels of attachment of the three carpels in *Berberis* indicate spiral arrangement, evidence of ranalian relationships, though a high type of syncarpy is present. (Syncarpy in gynoecia with few spirally placed carpels is probably rare.)

The genera *Berberis* and *Mahonia* stand apart from the other genera in their woody habit. In their stem anatomy, they show evidence of derivation from herbaceous ancestors. The woody stem consists of a compacted cylinder of vascular strands, separated by broad rays, like those of many herbs. In histological characters, the xylem resembles that of herbs with well-developed woody cylinders: vessel elements, with few exceptions, short and simply porous; parenchyma absent; fibers libriform, short, sometimes septate; intervascular pitting alternate.

Although highly specialized in habit and in gynoecial structure, the ovules (*Berberis*) show some primitive characters—a massive nucellus and multiple archesporial cells. The embryo resembles that of the Ranunculaceae.

The family is apparently closely related to the Ranunculaceae. It is heterogeneous, with subfamilies probably derived independently from herbaceous stock of primitive ranalian type.

Nymphaeaceae and Cabombaceae

The Nymphaeaceae, together with the probably related Cabombaceae, form a complex taxon, remarkable for retention in the flowers of primitive characters combined with specialized structure. Morphologically, they are especially important, because they provide examples of primitive structure in herbaceous ranalian taxa similar to those of the woody taxa—in perianth, stamens, and placentation especially. In the perianth (calyx and corolla distinct) and in the androecium and gynoecium, there is advance from numerous, spiral, to few, cyclic organs. The stamens range from laminar forms to those with well-differentiated anther and filament. Laminar placentation is characteristic of the Nymphaeaceae, but there is reduction to the puzzling types of *Cabomba, Brasenia,* and *Nelumbo* (Fig. 85). The Cabombaceae, as compared with the Nymphaeaceae, are advanced and provide, in some characters, examples of late stages in specialization.

The perianth shows stages in the development of distinct calyx and corolla. In some genera, especially in *Nymphaea,* there are many transitional forms between stamens and petals, with evidence that the corolla in this family represents sterile stamens, morphologically (Fig. 54). Other, less primitive genera have no staminodia; the distinction between stamen and petal is obvious. But in the polymerous taxa, a line is often not distinct; the nature of the sepals is not clear. In *Nuphar,* specialization in the perianth has progressed further than in the other Nymphaeaceae; the sepals have become petaloid, and the petals reduced and somewhat bractlike. In the Cabombaceae, calyx and corolla are sharply distinct; the differentiation of a corolla has accompanied the establishment of cyclic arrangement of organs.

The stamens in these two families show many stages in evolutionary modification (Figs. 54 and 55). The most primitive type is similar to that of the more primitive members of the Magnoliaceae—flat and broad stamens, without distinction of anther and filament (Fig. 55*B* to *D*); wall-less, elongate, median sporangia, deeply sunken in the lamina and remote from the vascular supply; and prominent distal appendage. This primitive type is characteristic of *Victoria* and some species of *Nymphaea.* Other taxa show forms transitional to the stamens of *Cabomba, Brasenia* (Figs. 54 and 55), and species of *Euryale* and *Barclaya* (Fig. 49), which have well-differentiated anther and filament, no sterile, distal appendage, and protuberant sporangia. The Nymphaeaceae and *Brasenia* are alike in having introrse dehiscence; *Cabomba* has extrorse dehiscence. The significance of this difference is discussed in Chap. 4. The water lilies, like the magnolias, demonstrate the evolutionary advance of the stamen in two independent lines of the Ranales.

The morphology of the gynoecium in these families has been much discussed. Connation and adnation, complicated by spiral arrangement of the numerous appendages, has made the perigyny and epigyny of the Nymphaeaceae difficult of anatomical interpretation. *Nelumbo,* with its isolated carpels, "sunken in the receptacle," increases the anatomical complexity. No broad and critical study of the gynoecial structure of these families has been made.

The morphology of the perigyny and epigyny in the Nymphaeaceae has been interpreted in two ways: as the result of the adnation of the outer, lower organs to the gynoecium; and as a result of the envelopment of the bases of the appendages by the receptacle. The ontogenetic growth of receptacular tissue about the free carpels in *Nelumbo* has been considered evidence that perigyny and epigyny in the other genera have come about in the same way. But ontogeny does not support this interpretation. This family, like the Rosaceae, shows all stages in the development of epigyny—here complicated by spiral arrangement of appendages. Evidence from the vascular skeleton is essential to the morphological interpretation of these flowers. (The growth of the receptacle about the carpels in *Nelumbo* is not evidence that, in other genera, the receptacle has similarly surrounded the bases of all appendages.)

Comparison is frequently made between the gynoecium of *Nelumbo* and that of *Eupomatia*, a primitive woody ranalian genus; both genera are described as having carpels sunken in the gynoecium. This comparison is invalid, because the carpels of *Eupomatia* are connate and intimately fused, basally (Fig. 72A).

The Nymphaeaceae and Cabombaceae provide examples of a rare type of placentation (Fig. 85)—few or solitary ovules, representing reduction directly from typical laminar placentation with many ovules. Solitary and paired ovules in most families are survivors of submarginal placentation, a stage intermediate between laminar and solitary arrangement. In laminar placentation, ovule traces are derived from the meshwork of small veins and also directly from the midvein and major lateral veins; in submarginal placentation, the ovule traces are derived directly from the ventral bundles only. (The reduced placentation in these families is described in Chap. 6.)

The relationships of the Nymphaeaceae have long been in question. The family has been considered, in some characters, transitional to the monocotyledons, especially in the possession of scattered vascular bundles in the stem and of laminar placentation. The water lilies have even been called monocotyledons; the presence in *Nelumbo* of a so-called single cotyledon has been used as partial evidence for this classification. But this cotyledon has been shown to consist of two cotyledons, connate at the base. In flower characters, general resemblance has been seen between the Nymphaeaceae and the Helobiales, and the suggestion made that the water lilies should be placed either in this monocotyledonous order or considered transitional to the monocotyledons.

The Nymphaeaceae and Cabombaceae form an isolated, heterogeneous ranalian assemblage, showing progressive stages in specialization in both external and internal structure.

Lauraceae

The Lauraceae are relatively more advanced than most of the ranalian families but belong with these families in a comparative and phylogenetic discussion. The family consists largely of tropical and subtropical trees and shrubs. In both external and internal characters, it is intermediate between most of the Ranales and more advanced orders.

The flowers show stages in transition from bisexual to unisexual. The simple, inconspicuous perianth of six members has been interpreted as uniseriate, and as biseriate with corolla reduced. The trimerous androecium is, morphologically, perhaps the most important part of the flower. There are four, sometimes five, whorls (Fig. 47). The outer two are introrse; the third, extrorse, with associated nectaries; the innermost whorl or whorls are represented by nectaries or lanceolate staminodia (Fig. 45). The nectaries, one on each side of the filament, are usually stalked, but sometimes sessile. The stalked form with enlarged tip is borne on the receptacle at or near the base of the filament, often united with the filament just above the base; secretion is from the enlarged terminal part. The sessile form is borne on the filament at various levels.

There has been disagreement about the morphology of these nectaries. The general opinion, based on comparative and anatomical studies, is that the nectaries are modified stamens, but they have been called mere "glandular protuberances." The interpretation of the nectaries as morphologically stamens is an important element in phylogenetic considerations. Evidence of several kinds supports this interpretation of nectaries derived from stamens. The lauraceous stamen has four sporangia in pairs, one above the other, with valvular dehiscence. (Some genera, such as *Lindera,* have two functional sporangia, two others abortive.) Stamenlike characters that persist in the stalked nectaries are the filamentlike stalk and the more or less anther-shaped tip (*Persea*); the occasional formation of pollen by some nectaries; in some genera, attachment of nectaries on the receptacle free from the stamen; the resemblance of early primordial stages of the nectary to anther primordia; in pistillate flowers, the similarity of the nectaries to the vestigial stamens. Vascular anatomy strongly supports the staminal nature of these nectaries. Though their vascular structure is complex within the secretory region—with many minute strands, as always in nectariferous tissues—the origin and course of the vascular supply is the same as that of the stamens. Where the nectary stalk is free, or nearly so, at the base, one trace, independent in the torus, passes into and through the stalk to the enlarged tip, as in a stamen—*Lindera;* where the nectaries are adnate to the stalk, the filament has three bundles (in some genera, five—*Persea*), and the outer, lateral bundles arise independently in the receptacle by the division of a single bundle of the floral stele and pass independently through the filament to the glands—*Umbellularia* (Fig. 45). Less commonly, the three bundles remain united in origin and part way through the filament.

The location of the nectaries, individually beside the stamens and in groups with a common vascular supply, supports the interpretation of the androecium of this family as fasciculate; each stamen, with its associated nectaries, represents a fascicle of three stamens, probably five in those genera where the stamen base has five bundles. "Stamen fascicles" are frequent in taxonomic descriptions of the family and of certain genera. The fasciculate condition is doubtless ancestral for the family but, in living forms, is greatly reduced. The innermost whorl may consist of nectaries only, or of lanceolate staminodes. Anatomical evidence that the cluster is a fascicle is seen in *Laurus*, where a single bundle, a "trunk bundle," divides into three trace bundles—the outer one normally oriented, the inner two inverted. The inverted bundles, in their course to the base of the stamen, where they enter the nectary stalk, twist and become normally oriented. (A similar vascular supply of stamen fascicles is present in other taxa—Dilleniaceae, Guttiferae.) Where there are five whorls, as in *Umbellularia*, the innermost fascicles are reduced to lanceolate staminodes; yet these minute organs still retain the three-trace supply of a fascicle.

A fasciculate androecium is also present in some of the Monimiaceae and, with valvate dehiscence, strengthens other evidence for fairly close relationship of the Monimiaceae, Gomortegaceae, Hernandiaceae, and Lauraceae.

The gynoecium, with its apparently solitary carpel and a single ovule, is more advanced than the androecium, with its four or five whorls of sporophylls. But the gynoecium is basically tricarpellate. *Umbellularia* frequently shows additional carpels, and *Cassytha* is shown by anatomy to be tricarpellate. The solitary ovule is anatropous and attached distally.

The xylem has a mixture of primitive and advanced characters: both scalariform and simply perforate vessel elements; fiber tracheids and libriform fibers; heterogeneous rays and alternate intervascular pitting. Wood structure supports the high rank within the Ranales commonly given the Lauraceae.

Some Families Less Well Known Morphologically, Commonly Placed in the Ranales

Little detailed information about the morphology of the small tropical families Hernandiaceae and Canellaceae is available. Both have been considered related to the Myristicaceae. The Canellaceae are advanced in the connation of their numerous stamens and their apparently syncarpous gynoecium—"one-loculate with several parietal placentae"—and seem hardly ranalian. The presence of lateral appendages on the stamens of the Hernandiaceae is perhaps evidence of relationship to the Myristicaceae or to the Lauraceae.

The Trimeniaceae, a small family, recently segregated from the Monimiaceae, show primitive characters in the lack of distinction in the floral axis of pedicel and receptacle; in the intergrading of bracteoles and tepals; in their numerous stamens; in their two-trace nodes and scalariform vessels. But they have advanced characters: flowers that show transitions from bisexual to unisexual; polyporate pollen; solitary carpels; and septate wood fibers, which replace parenchyma. This family is an excellent example of the mixture of primitive and advanced characters found in most ranalian families.

Discussion and Summary of the Ranales

In the twentieth century, the increasing acceptance of the Ranales as the most primitive living dicotyledons has stimulated critical morphological study throughout the order. The studies have been broad, with conclusions based on evidence from many fields. The value of palynology and the anatomy of wood in phylogenetic comparisons has been recognized. Doubtless, no other major taxon in the angiosperms is now so well known, structurally.

Phylogenetically, the Ranales now appear as a somewhat heterogeneous group, held together loosely by the possession of several primitive characters. Relationships within the order are difficult to determine; the families seem to be an assortment of relic types, representing several ancient lines rather than a single basic stock (Fig. 147). Archaic characters have been retained but are combined with well, even highly, advanced structure—for example, unsealed, stipitate, styleless carpels and vesselless xylem are associated with unisexuality and syncarpy (Winteraceae); laminar stamens with solitary carpels (Degeneriaceae); beetle pollination and the most primitive scalariform vessel elements with connation of carpels, perigyny, and multilacunar nodes (Eupomatiaceae). Obviously, these most primitive angiosperms are highly specialized in some characters (evidence supporting the view that the flowering plants are very old). Living angiosperms with *all* characters primitive are not to be expected.

The Plant Body. The Ranales strongly support the view that the arborescent habit is primitive in angiosperms, for it appears in correlation with primitive characters in nodal structure, xylem, phloem, pollen, ovule, nucellus, archesporial tissue. The herbaceous taxa have the advanced characters: trilacunar or multilacunar nodes, simply perforate vessel elements, libriform and septate fibers, tricolpate pollen. The dominance in the Ranales of the simple, pinnately veined leaf is strong evidence that this is a primitive leaf type in the angiosperms. Large and palmate leaves are restricted chiefly to the herbaceous Ranunculaceae. (The simple leaves of *Tetracentron* and *Cercidiphyllum* are palmately veined.) The Ranales lend no support to the view—a part of the durian theory of the nature of the plant body of primitive angiosperms—that the primitive angiosperm leaf was large and pinnately compound. Further, the freely branching, arborescent members of this order show no

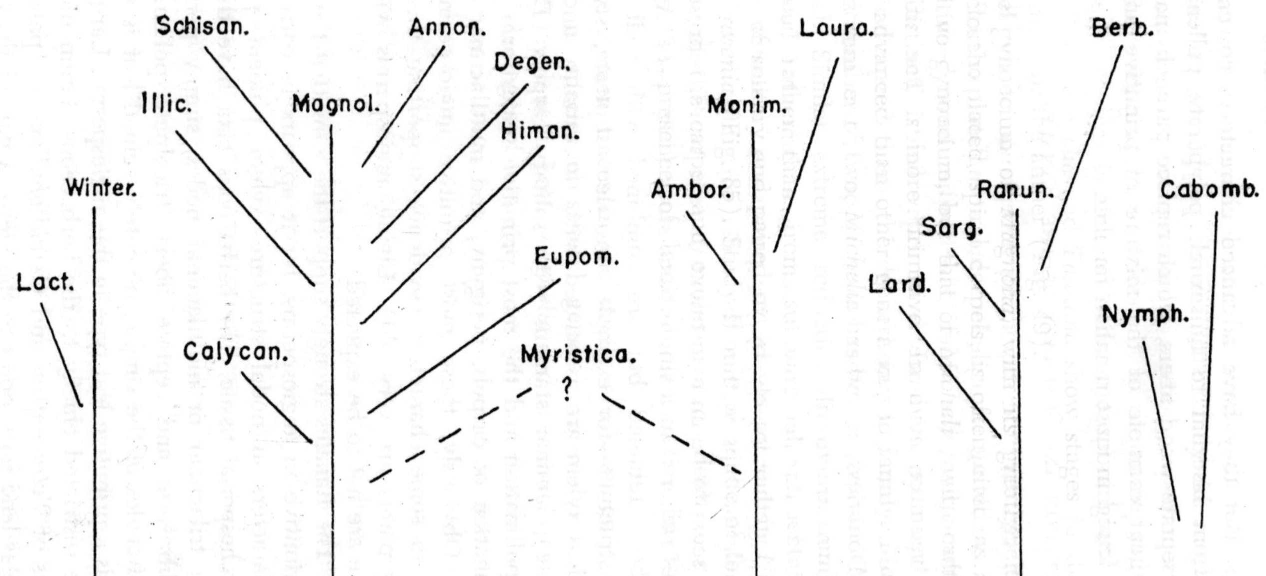

Fig. 147. Chart showing suggested relationship among some ranalian families based on consideration of all available characters. Omitted are *Austrobaileya*, Trochodendraceae, Tetracentraceae, Cercidiphyllaceae, Ceratophyllaceae, Eupteleaceae, Hernandiaceae, Canellaceae, Trimeniaceae, each of which seems to form an independent line.

examples or suggestions of the unbranched "durian" habit, as might be expected in this large and diverse primitive order if the durian theory were valid.

In leaf structure, some attention has been given to differences in form and arrangement of the subsidiary cells of the stomata, with the view that these differences—"haplocheile" versus "syndetocheile"—may be of phylogenetic significance, as they have been shown to be in the gymnosperms. But subsidiary-cell patterns seem not to be clear in angiosperms; some families appear to have no pattern, and, in other families—even within the genus *Magnolia*—more than one pattern can be distinguished. There may have been evolutionary advance in angiosperms from a simple subsidiary-cell pattern to a more complex one, and to a loss of distinction from surrounding epidermal cells, but there is, as yet, insufficient information to determine this.

The unilacunar node is characteristic of many of the ranalian families, and the origin of the trilacunar and multilacunar types is well shown in *Austrobaileya* and other genera. The origin of the odd-number trace system, long supposed characteristic and unique in angiosperms, is well shown in this order. The basic number of traces for appendages is clearly two, from one gap (Fig. 5). (Cotyledons, carpels, and stamens also show the basic number two in various orders.) Within the Ranales, nodal structure shows stages in advance from unilacunar to trilacunar and multilacunar and is rarely reduced to a secondary unilacunar with one trace.

In the shoot apex, the two-layered tunica is characteristic of several families and seems to be the primitive form in angiosperms.

Xylem and Phloem. Four ranalian families have vesselless xylem and accompanying highly primitive tissue structure and cell arrangement: the Winteraceae (Fig. 25), Trochodendraceae, Tetracentraceae, and Amborellaceae (Fig. 24). (Outside the order, only *Sarcandra* (Fig. 23) in the Chloranthaceae is known to be vesselless.) Only a step in histological advance above these taxa are the Eupomatiaceae and Cercidiphyllaceae; their tracheary cells differ little from those of the vesselless taxa except in the presence of perforations in the scalariform-pitted end walls. Other families have only scalariform or both scalariform and simply perforate vessel elements. The various families show progressive shortening of the scalariform-perforate end wall of the vessel element; reduction in number and increase in size of openings lead to the short, simply perforate element of the Berberidaceae and Ranunculaceae, with the exception of *Hydrastis*, which probably does not belong in this family. (*Paeonia* is now generally recognized as a dillenialian taxon.) Similar progressive modification is present in the tracheid-fiber and vascular-ray series.

The secondary phloem of the Ranales is little known; only that of the Calycanthaceae and Austrobaileyaceae (Fig. 26) has been critically studied. In several families, it has been loosely described as "soft," "nonfibrous," or "unstratified." Histological simplicity characterizes the secondary phloem in most families, but perhaps only in the Calycanthaceae, Eupomatiaceae, Austrobaileyaceae, and Lardizabalaceae is this tissue without sclerenchyma. Most of the families have "scattered fibers," or "patches of sclerenchyma." Only the Himantandraceae and Tetracentraceae are described as having "stratified phloem," as have most woody higher orders. The Ranales seem to show all stages in the specialization of secondary phloem as a complex tissue.

Detailed histological structure of the secondary phloem of the Ranales is not well known. In some families, the sieve-tube elements are of the more primitive, elongate, tapering type. The Calycanthaceae have sieve cells of high type—short, with transverse end wall; in this character, they accompany in the family the high types of vessel element and seed structure. This is a surprising group of advanced characters for a ranalian family with primitive perianth, stamens, and pollination. The phloem of *Austrobaileya* (Fig. 26) is remarkable in its lack of companion cells, a gymnosperm character, probably unknown elsewhere in angiosperms.

The Flower. Ranalian flowers, in addition to the generally recognized polymery and absence of fusion, show much of the probable nature and structure of the early angiosperm flower. The perianth is shown to have arisen independently in different phyletic lines and, morphologically, in different ways. In the Trochodendraceae, the flowers lack a perianth, and there is no evidence of lost organs. Where a perianth is present, the modified appendages surrounding the sporophylls are of two basic types, vegetative and reproductive. Vegetative types are represented by deciduous, calyptralike bud scales—one in *Eupomatia* (Figs. 143 and 144) and *Himantandra*—and by bracts, which range in form from minute scales to leaflike, from deciduous to persistent, from spiral—often connecting the phyllotaxy of the leaves with that of the androecium—to whorled. Leaflike tepals are prominent in some families; petaloid in others. *Liriodendron* and the more advanced species of *Magnolia* and *Asimina* show stages in the differentiation of a corolla in the form of whorls of three "petals" from the uppermost spirally borne tepals. The flowers of *Trochodendron* are described as without perianth, but possess scalelike organs below the flowers. Some flowers without a perianth, except in the form of protective bud scales, are probably primitively so—*Eupomatia* (frontispiece and Fig. 143), *Trochodendron;* others without a perianth—*Cercidiphyllum* (Fig. 146), *Euptelea*—have perhaps lost their perianth.

The corolla in the Ranales may consist of one, two, three, or even more series; these series are subwhorled and more or less distinct in form and color. Reduction in petal whorls is apparent in the Nymphaeaceae-Cabombaceae and Magnoliaceae-Annonaceae. Stages in the reduction in petal number are present in the Winteraceae, Magnoliaceae, and Annonaceae. In *Drimys*, the petals are reduced from several or many to two.

The origin of the corolla from stamens is apparent in many families by the presence of transitional organs and by anatomical structure. (The number of traces to petals is usually the same as that to stamens and unlike that to sepals.) The outer tepals of the Nymphaeaceae are perhaps, morphologically, bracts and the inner are stamens, with many transitional forms. (The somewhat similar petals of *Paeonia*, in the Dilleniales, are all bracteoid; there are no staminodes.) The conspicuous, petaloid staminodes of *Eupomatia* form a pseudocorolla, a corolla *above* the stamens (frontispiece). (*Himantandra* has staminodes both above and below the stamens.) Perhaps the upper position of petaloid organs in these two genera represents an ancient type of corolla, adapted to pollination by beetles only.

Stamens. The androecium of the Ranales shows in the laminar stamens —especially in *Degeneria* (Fig. 145), *Himantandra*, and the Nymphaeaceae (Fig. 54)—convincing evidence of the leaf nature of this sporophyll —form, origin by marginal and apical growth, vascular supply. All transitional stages from the simple, sessile blade to distinct anther and filament are present. The paired arrangement of the sporangia—two on each side of the midvein—is already established in the most primitive stamens, as is the position of the pairs—on the blade between the midrib and margin, without direct relation to the major vascular supply. The arrangement in two pairs—in strong contrast to the variety in ovule number and arrangement—persists throughout higher angiosperms. Other highly important characters are the sunken position of the sporangia and the absence of a sporangial wall. There are many stages in the protrusion of the sporangia and their protection by anther-sac walls built of laminar tissue. The value of the protruding connective tip as a primitive character in angiosperms is shown by its prominence and nearly constant presence in the ranalian families.

Interpretation, with convincing evidence, that the nectary-bearing stamens of the Lauraceae, Gomortegaceae, and some of the Monimiaceae represent modified stamen fascicles brings fasciculate stamens into the picture of the relationship of the Ranales to the Dilleniales. Separation of these two orders has been based on few characters, among them the presence of fasciculate stamens in the Dilleniales and their absence in the Ranales. The fascicles of the two orders differ, in that those of the Dilleniales have numerous stamens (reduced in *Hibbertia*) and those of the Ranales only three, or perhaps five, without evidence that these numbers represent reduction. Fasciculate stamens appear to have arisen independently in the two orders.

Carpels. The carpels of the Ranales show primitive laminar structure comparable with that of the stamen and with similar stages in elaboration of form related to function—in the stamen, anther, and filament; in the carpel, stigma, style, and ovary. The most primitive carpel, well shown in *Drimys* (Fig. 72*C*) and *Degeneria* (Fig. 72*F*), is folded, but unsealed; pollen-receiving areas are on marginal bands of the external and internal surfaces throughout the length of the lamina, and the pollen tubes enter between the marginal lips of the lamina. The two marginal strips projecting at the tip of the folded carpel show the origin of the two-lobed (unicarpellate) stigma of some higher orders. (The bilobed stigma is not the result of a basic dichotomy in the sporophyll.) The frequent presence of a stipe in the Ranales supports the view that the ancient carpel was stipitate, resembling a petiolate leaf.

The retention of sporophylls so leaflike and undifferentiated as those of *Degeneria, Drimys, Magnolia, Nymphaea, Lactoris* is remarkable in an order with many highly advanced characters.

The families of the Ranales show different methods of derivation of the isolated terminal stigma from the primitive stigmatic ridge, which, in *Degeneria* and many species of *Drimys*, extends the full length of the carpel. Progressive acropetal shortening and distortion of carpel form brought the stigmatic area to a pseudoterminal position in *Drimys;* sterilization and constriction of the distal part of the carpel formed a terminal stigma in *Himantandra;* connation of many carpels en masse in *Eupomatia* restricted the stigmatic area to the tips of the carpels. Other taxa show the style as an elongation of the distal part of the carpel.

The restriction of ovules to areas directly below the shortened stigmatic crest is perhaps evidence that, in early angiosperms, pollen tubes were short—as in *Degeneria* and the Winteraceae, where the pollen grains germinate close to the ovules—and that the long tube is correlated with the presence of a style.

The stipe—perhaps homologous with the petiole—characteristic of the primitive carpel in the Ranales, is present in other primitive orders, especially the Rosales; it is greatly reduced or lost in some ranalian families.

Placentation. In placentation, the history of evolutionary modification is one of progressive reduction in ovule number and in fertile wall area. Accompanying these changes has been modification of the relation of ovule traces to the vascular system of the carpel lamina. The ranalian families show that different lines have been followed in the reduction, although some of the placental types are difficult to interpret from descriptions based on herbarium material where the carpels are shrunken. The terms "median," "lateral," "septal," "marginal," "on the external face of the locule" have little certain morphological meaning.

Typical laminar placentation is present in the Nymphaeaceae and Lardizabalaceae. One modification of this seems to be that of *Degeneria* and the species of *Drimys* that have the decurrent stigmatic crest, where the ovules are borne in two longitudinal bands, about midway in the wings of the lamina. From this type has perhaps been derived the narrower, submarginal rows, characteristic of the follicles of higher orders. Another type of ovule reduction is seen in the Cabombaceae, where the very few surviving ovules are scattered over the locule wall (Fig. 85).

Modification of the relation of the ovule traces to the vascular system of the sporophyll is of different types within the Ranales. In "all-over" placentation, ovule traces are derived from all parts of the vascular network of laminar veins, including the major veins; the majority of traces depart from anastomoses of the veinlets. With reduction in ovule number, traces from the midvein first drop out. Where there are two longitudinal bands of ovules, the traces of some of the ovules come from anastomoses of branches of both midvein and major lateral veins. With reduction of ovule number in the rows, all the traces arise from strengthened lateral (ventral) bundles—the condition in submarginal placentation, the type characteristic of follicles. In *Degeneria* and some species of *Drimys,* the ovule traces are derived, in part, from anastomoses of branches of the dorsal and lateral veins (Fig. 83). In the Cabombaceae, the traces to the few ovules are derived from the network of small veins at points remote from the major veins (Fig. 85); this trace origin is evidence of derivation directly from laminar distribution.

The remarkable two-trace ovule supply in *Schisandra* and *Magnolia* is derived from branches of both the midvein and one of the ventral veins. Ovule number in the family has been greatly reduced to one or two, and the large ovule has "captured" two branchlets from the network of veins. (In higher orders, solitary basal ovules may have two traces—one from each ventral vein, or even one from adjacent veins of two carpels.) Placental reduction may be by localization. *Trochodendron* has numerous ovules but they are restricted "to the distal part of the locule wall." In *Exospermum, Pseudowintera,* and *Belliolum,* the fertile area is restricted to an area directly below the stigma. Similarly, in species of *Drimys* where the stigmatic ridge is greatly shortened and becomes pseudoterminal, the ovules lie below the stigmatic area. In

Exospermum and *Zygogynum*, the ovules are borne on the dorsal face of the locule wall (the basis for the name *Exospermum*) in two longitudinal rows, and their traces are derived from the dorsal vein. This ovule position was called "extremely rare" in 1900, and probably no other examples have been described. The two genera appear unique among angiosperms in their placentation. [In some achenes, ovule traces appear to be derived from dorsal bundles, but, in these fruits, the dorsal and lateral veins are greatly fused and condensed, and the ovule trace is derived from the united veins (Fig. 42).]

The Ovule. The primitiveness of the anatropous ovule is strongly supported by the Ranales, where it is apparently present in all families. Ovule type is perhaps the most constant character in the order.

Nectaries. The Ranales show stages in the development of nectaries from poorly defined areas of secretion to prominent secretory organs that represent stamens or petals. Where pollination is by chewing insects, beetles—*Eupomatia* and *Calycanthus*—food bodies (Fig. 69) take the place of secretory structures.

Pollen and Pollination. Pollen in the Ranales is dominantly monocolpate or of types (dicolpate and acolpate) derived phylogenetically from monocolpate, the form characteristic of lower seed plants. The higher, tricolpate type is present in families that are most specialized in habit—Ranunculaceae—or in flower structure—Berberidaceae. Pollen in permanent tetrads—generally considered an advanced character—is present in the Winteraceae, Lactoridaceae, and some of the Annonaceae.

Within the Ranales, there are all methods of pollination, from the most primitive, that by beetles only—by a single species of beetle in *Eupomatia*—to pollination by various advanced methods. The Ranales support the theory that pollination by wind is associated with the development of unisexuality. Pollination types and the related nectary morphology have been largely neglected in phylogenetic studies. The Ranales show the primitiveness of beetle pollination by their structural adaptations to this type, and by transitions in structure to pollination by other insects and by wind. The food bodies of *Eupomatia* and *Calycanthus* are replaced by the simplest of nectaries—mere areas of diffusion or secretion; these areas are replaced by emergences, and these emergences by transformed organs or parts of organs.

Receptacle. Modifications in the form of the receptacle that are steps to the complex gynoecial forms in higher orders are already present in the Ranales. The large flowers of the Magnoliaceae and Annonaceae, with many sporophylls, have the primitive elongate receptacle. Accompanying the reduction in number of appendages and change in arrangement from spiral to cyclic is progressive shortening of the receptacle by "telescoping"—shortening and elimination of internodes.

This reduction is well shown in the Magnoliaceae, Annonaceae, and Winteraceae. Condensation of the flower has come about also by restriction of apical growth in the floral meristem, while growth continues on the "shoulders" below. The Ranales show stages in this modification: a flattened receptacle in *Eupomatia* and in the Monimiaceae (where lateral expansion may be exaggerated); a slightly concave receptacle in other families and a deeply concave one in *Calycanthus*. In the Ranunculaceae, evidence of the shortening of the receptacle is frequently present in the form of a sterile vestigial tip between the carpel bases. Where the carpel is solitary, the tip may be pushed aside in ontogeny and the carpel take a falsely terminal position.

Embryo and Endosperm. Embryo and endosperm types are remarkably constant throughout the order. The embryo is very small, often undifferentiated at time of seed shedding, and is embedded in abundant endosperm. After-ripening, in the form of morphological maturation within the seed, is common in this order. (*Calycanthus* is an exception; it has large, well-developed embryos and little or no endosperm in the mature seed.)

Chromosome Number. Chromosome number varies considerably in the order but seems constant for some families; 19 is apparently basic for several families—Magnoliaceae, Cercidiphyllaceae, Trochodendraceae, Tetracentraceae, Winteraceae; 14, for other families—Illiciaceae, Schisandraceae, Eupteleaceae.

Relationships within the Ranales. Morphology, even with the aid of palynology, cytology, and serology, provides evidence of probable relationships for rather few ranalian families. The totality of evidence points, first of all, to the certainty that this order includes families with more primitive characters than any other dicotyledonous order; secondarily, that these families represent several or many lines of evolutionary development (Fig. 147), each represented by an individual combination of primitive and advanced characters. Some of the lines are represented by unigeneric families. The order appears to be an assemblage of relic products of ancient stocks, held together by the retention of many primitive characters.

A few characters—pollen type, nodal structure, possession of "ethereal oil cells," fasciculate stamens—seem important, but none of these can be used alone to indicate relationship.

Monocolpate pollen, together with pollen of types derived phylogenetically by modification of monocolpate pollen, characterizes the families which seem to have the largest number of primitive characters: Winteraceae, Magnoliaceae, Austrobaileyaceae, Eupomatiaceae, Himantandraceae, Degeneriaceae. Tricolpate pollen and pollen of modified tricolpate types characterize the Trochodendraceae, Tetracentraceae,

Eupteleaceae, Cercidiphyllaceae, Illiciaceae, Schisandraceae, Annonaceae, Ranunculaceae, Berberidaceae, Lardizabalaceae.

Secretory cells of the "ethereal oil" type are characteristic of many ranalian families and probably represent structure inherited from a common ancestral stock. They are present in the Winteraceae, Degeneriaceae, Eupomatiaceae, Himantandraceae, Magnoliaceae, Annonaceae, Myristicaceae, Monimiaceae, Gomortegaceae, Hernandiaceae, Lauraceae. The oil cells, when abundant, give an aromatic character to all parts of the plant. (They are present also in the Piperales.)

The primitive nodal structure, unilacunar with two traces, is present in the Austrobaileyaceae, Amborellaceae, Monimiaceae, Lactoridaceae, Illiciaceae, Schisandraceae, Chloranthaceae, Lauraceae, Calycanthaceae. The derived, trilacunar node and its modifications characterize the remaining families.

Two groups of families that seem closely related are the magnolian line—the Magnoliaceae, Annonaceae, Schisandraceae, Illiciaceae with perhaps the Eupomatiaceae, Degeneriaceae, and Himantandraceae—and the lauralian line—Hernandiaceae, Myristicaceae, Monimiaceae, Gomortegaceae, Lauraceae, with perhaps the Berberidaceae. The magnolian line is held together by large, spirally built, polymerous flowers with well-developed perianth, laminar stamens, and tri- or multilacunar nodes. The Degeneriaceae, Himantandraceae, and Eupomatiaceae have been called a "fairly natural group," allied to the Magnoliaceae. The lauralian families have in common unusual androecial structure, closely similar wood structure, and unilacunar nodes. The stamens are nectary-bearing (the Myristicaceae and some of the Monimiaceae lack the nectaries), borne in small fascicles, and have distinctive valvular dehiscence.

Isolated among the Ranales are the Cercidiphyllaceae, *Ceratophyllum*, Tetracentraceae, Trochodendraceae, Eupteleaceae, and Austrobaileyaceae. The Winteraceae also stand apart from all other families, with the possible exception of the Lactoridaceae. The Calycanthaceae show resemblances to the Magnoliaceae in perianth and cortical vascular system and to the Eupomatiaceae in pollen type and in remarkable adaptation to beetle pollination. But staminodes are abundant in *Eupomatia* and *Calycanthus* and absent in the Magnoliaceae. The presence of staminodes above the stamens in *Eupomatia* and above and below in *Himantandra* seems to bring these two genera together, but places them far from the magnolian line. The staminodes are below in *Calycanthus*.

The Myristicaeae, Eupomatiaceae, and Annonaceae have ruminate endosperm, a character claimed to tie these families together, but the

Magnoliaceae, surely close to the Annonaceae, do not have this type of endosperm.

The absence of vessels in the wood was at one time used as a basis for the setting apart of the taxon Homoxyleae. It has become apparent that this character is valueless in phylogenetic considerations, but that stages in the evolution of the vessel, when used together with other characters showing stages in evolutionary advance, are important. In the Ranales, the presence of scalariform vessels in most families and the existence, within the order, of stages in the reduction in number of bars and in the increase in size of the perforations are outstanding characters. Simply perforate vessel elements are present in the Lactoridaceae, Calycanthaceae, Ranunculaceae, Annonaceae, Berberidaceae, Lardizabalaceae, Himantandraceae, Lauraceae. The inclusion in this list of advanced ranalian families of the more primitive Lactoridaceae and Himantandraceae is an example of the unequal advance in rate of specialization in flower and vascular anatomy. Simply perforate vessel elements are present also in the aquatic families Nymphaeaceae, Cabombaceae, and Ceratophyllaceae, in which the vascular tissues are highly specialized. In these families, reduction has doubtless removed much evidence of relationship, and no affinities with other families are apparent. *Ceratophyllum* perhaps does not belong in the Ranales, though its stamens seem to place it there. The supposed resemblance of the gynoecium of *Nelumbo* to that of *Eupomatia* is in error; the carpels in *Eupomatia* are intimately connate, not sunken individually in the receptacle, as commonly described.

A possible correlation of basic chromosome number with nodal structure has been suggested. The genera studied that have tri- or multilacunar nodes—in the Winteraceae, Magnoliaceae, Trochodendraceae, Tetracentraceae, and Cercidiphyllaceae—have 19 small, short-rod-like chromosomes; the genera studied which have unilacunar nodes—in the Illiciaceae, Schisandraceae, and Eupteleaceae—have 14 much larger chromosomes. But differences in chromosomes, as shown in illustrations, are not sufficiently clear, and the nature of some of the unilacunar nodes—primitive or derived from trilacunar—is uncertain.

The Ranales show a variety of important ancestral characters: the primitive flower form—bisexual, polymerous, without perianth, appendages spirally arranged; stamens laminar, with sporangia paired, medianly placed, wall-less, and embedded in the sporophyll; megasporangia with multicellular archesporium embedded in projections of sporophyll tissue (ovules); vesselless xylem and fiberless phloem; unilacunar nodes, with two traces. Associated with these primitive characters and scattered among the various families are many of the most advanced characters of higher orders. The Ranales have retained many archaic characters, while progressing far in other characters. The variety of combinations of these archaic and advanced characters—almost as great as the number of families—suggests highly specialized remnants of ancient stocks. No one family can be selected as the most primitive. Both the Winteraceae and the Magnoliaceae have been so considered, but both have advanced genera with syncarpy and unisexuality.

DILLENIALES

The Dilleniales, closely resembling the Ranales in morphological characters, have, in some classifications, been merged with them. The orders have been held apart by the presence of fasciculate arrangement of stamens in the Dilleniales (stages in development are present in the Dilleniaceae and Paeoniaceae); of centrifugal sequence in stamen ontogeny; of seeds with an aril (except in Paeoniaceae, where the ovule, with massive integuments, is borne on placental projections); of tricolpate pollen. These distinctions are weak, because the Winteraceae (*Drimys*) have recently been shown to have centrifugal stamen development, and fasciculate stamens are present in the Monimiaceae and Lauraceae. The fasciculate stamens of the two orders are, however, apparently of somewhat different nature and of independent origin as clusters. The fascicles of the ranalian taxa consist of three, or perhaps five, stamens and show reduction in number by transformation into nectaries; the fascicles of the Dilleniales consist of numerous stamens, reduced in number in *Hibbertia,* but without transformation into nectaries. The seeds of peonies approach the arillate type in the massive integuments and placental enlargement. A third weakness in this distinction between the Ranales and the Dilleniales is the presence of tricolpate pollen in a few (the more advanced) ranalian families. From a morphological viewpoint, the maintenance of the order Dilleniales is justified. The families constituting the order—Dilleniaceae, Paeoniaceae, Crossosomataceae—form a related group, well apart from the ranalian families. The presence of fasciculate stamens in the Dilleniales supports the theory that the order represents an ancestral stock for the Guttiferae and related families.

Dilleniaceae

The Dilleniaceae, chiefly an Australasian family, have many of the primitive characters of the Ranales. They stand apart from the Ranales chiefly in their centrifugal and, in most taxa, fasciculate stamens. In habit, the Dilleniaceae are trees, shrubs, vines, and, rarely, herbs. *Hibbertia,* a large genus of about 110 species, chiefly of temperate

Australia, is of major morphological interest, because it shows in its androecium the origin and modification of stamen fascicles. The various species show all stages in the fusion of the numerous stamens to form five fascicles, each with many stamens, and the reduction of the stamens within each fascicle to few or one, with simultaneous reduction in fascicle number until only one fascicle with few stamens remains. Reduction in fascicle number gives strong zygomorphy to the androecium, but the corolla is only slightly modified. The bractlike nature of the calyx is marked in some species where the lowest of the spirally arranged, unequal sepals is leaflike, the uppermost sepaloid.

In wood structure, the Dilleniaceae show many primitive characters. The vessels are solitary and largely scalariform (simply perforate in a few advanced genera). The scalariform vessel elements, resembling those of *Eupomatia*, are highly primitive, very long and long-tapering, with many bars—up to 130. The wood parenchyma is mostly diffuse. The fibers have bordered pits. The rays are heterogeneous and both wide and narrow.

Paeoniaceae

The taxonomic position of *Paeonia*, as a member of the Ranunculaceae, long was in question. The genus has been placed in the Berberidaceae and in the Magnoliaceae. The suggestion that it should be segregated as an independent, unigeneric family was made in 1830 and again, seventy years later. In the twentieth century, evidence from cytology, floral anatomy, and ontogeny has given this change strong support and shown that the new family belongs in the Dilleniales.

In general flower structure, *Paeonia* differs from the genera of the Ranunculaceae and the removal of the genus from this family is justified. In gross structure, the flower of *Paeonia* differs from that of all genera in the Ranunculaceae. The perianth is not sharply limited. The phyllotactic spiral is continuous from leaves through bracts, sepals, petals, stamen trunks, and carpels (in herbaceous species, reduction of the carpels to very few may break the series). The receptacle is somewhat concave, in contrast to the convex apex in other genera, and a prominent, lobed disc surrounds the gynoecium. The fasciculate stamens mature centrifugally. The ovules are large, with a massive outer integument, and are borne on placental projections. The nucellus is absorbed before flowering, so that the inner integument encloses the embryo sac. The micropyle is closed, and the seed coat has three major layers. (Other genera have one or two layers.) Pollination is largely—in woody species, perhaps wholly—by beetles. Beetles lick the lobes of the prominent fleshy disc. The disc is, anatomically, a part of the androecium; the lobes receive vascular supply from the stamen trunks, as do the stamens (Fig. 46). The disc perhaps corresponds, morphologically, to the staminodia of *Eupomatia* and *Calycanthus* where inner staminodia are also modified as adaptations to beetle pollination. (In *Paeonia,* the staminodia are secretory and the "food" is fluid.)

Other characters of *Paeonia* supporting the removal of the Paeoniaceae from the Ranales are: pollen unlike that of any of the ranalian families; persistent sepals; arillate seeds; hypogeal seed germination, in contrast to that of true ranalian families, which have epigeal (except in part of the genus *Clematis*); petals that are, in petal-trace number, bracts in *Paeonia,* whereas those of the Ranunculaceae are sterile stamens.

Anatomically, *Paeonia* differs from its former sister genera chiefly in the type of vascular bundles and wood structure. The vascular bundles of ranunculaceous genera approach the amphivasal type; the xylem is V-shaped in cross section, with the phloem between the arms of the V. The bundles of *Paeonia* approach the amphicribral type; the xylem is unforked, the phloem is distal and lateral, spreading and overlapping the xylem. The vessels of *Paeonia* are scalariform.

Differences in the traces of the floral organs also set *Paeonia* apart from the Ranunculaceae. The sepals and petals have few to several traces; in other ranalian genera, the sepals usually have three traces and the petals one. A stamen-trunk system is characteristic of *Paeonia.*

The vessels of *Paeonia* are small, solitary, and scalariform, in contrast with the large, usually clustered, simple vessels of the Ranunculaceae. The fibers have distinctly bordered pits, in contrast to the libriform fibers of the Ranunculaceae. The phloem has little or no sclerenchyma; that of the Ranunculaceae has many fibers and stone cells. Accessory, cortical, vascular bundles, suggesting those of *Calycanthus* and those in flowers of *Magnolia,* are present in *Paeonia,* absent in the Ranunculaceae. The Paeoniaceae are a good example of the value, in phylogenetic classification, of characters other than those commonly used in taxonomic treatments. The basic number of chromosomes also sets *Paeonia* (five) far from that of the Ranunculaceae (seven to nine).

Relationships of the Paeoniaceae have long been in question; resemblances to various ranalian families have been seen. General flower structure suggests the Magnoliaceae, Annonaceae, and Calycanthaceae. The concave receptacle, cortical bundles in the stem, beetle pollination, and especially the peculiar staminal disc, make resemblance to the Calycanthaceae, perhaps, the strongest. Resemblance has been seen in leaf form to *Hydrastis, Actaea,* and *Cimicifuga* and, in habit of the herbaceous species, to *Podophyllum* and other berberidaceous genera. But the Paeoniaceae clearly belong in the Dilleniales, though doubtfully near the Crossosomataceae, as has been suggested. The wood of *Crossosoma,* with its simply perforate vessels, is unlike that of *Paeonia.* All the dillenialian families have centrifugal stamens, but the value of this character in determining relationships of major taxa has been, perhaps, overemphasized.

The Paeoniaceae, like the Nymphaeaceae, are an example of herbs with some highly primitive characters. The peonies show a series from woody shrubs to typical perennial herbs. Within the genus *Paeonia*, the section Moutan (woody shrubs) has ten to fifteen carpels, and primitive vessels and fibers; the herbaceous species have two to four carpels (with additional, abortive carpels in some species) and advanced vessel and fiber types. The section Moutan is of special interest as representing ancestral peony stock surviving in the mountains of China, "a region noted for its wealth of relic types."

Crossosomataceae

The Crossosomataceae, though more specialized in wood structure than the Paeoniaceae and Dilleniaceae, resemble these families, especially the peonies, in many characters and, doubtless, are well placed as an advanced family in the Dilleniales.

PIPERALES

Chloranthaceae

The number of known vesselless angiosperm taxa was increased in 1950 by the addition of the genus *Sarcandra* in the Chloranthaceae. The Chloranthaceae are a small tropical and subtropical family, not well known morphologically. Vesselless wood in this family—in contrast with that in the Trochodendraceae, Tetracentraceae, and Amborellaceae—accompanies flowers of advanced structure. The flowers, however, show primitive anatomy—stamens with two traces and carpels with double, partially fused, dorsal bundles. The vesselless wood and the primitive vascular structure of the floral organs of *Sarcandra* (Figs. 23 and 82) hardly form a sufficient basis for family rank for this genus.

VERTICILLATAE (Casuarinales)

Casuarinaceae

The Casuarinaceae have been prominent in most of the older, as well as the modern, classifications as the basic family of the dicotyledons —a member of the Apetalae, Verticillatae, or Amentiferae. In the twentieth century, they have been considered both highly specialized and so primitive as to be excluded from angiosperms. A combination of very simple, unisexual flowers and strange, coniferlike inflorescences, with highly specialized vegetative habit—whorled scalelike leaves and deciduous, photosynthetic twigs—has been responsible for a great range in interpretations of the phyletic position of the family.

The family is unigeneric. The only genus, *Casuarina,* consists of about fifty species of trees and shrubs of xerophytic habitats in Australasia, often on sea beaches at tidal limits. It is not well understood morphologically or taxonomically; species apparently undescribed are in cultivation, and little attention has been given to the more primitive species, in which the pistillate inflorescences are less specialized and are borne on vegetative short-shoots, like the staminate.

The simple, perianthless flowers—the staminate consisting of a single "terminal" stamen and the pistillate of two carpels, one vestigial—and their apparent aggregation in a unisexual catkin or conelike inflorescence long kept the family in the Amentiferae. Support for this classification was found in resemblances to conifers in the ovulate inflorescences, in anemophily, and needlelike, deciduous twigs. On the basis of similarities in whorled, scalelike leaves and "jointed" stems, even more distant relationships have been suggested, such as to the horsetails and calamites and to *Ephedra.* But critical morphological comparison of the Casuarinaceae with angiosperms in general demonstrates that this family is undoubtedly angiospermous. *Casuarina* is still insufficiently known, taxonomically and morphologically. Both its vegetative and reproductive structures represent adaptation accompanied by strong reduction. (The resemblance to *Ephedra* and *Equisetum* is in the jointed stems, with photosynthetic cortex, and the whorled, scalelike leaves.) But stelar and nodal structure are those of specialized angiosperms—the node unilacunar, with one trace. Histologically, the vascular tissue is angiospermous. The phloem has companion cells (not present in the Gnetales or lower taxa). The wood is typically angiospermous; its scalariform vessels are of the angiosperm type—not of the *Ephedra* (Gnetalian) type—with round perforations.

The flowers are extremely simple; the simplicity is clearly the result of reduction, as shown by anatomical evidence and from comparison with similar simple flowers in highly specialized families. The staminate flower consists of a single, "terminal" stamen, like that of *Euphorbia,* which is generally recognized as reduced, on evidence of many kinds. The stamens are advanced in form and structure, with filaments terete and slender and anthers sharply set apart. The gynoecium is also highly specialized; it consists of two completely connate carpels, one vestigial. The ovule is solitary. The stigma is threadlike, the specialized form of wind-pollinated plants. The structure of the ovule is insufficiently known, especially that of the more primitive species, but it is angiospermous. It has two integuments and a typical 8-nucleate embryo sac. There are several archesporial cells (Fig. 111), and sometimes an extra gametophyte is formed—a structure sometimes present in higher families.

Pollen-grain form gives no support to the theory that *Casuarina* is not an angiosperm; the pollen closely resembles that of the Betulaceae and Myricaceae and is wholly unlike the polyplicate pollen of *Ephedra.* (A position close to the Betulaceae was once suggested for the Casuarinaceae.)

Chalazogamy, first known in *Casuarina* (Fig. 103), was considered gymnospermlike, and the genus was, therefore, believed transitional to the gymnosperms. But chalazogamy was soon found in other families, and its importance in phylogeny discounted.

Little is known of the anatomy of the flower. The stamen is described as having two traces, which unite in the filament, then separate in the connective. The filament is described as "having a tendency to split," a character considered, in the classification that removes *Casuarina* from the angiosperms, evidence of dichotomy and support for the protangiosperm theory. The two-trace stamen is indeed primitive but occurs in many other families, as in the Proteaceae and Betulaceae—an example of retention of the primitive nodal structure of angiosperms. The ovule is described as "lateral," but the origin of its trace is apparently unknown.

The Casuarinaceae are unquestionably angiosperms, a highly specialized family derived from primitive stock. Only a complete disregard of morphological characters, external and internal, makes possible the placing of this family with the Gnetales in "Protangiosperms." Pollen and flower morphology suggest the Betulaceae as the near relatives of the Casuarinaceae; wood structure shows many similarities to that of the Hamamelidaceae. The casuarinas remain an isolated family; the probability that they are nearest to some of the amentiferous families is of little significance, since the Amentiferae have been shown in recent years to be a highly artificial assemblage of diverse taxa, characterized by reduction in inflorescence and flowers, in association with the adoption of unisexuality and anemophily.

HELOBIALES

Among monocotyledons, the Helobiales have been generally accepted as the most primitive, because of the polymery of the flowers, which have little or no fusion.

The flowers of the Helobiales have numerous free stamens and carpels with transitions from spiral to whorled. (Some taxa have slight basal fusion in the gynoecium, and carpel number may be reduced to few or one.) The stamens have fairly primitive form, with flattened filaments and basifixed anthers. Some highly specialized stamens—those of *Potamogeton, Posidonia, Lilaea, Phyllospadix*—are described as having an expanded laminalike connective or filament. The carpels are basically follicles—reduced to achenes in many genera—with a sessile or nearly sessile stigma.

The morphological story in this order is one of reduction throughout the plant, in adaptation to moist or aquatic habitats. In the flowers, the numerous spiral sporophylls are reduced to trimerous whorls, like those of the perianth, and the whorls are progressively reduced to one and to a solitary, "terminal" carpel or stamen. (The gynoecium of the most reduced forms closely resembles that of the dicotyledon *Ceratophyllum.*)

In monocotyledons, two types of perianth have been distinguished: one, with calyx and corolla free and unlike in form, structure, and color; the other, with the two whorls closely alike in every way and tending to unite. Both these types are present in the Helobiales. (The first type is rare elsewhere in the monocotyledons—in some primitive members of the Liliaceae, the Trillieae.) In the aquatic taxa, the perianth is greatly reduced and often lost. It may persist in vestigial form, even in submersed flowers—*Wisneria*, a highly specialized alismataceous genus, with submersed flowers. Reduction of the stamen whorls brings about the formation of one, rarely two, outer whorls, in which the stamens stand in pairs (Fig. 48). (Morphologically, these pairs have been formed by telescoping of the whorls, not by *dédoublement.*) Lateral enlargement of the receptacle in *Thalassia* and *Cymodocea*, enclosing the carpels, forms a false inferior ovary. Ventral adnation of the carpels to the enlarged receptacle forms pseudosyncarpy in the Alismataceae, Aponogetonaceae, Scheuchzeriaceae.

The placentation is fundamentally laminar—Butomaceae, Hydrocharitaceae—with many ovules. The Alismataceae show, in *Damasonium*, a series of stages in reduction from laminar to basal placentation: *D. polyspermum* has several ovules; *D. stellatum*, two ovules; and *D. californicum*, a solitary ovule. The vascular anatomy of these carpels is apparently unknown, and the relation of this placentation type to that of *Cabomba* and *Brasenia*—which shows similar reduction—is uncertain. Laminar placentation in the monocotyledons perhaps differs somewhat from that in the dicotyledons; absence of ovules along the midrib line has not been reported in laminar placentation in the monocotyledons.

The ovules are anatropous, as in primitive dicotyledons, with possible exceptions among solitary ovules; some basal ovules are described as "erect" and some suspended ovules as "doubtfully anatropous." Emphasis has been placed on the presence of several archesporial cells in the ovules of some genera as indicative of primitiveness and as suggesting derivation of the Helobiales from the Ranunculaceae, but multiple archesporial cells are present in some high families. The absence of endosperm in the seeds is an important character.

The leaves of the Helobiales vary greatly in type; those with cordate or sagittate blades are prominent. Probably the only free, paired stipules in the monocotyledons are in the Potamogetonaceae. In other orders, the stipules are merged with the leaf sheath; in the Scheuchzeriaceae, the stipules are still prominent as distal appendages of the sheath.

The vascular structure of the Helobiales is little known; its reduced condition makes interpretation difficult. The vessel elements of the Alismataceae are of the simply perforate type, a type, perhaps, not to be expected in this primitive family, but this type of vessel is characteristic of herbaceous, and especially of aquatic, plants.

Within the Helobiales, the Alismataceae and Butomaceae are generally considered the most primitive families. Though primitive in apocarpy and general absence of fusion in floral organs, they are advanced in the possession, in some taxa, of achenes, unisexuality, whorled perianth (with clear distinction of calyx and corolla), whorled androecium, gynoecium ranging from spiral to whorled, solitary ovules, and simply perforate vessels. As compared with the most primitive dicotyledons, these families have fewer primitive characters.

The genera of Alismataceae are closely interrelated; *Alisma* is most primitive on cytological evidence—$7n$ chromosomes, in contrast with much higher numbers in other genera. The pollen grains of this family—multiporate with elaborate apertures—are markedly unlike the monosulcate type characteristic of most of the higher monocotyledons. In the Butomaceae, all the genera, except *Butomus*, have pollen resembling that of the Alismataceae, but *Butomus* has monosulcate pollen grains. On the basis of pollen morphology, it has been suggested that *Butomus* alone be retained in the Butomaceae and the other genera placed in the Alismataceae. Relationships among these aquatic or marsh-living taxa are difficult to determine because of their great reduction.

The carpels of *Alisma* perhaps show the origin of septal nectaries. Between the bases of adjacent carpels, nectar is secreted from simple, unspecialized areas of the carpel walls. With connation of the carpels in higher taxa, the secretory areas are enclosed, and a specialized aperture developed.

The Hydrocharitaceae are important in showing, with the Butomaceae, primitive laminar placentation. The morphology of the gynoecium in this family needs anatomical study; the relation of the carpels to the receptacle is obscure and has been variously interpreted.

The other families in the Helobiales show specialization of various types, including a reduction series in flower structure—Aponogetonaceae,

Potamogetonaceae, Najadaceae. Family lines are difficult to draw, morphologically, in this order, because of its greatly reduced floral structure and the questionable interpretations of some of the floral organs, especially in those families that have been described as possessing stamens with remarkable, more or less sepallike, expansions of the connective—Potamogetonaceae *sensu lato*, Lilaeaceae, Scheuchzeriaceae. Anatomy shows, however, that these peculiar stamens have been misinterpreted, morphologically; the pseudoperianth parts of the potamogetons and their close relatives are compound organs, stamens with adnate sepals, the adnation varying in degree. The sepal and the stamen have separate vascular supplies and are, morphologically, independent organs.

The interpretation of these peculiar organs has been extended to cover the flowers of the Scheuchzeriaceae (Juncaginaceae), *Triglochin* and its closely related genera, where sepal and stamen are closely associated but structurally independent. The "flower" of these genera is an inflorescence, and it seems most doubtful that *Triglochin* and its related genera belong in the Potamogetonaceae. The association of *Triglochin* with *Scheuchzeria* seems unnatural. The "perianth" parts of *Triglochin* are bracts; those of *Scheuchzeria* are tepals. The leaves of *Scheuchzeria* are bifacial, those of *Triglochin* are terete. The rhizome of *Scheuchzeria* has a fibrous sheath; that of *Triglochin* has a sheath of stone cells.

Resemblances to the Ranunculaceae and to the Nymphaeaceae have been seen in the Helobiales, and these similarities used as evidence of derivation of the monocotyledons from the dicotyledons. But the similarities are largely those of general primitiveness and reduction related to habitat. The sharply distinct calyx and corolla suggest the similar differentiation in the Ranunculaceae, but this character alone can hardly be significant. The presence, in the Alismataceae, of laticiferous cells, and the absence of these cells in the Ranunculaceae is evidence against relationship of these two families.

Among the major primitive monocotyledonous taxa, the well-differentiated calyx and corolla set the Helobiales apart from the Palmae and the primitive Liliales.

LILIALES

The Primitive Liliaceae

The Liliaceae are generally considered, in taxonomic treatments, to be one of the primitive monocotyledonous families, an interpretation supported by morphology, anatomy, and palynology. Most primitive in the family are the Melanthioideae, a subfamily with genera rather diverse in combinations of primitive and advanced characters. Its various genera show stages in the change from spiral to whorled arrangement of the floral organs; in the reduction of the receptacle tips and the carpellary stipes; in the occasional presence of more than six (nine to twelve) stamens; in the connation of the carpels, laterally and ventrally; and in the closure of the carpels. The origin and elaboration of the septal nectary is well shown.

The Liliaceae, as a whole, show steps in the union of the three styles and stigmas to form a single, simple, median structure. No genus shows all the more primitive characters; *Veratrum* and *Tofieldia* probably best show general primitive structure. In *Veratrum*, the "whorls" of all floral organs are shown by anatomy to be "tight" spirals, and there is a vestigial receptacular tip between the carpels. The carpels are incompletely connate; their distal parts are free and divergent. The carpel margins are merely appressed, not histologically fused; the styles are structurally unspecialized, resembling the fertile part of the carpel in shape, and are open—horseshoe-shaped in cross section—distally. The three traces of the carpels arise independently from the receptacular stele—in contrast with those of most higher taxa in the family, where they arise fused. In advanced characters, the flower shows slight perigyny—the stamens and perianth are adnate to the carpel bases.

Primitive characters are seen in *Tofieldia;* there may be 9 to 12 stamens, and the carpels, varying from three to five, are stipitate and free of connation for some distance from the base, though often united distally. In *T. glutinosa,* the carpels are almost free of lateral connation. There is no perigyny. In advanced characters, the flower of *Tofieldia* shows appendages whorled, the receptacle not prolonged, and carpels with a single trace.

Absence of septal glands—highly specialized structures—characterizes the Melanthioideae, primitive members of the Liliaceae. (*Tofieldia* perhaps shows an early stage in these glands, but the epithelial lining of the intercarpellary spaces is unlike that of the glands of higher subfamilies.)

The gynoecia of the Melanthioideae have been considered to resemble those of *Butomus* (Helobiales) and *Scheuchzeria* (Alismatales) —part of the evidence that these taxa belong to the same ancestral line. There is surely close resemblance in the carpels, but this resemblance may represent merely the same level of carpel evolution. The sessile stigma of *Scheuchzeria* shows stages in the evolution of a style, stages similar to those in the more primitive lilies. Fusion of the carpels in the Melanthioideae is ontogenetic; that in the more advanced genus, *Tricyrtis*, is congenital, as in most of the Liliaceae. In the primitive Liliaceae, placentation also shows steps in evolutionary advance. The ovules are chiefly in submarginal rows, with traces from the ventral bundles, but some ovules derive their traces from the laminar meshwork between the median and lateral veins. This vascularization is evidence that the submarginal placentation in the Liliaceae represents reduction from laminar, as is shown for the dicotyledons in the Degeneriaceae and Winteraceae. The laminar placentation, as seen in the Helobiales (*Butomus*), is reduced to submarginal in the Liliaceae and to basal (one or two ovules) in Alismatales (*Scheuchzeria*). The position of the solitary ovule of *Aphyllanthes* (Liliaceae) is described, obscurely, as "marginal median," perhaps between the median and lateral veins, as in *Cabomba*. The story of placentation reduction is seen in both dicotyledons and monocotyledons.

PALMAE

The palms, one of the largest families of the monocotyledons—over two hundred genera and four thousand to five thousand species, are remarkable for range in form and for combinations of primitive and advanced characters. Unfortunately, the family is not well known morphologically, because of its large size and the difficulty of obtaining material. There has been comparatively little anatomical study of the flowers and inflorescences.

Habit

In habit, the palms range from small, essentially herbaceous—acaulescent—types to large, unbranched trees—some, 100 feet tall!—and there are woody vines and clustered, much-branched "shrubs." Many of the arborescent taxa branch occasionally. The great range in habit, together with anatomical structure, suggests that the arborescent, unbranched habit has been derived from the herbaceous, rhizomatous, freely branching habit. (The bamboos seem to represent a similar state in the grasses.)

The trunks of many palm trees consist entirely of primary tissues, built up by the huge, terminal meristem; in other genera, there is increase in trunk diameter by secondary growth. This increase in diameter is acquired after increase in length of the region has ceased and, histologically, resembles primary growth rather than that brought about by a cambium. New-cell formation in the inner cortex adds new tissues—vascular bundles, fiber strands, and parenchyma, which resemble the central primary tissues. (A small amount of "delayed" primary increase in diameter occurs in some palms after the tightly sheathing leaf base is loosened or lost. This increase consists of enlargement in cell size and change in shape, without new-cell formation, and is always of brief duration.) The unbranched trunk habit is typical of most arborescent genera. Branching may be the result of injury to the terminal meristem or of the replacing of axillary inflorescences by vegetative shoots. One species of *Hyphaene* branches sparsely in a way that resembles dichotomy. This has been called true dichotomy "persisting in the angiosperms," but other species of the genus are unbranched, or rather freely branched without suggestion of dichotomy. It seems improbable that within one genus both ancient and modern branching types could exist. (Superficial form is not sufficient in determining dichotomy; only ontogeny and anatomy can determine the nature of this branching.)

When genera with a variety of habits are compared, the unbranched-tree habit appears to be a specialized form. Similar conditions are present in some woody Liliales. (Trees of simple form have existed in all major taxa of land plants. In gross habit, simplicity in form may represent reduction from complexity. Examples are seen in the lepidodendrids, cycads, calamites, ferns.)

The Leaf

The leaves of the palms are probably the largest of all seed plants, attaining lengths, in pinnate genera, of up to 50 feet (*Raphia*) and, in palmate genera, of up to 30 feet (*Corypha*). In form, palm leaves range widely. Simple leaves characterize a few taxa and the early leaves of many genera, but compound leaves are typical for the family. The two well-known compound forms, pinnate and palmate, are connected by many transitional, "costa-palmate" types; these, although palmate in general outline, have a strong midrib, which is a remnant of the rachis of an ancestral pinnate type. The simple leaf is apparently primitive; the compound leaf has been derived by dissection of the simple lamina (Chap. 1). Leaflet formation in the compound leaf is by ontogenetic dissection at a very early stage; leaflets in other angiosperm families (except the related Cyclanthaceae) are formed independently by lateral lobes of the foliar primordium (Fig. 2). All palm leaves, except the simple types, are characterized by a "folding" of the blade in early stages, a folding that seems unique in the palms and the related Cyclanthaceae (Figs. 18, 19, 20, and 148).

The method of compound-leaf formation by ontogenetic dissection (Fig. 17) involves the freeing of marginal strips of the lamina, which, as the "reins" or lorea, are a prominent part of the young leaf in many taxa (Fig. 148). Like other parts of the leaf, the reins have undergone strong modification, a reduction from prominent ribbonlike bands to ephemeral filaments in the specialized palmate leaf.

The remarkable ontogeny of the compound palm leaf is evidence that the palms are only distantly related to other monocotyledons. It provides support for the theory that the primitive angiosperm leaf was simple, as shown by the Ranales.

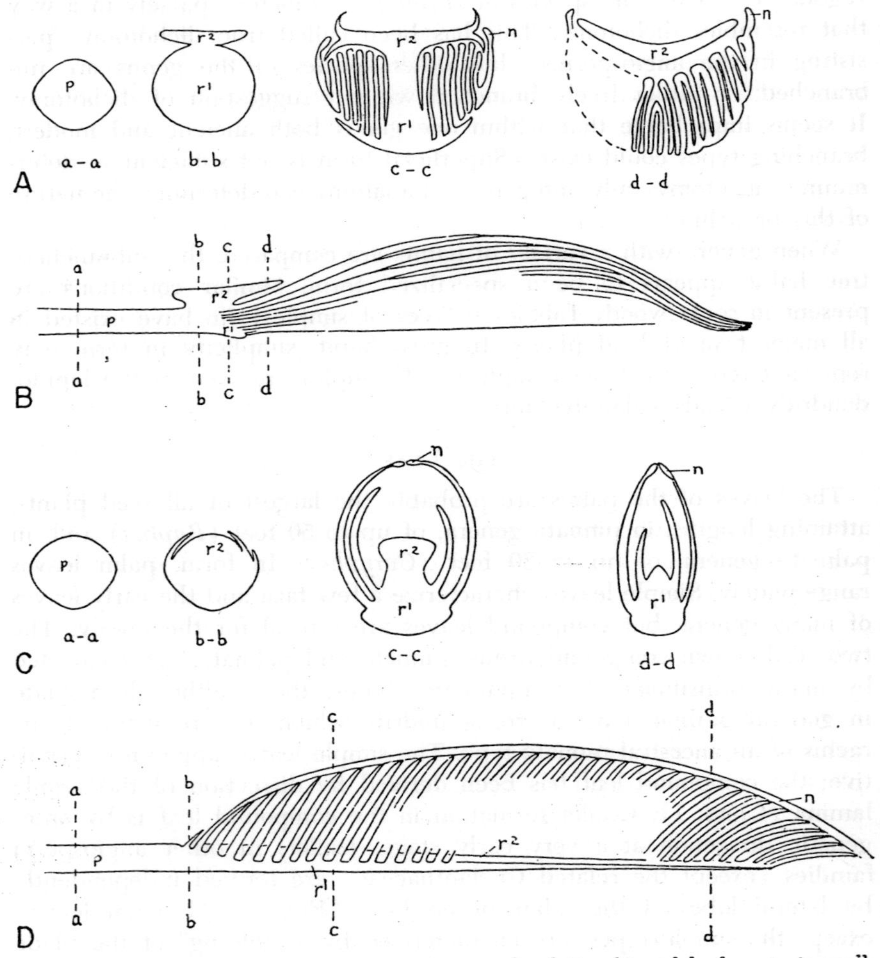

Fig. 148. Diagrams showing gross structure and relationship of leaf parts in weil-developed but still folded leaves of two types. *A*, *B*, palmate type; *C*, *D*, pinnate type. *B*, *D*, longitudinal; *A*, *C*, cross sections at levels indicated in longitudinal diagrams. *n*, reins; *p*, petioe; r^1, dorsal lobe of rachis; r^2, ventral lobe of rachis (hastula). (*From Eames*, 1953.)

Within the family, the palmate type is clearly the advanced type, as shown by the presence in the palmate forms of a strong midrib in all stages of reduction, the costa. The costa shows evidence, not only in its large size but in its form and vascular structure, that it represents the "telescoped" rachis of the pinnate leaf. In advanced palmate genera, the hastula, or so-called ligule, at the base of the blade, often showing a "crumpling" of the fibrous strands of the rachis, represents the end result of the telescoping (Fig. 148).

Inflorescence

The inflorescences, mostly large to huge, range from simple to complex. (The inflorescence of the talipot palm, *Corypha umbraculifera*, is cone-shaped and may be 30 feet high and 50 feet in diameter at the base. Inflorescences of *Raphia* may be 10 to 15 feet long.) The basic form of the inflorescence seems to be a large, racemose panicle. This panicle has been modified in many ways: by condensation and reduction; by fusion of branchlets, and by partial sinking of flowers or flower clusters in thickened axes (Fig. 30). Simple spiral arrangement of flowers is replaced in most genera by more complex spirals and by a two-rowed or an irregular arrangement. Reduction of the complex branch system has brought about extraordinary modifications. Branchlets have become connate laterally with the mother axis on which they seem to be borne. Part of the story of condensation of the complex panicle is the formation of small clusters of flowers—"sessile" or partly sunken in the axis. These clusters consist of "triads" (Fig. 30*D*)—a pistillate flower, with two staminate flowers, one above and one below —or "diads," one pistillate and one staminate. It has been suggested that these clusters indicate that the inflorescence is basically cymose, but these "cymose" clusters seem to represent small, reduction groups. The flower position is obscured.

The inflorescences are borne in various ways: terminal on the trunk; in the leaf axils, among or below the leaves; even on old trunks. Interfoliar is the more common type. In some monocarpic genera, where the plant flowers once and then dies, they are terminal. In the fishtail palms, the inflorescences mature in leaf-scar axils progressively down the trunk—an example of centrifugal succession in development like that in some androecia.

Extreme condensation and connation in the inflorescence is seen in the subfamily Lepidocaryoideae, in which the pistillate flowers are enclosed in a lorica, or "armor," of scalelike structures (Fig. 34). (The term lorica has also been applied to the individual members of this sheath.) The morphological nature of the lorica has been uncertain; it has been interpreted as a sheath of "excrescences" or of emergences of the ovary wall, as an "axillary cup" with "secondary leaves," as the connate bracts of suppressed inflorescence branchlets. The arrangement and anatomy of the "scales," together with the structure of the "ovary" as a whole, demonstrate that the lorica consists of leaflike organs, bracts that subtended lost branchlets. The bracts are connate and form a protective shell for a pistillate flower. The acorn cup of *Quercus* is, morphologically, a closely similar structure.

The scales of the lorica show a phyllotactic arrangement similar to that of the leaves. Though the scales have been described as lacking vascular bundles, bundles are shown in some illustrations. The presence of these bundles is evidence that the scales are not emergences. (It is probable that, commonly, the vascular tissue in these organs may not have matured at the flowering stage, the stage usually studied.)

It has been argued that the scales of the lorica are lobes of the ovary wall and, therefore, cannot represent bracts. There has been little histological study of the structure of the lorica wall, but stages in the adnation of the scales to the carpellary wall have been described. The carpels may be completely free from the lorica; they may be adnate but histologically distinct (*Laccosperma*); in other genera, the fusion of scales and ovary wall seems complete. (Further anatomical study of palm flowers of this type should be made.) Loricate flowers represent, without doubt, one of the many forms of inflorescence reduction in the palms.

Various genera show stages in the evolutionary development of the lorica. In *Zalacca,* partial inflorescences are clustered in a head or club-shaped mass. In *Latania,* the flowers are borne on very slender branchlets, each flower subtended by an adnate bract. In *Eugeissona tristis,* the lorica is weakly developed; the scale units are bractlike in form.

THE FLOWER

In general structure, the flower of the palms is of the monocotyledonous type, with trimerous whorls and various degrees of connation and adnation. It ranges from bisexual to unisexual, and the unisexual from monoecism to dioecism. Only a few genera have bisexual flowers—most of the Sabal palms and some of the Lepidocaryoideae. The subfamily Ceroxyloideae shows all transitional stages from bisexual to unisexual flowers, and, in the location of unisexual flowers, from both sexes in the same inflorescence to separation of the sexes in different inflorescences in the same plant and on different plants. Dioecism is uncommon or rare—*Raphia, Phoenix, Phytelephas.* Monoecious taxa are of two types: those with staminate and pistillate flowers in the same inflorescence, even side by side; and those with the two floral types in different parts of the same inflorescence. Nearly all unisexual flowers have vestigial sporophylls of the reduced sex and show all stages of reduction to complete disappearance of these abortive organs. Only in strongly dioecious taxa are there no vestiges of abortive sporophylls.

The staminate flowers of *Nipa, Zalacca,* and the Mauritiae, for example, are described as lacking all traces of a gynoecium.

The Perianth. The perianth is usually inconspicuous; staminate flowers are sometimes showy; calyx and corolla are much alike but distinguishable as inner and outer series of three organs. Sepals and petals are persistent, enlarging after flowering and becoming leathery or hard, sometimes partly fleshy. In the fruit, they form, with bracts, a basal cupule. Rarely, the perianth is rudimentary or absent—*Nipa.*

The Androecium. The stamens are commonly six, in two series, but there is considerable variation; when more—thirty-five to fifty, up to one hundred, and more than one hundred—they have been described as in whorls of three, but other descriptions are "in a fascicle" and "in one series." Where there are numerous stamens, many may be sterile. Rarely, there are only three stamens, as in *Nipa, Areca triandra.* In androecia with only three stamens, the inner of the two ancestral whorls appears to have been lost in reduction. In dicotyledons, similar reduction in stamen whorls is by loss of the outer whorl. The filaments range from narrowly laminar to terete and slender, and are often broad at the base. The anthers are chiefly introrse. Connation of filaments is frequent, and the whorls may be independent or united. Adnation of stamens to perianth and to gynoecium is frequent.

The Gynoecium. The gynoecium consists of three—rarely four to seven—carpels, free or connate in various degrees. Complete apocarpy is rare—*Chamaerops, Trachycarpus*—and perhaps present only in fully bisexual flowers. (Vestigial carpels are commonly free.) In syncarpy, connation varies in extent and in place of fusion; in some taxa, fusion is distal or proximal only. Frequently, the tip of the receptacle projects between the carpel bases, and the carpel bases may be fused to it laterally, as in some dicotyledonous families. All steps in the evolution of septal nectaries are shown—from mere secretory areas to enclosed glands with "canals" to apertures on the upper surface of the ovary. (The story here parallels that in the Liliaceae; septal glands have developed independently, perhaps several times.) The stigmas are usually sessile; a few genera have short styles.

The Ovule. Each of the carpels bears one "basal" ovule, but only one ovule becomes a seed, even in apocarpous taxa. In syncarpy, two of the three ovules begin to degenerate at flowering time, and, as the fruit develops, the sterile carpels are compressed and enter into the formation of the wall of the fruit, with the position of their locules apparent in the wall, as in the coconut (Fig. 93). The ovules are basically anatropous; the massive funicle and integuments and crowding in the small ovarian cavity distort their form. The ovules are subbasal and described as borne "opposite the median trace" of the carpel. They seem to represent survivors of submarginal placentation, though described under the peltate theory of carpel form as "laminar." The ovule has a strong vascular trace, which branches freely in the chalaza and enters the outer integument. Differentiation of funicle, raphe, and integuments is difficult in most taxa and is often complicated by adnation of the ovule to the ovary wall; the adnate condition is the basis for the description "ovule sessile." The integuments show various degrees of fusion; often they are distinct only near the micropyle. *Attalea* is described as having two integuments; *Phoenix,* only one.

The Pollen and Pollination. The pollen grains are of various sulcate types and most resemble those of the Cyclanthaceae, Liliaceae, and Amaryllidaceae. Pollination is by various agents, perhaps chiefly by wind, but also by insects—wasps, in part, in *Cocos.* It has been reported as by birds in *Eupritchardia.* Pollination seems to be in transition from entomophilous to anemophilous, as in *Salix* and *Acer,* accompanying development of unisexuality. Pollination by beetles should be looked for in the primitive genera.

The Fruit

The fruits vary in type but are mostly drupaceous, ranging to nonfleshy types. In *Nipa,* the closely aggregated fruits form a syncarpium.

Discussion

The palms are primitive monocotyledons that have progressed far in specialization in various characters. Habit, leaves, inflorescences, flowers, pollination, fruits, all give evidence, in their diversity of form, of a long period of evolutionary modification. Variety in habit is great. Whatever the interpretation of the evolutionary history of the unbranched tree, the range of form from rhizomatous "herbs" to shrubs, woody vines, and tall trees is remarkable for a single family.

The history of leaf form in the palms is probably unique in angiosperms (see Chap. 1). The derivation of the compound leaf from the simple—doubtless an evolutionary story in other families also—is, in this family, by dissection of a type probably not present in other angiosperms. The remarkable ontogeny of the compound leaf of the palms sets the palms well apart from other angiosperms. The compound leaf must have arisen early in the history of the angiosperms and independently in the palms. The primitive palm leaf was, doubtless, simple, with pinnately parallel venation.

The history of leaf modification alone is sufficient evidence that the palms are a very old family. Fossil leaves support this record. *Pro-*

palmophyllum, fossil leaves from the Lower Jurassic, accompanying a typical Jurassic flora, have been generally accepted as those of palms, and the fossil leaves of *Sanmiguelia* from the Triassic of Colorado closely resemble palm leaves. The leaves of *Sanmiguelia* are simple, those of *Propalmophyllum* are costapalmate, a type intermediate between pinnate and palmate. If these leaves are correctly interpreted, the palms were already specialized in leaf structure by late Jurassic.

The inflorescences show clear evidence of reduction from large panicles to reduced, compacted, and fused smaller inflorescences, with greatly modified phyllotaxy and the sinking of flowers in branches of the axis. The range in position of inflorescences, from axillary among the leaves to terminal on the tree and to axillary on the trunk—both positions associated with monocarpy—is unusual. A remarkable specialization is the basipetal sequence in maturation of the inflorescences on the trunks of the fishtail palms and in the bractlike scales of the lorica.

The flowers show a full series in reduction from bisexuality to unisexuality. In typical unisexual genera, there are no vestiges of the lost sporophylls; the staminate flowers of *Phytelephas* and the Mauritieae have no vestiges of a gynoecium. Supporters of the theory that the unisexual flower is of two types—primitive and derived from the bisexual flower—accept the interpretation of reduction where vestigial organs are present, but deny it where they are absent. They should consider the flowers of the palm family as a whole. Even within a genus, vestigial carpels may be present or absent. Unisexuality has doubtless arisen many times in different lines, but the existence of two types of widely different origin within a family is unacceptable morphologically.

The perianth is remarkable for absence of specialization; it usually remains simple, with minor elaborations.

The story of the androecium is one of reduction in stamen number, with the establishment of whorled arrangement from indefinite or spiral position.

The number of carpels, three, is constant throughout the family, with stages in connation and in reduction of fertility in two of them. Rarely, even in apocarpous taxa, do all three carpels develop seeds. In the sterile carpels, ovules may develop apparently normally until flowering time; in other families, ovules that abort rarely develop beyond early stages. In syncarpous genera, as in *Cocos,* the sterile carpels collapse after fertilization and, in flattened condition, form part of the wall of the fruit (Fig. 93).

Syncarpy varies in method of carpel union; typically, the carpels are closed and the placentation is axillary basal, but in Asiatic members of the Calamineae, the partly open carpels are united by their margins to form a unilocular ovary. The palms seem to show that syncarpy has arisen by the union of both open and closed carpels.

The carpels are primitive in their sessile or subsessile stigmas (a few have short styles) but advanced in their solitary, subbasal ovule. The ovule is specialized in its massive structure, filling the ovarian cavity, and in its frequent adnation to the carpellary wall. The two integuments show steps in fusion. Fusion among the floral organs is common and of many types, but there are no inferior ovaries.

The palms give evidence of great age; they are a basically primitive taxon that has become greatly diversified and advanced in many characters, each character giving evidence of long specialization. There is fossil evidence of great age—evidence of their certain presence in the Jurassic and probable presence in the Triassic. And the leaves of *Propalmophyllum* are costapalmate, an advanced type; the palms seem to have advanced far in leaf form by Jurassic time—from simple to compound, and from pinnately compound to costapalmate.

BIBLIOGRAPHY

Ranales

Winteraceae

Bailey, I. W., and C. G. Nast: The comparative morphology of the Winteraceae. I. Pollen and stamens, *Jour. Arnold Arb.*, **24**: 340–346, 1943; II. Carpels, *Jour. Arnold Arb.*, **24**: 472–481, 1943; VII. Summary and conclusions, *Jour. Arnold Arb.*, **26**: 37–47, 1945.

Dandy, J. E.: The Winteraceae of New Zealand, *Jour. Bot.*, **71**: 119–122, 1933.

Hotchkiss, A. T.: Chromosome numbers and pollen-tetrad size in the Winteraceae, *Proc. Linn. Soc. N.S.W.*, **80**: 49–53, 1955.

Hutchinson, J.: The family Winteraceae, *Kew Bull.*, **1921**: 185–190.

Nast, C. G.: The comparative morphology of the Winteraceae. VI. Vascular anatomy of the flowering shoot, *Jour. Arnold Arb.*, **25**: 454–466, 1944.

Smith, A. C.: Taxonomic notes on the Old World species of Winteraceae, *Jour. Arnold Arb.*, **24**: 119–164, 1943.

———: Studies of Papuasian plants. V, *Jour. Arnold Arb.*, **23**: 417–443, 1942.

———: The American species of *Drimys*, *Jour. Arnold Arb.*, **24**: 1–33, 1943.

Tucker, S. C.: Ontogeny of the inflorescence and the flower in *Drimys winteri* var. *chilensis*, *Univ. Calif. Publ. Bot.*, **30**: 257–336, 1959.

Lactoridaceae

Challenger Expedition: Rpt. Sci. Results Bot., III. Rpt. Bot. Juan Fernandez and Masafuera, 1884.

Engler, A.: Über Familie der Lactoridaceae, *Bot. Jahrb.*, **8**: 53–56, 1886.

McLaughlin, R. P.: Systematic anatomy of the woods of the Magnoliales, *Trop. Woods*, **34**: 3–39, 1933.

Philippi, R. A.: Über zwei neue Pflanzen-Gattungen, *Verhl. Zool.-Bot. Ges. Wien*, **15**: 517–524, 1805.

Magnoliaceae

Canright, J. E.: The comparative morphology and relationships of the Magnoliaceae. I. Trends of specialization in the stamens, *Am. Jour. Bot.*, **39**: 484–497, 1952; II. Significance of the pollen, *Phytomorph.*, **3**: 355–365, 1952; III. Carpels, *Am. Jour. Bot.*, **47**: 145–155, 1960; IV. Wood and nodal anatomy, *Jour. Arnold Arb.*, **36**: 119–139, 1955.

Dandy, J. E.: A survey of the genus *Magnolia* together with *Manglietia* and *Michelia*, Rpt. of the Camellias and Magnolias Conference, **1950**: 65–81.

———: The genera of Magnolieae, *Bull. Misc. Inform. Kew*, **1927**: 257–264.

Good, R. D'O.: The past and present distribution of the Magnoliaceae, *Ann. Bot.*, **39**: 409–430, 1925.

Howard, R. A.: The morphology and systematics of the West Indian Magnoliaceae, *Bull. Torrey Bot. Club*, **75**: 335–357, 1948.

Johnson, M. A.: Relationship in the Magnoliaceae as determined by the precipitin reaction, *Bull. Torrey Bot. Club*, **80**: 349–350, 1953.

———: The precipitin reaction as an index of relationship in the Magnoliaceae, *Bull. Ser. Mus.*, **13**: 1–5, 1954.

Lemesle, R.: De l'ancienneté des caractères anatomiques des Magnoliacées, *Rev. Gén. Bot.*, **45**: 341–355, 1933.

———: Les caractères histologiques du bois secondaire des Magnoliales. *Phytomorph.*, **3**: 430–446, 1953.

McLaughlin, R. P.: Systematic anatomy of the woods of the Magnoliales, *Trop. Woods*, **34**: 3–39, 1933.

Whitaker, T. W.: Chromosome number and relationship in the Magnoliales, *Jour. Arnold Arb.*, **14**: 376–385, 1933.

Annonaceae

Corner, E. J. H.: The annonaceous seed and its four integuments, *New Phyt.*, **48**: 332–364, 1949.

Diels, L.: Die Gliederung der Annonaceen und ihre Phylogenie, *Sitzungsber. Preuss. Akad. Wiss.*, **1932**: 77–85.

Hutchinson, J.: Contributions toward a phylogenetic classification of angiosperms. II. Anonaceae and related forms, *Kew Bull.*, **1923**: 241–261.

Ingle, H. D., and H. E. Dadswell: The anatomy of the timbers of the south-west Pacific area. II. Apocynaceae and Annonaceae, *Aust. Jour. Bot.*, **1**: 1–26, 1953.

Safford, W. E.: *Raimondia*, a new genus of Anonaceae from Colombia, *Contr. U.S. Nat. Herb.*, **16**: 217–219, 1913.

———: *Anona sericea* and its allies, *Contr. U.S. Nat. Herb.*, **16**: 263–275, 1913.

———: Classification of the genus *Anona*, *Contr. U.S. Nat. Herb.*, **18**: 1–68, 1914.

Samuelsson, G.: Über die Pollenentwicklung von *Anona* und *Aristolochia* und ihre systematische Bedeutung, *Svensk Bot. Tidskr.*, **8**: 181–189, 1914.

Smith, G. H.: Vascular anatomy of Ranalian flowers. II. Menispermaceae, Calycanthaceae, Anonaceae, *Bot. Gaz.*, **85**: 152–177, 1928.

Wyk, Van der, R. W., and J. E. Canright: The anatomy and relationships of the Annonaceae, *Trop. Woods*, **104**: 1–24, 1956.

Eupomatiaceae

Baillon, H.: Recherches organogéniques sur les *Eupomatia*, *Adansonia*, **9**: 22–28, 1868–1870.

Brown, R.: "Misc. Bot. Works," Atlas of Plates, 1868.

Diels, L.: Über primitive Ranalen der australischen Flora, *Bot. Jahrb.*, **48**, *Beibl.* **107**: 7–13, 1912.

Hamilton, A. G.: On the fertilization of *Eupomatia laurina* R. Br., *Proc. Linn. Soc. N.S.W.*, **22**: 48–55, 1897.

Hotchkiss, A. T.: Geographical distribution of the Eupomatiaceae, *Jour. Arnold Arb.*, **36**: 385–396, 1955.

———: Pollen and pollination in the Eupomatiaceae, *Proc. Linn. Soc. N.S.W.*, **83**: 86–91, 1958.

Lemesle, R.: Les vaissaux à perforation scalariformes de l'*Eupomatia* et leur importance dans le phylogénie des Polycarpes, *Compt. Rend. Acad. Sci. Paris*, **203**: 1538–1540, 1936.

———: Contribution à l'étude de l'*Eupomatia* R. Br., *Rev. Gén. Bot.*, **50**: 692–712, 1938.

Himantandraceae

Bailey, I. W., C. G. Nast, and A. C. Smith: The family Himantandraceae, *Jour. Arnold Arb.*, **24**: 190–206, 1943.

Diels, L.: Über primitive Ranales der australischen Flora, *Bot. Jahrb.*, **48**, *Beibl.* **107**: 11–13, 1912.

———: Ueber die Gattung *Himantandra,* ihre Verbreitung und ihre systematische Stellung, *Bot. Jahrb.*, **55**: 126–134, 1919.

Sprague, T. A.: *Galbulimima baccata,* F. M. Bailey, *Hook. Icon. Plant.*, vol. 31, pl. 3001, 1915.

Degeneriaceae

Bailey, I. W., and A. C. Smith: Degeneriaceae: a new family of flowering plants from Fiji, *Jour. Arnold Arb.*, **23**: 356–365, 1942.

Lemesle, R., and A. Duchaigne: Contribution à l'étude histologique et phylogénetique du *Degeneria vitiensis,* I. W. Bailey and A. C. Smith, *Rév. Gén. Bot.,* **62**: 1–12, 1955.

Swamy, B. G. L.: Further contributions to the morphology of the Degeneriaceae, *Jour. Arnold Arb.,* **30**: 10–38, 1949.

Myristicaceae

Corner, E. J. H.: The durian theory or the origin of the modern tree, *Ann. Bot.*, **13**: 367–415, 1949.

Garratt, G. A.: Systematic anatomy of the woods of the Myristicaceae, *Trop. Woods,* **35**: 6–48, 1933.

———: Bearing of wood anatomy on the relationships of the Myristicaceae, *Trop. Woods,* **36**: 20–44, 1933.

Joshi, A. C.: A note on the development of pollen of *Myristica fragrans* and the affinities of the family Myristicaceae, *Jour. Indian Soc. Bot.,* **25**: 139–143, 1946.

Nair, N. C., and P. N. Bahl: Vascular anatomy of the flower of *Myristica malabarica* Lamk., *Phytomorph.,* **6**: 127–134, 1956.

Smith, A. C., and R. P. Wodehouse: The American species of Myristicaceae, *Brittonia,* **2**: 393–510, 1938.

Sporne, K. R.: A new approach to the problem flower, *New Phyt.,* **48**: 259–276, 1949.

Illiciaceae and Schisandraceae

Bailey, I. W., and C. G. Nast: Morphology and relationships of *Illicium, Schisandra* and *Kadsura.* I. Stem and leaf, *Jour. Arnold Arb.,* **29**: 77–89, 1948.

Lemesle, R.: Les ponctuations aréolées des fibres des genres *Schizandra, Kadsura, Illicium* et leurs rapports avec la phylogénie, *Compt. Rend. Acad. Sci. Paris*, **221**: 113, 1945.

Ozenda, P.: Sur l'anatomie libéroligneuse des Schizandrées, *Compt. Rend. Acad. Sci. Paris*, **223**: 207–209, 1946.

———: Récherches sur les dicotyledones apocarpiques, *Publ. Lab. École Norm. Sup.*, sér. biol. II, Paris, 1949.

Smith, A. C.: The families Illiciaceae and Schisandraceae, *Sargentia*, **7**: 1–224, 1947.

Calycanthaceae

Cheadle, V. I., and K. Esau: Secondary phloem of Calycanthaceae, *Univ. Calif. Publ. Bot.*, **29**: 397–510, 1958.

Daumann, E.: Das Blütennektarium von *Magnolia* und die Fütterkörper in der Blüte von *Calycanthus, Planta*, **11**: 108–116, 1930.

Diels, L.: Käferblumen bei den Ranales und ihre Bedeutung für die Phylogenie der Angiospermen, *Deutsch. Bot. Ges.*, **34**: 758–774, 1916.

Fahn, A., and I. W. Bailey: The nodal anatomy and the primary vascular cylinder of the Calycanthaceae, *Jour. Arnold Arb.*, **38**: 107–117, 1957.

Grant, V.: The pollination of *Calycanthus occidentalis, Am. Jour. Bot.*, **37**: 294–297, 1950.

Lignier, O.: Recherches sur l'anatomie comparée des Calycanthacées, des Mélastomacées et des Myrtacées, *Arch. Bot. du Nord de la France*, **2**: 1–455, 1886.

Mattfeld, J.: Das morphologische Wesen und die phylogenetische Bedeutung der Blumenblätter, *Ber. Deutsch. Bot. Ges.*, **56**: 86–116, 1938.

Peter, J.: Zur Entwicklungsgeschichte einiger Calycanthaceae, *Beitr. zur Biol. der Pfl.*, **14**: 59–84, 1900.

Schaeppi, H.: Morphologische Untersuchungen an den Karpellen der Calycanthaceae, *Phytomorph.*, **3**: 112–120, 1953.

Smith, G. H.: Vascular anatomy of Ranalian flowers. II. Menispermaceae, Calycanthaceae, Anonaceae, *Bot. Gaz.*, **85**: 152–177, 1928.

Van Tieghem, P.: Structure de la tige des Calycanthaceae, *Ann. Sci. Nat. Bot.*, 8 sér., **19**: 305–320, 1904.

Worsdell, W. C.: A study of the vascular system in certain orders of the Ranales, *Ann. Bot.*, **22**: 651–682, 1908.

Trochodendraceae and Tetracentraceae

Bailey, I. W., and W. P. Thompson: Additional notes upon the angiosperms *Tetracentron, Trochodendron,* and *Drimys, Ann. Bot.*, **37**: 503–512, 1918.

——— and C. G. Nast: Morphology and relationships of *Trochodendron* and *Tetracentron*. I. Stem, root and leaf, *Jour. Arnold Arb.*, **26**: 143–154, 1945.

Bondeson, W.: Entwicklungsgeschichte und Bau der Spaltöffnungen bei den Gattungen *Trochodendron* Sieb. et Zucc., *Tetracentron* Oliv. und *Drimys* J. R. et G. Forst., *Acta Horti Berg.*, **16**: 169–217, 1952.

Croizat, L.: *Trochodendron, Tetracentron*, and their meaning in phylogeny, *Bull. Torrey Bot. Club*, **74**: 60–76, 1947.

Nast, C. G., and I. W. Bailey: Morphology and relationships of *Trochodendron* and *Tetracentron*. II. Inflorescence, flower and fruit, *Jour. Arnold Arb.*, **26**: 265–276, 1945.

Smith, A. C.: A taxonomic review of *Trochodendron* and *Tetracentron, Jour. Arnold Arb.*, **26**: 123–142, 1945.

Austrobaileyaceae

Bailey, I. W., and B. G. L. Swamy: The morphology and relationships of *Austrobaileya, Jour. Arnold Arb.,* **30**: 211–226, 1949.

Croizat, L.: Notes on the Dilleniaceae and their allies: Austrobaileyeae, subfam. nov., *Jour. Arnold Arb.,* **21**: 397–404, 1940.

———: New families, *Jour. Cactus and Succulents Soc. Amer.,* **15**: 64, 1943.

White, C. T.: Ligneous plants collected for the Arnold Arboretum in North Queensland by S. F. Kajewsky in 1929, *Contrib. Arnold Arb.,* **4**: 1–113, 1933.

Cercidiphyllaceae

Bailey, I. W., and C. G. Nast: Morphology and relationships of *Trochodendron* and *Tetracentron.* I. Stem, root, and leaf, *Jour. Arnold Arb.,* **26**: 143–154, 1945.

Brown, R. W.: Fossil leaves, fruits, and seeds of *Cercidiphyllum, Jour. Paleontol.,* **13**: 485–499, 1939.

Croizat, L.: *Trochodendron, Tetracentron* and their meaning in phylogeny, *Bull. Torrey Bot. Club,* **74**: 60–76, 1947.

Lemesle, R.: Contribution à l'étude morphologique et phylogénique des Eupteléacées, Cercidiphyllacées, Eucommiacées, *Ann. Sci. Nat. Bot.,* 11 sér., **7**: 41–52, 1946.

McLaughlin, R. P.: Systematic anatomy of the woods of the Magnoliales, *Trop. Woods,* **34**: 3–39, 1933.

Nast, C. G., and I. W. Bailey: Morphology and relationships of *Trochodendron* and *Tetracentron.* II. Inflorescence, flower, and fruit, *Jour. Arnold Arb.,* **26**: 265–276, 1945.

Solereder, H.: Zur Morphologie und Systematik der Gattung *Cercidiphyllum* Sieb. und Zucc., mit Berücksichtigung der Gattung *Eucommia, Ber. Deutsch. Bot. Ges.,* **17**: 387–405, 1900.

Swamy, B. G. L., and I. W. Bailey: The morphology and relationships of *Cercidiphyllum, Jour. Arnold Arb.,* **30**: 187–210, 1949.

Monimiaceae

Garratt, G. A.: Bearing of wood anatomy on the relationships of the Myristicaceae, *Trop. Woods,* **36**: 20–44, 1933.

———: Systematic anatomy of the woods of the Monimiaceae, *Trop. Woods,* **39**: 18–44, 1939.

Hobein, M.: Beitrag zur anatomischen Charakteristik der Monimiaceen unter vergleichender Berücksichtigung der Lauraceae, *Bot. Jahrb.,* **10**: 51–74, 1888.

Money, L., I. W. Bailey, and B. G. L. Swamy: The morphology and relationships of the Monimiaceae, *Jour. Arnold Arb.,* **31**: 372–404, 1950.

Tippo, O.: Comparative anatomy of the Monimiaceae and their allies, *Bot. Gaz.,* **100**: 1–99, 1938.

Amborellaceae

Bailey, I. W.: Additional notes on the vesselless dicotyledon, *Amborella trichopoda* Baill., *Jour. Arnold Arb.,* **38**: 374–378, 1957.

——— and B. G. L. Swamy: *Amborella trichopoda* Baill.: A new morphological type of vesselless dicotyledon, *Jour. Arnold Arb.,* **29**: 245–254, 1948.

Eupteleaceae

Lemesle, R.: Contribution à l'étude morphologique et phylogénique des Eupteléacées, Cercidiphyllacées, Eucommiacées, *Ann. Sci. Nat. Bot.,* 11 sér., **7**: 41–52, 1946.

———: Les caractères histologiques du bois secondaire des Magnoliales, *Phytomorph.*, **3**: 430–446, 1953.

Nast, C. G., and I. W. Bailey: Morphology of *Euptelea* and comparison with *Trochodendron, Jour. Arnold Arb.*, **27**: 186–192, 1946.

Smith, A. C.: A taxonomic review of *Euptelea, Jour. Arnold Arb.*, **27**: 175–185, 1946.

Ceratophyllaceae

Eckardt, T.: Untersuchungen über Morphologie, Entwicklungsgeschichte und systematische Bedeutung des pseudomonomerous Gynoeceums, *Nova Acta Leopold.*, **5**: 1–112, 1937.

Klercker, F. de: Sur l'anatomie et le développement de *Ceratophyllum, Bihang. Svensk. Vet. Akad. Handl.*, **9**: 1–23, 1885.

Schleiden, M. J.: Beiträge zur Kenntniss der Ceratophylleen, *Linnaea*, **11**: 513–542, 1837.

Strasburger, E.: Ein Beitrag zur Kenntnis von *Ceratophyllum submersum* und phylogenetische Erörterungen, *Jahrb. Wiss. Bot.*, **37**: 477–526, 1902.

Ranunculaceae

Baillon, H.: "The Natural History of Plants," London, 1871.

Bessey, E. A.: The comparative morphology of the pistils of the Ranunculaceae, Alismaceae and Rosaceae, *Bot. Gaz.*, **26**: 297–313, 1898.

Brouland, M.: Recherches sur l'anatomie florale des Renonculacées, *Le Botaniste*, **27**: 1–278, 1935.

Chute, H. M.: The morphology of the achene, *Am. Jour. Bot.*, **17**: 703–723, 1930.

Fraser, M. S.: A study of the vascular supply to the carpels in the follicle-bearing Ranunculaceae, *Trans. Roy. Soc. Edinburgh*, **59**: 1–56, 1937.

Goffart, J.: Recherches sur l'anatomie des feuilles dans les Renonculacées, *Archiv Inst. Bot. Univ. Liège*, **3**: 1–190, 1901.

Gregory, W. C.: Phylogenetic and cytological studies in the Ranunculaceae, *Trans. Am. Phil. Soc.*, n.s., **31**: 443–501, 1941.

Hammond, H. D.: Systematic serological studies in Ranunculaceae, *Ser. Mus. Bull.*, **14**: 1–3, 1955.

Janchen, E.: Die systematische Gliederung der Ranunculaceae und Berberidaceae, *Denkschr. Akad. Wiss. Wien Math.-Natur. Kl.*, **108**: 1–82, 1949.

Kumazawa, M.: Morphology and biology of *Glaucidium palmatum* Sieb. et Zucc. with notes on affinities to the allied genera *Hydrastis, Podophyllum* and *Diplophylleia, Jour. Fac. Sci. Imp. Univ. Tokyo*, sec. 3, *Bot.*, **2**: 345–380, 1930.

———: Systematic and phylogenetic consideration of the Ranunculaceae and Berberidaceae, *Bot. Mag. Tokyo*, **52**: 9–15, 52–53, 1938.

Lemesle, R.: Position phylogénétique de l'*Hydrastis canadensis* L. et du *Crossosoma californica* Nutt. d'après les particularités histologiques du xylème, *Compt. Rend. Acad. Sci. Paris*, **227**: 221–223, 1948.

———: Contribution à l'étude de quelques familles de dicotylédones considérées comme primitives, *Phytomorph.*, **5**: 11–45, 1955.

Marié, P.: Recherches sur la structure des Renonculacées, *Ann. Sci. Nat. Bot.*, 6 sér., **20**: 5–180, 1885.

Schöffel, K.: Untersuchungen über den Blütenbau der Ranunculaceen, *Planta*, **17**: 315–371, 1932.

Sterckx, R.: Recherches anatomiques sur l'embryon et les plantules dans la famille de Renonculacées, *Archiv Inst. Bot. Univ. Liège*, **2**: 1–100, 1900.

Troll, W.: Beiträge zur Morphologie des Gynaeceums. III. Uber das Gynaeceum von *Nigella* und einiger anderer Helloboreen, *Planta,* **21**: 266–291, 1933.

Lardizabalaceae

Decaisne, J.: Mémoire sur la famille des Lardizabalacées, *Ext. Archiv Mus. Nat. Hist. Paris,* 1859.

Kumazawa, M.: Pollen grain morphology in Ranunculaceae, Lardizabalaceae and Berberidaceae, *Jap. Jour. Bot.,* **8**: 19–46, 1936.

Sargentodoxaceae

Lemesle, R.: Les divers types de fibres à ponctuations areolées chez les dicotylédones apocarpiques les plus archaïques et leur role dans la phylogénie, *Ann. Sci. Nat. Bot.,* 11 sér., **7**: 19–39, 1946.

Stapf, O.: *Sargentodoxa cuneata, Curtis's Bot. Mag.,* **151**: 9111–9112, 1926.

Berberidaceae

Chapman, M.: Carpel morphology in the Berberidaceae, *Am. Jour. Bot.,* **23**: 340–348, 1936.

Eckardt, T.: Untersuchungen uber Morphologie, Entwicklungsgeschichte und systematische Bedeutung des pseudomonomerous Gynoeceums, *Nova Acta Leopold,* N.F., **5**: 1–108, 1937.

Himmelbaur, W.: Die Berberidaceen und ihre Stellung im System, *Denkschr. Akad. Wiss. Wien. Math.-Natur. Kl.,* **89**: 733–795, 1914.

Johri, B. M.: The gametophytes of *Berberis nepalensis* Spreng, *Proc. Indian Acad. Sci.,* **1**(B): 640–649, 1935.

Kumazawa, M.: Systematic and phylogenetic considerations of the Ranunculaceae and Berberidaceae, *Bot. Mag. Tokyo,* **52**: 9–15, 1938.

Leinfellner, W.: Zur Morphologie des Gynoezeums von *Berberis, Oesterr. Bot. Zeitschr.,* **103**: 600–612, 1957.

Nymphaeaceae and Cabombaceae

Chifflot, J. B. J.: Contributions à l'étude de la classe des Nymphéinées, *Ann. Lyons Univ.,* n.s., **10**: 1–294, 1902.

Conard, H. S.: The waterlilies: Taxonomy and bibliography, *Carnegie Inst. Publ.* **5**, Washington, D.C., 1905.

Gray, A.: Genera Florae Americanae Boreali-Orientalis Illustrata, **1**: 91–96; pls. 38, 39, 42, 44, 1848.

Leinfellner, W.: Die blattartig flachen Staubblätter und ihre gestaltlichen Beziehungen zum Bautypus des Angiospermen Staubblattes, *Oesterr. Bot. Zeitschr.,* **103**: 247–290, 1956.

Leonhardt, R.: Phylogenetisch-systematische Betrachtungen. II. Gedanken zur systematischen Stottung, bzw. Gliederung einger Familien der Choripetalae, *Oesterr. Bot. Zeitschr.,* **98**: 1–43, 1951.

Li, H. L.: Classification and phylogeny of the Nymphaeaceae and allied families, *Am. Midl. Nat.,* **54**: 33–41, 1955.

Mattfeld, J.: Das morphologische Wesen und die phylogenetische Bedeutung der Blumenblätter, *Ber. Deutsch. Bot. Ges.,* **56**: 86–116, 1938.

Moseley, M. F., Jr.: Morphological studies in the Nymphaeaceae. I. The nature of the stamens, *Phytomorph.,* **8**: 1–29, 1958.

Raciborski, M.: Die Morphologie der Cabombeen und Nymphaeaceen, *Flora,* **78**: 244–279, 1894.

Saunders, E. R.: Some morphological problems presented by the flower of the Nymphaeaceae, *Jour. Bot.,* **74:** 217–221, 1936.

Strasburger, E.: "Die Angiospermen und die Gymnospermen," Jena, 1879.

Troll, W.: Beiträge zur Morphologie des Gynaeceums. IV. Über das Gynaeceum der Nymphaeaceen, *Planta,* **21:** 447–485, 1934.

Lauraceae

Brown, W. H.: The bearing of nectaries on the phylogeny of flowering plants, *Proc. Am. Phil. Soc.,* **79:** 549–595, 1937.

Coy, G. V.: Morphology of *Sassafras* in relation to phylogeny of angiosperms, *Bot. Gaz.,* **86:** 149–170, 1928.

Dadswell, H. E., and A. M. Eckersley: The wood anatomy of some Australian Lauraceae with methods for their identification, *CSIRO, Div. Forest Products, Tech. Paper* 34, 48 pp., 1940.

Daumann, E.: Zur morphologischen Wertigkeit der Blütennektarien von *Laurus, Beih. Bot. Centralbl.,* **48:** 183–208, 209–213, 1931.

Garratt, G. A.: Bearing of wood anatomy on the relationships of the Myristicaceae, *Trop. Woods,* **36:** 20–45, 1933.

Hobein, M.: Beitrag zur anatomischen Charakteristik der Monimiaceen unter vergleichender Berücksichtigung der Lauraceae, *Bot. Jahrb.,* **10:** 51–74, 1888.

Kasapligil, B.: Morphological and ontogenetic studies of *Umbellularia californica* Nutt. and *Laurus nobilis* L., *Univ. Calif. Publ. Bot.,* **25:** 115–240, 1951.

Mirande, M.: Sur l'origine pluricarpellaire du pistil des Lauracées, *Compt. Rend. Acad. Sci. Paris,* **145:** 570–572, 1907.

Reece, P. C.: The floral anatomy of the avocado, *Am. Jour. Bot.,* **26:** 429–433, 1939.

Sastri, R. L. N.: Studies in the Lauraceae. I. Floral anatomy of *Cinnamomum iners* and *Cassytha filiformis, Jour. Indian Bot. Soc.,* **31:** 240–246, 1952.

Schroeder, C. A.: Floral development, sporogenesis, and embryology in the avocado, *Persea americana, Bot. Gaz.,* **113:** 270–278, 1952.

Stern, W. L.: Comparative anatomy of xylem and phylogeny of Lauraceae, *Trop. Woods,* **100:** 1-73, 1954.

Wood, C. E., Jr.: The genera of the woody Ranales of the southeastern United States, *Jour. Arnold Arb.,* **39:** 296–346, 1958.

General for the Ranales

Axelrod, D. I.: A theory of angiosperm evolution, *Evol.,* **6:** 49–60, 1952.

Bailey, I. W.: Nodal anatomy in retrospect, *Jour. Arnold Arb.,* **37:** 269–287, 1956.

Brown, W. H.: The bearing of nectaries on the phylogeny of flowering plants, *Proc. Am. Phil. Soc.,* **79:** 549–594, 1938.

Diels, L.: Käferblumen bie den Ranales und ihre Bedeutung für die Phylogenie der Angiospermen, *Ber. Deutsch. Bot. Ges.,* **34:** 758–774, 1916.

Gifford, E. M.: The structure and development of the shoot apex in certain woody Ranales, *Am. Jour. Bot.,* **37:** 595–611, 1950.

Hsü, J., and M. N. Bose: Further information on *Homoxylon rajmahalense* Sahni, *Jour. Indian Bot. Soc.,* **31:** 1–12, 1952.

Kumazawa, K.: Pollen grain morphology in Ranunculaceae, Lardizabalaceae and Berberidaceae, *Jap. Jour. Bot.,* **8:** 19–46, 1936.

Lemesle, R.: Les divers types des fibres à ponctuations aréolées chez les dicotoylédones apocarpiques et leur rôle dans la phylogénie, *Ann. Sci. Nat. Bot.,* 11 sér., **7:** 19–39, 1946.

———: Contribution à l'étude morphologique et phylogénétique des Eupteléacées, Cercidiphyllacées, Eucommiacées (ex-Trochodendracées), *Ann. Sci. Nat. Bot.*, 11 sér., **7**: 41–52, 1946.

Marsden, M. P. F., and I. W. Bailey: A fourth type of nodal anatomy in dicotyledons illustrated by *Clerodendron trichotomum* Thunb., *Jour. Arnold Arb.*, **36**: 1–50, 1955.

Ozenda, P.: Recherches sur les dicotylédones apocarpiques, *Publ. Lab. École Norm. Sup.*, sér. biol. II, Paris, 1949.

Petersen, A. E.: A comparative study of the secondary xylem elements of certain species of the Amentiferae and Ranales, *Bull. Torrey Bot. Club*, **80**: 365–385, 1953.

Sahni, B.: *Homoxylon rajmahalense* gen. et sp. nov., a fossil angiospermous wood devoid of vessels from the Rajmahal Hills, *Behar. Mem. Geol. Surv. India, Palaeontologica Indica*, n.s., vol. 20, no. 2, 1932.

Sterckx, R.: Recherches anatomiques sur l'embryon et les plantules dans la famille de Renonculacées, *Archiv Inst. Bot. Univ. Liège*, **2**: 1–100, 1900.

Swamy, B. G. L.: Some aspects in the embryogeny of *Zygogynum Bailloni* v. Tiegh., *Proc. Nat. Inst. Sci. India*, **17**: 399–406, 1953.

Tippo, O.: The role of wood anatomy in phylogeny, *Am. Midl. Nat.*, **36**: 367–372, 1946.

Van Tieghem, P.: Sur les dicotylédones du groups Homoxylées, *Jour. Bot.*, **14**: 259–297, 330–361, 1900.

Whitaker, T. W.: Chromosome number and relationship in the Magnoliales, *Jour. Arnold Arb.*, **14**: 376–385, 1933.

Wodehouse, W. P.: Evolution of pollen grains, *Bot. Rev.*, **2**: 67–84, 1936.

Wood, C. E., Jr.: The genera of the woody Ranales in the southeastern United States, *Jour. Arnold Arb.*, **39**: 296–346, 1958.

Worsdell, W. C.: A study of the vascular system in certain orders of the Ranales, *Ann. Bot.*, **22**: 651–682, 1908.

Dilleniales

Dilleniaceae

Brown, W. H.: The bearing of nectaries on the phylogeny of flowering plants, *Proc. Am. Phil. Soc.*, **17**: 549–594, 1935.

Diels, L.: Dilleniaceae, *Bot. Jahrb.*, **57**: 436–459, 1922.

Ozenda, P.: Recherches sur les dicotylédones apocarpiques, *Publ. Lab. École Norm. Sup.*, sér biol. II, 1–183, 1949.

Vestal, P. A.: The significance of comparative anatomy in establishing the relationship of the Hypericaceae to the Guttiferae and their allies, *Philip. Jour. Sci.*, **64**: 199–256, 1937.

Paeoniaceae

Brouland, M.: Recherches sur l'anatomie florale des Rénonculacées, *Le Botaniste*, **27**: 1–280, 1935.

Corner, E. J. H.: Centrifugal stamens, *Jour. Arnold Arb.*, **27**: 423–437, 1946.

Eames, A. J.: Floral anatomy as an aid in generic limitation, *Chron. Bot.*, **14**: 126–132, 1953.

Gregory, W. C.: Phylogenetic and cytological studies in the Ranunculaceae, *Trans. Am. Phil. Soc.*, **31**: 443–500, 1941.

Kumazawa, M.: The structure and affinities of *Paeonia*, *Bot. Mag. Tokyo*, **49**: 306–315, 1935.

———: On the ovular structure in the Ranunculaceae and Berberidaceae, *Jour. Jap. Bot.*, **14**: 10–25, 1938.

———: Systematic and phylogenetic considerations of the Ranunculaceae and Berberidaceae, *Bot. Mag. Tokyo*, **52**: 9–15, 1938.

Lemesle, R.: Trachéides aréolées du type cycadéen dans le genre *Paeonia*, leur intérêt au point de vue systématique et phylogénetique, *Compt. Rend. Acad. Sci. Paris*, **226**: 2172–2173, 1948.

Ozenda, P.: Recherches sur les dicotylédones apocarpiques, *Publ. École Sup. Lab.*, sér. biol. II, Univ. Neu Chapelle, Switzerland, 1949.

Schöffel, K.: Untersuchungen über den Blütenbau der Ranunculaceen, *Planta*, **17**: 315–371, 1932.

Stebbins, G. L., Jr.: Notes on some systematic relationships in the genus *Paeonia*, *Univ. Calif. Publ. Bot.*, **19**: 245–266, 1939.

Tippo, O.: The role of wood anatomy in phylogeny, *Am. Midl. Nat.*, **36**: 367–372, 1946.

Vesque, J.: Les tissues vegétaux appliqués à la classification, *Nouv. Arch. Mus. Hist. Nat.*, sér. 2, **4**: 22–29, 1881.

Worsdell, W. C.: The affinities of *Paeonia*, *Jour. Bot.* (*London*), **46**: 114–116, 1908.

Crossosomataceae

Lemesle, R.: Position phylogénétique de l'*Hydrastis canadensis* L. et du *Crossosoma californica* Nutt. d'après les particularitiés histologiques du xylème, *Compt. Rend. Acad. Sci. Paris*, **227**: 221–223, 1948.

See also Bibliography for Ranunculaceae.

Piperales (Chloranthaceae)

Armour, H. M.: On the morphology of *Chloranthus*, *New Phyt.*, **5**: 49–55, 1906.

Swamy, B. G. L., and I. W. Bailey: *Sarcandra*, a vesselless genus of the Chloranthaceae, *Jour. Arnold Arb.*, **31**: 117–129, 1950.

———: The morphology and relationships of the Chloranthaceae, *Jour. Arnold Arb.*, **34**: 375–408, 1953.

Verticillatae (Casuarinaceae)

Arber, E. A. N.: The relationship of the angiosperms to the Gnetales, *Ann. Bot.*, **22**: 489–515, 1908.

Bailey, I. W., and E. W. Sinnott: Investigations on the phylogeny of the angiosperms. II. Anatomical evidence of reduction in the Amentiferae, *Bot. Gaz.*, **58**: 36–60, 1914.

Benson, M.: Contributions to the embryology of the Amentiferae. I, *Trans. Linn. Soc.*, (II) **3**: 409–424, 1894.

Frye, T. C.: The embryo sac of *Casuarina stricta*, *Bot. Gaz.*, **36**: 101–113, 1903.

Gaussen, P.: L'évolution pseudocyclique, *Année Biol.*, (III) **56**: C207–C220, 1952.

Hjelmquist, H.: Studies on the floral morphology and phylogeny of the Amentiferae, *Bot. Not. Suppl.* **2**: 1-171, 1948.

Ivancich, A.: Der Bau der Filamente der Amentaceen, *Oesterr. Bot. Zeitschr.*, **56**: 385–394, 1906.

Lam, H. J.: Classification and the new morphology, *Acta Biotheor.*, **8**: 107–154, 1948.

———: Stachyospory and phyllospory as factors in the natural system of the Cormophyta, *Svensk Bot. Tidskr.*, **44**: 517–534, 1950.

Moseley, M. F., Jr.: Comparative anatomy and phylogeny of the Casuarinaceae, *Bot. Gaz.*, **110**: 231–280, 1948.

Moss, C. E.: Modern systems of classification of the angiosperms, *New Phyt.*, **11**: 206–213, 1912.

Neumayer, H.: Die Geschichte der Blüte, *Abh. Zoo.-Bot. Ges. Wien*, **14**: 3–111, 1924.

Poisson, J.: Recherches sur les *Casuarina*, *Nouv. Arch. Mus. Hist. Nat., Paris*, sér. 1, **10**: 59–111, 1874.

Porsch, O.: Die Spaltöffnungsapparat von *Casuarina* und seine phyletische Bedeutung, *Oesterr. Bot. Zeitschr.*, **54**: 7–17, 41–51, 1904.

Robertson, C.: The structure of the flowers and the mode of pollination of the primitive angiosperms, *Bot. Gaz.*, **37**: 294–298, 1904.

Swamy, B. G. L.: A contribution to the life history of *Casuarina*, *Proc. Am. Acad. Arts and Sci.*, **77**: 1–34, 1948.

Tippo, O.: Comparative anatomy and morphology of the Moraceae and their presumed allies, *Bot. Gaz.*, **100**: 1–99, 1938.

Treub, M.: Sur les Casuarinées et leur place dans le système naturel, *Ann. Jard. Bot. Buitenzorg*, **10**: 145–231, 1891.

Helobiales

Arber, A.: On the "squamae intravaginales" of the Alismataceae and Butomaceae. *Ann. Bot.*, 39: 169–173, 1925.

Brown, W. V.: Cytological studies in the Alismataceae, *Bot. Gaz.*, **108**: 263–267, 1946.

Chrysler, M. A.: The structure and relationships of the Potamogetonaceae and allied families, *Bot. Gaz.*, **44**: 161–188, 1907.

Eber, E.: Karpelbau und Plazentationverhältnisse in der Reihe der Helobiae: Mit einem Anhang über die verwandschaftlichen Beziehungen zwischen Ranales und Helobiae, *Flora*, N.F., **127**: 273–330, 1934.

Holmgren, I.: Zur Entwicklungsgeschichte von *Butomus umbellatus*, *Svensk Bot. Tidskr.*, **7**: 58–77, 1913.

Johri, B. M.: Studies in the family Alismaceae. I. *Limnophyton obtusifolium* Miq., *Jour. Indian Bot. Soc.*, **14**: 49–66, 1935; II. *Sagittaria sagittifolia* L., *Proc. Indian Acad. Sci.*, **B1**: 340–348; III. *Sagittaria guayanensis* H.B.K. and *S. latifolia* Willd., *ibid.*, **B2**: 33–48, 1935; IV. *Alisma Plantago* L., *A. plantago-aquatica* L., and *Sagittaria graminea* Mich., *ibid.*, **B4**: 128–138, 1936.

———: The life history of *Butomopsis lanceolata* Kunth., *Proc. Indian Acad. Sci.*, **B4**: 139–162, 1936.

———: The embryo sac of *Limnocharis emarginata* L., *New Phyt.*, **37**: 279–285, 1938.

———: The embryo sac of *Hydrocleis nymphoides* Buch., *Beih. Bot. Centralbl.*, **48A**: 165–172, 1938.

Markgraf, F.: Blütenbau und Verwandschaft bei den einfachsten Helobiae, *Ber. Deutsch. Bot. Ges.*, **54**: 191–229, 1936.

Mayr, F.: Beiträge zur Anatomie der Alismataceen, *Beih. Bot. Centralbl.* Abt. A. *Morph. v. Phyl. der Pflanzen*, **62**: 61–77, 1943.

Meyer, F. J.: Die Verwandschaftsbeziehungen der Alismataceen zu den Ranales im Lichte der Anatomie, *Bot. Jahrb.*, **65**: 53–59, 1932.

Nitzschke, J.: Bieträge zur Phylogenie der Monokotylen, gegründet auf der Embryosackentwicklung apocarper Nymphaeaceen und Helobien, *Beitr. Biol. Pfl.*, **12**: 223–267, 1914.

Salisbury, D. S. C.: Floral constitution in the Helobiales, *Ann. Bot.*, **40**: 419–445, 1926.

Wodehouse, R. P.: Pollen grains in the identification and classification of plants. VIII. The Alismataceae, *Am. Jour. Bot.*, **23**: 535–539, 1936.

Witmer, S. W.: Morphology and cytology of *Vallisneria spiralis* L., *Am. Midl. Nat.*, **18**: 309–333, 1937.

Liliales

Buchenau, F.: Zur Naturgeschichte von *Narthecium ossifragum*, *Bot. Zeit.*, **24**: 349–355, 1866.

Buxbaum, F.: Die Entwicklungslinien der Lilioideae, *Bot. Archiv*, **38**: 242–293, 1936.

El-Hamidi, A.: Vergleichend-morphologische Untersuchungen am Gynoeceum der Unterfamilien Melanthioideae und Asphodeloideae der Liliaceae, *Arb. Inst. Allgem. Bot. Univ. Zürich*, **A4**: 1–49, 1952.

Markgraf, F.: Blütenbau und Verwandtschaft bei den einfachsten Helobiae, *Ber. Deutsch. Bot. Ges.*, **54**: 191–229, 1936.

Schnarf, K.: Die Embryologie der Liliaceae und ihre systematische Bedeutung, *Sitzungsberichte Akad. Wiss. Wien Math.-Nat. Kl.*, Abt I, **138**: 69–72, 1929.

Souèges, R.: Recherches sur l'embryogenie des Liliacées, *Bull. Soc. Bot. France*, **78**: 662–682, 1931.

Wunderlich, R.: Vergleichende Untersuchungen von Pöllenkornern einiger Liliaceen und Amaryllidaceen, *Oesterr. Bot. Zeitschr.*, **85**: 30–55, 1936.

Palmae

Al-Rawi, A.: Blütenmorphologische und zytologische Untersuchungen an Palmen der Unterfamilie der Ceroxylidae, *Arb. Inst. Algem. Bot. Univ. Zürich*, III ser., no. 6, 1945.

Bailey, L. H.: Palms and their characteristics, *Gentes Herb.*, **3**: 3–29, 1933.

———: Certain palms of Panama, *Gentes Herb.*, **3**: 33–116, 1933.

Bosch, E.: Blütenmorphologische und zytologische Untersuchungen an Palmen, *Ber. Schweiz. Bot. Ges.*, **57**: 37–100, 1947.

Brown, R. W.: Palmlike Plants from the Dolores Formation (Triassic), Southwestern Colorado, *U.S. Geol. Surv. Prof. Paper*, 274-H, 1956.

Drude, O.: Ausgewählte Beispiele zur Erläuterung der Fruchtbildung bei den Palmen, *Bot. Zeit.*, **35**: 601–613, 1877.

Eames, A. J.: Neglected morphology of the palm leaf, *Phytomorph.*, **3**: 172–189, 1953.

Eckhardt, T.: Kritische Untersuchungen über das primäre Dickenwachstum bei Monokotylen, mit Ausblick auf dessen Verhältnis zur secundären Verdickung, *Bot. Archiv*, **42**: 289–334, 1941.

Lignier, O.: Nouvelles recherches sur le *Propalmophyllum liasinum*, *Mém. Soc. Linn. Normandie*, **23**: 1–15, 1908.

Martius, C. F. P. de: "Historia Naturalis Palmarum," München, 1823–1850.

Micheels, H.: Contribution à l'étude anatomique sur les organes végétatifs et floraux chez *Carludovica plica*, *Archiv Inst. Bot. Liège*, **2**: 1–86, 1900.

Morris, D.: On the phenomena concerned in the production of forked and branched palms, *Jour. Linn. Soc. London*, **29**: 281–298, 1893.

Naumann, A.: Beiträge zur Entwicklungsgeschichte der Palmenblätter, *Flora*, **70**: 193–202, 209–218, 227–242, 250–257, 1887.

Schoute, J. C.: Über das Dickenwachstum der Palmen, *Ann. Jard. Bot. Buitenzorg*, 2 sér., **11**: 1–209, 1912.

——— and L. Algera: Über den Morphologischen Wert der Schuppen der Lepidocaryineenfrucht, *Ber. Deutsch. Bot. Ges.*, **46** (6): 82–106, 1928.

Skutch, A.: The pollination of the palm *Archontophoenix Cunninghamii*, *Torreya*, **32**: 29–37, 1932.

Chapter 12

PHYLOGENY OF THE ANGIOSPERMS

Relation of Monocotyledons to Dicotyledons

Division of the angiosperms into two major groups, monocotyledons and dicotyledons, has long been maintained, although there is, morphologically, no clear-cut separation. Primarily, the bases for the distinction have been differences in cotyledon number, leaf venation, and vascular structure of the stem. The dicotyledons have two cotyledons, a terminal plumule, netted venation, and steles with cylinders of vascular tissue; the monocotyledons have one cotyledon, a so-called lateral plumule, parallel venation, and steles with scattered vascular bundles. But there are many exceptions. Cotyledon number is not constant; taxa with one cotyledon are frequent among dicotyledons, and some monocotyledons have a second vestigial cotyledon (Chap. 9). In the dicotyledons, there are examples of parallel venation—the Epacridaceae, especially *Dracophyllum;* in the monocotyledons, some of the Araceae and Liliaceae have netted venation. A difference of detail in venation, the presence of free vein endings in the dicotyledons, and closed venation —absence of free vein tips—in the monocotyledons has been emphasized. But there is closed venation in many dicotyledons.

Characters of less morphological significance are helpful in the separation of dicotyledons and monocotyledons. In general habit, the monocotyledons stand apart, because of their shortened internodes and frequent bulbous, cormose, or rhizomatous form, linear leaves, and hypogeal germination. Rather few dicotyledons have seeds in which the cotyledons remain underground and serve for food storage. *Peperomia* shows stages in the development of this character; in some species, one cotyledon remains underground (Fig. 136).

The monocotyledon commonly has many adventitious roots that largely take over the function of the primary root and sometimes replace it, structurally, in the seedling. The dicotyledon has few adventitious roots in the seedling, and a primary root is present, except in parasites, saprophytes, and some other highly specialized taxa.

There are fairly constant differences in flower structure. Spiral arrangement of floral appendages is prominent in many of the lower dicotyledonous families; it is present in only the lowest monocotyledonous families. The floral whorls of the dicotyledons are predominantly pentamerous; those of the monocotyledons are trimerous. In the dicotyledons, calyx and corolla are usually unlike in appearance, as well as morphologically; in the monocotyledons, calyx and corolla are usually alike in appearance and tend to unite to form a pseudowhorl.

The absence of a typical cambium has been emphasized as a distinctive character of the monocotyledons, but weakly developed cambium is present in the free vascular bundles of several monocotyledonous families, a cambium like that of similar free bundles in the stems of herbaceous dicotyledons.

Differences have been seen in root anatomy; the dicotyledons have small numbers of xylem poles, two or four, basically; the monocotyledon root is typically polyarch. Exceptions are numerous.

The nectaries of the monocotyledons are chiefly of the septal type; the characteristic type of the dicotyledons is the transformed stamen.

But the dicotyledons and monocotyledons have many close morphological resemblances, in spite of their obvious differences in general habit and gross anatomy.

Cotyledon number has little value in separating the major groups of angiosperms. Among the dicotyledons there are scattered genera and species that have a single cotyledon, and some of the monocotyledons have a second, smaller, usually vestigial cotyledon.

In anatomical structure—in steles, nodes, leaf traces, meristems, vascular tissues—there are no basic differences. For example, the vessel and the sieve tube have arisen independently in both monocotyledons and dicotyledons—several to many times in each—and the histological process and sequence are the same. (The vessels of the Gnetales also have independent origin.) The scattered-bundle stele and the V-shaped, closed vascular strand, characteristic of the monocotyledons, are present in some of the Ranunculaceae.

Differences in type of root system are those of degree only. The root system of the monocotyledons is a modification of that of the dicotyledons. Arrested growth of the primary root of the monocotyledon embryo and seedling is present in all stages, especially in the woody types; similar arrest is present in dicotyledons—Nymphaeaceae, Umbelliferae.

Close similarities of the Ranunculaceae and the Nymphaeaceae to the Alismataceae are seen as evidence of a monophyletic origin of the angiosperms, but there are also prominent differences between these families, especially in embryology.

Though the phyletic relationships between monocotyledons and dicotyledons have been much discussed, no general agreement has been reached as to which represents the more primitive stock. Each has been considered the older, ancestral taxon, and they have been interpreted as parallel groups, derived from the same ancestral stock. The presence of both woody and herbaceous forms in the primitive families of angiosperms is the basis for the theory that there were, probably, in the ancestral stock of angiosperms, or in the earliest angiosperms, both woody and herbaceous types. (The division of the dicotyledons into the Lignosae and Herbaceae, setting apart two major lines, based, in part, on habit, seems doubtfully valid as a basis for separating natural taxa. Much more information from various fields of study is necessary to determine the relations of the more primitive families.)

The monocotyledons may have been derived from dicotyledonous stock early in the history of the angiosperms. The presence of apparently only distantly related or unrelated lines among the more primitive orders—Butomales, Alismatales, Helobiales, Liliales—suggests independent origin of these orders. Variety of habit and structure in the primitive families suggests a very long period of specialization since angiosperms arose. If the monocotyledons arose from early dicotyledonous stock, the origin must have been before vessels arose in dicotyledons, because vessels have been shown to have arisen at least several times in the monocotyledons.

The Stachyospory-Phyllospory Theory. On the basis of supposed differences in ovule position—borne on stem (stachyosporous) or on leaves (phyllosporous)—the angiosperms are divided, under this theory, into two new major divisions—the Phyllosporeae and the Stachyosporeae (see Chaps. 6 and 7). By this grouping, the monocotyledons and dicotyledons and some families, even some genera, are divided. To cover taxa so divided, a third group was later proposed, one intermediate between the two. The theory is valueless, because comparative morphology and anatomy show that, in the angiosperms, all ovules are borne on appendages.

Origin of the Angiosperms

The problem of the origin of the angiosperms has received most attention since the beginning of the twentieth century, as increasing information made older theories untenable. Several major taxa were formerly suggested as possibly ancestral to the angiosperms.

The Isoëtes-Monocotyledon Theory. When the monocotyledons were generally accepted as the most primitive angiosperms, resemblances were seen in the linear leaves and cormous habit of *Isoëtes* to the simpler, aquatic monocotyledons. But the microphyllous vascular cryptogams seemed a most improbable ancestral stock.

The Coniferales-Amentiferae Theory. Attention then turned to the highest gymnosperms. Resemblances of the angiosperms to the conifers were seen in the Amentiferae, considered in prominent taxonomic treatments as the most primitive angiosperms. The inflorescences of the Amentiferae, with their simple, naked flowers and wind pollination, suggested the cones and pollination of the conifers. The angiosperm stamen and its two pollen sacs were considered the equivalent of the bisporangiate microsporophyll of many conifers. But the flowers and inflorescences of the Amentiferae have been shown to be specialized, rather than primitive. The simplicity of the flowers of the Salicaceae, Betulaceae, Myricaceae, Casuarinaceae, and other families is the simplicity of reduction, rather than of primitiveness. Anatomical structure, especially that of the wood, strongly supports the advanced position of these families.

Gymnosperm-Gnetales-Angiosperm Theory. The Gnetales (*sensu lato*), commonly called the highest gymnosperms, have been interpreted as transitional from gymnosperms to angiosperms. They combine naked seeds with vessel-bearing wood. *Gnetum* has an angiospermlike leaf and a female gametophyte that, in some characters, suggests the gametophyte of angiosperms. The vessels of the Gnetales are, histologically, unlike those of angiosperms; their perforations have been derived from round-pitted tracheids, rather than scalariform-pitted tracheids, as have the vessels of the angiosperms. The ovules are unlike in general structure, and the female gametophyte of *Gnetum* is unique.

The Anthostrobilus (Bennettitalean) Theory. The anthostrobilus theory is founded on resemblances in the angiosperm flower (magnolian type) to the cone of the bennettitalean cycads (cycadeoides), a Mesozoic taxon. The *anthostrobilus* is a hypothetical, conelike reproductive structure that differs from the strobili of lower vascular taxa in its amphisporangiate structure, with megasporophylls above the microsporophylls, and in the presence of a distinct perianth.

The prominence of the Bennettitales in the Cretaceous, at the time when angiosperm fossils are first abundant, seemed to support the theory of a phyletic relationship between the two taxa. In gross structure, the flower and the cycadeoid cone are indeed alike, but sporophyll types are wholly unlike. The microsporophyll of the Bennettitales is a large, pinnately compound organ; that of the angiosperms is basically a simple, laminar organ. The megasporophyll of the Bennettitales is stalklike, with a terminal ovule; that of the angiosperms is laminar, with numerous ovules on the adaxial surface. Separating the megasporophylls of the cycadeoides are many scales, appendages unknown in angiosperms. The bennettitalean stem has a large pith, thin vascular cylinder, and thick cortex, with strongly girdling leaf traces; the angiosperm stem has a small pith, thick vascular cylinder, and thin cortex, with leaf traces not girdling or lateral ones only slightly girdling. There are important differences also in wood structure. The wood rays of the Bennettitales are simple, consisting of one kind of cell only, and lacking marginal cells; the wood rays of the angiosperms are made up of two major types of cells, and marginal cells are prominent. The bennettitalean cone and the angiosperm flower seem to represent parallel development in unrelated lines.

The Caytonialian Theory. The Caytoniales, a small group of Mesozoic fossil gymnosperms, known chiefly by their fruiting structures, have been considered possible ancestors of the angiosperms, because their seeds are enclosed in a carpellike sheath. They are described and their morphology is discussed in Chap. 6. Possessing closed carpels, they can hardly be considered ancestral to the angiosperms, in which the carpels are still incompletely closed in many genera.

The Pteridosperms. The seed ferns are frequently suggested as possible ancestors to the angiosperms. But, in addition to ovule differences between the two groups discussed in Chap. 7, wood structure does not uphold this seed-fern origin of the angiosperms. Scalariform-pitted tracheids in secondary xylem, prominent in the wood of primitive angiosperms, are absent in the pteridosperms.

The Prephanerogamae and Chlamydospermae. The terms Prephanerogamae and Chlamydospermae have both been applied to seed plants that are intermediate in some characters between gymnosperms and angiosperms. Authors differ in the use of the two terms and of their coverage—the Gnetales alone; the Gnetales and *Ginkgo;* and *Casuarina* and *Sarcopus,* sometimes called the "Protangiosperms." The various bases for classification are so confused and artificial that the terms should be discarded.

Angiosperms, Polyphyletic or Monophyletic. Regardless of the ancestral stock from which the angiosperms arose, the question is frequently raised—are they monophyletic or polyphyletic? In the primitive orders of dicotyledons and monocotyledons, there are groups of families or single families that seem to have no close relationship to other families. In the Ranales, there are about ten of these lines (Fig. 147); in the monocotyledons, there seem to be three or four. This great diversity in the primitive families suggests a polyphyletic origin, a long specialization of a primitive stock, or origin en masse from a diverse ancestral stock. But absence of close relationship among these primitive families need not indicate independence of origin; differences may be the result of modification over long periods.

Support can perhaps be found for the hypothesis of a polyphyletic or en masse origin of the angiosperms (1) in the variety of perianth in the more primitive families—a perianth of one, two, or several bud scales, with or without petaloid organs; (2) in the evidence that the closed carpel was not a characteristic of the earliest angiosperms. But carpel closure, like the development of the filament-anther type of stamen, is clearly the result of a general trend in specialization. The microsporangiate characters—four sporangia in two pairs—seem to be a basic and most important character that became fixed very early in the angiosperm line.

The strongest evidence in support of the theory of monophyletic origin is the common possession of the 8-nucleate embryo sac and endosperm of "secondary nature"—formed by the union of a male nucleus and two female nuclei. Other supporting evidence is seen in the similarity in basic structure of the flowers—position, arrangement, and type of appendages, which are modified in similar ways: in the stamens, throughout, with four sporangia in two pairs, one pair on each side of the midvein; in the follicular carpel and its anatropous ovules, the primitive type in both lines; and in similar pollen, simple and monocolpate in primitive families of both monocotyledons and dicotyledons.

Age of the Angiosperms*

In the middle of the twentieth century, the opinion that the angiosperms arose in a fairly recent geologic age began to be replaced by the opinion that the origin was one much older than the long-accepted Lower Cretaceous. The basis for this change is the gradual accumulation of evidence of greater age obtained largely in two fields of botany. Evidence from paleobotany in the discovery of angiosperm fossils in the Jurassic and Triassic has naturally been most convincing; and that from morphology, although less readily recognized, has become impressive, as the more primitive living taxa have been broadly and critically studied.

Evolutionary change in form has been, first of all, toward increasing complexity. Simplicity has been commonly interpreted as evidence of primitiveness, but the importance of retrogressive change—reduction after complexity—is now recognized as prominent in many taxa. Simplicity must be critically examined; it is as likely to represent high specialization as primitiveness; more likely, in advanced taxa. Changes in the interpretation of simple form have brought about major changes in views of the phylogenetic relationships of some taxa and, therefore, of their probable age. The Amentiferae are no longer considered examples of the earliest angiosperms and indicators of the time of origin of the angiosperms.

* This discussion contains extracts from an invitation paper presented by the author at the Ninth International Botanical Congress, Montreal, 1959, entitled "The Morphological Basis for a Paleozoic Origin of the Angiosperms."

The necessity of considering the nature of simplicity in form is, of course, only a part of the understanding of the entire story of evolutionary change, but the nature of simplicity needs emphasis, because its neglect has been a major element in misinterpretations. In the search for a time of origin and an ancestral stock for the angiosperms, it is necessary to know, as completely as possible, the story of evolutionary modification of all organs and tissues. On this basis, the evolutionary stage of a fossil can be interpreted, and the age of angiosperms, at that time, perhaps be estimated.

Since 1920, thorough morphological studies of the Ranales and of some of the more primitive monocotyledons have provided a basis for an estimate of the relative advance of a taxon (Chap. 11). These studies have shown that a decision as to the level of advance must rest on all possible characters, not on one or two; uneven advance in structural specialization is common, and the *sum total* of modifications is needed to estimate the stage of evolution.

Assessment of the general level of morphological advance in prominent genera of Lower and Middle Cretaceous floras should give a basis for an opinion on the probable age of those floras. *Populus, Sassafras,* and *Platanus*—well-known Cretaceous genera of unquestioned identity —give evidence that the angiosperms, as early as the Cretaceous were already highly specialized, for the three genera show many advanced characters. In general flower structure, simplicity has replaced complexity. All are dioecious; two are anemophilous. The stamens have terete filaments and specialized anthers; carpel number is reduced to two in *Populus,* and one in *Sassafras* and *Platanus.* The gynoecium in *Populus* is syncarpous. The ovules are solitary in *Sassafras* and *Platanus.* There is little evidence of the primitive follicle in the achene of *Platanus* or the drupe of *Sassafras.* In nodal structure, no evidence remains of the primitive two-trace structure. The xylem of all genera is specialized. The angiosperms had had a long history before Cretaceous times.

By Middle Cretaceous times, the fossil record shows great diversity among angiosperm families, and many families now considered basic had become cosmopolitan. The Cretaceous flora of Australia is not different in its basic families from floras of other continents. After early differentiation of fundamental families, the angiosperms had spread throughout the world. This widespread flora was a "mature" flora; it can hardly be considered young in the Cretaceous.

In a discussion (1960) of the age of the angiosperms, a probable paleozoic origin for these plants is questioned. But the evidence of high specialization of many of the Lower and Middle Cretaceous genera is not considered. Morphology of living plants strongly supports the evidence accumulating from the fossil record that the angiosperms had their origin much further back in geological time than has been commonly assumed, perhaps to the Permian.

In the search for possible ancestors of the angiosperms, morphologists have been neglecting three aspects of stamen structure that seem of major importance: (1) the laminar form of the microsporophyll; (2) the wall-less, sunken position of the microsporangium; (3) the type of pollination in families with primitive stamens. The laminar stamen is surely primitive, as shown by *Degeneria, Himantandra,* the Magnoliaceae and Nymphaeaceae, and by the laminar form of the carpel. The wall-less character of the sporangium—clearly evident in the laminar sporophylls—is concealed, in typical anthers, by the interpretation of the anther-sac walls as sporangium walls. The megasporangium also is wall-less and sunken, if the nucellus is interpreted as an emergence of the lamina, as it seems to be, from comparison of micro- and megasporophylls.

Important in the evolution of angiosperm flowers have been the *biological* aspects, especially those aspects related to pollination. The history of pollination in the angiosperms has surely played a prominent part in major changes in the structure of the flower and the inflorescence. The story is one of gradual modification, accompanying the change from pollination by beetles to pollination by higher insects, birds, bats, wind, and to self-pollination. Note the major change from the nectarless *Eupomatia* and *Calycanthus,* with their supply of food bodies for chewing insects and their heavy, penetrating odors—attractive to beetles only—to nectary-bearing, fragrant flowers—attractive to bees—and to nectarless, odorless, wind-pollinated flowers. The development of nectaries and other early changes parallel the evolution of the insects. Beetles first appear in the fossil record in the Permian; other insects in later periods.

A line between the earliest angiosperms and the stock from which they arose perhaps cannot be drawn. But from a general survey of what appear to be the more conservative characters in living taxa, some suggestions can be made as to the characters of a possible ancestral stock.

General body habit, especially anatomy, suggests that angiosperms belong in the fern line of vascular plants. In stelar structure and vascular histology, they are fernlike, rather than cycadlike. Amphiphloic steles are frequent, and the primitive vessel elements are closely like those of *Pteridium.* Basic nodal structure is two traces to one gap. Laminar placentation suggests the soral distribution of ferns. (The distribution of microsporangia—four, in two pairs, median on the sporophyll, not known in ferns—is perhaps a critical character, one that must have been established very early in angiosperm history, or was already present in ancestral stock.) Sporangial development is eusporangiate in angiosperms and the more primitive ferns.

The description, in 1956, of apparently bisexual fructifications in the fernlike Mesozoic genus *Glossopteris*, commonly considered a pteridosperm, may be important to the theory of angiosperm derivation from fern stock. Though details of the reproductive structures are still unknown, there seems no question but that there are two kinds in each fruiting structure, and resemblance of the *Glossopteris* leaf to simple fern leaves is close. *Glossopteris* is perhaps closer to the eusporangiate ferns than to the pteridosperms; if so, it may form a link between the ferns and the angiosperms.

Ancestry for the angiosperms should probably be sought far back in the eusporangiate line, as far as the Permian, when beetles are first known. (The Permian flora was rich in pteridophytes and pteridosperms.) Seed bearing is perhaps not an essential character in the ancestral forms. (Seedlike structures have developed independently in several major vascular lines.) It is difficult to interpret the simple ovule of the angiosperms in terms of the complex pteridosperm seed, and the pteridosperms should perhaps be dropped as a possible ancestral stock of the angiosperms.

Further evidence for great age in the angiosperms comes from ontogeny, from the story of carpel closure and sealing. The living angiosperms show all stages in the closure of the carpels; many that appear closed are unsealed or sealed only in the late stages of carpel maturation. In many taxa, the carpels are congenitally fused; they develop from ring-shaped primordia. Carpel sealing, with the establishment of a primordium of an entirely different shape, is apparently a slow process, slower than reduction in ovule number and the development of the achene, because the achenes of *Sparganium* and *Platanus* are still unsealed at flowering.

The search for the ancestral stock of the angiosperms must continue. Progress has been made in the twentieth century, though it is chiefly in the elimination of many suggested ancestral lines.

BIBLIOGRAPHY

Phylogeny

Andrews, H. N.: On the stelar anatomy of the pteridosperms, with particular reference to the secondary wood, *Ann. Mo. Bot. Gard.*, **27**: 51–118, 1940.

Arber, A.: The interpretation of the flower: a study of some aspects of morphological thought, *Biol. Rev.*, **12**: 157–184, 1937.

Arber, E. A. N., and J. Parkin: The origin of angiosperms, *Jour. Linn. Soc. Bot.*, **38**: 29–80, 1907.

——— and ———: Studies on the evolution of the angiosperms, *Ann. Bot.*, **22:** 489–515, 1908.

Axelrod, D. I.: A theory of angiosperm evolution, *Evol.*, **6:** 29–60, 1952.

Bailey, I. W.: The development of vessels in angiosperms and its significance in morphological research, *Am. Jour. Bot.*, **31:** 421–428, 1944.

———: Origin of the angiosperms: The need for a broadened outlook, *Jour. Arnold Arb.*, **30:** 64–70, 1949.

——— and E. W. Sinnott: Investigations on the phylogeny of the angiosperms. II. Anatomical evidences of reduction in certain of the Amentiferae, *Bot. Gaz.*, **58:** 36–60, 1914.

Bancroft, H.: A review of literature concerning the evolution of monocotyledons, *New Phyt.*, **13:** 185–308, 1914.

Beauverie, J., and M. Durand: L'ancienété et la phylogénie des plantes à fleurs, *Rev. Gén. Sci. Pur. App.*, **46:** 269–334, 1930.

Bessey, C. E.: The phylogenetic taxonomy of flowering plants, *Ann. Mo. Bot. Gard.*, **2:** 109–164, 1915.

Brown, W. H.: The bearing of nectaries on the phylogeny of flowering plants, *Proc. Am. Phil. Soc.*, **79:** 549–594, 1935.

Calestani, V.: Le origini e la classificazione delle angiosperme, *Archivo Bot.*, **9:** 274–311, 1933.

Campbell, D. H.: The phylogeny of the angiosperms, *Bull. Torrey Bot. Club*, **55:** 479–497, 1928.

Chalk, L.: The phylogenetic value of certain anatomical features of dicotyledonous woods, *Ann. Bot.*, n.s., **1:** 409–428, 1937.

Constance, L.: "The systematics of the angiosperms." In "A Century of Progress in the Natural Sciences, 1853–1953," California Academy of Science, San Francisco, 1955.

Cuénod, A.: Hypothèse relative à la place des monocotylédones dans la classification naturelle, *Bull. Soc. Bot. France*, **79:** 365–393, 1932.

Eames, A. J.: Again: "The New Morphology," *New Phyt.*, **50:** 17–35, 1951.

Eckardt, T.: Über zweizählige Wirtelstellungen bei den Monokotylen und die Bedeutung der Symmetrieverhältnisse für ihr Verständnis, *Bot. Archiv*, **42:** 44–49, 1941.

———: Kritische Untersuchungen über das primäre Dickenwachstum bei Monokotylen, mit Ausblick auf dessen Verhältnis zur secundären Verdickung, *Bot. Archiv*, **42:** 289–334, 1941.

Emberger, L.: Les Préphanérogames, *Ann. Sci. Nat. Bot.*, 11 sér., **10:** 131–144, 1950.

Fisher, M. J.: The morphology and anatomy of the flowers of the Salicaceae, *Am. Jour. Bot.*, **15:** 307–326, 372–394, 1928.

Florin, R.: On female reproductive organs in the Cordaitineae, *Acta Hort. Berg.*, **15:** 111–134, 1950.

Gagnepain, F., and E. Boureau: Une nouvelle famille de Gymnospermes: Les Sarcopodacées, *Bull. Soc. Bot. France*, **93:** 313–320, 1946.

Gatin, C. L.: La morphologie de la germination et ses rapports avec la phylogénie, *Rev. Gén. Bot.*, **21:** 147–158, 1909.

Gerasimova-Navasina, E. N.: Die Entwicklung des Embryosacks, die doppelte Befruchtung und die Frage nach der Ableitung der Angiospermen, *Bot. Zeit.*, **39:** 655–680, 1954.

Haskell, G.: Pleiocotyly and differentiation within angiosperms, *Phytomorph.*, **4:** 140–152, 1954.

Henslow, G.: A theoretical origin of endogens from exogens, through self-adaptation to an aquatic habitat, *Jour. Linn. Soc. Bot. London,* **29**: 485–528, 1893.

Hirmer, M.: Die Pteridospermae, insbesondere die Caytoniales und die Entwicklung der Angiospermae, *Proc. Zesda Int. Bot. Congr. Amsterdam,* II: 231–234, 1935.

Hjelmqvist, H.: Studies in the floral morphology and phylogeny of the Amentiferae, *Bot. Not. Suppl.,* **2**: 1–171, 1948.

Jacob de Cordemoy, H.: "Recherches sur les Monocotylédones à Accroissement Sécondaire," Lille, 1894.

Janchen, E.: Die Herkunft der Angiospermen-Blüte und die systematische Stellung der Apetalen, *Oesterr. Bot. Zeitschr.,* **97**: 129–167, 1950.

Joshi, A. C.: Systematic distribution of the Fritillaria-type of embryo sac and the mono- or polyphyletic origin of the angiosperms, *Chron. Bot.,* **4**: 507–508, 1938.

Just, T.: Gymnosperms and the origin of the angiosperms, *Bot. Gaz.,* **110**: 91–103, 1948.

———: Fifty years of paleobotany, *Am. Jour. Bot.,* **44**: 93–99, 1957.

Lam, H. J.: Classification and the New Morphology, *Acta Biotheoretica,* **8**: 107–154, 1948.

———: Stachyospory and phyllospory as factors in the natural system of the Cormophyta, *Svensk Bot. Tidskr.,* **44**: 517–534, 1950.

Martens, P.: Les préphanerogams et le problème de la graine, *La Cellule,* **54**: 105–132, 1951.

McNair, J. B.: Angiosperm phylogeny on a chemical basis, *Bull. Torrey Bot. Club,* **62**: 515–522, 1935.

Nemecj, F.: On the problem of the origin and phylogenetic development of the angiosperms, *Acta Mus. Nat. Prague,* **12**: 65–145, 1926.

Nitzschki, J.: Beiträge zur Phylogenie der Monocotyledonen, gegründet auf der Embryosackentwicklung apocarper Nymphaeaceen und Helobien, *Beitr. Biol. Pfl.,* **12**: 223–267, 1914.

Parkin, J.: The unisexual flower—a criticism, *Phytomorph.,* **2**: 75–79, 1952.

Plumstead, E. P.: Bisexual fructifications borne on *Glossopteris* leaves from South Africa, *Paleontographica,* Abt. B, **100**: 1–25, 1956.

Porsch, O.: Die Abstammung der Monokotylen und die Blütennektarien, *Ber. Deutsch. Bot. Ges.,* **31**: 580–590, 1913.

Sargant, E.: Evolution of the monocotyledons, *Bot. Gaz.,* **37**: 325–345, 1904.

———: The reconstruction of a race of primitive angiosperms, *Ann. Bot.,* **22**: 121–186, 1908.

Scott, R. A., E. S. Barghoorn, and E. B. Leopold: How old are the angiosperms? *Am. Jour. Sci.,* 258A: 284–299, 1960.

Soo, R. de: Die modernen Grundsetze der Phylogenie im neuen System der Blütenpflanzen, *Acta Biol. Hungar.,* **4**: 257–306, 1953.

Sporne, K. R.: Statistics and the evolution of dicotyledons, *Evol.,* **8**: 55–64, 1954.

———: The phylogenetic classification of angiosperms, *Biol. Rev.,* **31**: 1–19, 1956.

Stebbins, G. L., Jr.: Natural selection and the differentiation of angiosperm families, *Evol.,* **5**: 299–324, 1951.

———: Plant phylogeny and evolution, *Evol.,* **7**: 281–285, 1953.

Suessenguth, K.: The flora of Australia as an index of the antiquity of angiosperms, *Pacific Sci.,* **4**: 287–308, 1950.

——— and H. Merxmüller: Über die Herkunft der Angiospermen, *Phyton,* **4**: 1–18, 1952.

Takhtajian, A. L.: Origin of angiospermous plants, 1954 (translation, 68 pp.), *Washington Publ. Am. Inst. Biol. Sci.,* 1958.

Thomas, H. H.: Paleobotany and the origin of the angiosperms, *Bot. Rev.*, **2:** 397–419, 1936.

———: The old morphology and the new, *Proc. Linn. Soc. Bot. London,* **145:** 17–32, 1952.

Thomson, R. B.: Origin of the seed habit in plants, *Trans. Roy. Soc. Canada,* 3 ser., **21:** 229–272, 1927.

Turrill, W. B.: Taxonomy and phylogeny, *Bot. Rev.*, **8:** 247–270, 473–532, 655–707, 1942.

Walton, J.: L'évolution de téguments et de la protectrice du sporange, *Année Biol.*, **26:** 129–152, 1952.

General Bibliography

Arber, A.: The "law of loss" in evolution, *Proc. Linn. Soc. London,* **131**: 70–78, 1918–19.

———: "Monocotyledons," Cambridge, 1925.

———: The interpretation of the flower: A study of some aspects of morphological thought, *Biol. Rev.,* **12**: 157–184, 1937.

———: "The Natural Philosophy of Plant Form," Cambridge, 1950.

Bailey, I. W.: The use and abuse of anatomical data in the study of phylogeny and classification, *Phytomorph.,* **1**: 67–70, 1951.

———: "Contributions to Plant Anatomy," Waltham, Mass., 1954.

Bancroft, H.: A review of researches concerning floral morphology, *Bot. Rev.,* **1**: 77–99, 1935.

Bower, F. O.: "Size and Form in Plants with Special References to the Primary Conducting Tracts," London, 1930.

Boyden, A.: Homology and analogy: A critical review of the meaning and implications of these concepts in biology, *Am. Midl. Nat.,* **37**: 648–669, 1947.

Constance, L.: "The systematics of the angiosperms." In "A Century of Progress in the Natural Sciences, 1853–1953," California Academy of Science, San Francisco, 1955.

Coulter, J. M., and C. J. Chamberlain: "Morphology of the Angiosperms," New York, 1903.

Dorf, E.: Plants and the geological time scale, *Geol. Soc. Am. Spec. Paper,* **62**: 575–592, 1955.

Eames, A. J., and L. H. MacDaniels: "Introduction to Plant Anatomy," 2d ed., New York, 1947.

Eckardt, T.: "Neue Hefte zur Morphologie: Vergleichende Studie über die Morphologischen Beziehungen zwischen Fruchtblatt, Samenlage und Blütenachse bei einigen Angiospermen—zugleich als kritische Beleuchtung der 'New Morphology,'" Weimar, 1957.

Emberger, L.: "Les Plantes Fossiles dans leurs Rapports avec les Végétaux Vivants," Paris, 1944.

Erdtman, G.: "Pollen Morphology and Plant Taxonomy. Angiosperms," Stockholm and Waltham, Mass., 1952.

Esau, K.: "Plant Anatomy," New York, 1953.

Goebel, K. von: "Organographie der Pflanzen: Samenpflanzen," 3d ed., Jena, 1933.

Good, R.: "The Geography of Flowering Plants," New York, 1947.

Henslow, G.: A theoretical origin of endogens from exogens by self-adaptation to an aquatic habit, *Jour. Linn. Soc. London,* **29**: 485–528, 1893.

Hutchinson, J.: "The Families of Flowering Plants," vol. I, "Dicotyledons"; vol. II, "Monocotyledons," Oxford, England, 1959.

Kasapligil, B.: Morphological and ontogenetic studies of *Umbellularia californica* Nutt. and *Laurus nobilis* L., *Univ. Calif. Publ. Bot.,* **25**: 115–240, 1951.

Kozo-Poljanski, B.: On some "third" conceptions in floral morphology, *New Phyt.,* **35**: 479–491, 1936.

Lawrence, G. H. M.: "Taxonomy of Vascular Plants," New York, 1951.
Metcalfe, C. R.: An anatomist's view on angiosperm classification, *Kew Bull.*, **3**: 427–440, 1954.
——— and L. Chalk: "Anatomy of the Dicotyledons," Oxford, England, 1950.
Popham, R. A.: "Developmental Plant Anatomy," Columbus, Ohio, 1952.
Radforth, N. W.: The taxonomic treatment of the fern-pteridosperm complex, *Am. Midl. Nat.*, **36**: 325–330, 1946.
Sifton, H. B.: Developmental morphology of vascular plants, *New Phyt.*, **43**: 87–129, 1944.
Solereder, H.: "Systematic Anatomy of the Dicotyledons," Oxford, England, 1908.
——— and F. J. Meyer: "Systematische Anatomie der Monokotyledonen," Berlin, 1928, 1929, 1930.
Stebbins, G. L., Jr.: "Variation and Evolution in Plants," New York, 1950.
Strassburger, E.: "Die Angiospermen und die Gymnospermen," Jena, 1879.
Suessenguth, K.: "Neue Ziele der Botanik," München-Berlin, 1939.
Takhtajan, A. L.: Phylogenetic principles of the system of higher plants, *Bot. Rev.*, **19**: 1–45, 1953.
Tansley, A. G.: Reduction in descent, *New Phyt.*, **1**: 131–133, 1902.
Tippo, O.: A modern classification of the plant kingdom, *Chron. Bot.*, **7**: 203–206, 1942.
Troll, W.: "Vergleichende Morphologie der höheren Pflanzen," Berlin, vol. I, 1937; vol. II, 1939.
Turrill, W. B.: Taxonomy and phylogeny, *Bot. Rev.*, **8**: 247–270, 473–532, 655–707, 1942.
Van Tieghem, P.: "Traité de Botanique," Paris, 1891.
Velenovsky, J.: "Vergleichende Morphologie der Pflanzen," Prague, 1905–10.
Wardlaw, C. W.: "Phylogeny and Morphogenesis," London, 1952.
Wettstein, R.: "Handbuch der Systematischen Botanik," 4 Aufl., Wien-Leipzig, 1935.
Zimmermann, W.: "Die Phylogenie der Pflanzen," Jena, 1930.
———: "Geschichte der Pflanzen," Stuttgart, 1957.

Index

Page numbers in boldface type refer to illustrations